Lecture Notes in Artificial Intelligence 12979

Subseries of Lecture Notes in Computer Science

More information about this subseries at http://www.springer.com/series/1244

Yuxiao Dong · Nicolas Kourtellis ·
Barbara Hammer · Jose A. Lozano (Eds.)

Machine Learning and Knowledge Discovery in Databases

Applied Data Science Track

European Conference, ECML PKDD 2021
Bilbao, Spain, September 13–17, 2021
Proceedings, Part V

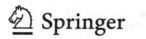 Springer

Editors
Yuxiao Dong
Facebook AI
Seattle, WA, USA

Nicolas Kourtellis
Torre Telefonica
Barcelona, Spain

Barbara Hammer
Bielefeld University, CITEC
Bielefeld, Germany

Jose A. Lozano ⓘ
Basque Center for Applied Mathematics
Bilbao, Spain

ISSN 0302-9743   ISSN 1611-3349 (electronic)
Lecture Notes in Artificial Intelligence
ISBN 978-3-030-86516-0   ISBN 978-3-030-86517-7 (eBook)
https://doi.org/10.1007/978-3-030-86517-7

LNCS Sublibrary: SL7 – Artificial Intelligence

This Springer imprint is published by the registered company Springer Nature Switzerland AG
The registered company address is: Gewerbestrasse 11, 6330 Cham, Switzerland

Preface

This edition of the European Conference on Machine Learning and Principles and Practice of Knowledge Discovery in Databases (ECML PKDD 2021) has still been affected by the COVID-19 pandemic. Unfortunately it had to be held online and we could only meet each other virtually. However, the experience gained in the previous edition joined to the knowledge collected from other virtual conferences allowed us to provide an attractive and engaging agenda.

ECML PKDD is an annual conference that provides an international forum for the latest research in all areas related to machine learning and knowledge discovery in databases, including innovative applications. It is the leading European machine learning and data mining conference and builds upon a very successful series of ECML PKDD conferences. Scheduled to take place in Bilbao, Spain, ECML PKDD 2021 was held fully virtually, during September 13–17, 2021. The conference attracted over 1000 participants from all over the world. More generally, the conference received substantial attention from industry through sponsorship, participation, and also the industry track.

The main conference program consisted of presentations of 210 accepted conference papers, 40 papers accepted in the journal track and 4 keynote talks: Jie Tang (Tsinghua University), Susan Athey (Stanford University), Joaquin Quiñonero Candela (Facebook), and Marta Kwiatkowska (University of Oxford). In addition, there were 22 workshops, 8 tutorials, 2 combined workshop-tutorials, the PhD forum, and the discovery challenge. Papers presented during the three main conference days were organized in three different tracks:

- Research Track: research or methodology papers from all areas in machine learning, knowledge discovery, and data mining.
- Applied Data Science Track: papers on novel applications of machine learning, data mining, and knowledge discovery to solve real-world use cases, thereby bridging the gap between practice and current theory.
- Journal Track: papers that were published in special issues of the Springer journals Machine Learning and Data Mining and Knowledge Discovery.

We received a similar number of submissions to last year with 685 and 220 submissions for the Research and Applied Data Science Tracks respectively. We accepted 146 (21%) and 64 (29%) of these. In addition, there were 40 papers from the Journal Track. All in all, the high-quality submissions allowed us to put together an exceptionally rich and exciting program.

The Awards Committee selected research papers that were considered to be of exceptional quality and worthy of special recognition:

- Best (Student) Machine Learning Paper Award: Reparameterized Sampling for Generative Adversarial Networks, by Yifei Wang, Yisen Wang, Jiansheng Yang and Zhouchen Lin.

- First Runner-up (Student) Machine Learning Paper Award: "Continual Learning with Dual Regularizations", by Xuejun Han and Yuhong Guo.
- Best Applied Data Science Paper Award: "Open Data Science to fight COVID-19: Winning the 500k XPRIZE Pandemic Response Challenge", by Miguel Angel Lozano, Oscar Garibo, Eloy Piñol, Miguel Rebollo, Kristina Polotskaya, Miguel Angel Garcia-March, J. Alberto Conejero, Francisco Escolano and Nuria Oliver.
- Best Student Data Mining Paper Award: "Conditional Neural Relational Inference for Interacting Systems", by Joao Candido Ramos, Lionel Blondé, Stéphane Armand and Alexandros Kalousis.
- Test of Time Award for highest-impact paper from ECML PKDD 2011: "Influence and Passivity in Social Media", by Daniel M. Romero, Wojciech Galuba, Sitaram Asur and Bernardo A. Huberman.

We would like to wholeheartedly thank all participants, authors, Program Committee members, area chairs, session chairs, volunteers, co-organizers, and organizers of workshops and tutorials for their contributions that helped make ECML PKDD 2021 a great success. We would also like to thank the ECML PKDD Steering Committee and all sponsors.

September 2021

Jose A. Lozano
Nuria Oliver
Fernando Pérez-Cruz
Stefan Kramer
Jesse Read
Yuxiao Dong
Nicolas Kourtellis
Barbara Hammer

Organization

General Chair

Jose A. Lozano — Basque Center for Applied Mathematics, Spain

Research Track Program Chairs

Nuria Oliver — Vodafone Institute for Society and Communications, Germany, and Data-Pop Alliance, USA
Fernando Pérez-Cruz — Swiss Data Science Center, Switzerland
Stefan Kramer — Johannes Gutenberg Universität Mainz, Germany
Jesse Read — École Polytechnique, France

Applied Data Science Track Program Chairs

Yuxiao Dong — Facebook AI, Seattle, USA
Nicolas Kourtellis — Telefonica Research, Barcelona, Spain
Barbara Hammer — Bielefeld University, Germany

Journal Track Chairs

Sergio Escalera — Universitat de Barcelona, Spain
Heike Trautmann — University of Münster, Germany
Annalisa Appice — Università degli Studi di Bari, Italy
Jose A. Gámez — Universidad de Castilla-La Mancha, Spain

Discovery Challenge Chairs

Paula Brito — Universidade do Porto, Portugal
Dino Ienco — Université Montpellier, France

Workshop and Tutorial Chairs

Alipio Jorge — Universidade do Porto, Portugal
Yun Sing Koh — University of Auckland, New Zealand

Industrial Track Chairs

Miguel Veganzones — Sherpa.ia, Portugal
Sabri Skhiri — EURA NOVA, Belgium

Award Chairs

Myra Spiliopoulou	Otto-von-Guericke-University Magdeburg, Germany
João Gama	University of Porto, Portugal

PhD Forum Chairs

Jeronimo Hernandez	University of Barcelona, Spain
Zahra Ahmadi	Johannes Gutenberg Universität Mainz, Germany

Production, Publicity, and Public Relations Chairs

Sophie Burkhardt	Johannes Gutenberg Universität Mainz, Germany
Julia Sidorova	Universidad Complutense de Madrid, Spain

Local Chairs

Iñaki Inza	University of the Basque Country, Spain
Alexander Mendiburu	University of the Basque Country, Spain
Santiago Mazuelas	Basque Center for Applied Mathematics, Spain
Aritz Pèrez	Basque Center for Applied Mathematics, Spain
Borja Calvo	University of the Basque Country, Spain

Proceedings Chair

Tania Cerquitelli	Politecnico di Torino, Italy

Sponsorship Chair

Santiago Mazuelas	Basque Center for Applied Mathematics, Spain

Web Chairs

Olatz Hernandez Aretxabaleta	Basque Center for Applied Mathematics, Spain
Estíbaliz Gutièrrez	Basque Center for Applied Mathematics, Spain

ECML PKDD Steering Committee

Andrea Passerini	University of Trento, Italy
Francesco Bonchi	ISI Foundation, Italy
Albert Bifet	Télécom ParisTech, France
Sašo Džeroski	Jožef Stefan Institute, Slovenia
Katharina Morik	TU Dortmund, Germany
Arno Siebes	Utrecht University, The Netherlands
Siegfried Nijssen	Université Catholique de Louvain, Belgium

Luís Moreira-Matias	Finiata GmbH, Germany
Alessandra Sala	Shutterstock, Ireland
Georgiana Ifrim	University College Dublin, Ireland
Thomas Gärtner	University of Nottingham, UK
Neil Hurley	University College Dublin, Ireland
Michele Berlingerio	IBM Research, Ireland
Elisa Fromont	Université de Rennes, France
Arno Knobbe	Universiteit Leiden, The Netherlands
Ulf Brefeld	Leuphana Universität Lüneburg, Germany
Andreas Hotho	Julius-Maximilians-Universität Würzburg, Germany
Ira Assent	Aarhus University, Denmark
Kristian Kersting	TU Darmstadt University, Germany
Jefrey Lijffijt	Ghent University, Belgium
Isabel Valera	Saarland University, Germany

Program Committee

Guest Editorial Board, Journal Track

Richard Allmendinger	University of Manchester
Marie Anastacio	Leiden University
Ana Paula Appel	IBM Research Brazil
Dennis Assenmacher	University of Münster
Ira Assent	Aarhus University
Martin Atzmueller	Osnabrueck University
Jaume Bacardit	Newcastle University
Anthony Bagnall	University of East Anglia
Mitra Baratchi	University of Twente
Srikanta Bedathur	IIT Delhi
Alessio Benavoli	CSIS
Viktor Bengs	Paderborn University
Massimo Bilancia	University of Bari "Aldo Moro"
Klemens Böhm	Karlsruhe Institute of Technology
Veronica Bolon Canedo	Universidade da Coruna
Ilaria Bordino	UniCredit R&D
Jakob Bossek	University of Adelaide
Ulf Brefeld	Leuphana Universität Luneburg
Michelangelo Ceci	Universita degli Studi di Bari "Aldo Moro"
Loïc Cerf	Universidade Federal de Minas Gerais
Victor Manuel Cerqueira	University of Porto
Laetitia Chapel	IRISA
Silvia Chiusano	Politecnico di Torino
Roberto Corizzo	American University, Washington D.C.
Marco de Gemmis	Università degli Studi di Bari "Aldo Moro"
Sébastien Destercke	Università degli Studi di Bari "Aldo Moro"
Shridhar Devamane	Visvesvaraya Technological University

Peer Kröger	Ludwig Maximilian University of Munich
Meelis Kull	University of Tartu
Michel Lang	TU Dortmund University
Helge Langseth	Norwegian University of Science and Technology
Oswald Lanz	FBK
Mark Last	Ben-Gurion University of the Negev
Kangwook Lee	University of Wisconsin-Madison
Jurica Levatic	IRB Barcelona
Thomar Liebig	TU Dortmund
Hsuan-Tien Lin	National Taiwan University
Marius Lindauer	Leibniz University Hannover
Marco Lippi	University of Modena and Reggio Emilia
Corrado Loglisci	Università degli Studi di Bari
Manuel Lopez-Ibanez	University of Malaga
Nuno Lourenço	University of Coimbra
Claudio Lucchese	Ca' Foscari University of Venice
Brian Mac Namee	University College Dublin
Gjorgji Madjarov	Ss. Cyril and Methodius University
Davide Maiorca	University of Cagliari
Giuseppe Manco	ICAR-CNR
Elena Marchiori	Radboud University
Elio Masciari	Università di Napoli Federico II
Andres R. Masegosa	Norwegian University of Science and Technology
Ernestina Menasalvas	Universidad Politécnica de Madrid
Rosa Meo	University of Torino
Paolo Mignone	University of Bari "Aldo Moro"
Anna Monreale	University of Pisa
Giovanni Montana	University of Warwick
Grègoire Montavon	TU Berlin
Katharina Morik	TU Dortmund
Animesh Mukherjee	Indian Institute of Technology, Kharagpur
Amedeo Napoli	LORIA Nancy
Frank Naumann	University of Adelaide
Thomas Dyhre	Aalborg University
Bruno Ordozgoiti	Aalto University
Rita P. Ribeiro	University of Porto
Pance Panov	Jozef Stefan Institute
Apostolos Papadopoulos	Aristotle University of Thessaloniki
Panagiotis Papapetrou	Stockholm University
Andrea Passerini	University of Trento
Mykola Pechenizkiy	Eindhoven University of Technology
Charlotte Pelletier	Université Bretagne Sud
Ruggero G. Pensa	University of Torino
Nico Piatkowski	TU Dortmund
Dario Piga	IDSIA Dalle Molle Institute for Artificial Intelligence Research - USI/SUPSI

Gianvito Pio	Università degli Studi di Bari "Aldo Moro"
Marc Plantevit	LIRIS - Université Claude Bernard Lyon 1
Marius Popescu	University of Bucharest
Raphael Prager	University of Münster
Mike Preuss	Universiteit Leiden
Jose M. Puerta	Universidad de Castilla-La Mancha
Kai Puolamäki	University of Helsinki
Chedy Raïssi	Inria
Jan Ramon	Inria
Matteo Riondato	Amherst College
Thomas A. Runkler	Siemens Corporate Technology
Antonio Salmerón	University of Almería
Joerg Sander	University of Alberta
Roberto Santana	University of the Basque Country
Michael Schaub	RWTH Aachen
Lars Schmidt-Thieme	University of Hildesheim
Santiago Segui	Universitat de Barcelona
Thomas Seidl	Ludwig-Maximilians-Universitaet Muenchen
Moritz Seiler	University of Münster
Shinichi Shirakawa	Yokohama National University
Jim Smith	University of the West of England
Carlos Soares	University of Porto
Gerasimos Spanakis	Maastricht University
Giancarlo Sperlì	University of Naples Federico II
Myra Spiliopoulou	Otto-von-Guericke-University Magdeburg
Giovanni Stilo	Università degli Studi dell'Aquila
Catalin Stoean	University of Craiova
Mahito Sugiyama	National Institute of Informatics
Nikolaj Tatti	University of Helsinki
Alexandre Termier	Université de Rennes 1
Kevin Tierney	Bielefeld University
Luis Torgo	University of Porto
Roberto Trasarti	CNR Pisa
Sébastien Treguer	Inria
Leonardo Trujillo	Instituto Tecnológico de Tijuana
Ivor Tsang	University of Technology Sydney
Grigorios Tsoumakas	Aristotle University of Thessaloniki
Steffen Udluft	Siemens
Arnaud Vandaele	Université de Mons
Matthijs van Leeuwen	Leiden University
Celine Vens	KU Leuven Kulak
Herna Viktor	University of Ottawa
Marco Virgolin	Centrum Wiskunde & Informatica
Jordi Vitrià	Universitat de Barcelona
Christel Vrain	LIFO – University of Orléans
Jilles Vreeken	Helmholtz Center for Information Security

Willem Waegeman	Ghent University
David Walker	University of Plymouth
Hao Wang	Leiden University
Elizabeth F. Wanner	CEFET
Tu Wei-Wei	4paradigm
Pascal Welke	University of Bonn
Marcel Wever	Paderborn University
Man Leung Wong	Lingnan University
Stefan Wrobel	Fraunhofer IAIS, University of Bonn
Zheng Ying	Inria
Guoxian Yu	Shandong University
Xiang Zhang	Harvard University
Ye Zhu	Deakin University
Arthur Zimek	University of Southern Denmark
Albrecht Zimmermann	Université Caen Normandie
Marinka Zitnik	Harvard University

Area Chairs, Research Track

Fabrizio Angiulli	University of Calabria
Ricardo Baeza-Yates	Universitat Pompeu Fabra
Roberto Bayardo	Google
Bettina Berendt	Katholieke Universiteit Leuven
Philipp Berens	University of Tübingen
Michael Berthold	University of Konstanz
Hendrik Blockeel	Katholieke Universiteit Leuven
Juergen Branke	University of Warwick
Ulf Brefeld	Leuphana University Lüneburg
Toon Calders	Universiteit Antwerpen
Michelangelo Ceci	Università degli Studi di Bari "Aldo Moro"
Duen Horng Chau	Georgia Institute of Technology
Nicolas Courty	Université Bretagne Sud, IRISA Research Institute Computer and Systems Aléatoires
Bruno Cremilleux	Université de Caen Normandie
Philippe Cudre-Mauroux	University of Fribourg
James Cussens	University of Bristol
Jesse Davis	Katholieke Universiteit Leuven
Bob Durrant	University of Waikato
Tapio Elomaa	Tampere University
Johannes Fürnkranz	Johannes Kepler University Linz
Eibe Frank	University of Waikato
Elisa Fromont	Université de Rennes 1
Stephan Günnemann	Technical University of Munich
Patrick Gallinari	LIP6 - University of Paris
Joao Gama	University of Porto
Przemyslaw Grabowicz	University of Massachusetts, Amherst

Eyke Hüllermeier	Paderborn University
Allan Hanbury	Vienna University of Technology
Daniel Hernández-Lobato	Universidad Autónoma de Madrid
José Hernández-Orallo	Universitat Politècnica de València
Andreas Hotho	University of Wuerzburg
Inaki Inza	University of the Basque Country
Marius Kloft	TU Kaiserslautern
Arno Knobbe	Universiteit Leiden
Lars Kotthoff	University of Wyoming
Danica Kragic	KTH Royal Institute of Technology
Sébastien Lefèvre	Université Bretagne Sud
Bruno Lepri	FBK-Irst
Patrick Loiseau	Inria and Ecole Polytechnique
Jorg Lucke	University of Oldenburg
Fragkiskos Malliaros	Paris-Saclay University, CentraleSupelec, and Inria
Giuseppe Manco	ICAR-CNR
Dunja Mladenic	Jozef Stefan Institute
Katharina Morik	TU Dortmund
Sriraam Natarajan	Indiana University Bloomington
Siegfried Nijssen	Université catholique de Louvain
Andrea Passerini	University of Trento
Mykola Pechenizkiy	Eindhoven University of Technology
Jaakko Peltonen	Aalto University and University of Tampere
Marian-Andrei Rizoiu	University of Technology Sydney
Céline Robardet	INSA Lyon
Maja Rudolph	Bosch
Lars Schmidt-Thieme	University of Hildesheim
Thomas Seidl	Ludwig-Maximilians-Universität München
Arno Siebes	Utrecht University
Myra Spiliopoulou	Otto-von-Guericke-University Magdeburg
Yizhou Sun	University of California, Los Angeles
Einoshin Suzuki	Kyushu University
Jie Tang	Tsinghua University
Ke Tang	Southern University of Science and Technology
Marc Tommasi	University of Lille
Isabel Valera	Saarland University
Celine Vens	KU Leuven Kulak
Christel Vrain	LIFO - University of Orléans
Jilles Vreeken	Helmholtz Center for Information Security
Willem Waegeman	Ghent University
Stefan Wrobel	Fraunhofer IAIS, University of Bonn
Min-Ling Zhang	Southeast University

Area Chairs, Applied Data Science Track

Francesco Calabrese	Vodafone
Michelangelo Ceci	Università degli Studi di Bari "Aldo Moro"
Gianmarco De Francisci Morales	ISI Foundation
Tom Diethe	Amazon
Johannes Fründkranz	Johannes Kepler University Linz
Han Fang	Facebook
Faisal Farooq	Qatar Computing Research Institute
Rayid Ghani	Carnegie Mellon Univiersity
Francesco Gullo	UniCredit
Xiangnan He	University of Science and Technology of China
Georgiana Ifrim	University College Dublin
Thorsten Jungeblut	Bielefeld University of Applied Sciences
John A. Lee	Université catholique de Louvain
Ilias Leontiadis	Samsung AI
Viktor Losing	Honda Research Institute Europe
Yin Lou	Ant Group
Gabor Melli	Sony PlayStation
Luis Moreira-Matias	University of Porto
Nicolò Navarin	University of Padova
Benjamin Paaßen	German Research Center for Artificial Intelligence
Kitsuchart Pasupa	King Mongkut's Institute of Technology Ladkrabang
Mykola Pechenizkiy	Eindhoven University of Technology
Julien Perez	Naver Labs Europe
Fabio Pinelli	IMT Lucca
Zhaochun Ren	Shandong University
Sascha Saralajew	Porsche AG
Fabrizio Silvestri	Facebook
Sinong Wang	Facebook AI
Xing Xie	Microsoft Research Asia
Jian Xu	Citadel
Jing Zhang	Renmin University of China

Program Committee Members, Research Track

Hanno Ackermann	Leibniz University Hannover
Linara Adilova	Fraunhofer IAIS
Zahra Ahmadi	Johannes Gutenberg University
Cuneyt Gurcan Akcora	University of Manitoba
Omer Deniz Akyildiz	University of Warwick
Carlos M. Alaíz Gudín	Universidad Autónoma de Madrid
Mohamed Alami	Ecole Polytechnique
Chehbourne Abdullah Alchihabi	Carleton University
Pegah Alizadeh	University of Caen Normandy

Reem Alotaibi	King Abdulaziz University
Massih-Reza Amini	Université Grenoble Alpes
Shin Ando	Tokyo University of Science
Thiago Andrade	INESC TEC
Kimon Antonakopoulos	Inria
Alessandro Antonucci	IDSIA
Muhammad Umer Anwaar	Technical University of Munich
Eva Armengol	IIIA-SIC
Dennis Assenmacher	University of Münster
Matthias Aßenmacher	Ludwig-Maximilians-Universität München
Martin Atzmueller	Osnabrueck University
Behrouz Babaki	Polytechnique Montreal
Rohit Babbar	Aalto University
Elena Baralis	Politecnico di Torino
Mitra Baratchi	University of Twente
Christian Bauckhage	University of Bonn, Fraunhofer IAIS
Martin Becker	University of Würzburg
Jessa Bekker	Katholieke Universiteit Leuven
Colin Bellinger	National Research Council of Canada
Khalid Benabdeslem	LIRIS Laboratory, Claude Bernard University Lyon I
Diana Benavides-Prado	Auckland University of Technology
Anes Bendimerad	LIRIS
Christoph Bergmeir	University of Granada
Alexander Binder	UiO
Aleksandar Bojchevski	Technical University of Munich
Ahcène Boubekki	UiT Arctic University of Norway
Paula Branco	EECS University of Ottawa
Tanya Braun	University of Lübeck
Katharina Breininger	Friedrich-Alexander-Universität Erlangen Nürnberg
Wieland Brendel	University of Tübingen
John Burden	University of Cambridge
Sophie Burkhardt	TU Kaiserslautern
Sebastian Buschjäger	TU Dortmund
Borja Calvo	University of the Basque Country
Stephane Canu	LITIS, INSA de Rouen
Cornelia Caragea	University of Illinois at Chicago
Paula Carroll	University College Dublin
Giuseppe Casalicchio	Ludwig Maximilian University of Munich
Bogdan Cautis	Paris-Saclay University
Rémy Cazabet	Université de Lyon
Josu Ceberio	University of the Basque Country
Peggy Cellier	IRISA/INSA Rennes
Mattia Cerrato	Università degli Studi di Torino
Ricardo Cerri	Federal University of Sao Carlos
Alessandra Cervone	Amazon
Ayman Chaouki	Institut Mines-Télécom

Paco Charte	Universidad de Jaén
Rita Chattopadhyay	Intel Corporation
Vaggos Chatziafratis	Stanford University
Tianyi Chen	Zhejiang University City College
Yuzhou Chen	Southern Methodist University
Yiu-Ming Cheung	Hong Kong Baptist University
Anshuman Chhabra	University of California, Davis
Ting-Wu Chin	Carnegie Mellon University
Oana Cocarascu	King's College London
Lidia Contreras-Ochando	Universitat Politècnica de València
Roberto Corizzo	American University
Anna Helena Reali Costa	Universidade de São Paulo
Fabrizio Costa	University of Exeter
Gustavo De Assis Costa	Instituto Federal de Educação, Ciência e Tecnologia de Goiás
Bertrand Cuissart	GREYC
Thi-Bich-Hanh Dao	University of Orleans
Mayukh Das	Microsoft Research Lab
Padraig Davidson	Universität Würzburg
Paul Davidsson	Malmö University
Gwendoline De Bie	ENS
Tijl De Bie	Ghent University
Andre de Carvalho	Universidade de São Paulo
Orphée De Clercq	Ghent University
Alper Demir	İzmir University of Economics
Nicola Di Mauro	Università degli Studi di Bari "Aldo Moro"
Yao-Xiang Ding	Nanjing University
Carola Doerr	Sorbonne University
Boxiang Dong	Montclair State University
Ruihai Dong	University College Dublin
Xin Du	Eindhoven University of Technology
Stefan Duffner	LIRIS
Wouter Duivesteijn	Eindhoven University of Technology
Audrey Durand	McGill University
Inês Dutra	University of Porto
Saso Dzeroski	Jozef Stefan Institute
Hamid Eghbalzadeh	Johannes Kepler University
Dominik Endres	University of Marburg
Roberto Esposito	Università degli Studi di Torino
Samuel G. Fadel	Universidade Estadual de Campinas
Xiuyi Fan	Imperial College London
Hadi Fanaee-T.	Halmstad University
Elaine Faria	Federal University of Uberlandia
Fabio Fassetti	University of Calabria
Kilian Fatras	Inria
Ad Feelders	Utrecht University

Songhe Feng	Beijing Jiaotong University
Àngela Fernández-Pascual	Universidad Autónoma de Madrid
Daniel Fernández-Sánchez	Universidad Autónoma de Madrid
Sofia Fernandes	University of Aveiro
Cesar Ferri	Universitat Politécnica de Valéncia
Rémi Flamary	École Polytechnique
Michael Flynn	University of East Anglia
Germain Forestier	Université de Haute Alsace
Kary Främling	Umeå University
Benoît Frénay	Université de Namur
Vincent Francois	University of Amsterdam
Emilia Gómez	Joint Research Centre - European Commission
Luis Galárraga	Inria
Esther Galbrun	University of Eastern Finland
Claudio Gallicchio	University of Pisa
Jochen Garcke	University of Bonn
Clément Gautrais	KU Leuven
Yulia Gel	University of Texas at Dallas and University of Waterloo
Pierre Geurts	University of Liège
Amirata Ghorbani	Stanford University
Heitor Murilo Gomes	University of Waikato
Chen Gong	Shanghai Jiao Tong University
Bedartha Goswami	University of Tübingen
Henry Gouk	University of Edinburgh
James Goulding	University of Nottingham
Antoine Gourru	Université Lumière Lyon 2
Massimo Guarascio	ICAR-CNR
Riccardo Guidotti	University of Pisa
Ekta Gujral	University of California, Riverside
Francesco Gullo	UniCredit
Tias Guns	Vrije Universiteit Brussel
Thomas Guyet	Institut Agro, IRISA
Tom Hanika	University of Kassel
Valentin Hartmann	Ecole Polytechnique Fédérale de Lausanne
Marwan Hassani	Eindhoven University of Technology
Jukka Heikkonen	University of Turku
Fredrik Heintz	Linköping University
Sibylle Hess	TU Eindhoven
Jaakko Hollmén	Aalto University
Tamas Horvath	University of Bonn, Fraunhofer IAIS
Binbin Hu	Ant Group
Hong Huang	UGoe
Georgiana Ifrim	University College Dublin
Angelo Impedovo	Università degli studi di Bari "Aldo Moro"

Nathalie Japkowicz	American University
Szymon Jaroszewicz	Institute of Computer Science, Polish Academy of Sciences
Saumya Jetley	Inria
Binbin Jia	Southeast University
Xiuyi Jia	School of Computer Science and Technology, Nanjing University of Science and Technology
Yuheng Jia	City University of Hong Kong
Siyang Jiang	National Taiwan University
Priyadarshini Kumari	IIT Bombay
Ata Kaban	University of Birmingham
Tomasz Kajdanowicz	Wroclaw University of Technology
Vana Kalogeraki	Athens University of Economics and Business
Toshihiro Kamishima	National Institute of Advanced Industrial Science and Technology
Michael Kamp	Monash University
Bo Kang	Ghent University
Dimitrios Karapiperis	Hellenic Open University
Panagiotis Karras	Aarhus University
George Karypis	University of Minnesota
Mark Keane	University College Dublin
Kristian Kersting	TU Darmstadt
Masahiro Kimura	Ryukoku University
Jiri Klema	Czech Technical University
Dragi Kocev	Jozef Stefan Institute
Masahiro Kohjima	NTT
Lukasz Korycki	Virginia Commonwealth University
Peer Kröger	Ludwig Maximilian University of Münich
Anna Krause	University of Würzburg
Bartosz Krawczyk	Virginia Commonwealth University
Georg Krempl	Utrecht University
Meelis Kull	University of Tartu
Vladimir Kuzmanovski	Aalto University
Ariel Kwiatkowski	Ecole Polytechnique
Emanuele La Malfa	University of Oxford
Beatriz López	University of Girona
Preethi Lahoti	Aalto University
Ichraf Lahouli	Euranova
Niklas Lavesson	Jönköping University
Aonghus Lawlor	University College Dublin
Jeongmin Lee	University of Pittsburgh
Daniel Lemire	LICEF Research Center and Université du Québec
Florian Lemmerich	University of Passau
Elisabeth Lex	Graz University of Technology
Jiani Li	Vanderbilt University
Rui Li	Inspur Group
Wentong Liao	Lebniz University Hannover

Jiayin Lin	University of Wollongong
Rudolf Lioutikov	UT Austin
Marco Lippi	University of Modena and Reggio Emilia
Suzanne Little	Dublin City University
Shengcai Liu	University of Science and Technology of China
Shenghua Liu	Institute of Computing Technology, Chinese Academy of Sciences
Philipp Liznerski	Technische Universität Kaiserslautern
Corrado Loglisci	Università degli Studi di Bari "Aldo Moro"
Ting Long	Shanghai Jiaotong University
Tsai-Ching Lu	HRL Laboratories
Yunpu Ma	Siemens AG
Zichen Ma	The Chinese University of Hong Kong
Sara Madeira	Universidade de Lisboa
Simona Maggio	Dataiku
Sara Magliacane	IBM
Sebastian Mair	Leuphana University Lüneburg
Lorenzo Malandri	University of Milan Bicocca
Donato Malerba	Università degli Studi di Bari "Aldo Moro"
Pekka Malo	Aalto University
Robin Manhaeve	KU Leuven
Silviu Maniu	Université Paris-Sud
Giuseppe Marra	KU Leuven
Fernando Martínez-Plumed	Joint Research Centre - European Commission
Alexander Marx	Max Plank Institue for Informatics and Saarland University
Florent Masseglia	Inria
Tetsu Matsukawa	Kyushu University
Wolfgang Mayer	University of South Australia
Santiago Mazuelas	Basque center for Applied Mathematics
Stefano Melacci	University of Siena
Ernestina Menasalvas	Universidad Politécnica de Madrid
Rosa Meo	Università degli Studi di Torino
Alberto Maria Metelli	Politecnico di Milano
Saskia Metzler	Max Planck Institute for Informatics
Alessio Micheli	University of Pisa
Paolo Mignone	Università degli studi di Bari "Aldo Moro"
Matej Mihelčić	University of Zagreb
Decebal Constantin Mocanu	University of Twente
Nuno Moniz	INESC TEC and University of Porto
Carlos Monserrat	Universitat Politécnica de Valéncia
Corrado Monti	ISI Foundation
Jacob Montiel	University of Waikato
Ahmadreza Mosallanezhad	Arizona State University
Tanmoy Mukherjee	University of Tennessee
Martin Mundt	Goethe University

Mohamed Nadif	Université de Paris
Omer Nagar	Bar Ilan University
Felipe Kenji Nakano	Katholieke Universiteit Leuven
Mirco Nanni	KDD-Lab ISTI-CNR Pisa
Apurva Narayan	University of Waterloo
Nicolò Navarin	University of Padova
Benjamin Negrevergne	Paris Dauphine University
Hurley Neil	University College Dublin
Stefan Neumann	University of Vienna
Ngoc-Tri Ngo	The University of Danang - University of Science and Technology
Dai Nguyen	Monash University
Eirini Ntoutsi	Free University Berlin
Andrea Nuernberger	Otto-von-Guericke-Universität Magdeburg
Pablo Olmos	University Carlos III
James O'Neill	University of Liverpool
Barry O'Sullivan	University College Cork
Rita P. Ribeiro	University of Porto
Aritz Pèrez	Basque Center for Applied Mathematics
Joao Palotti	Qatar Computing Research Institute
Guansong Pang	University of Adelaide
Pance Panov	Jozef Stefan Institute
Evangelos Papalexakis	University of California, Riverside
Haekyu Park	Georgia Institute of Technology
Sudipta Paul	Umeå University
Yulong Pei	Eindhoven University of Technology
Charlotte Pelletier	Université Bretagne Sud
Ruggero G. Pensa	University of Torino
Bryan Perozzi	Google
Nathanael Perraudin	ETH Zurich
Lukas Pfahler	TU Dortmund
Bastian Pfeifer	Medical University of Graz
Nico Piatkowski	TU Dortmund
Robert Pienta	Georgia Institute of Technology
Fábio Pinto	Faculdade de Economia do Porto
Gianvito Pio	University of Bari "Aldo Moro"
Giuseppe Pirrò	Sapienza University of Rome
Claudia Plant	University of Vienna
Marc Plantevit	LIRIS - Universitè Claude Bernard Lyon 1
Amit Portnoy	Ben Gurion University
Melanie Pradier	Harvard University
Paul Prasse	University of Potsdam
Philippe Preux	Inria, LIFL, Universitè de Lille
Ricardo Prudencio	Federal University of Pernambuco
Zhou Qifei	Peking University
Erik Quaeghebeur	TU Eindhoven

Tahrima Rahman	University of Texas at Dallas
Herilalaina Rakotoarison	Inria
Alexander Rakowski	Hasso Plattner Institute
María José Ramírez	Universitat Politècnica de València
Visvanathan Ramesh	Goethe University
Jan Ramon	Inria
Huzefa Rangwala	George Mason University
Aleksandra Rashkovska	Jožef Stefan Institute
Joe Redshaw	University of Nottingham
Matthias Renz	Christian-Albrechts-Universität zu Kiel
Matteo Riondato	Amherst College
Ettore Ritacco	ICAR-CNR
Mateus Riva	Télécom ParisTech
Antonio Rivera	Universidad Politécnica de Madrid
Marko Robnik-Sikonja	University of Ljubljana
Simon Rodriguez Santana	Institute of Mathematical Sciences (ICMAT-CSIC)
Mohammad Rostami	University of Southern California
Céline Rouveirol	Laboratoire LIPN-UMR CNRS
Jože Rožanec	Jožef Stefan Institute
Peter Rubbens	Flanders Marine Institute
David Ruegamer	LMU Munich
Salvatore Ruggieri	Università di Pisa
Francisco Ruiz	DeepMind
Anne Sabourin	Télécom ParisTech
Tapio Salakoski	University of Turku
Pablo Sanchez-Martin	Max Planck Institute for Intelligent Systems
Emanuele Sansone	KU Leuven
Yucel Saygin	Sabanci University
Patrick Schäfer	Humboldt Universität zu Berlin
Pierre Schaus	UCLouvain
Ute Schmid	University of Bamberg
Sebastian Schmoll	Ludwig Maximilian University of Munich
Marc Schoenauer	Inria
Matthias Schubert	Ludwig Maximilian University of Munich
Marian Scuturici	LIRIS-INSA de Lyon
Junming Shao	University of Science and Technology of China
Manali Sharma	Samsung Semiconductor Inc.
Abdul Saboor Sheikh	Zalando Research
Jacquelyn Shelton	Hong Kong Polytechnic University
Feihong Shen	Jilin University
Gavin Smith	University of Nottingham
Kma Solaiman	Purdue University
Arnaud Soulet	Université François Rabelais Tours
Alessandro Sperduti	University of Padua
Giovanni Stilo	Università degli Studi dell'Aquila
Michiel Stock	Ghent University

Lech Szymanski	University of Otago
Shazia Tabassum	University of Porto
Andrea Tagarelli	University of Calabria
Acar Tamersoy	NortonLifeLock Research Group
Chang Wei Tan	Monash University
Sasu Tarkoma	University of Helsinki
Bouadi Tassadit	IRISA-Université Rennes 1
Nikolaj Tatti	University of Helsinki
Maryam Tavakol	Eindhoven University of Technology
Pooya Tavallali	University of California, Los Angeles
Maguelonne Teisseire	Irstea - UMR Tetis
Alexandre Termier	Université de Rennes 1
Stefano Teso	University of Trento
Janek Thomas	Fraunhofer Institute for Integrated Circuits IIS
Alessandro Tibo	Aalborg University
Sofia Triantafillou	University of Pittsburgh
Grigorios Tsoumakas	Aristotle University of Thessaloniki
Peter van der Putten	LIACS, Leiden University and Pegasystems
Elia Van Wolputte	KU Leuven
Robert A. Vandermeulen	Technische Universität Berlin
Fabio Vandin	University of Padova
Filipe Veiga	Massachusetts Institute of Technology
Bruno Veloso	Universidade Portucalense and LIAAD - INESC TEC
Sebastián Ventura	University of Cordoba
Rosana Veroneze	UNICAMP
Herna Viktor	University of Ottawa
João Vinagre	INESC TEC
Huaiyu Wan	Beijing Jiaotong University
Beilun Wang	Southeast University
Hu Wang	University of Adelaide
Lun Wang	University of California, Berkeley
Yu Wang	Peking University
Zijie J. Wang	Georgia Tech
Tong Wei	Nanjing University
Pascal Welke	University of Bonn
Joerg Wicker	University of Auckland
Moritz Wolter	University of Bonn
Ning Xu	Southeast University
Akihiro Yamaguchi	Toshiba Corporation
Haitian Yang	Institute of Information Engineering, Chinese Academy of Sciences
Yang Yang	Nanjing University
Zhuang Yang	Sun Yat-sen University
Helen Yannakoudakis	King's College London
Heng Yao	Tongji University
Han-Jia Ye	Nanjing University

Kristina Yordanova	University of Rostock
Tetsuya Yoshida	Nara Women's University
Guoxian Yu	Shandong University, China
Sha Yuan	Tsinghua University
Valentina Zantedeschi	INSA Lyon
Albin Zehe	University of Würzburg
Bob Zhang	University of Macau
Teng Zhang	Huazhong University of Science and Technology
Liang Zhao	University of São Paulo
Bingxin Zhou	University of Sydney
Kenny Zhu	Shanghai Jiao Tong University
Yanqiao Zhu	Institute of Automation, Chinese Academy of Sciences
Arthur Zimek	University of Southern Denmark
Albrecht Zimmermann	Université Caen Normandie
Indre Zliobaite	University of Helsinki
Markus Zopf	NEC Labs Europe

Program Committee Members, Applied Data Science Track

Mahdi Abolghasemi	Monash University
Evrim Acar	Simula Research Lab
Deepak Ajwani	University College Dublin
Pegah Alizadeh	University of Caen Normandy
Jean-Marc Andreoli	Naver Labs Europe
Giorgio Angelotti	ISAE Supaero
Stefanos Antaris	KTH Royal Institute of Technology
Xiang Ao	Institute of Computing Technology, Chinese Academy of Sciences
Yusuf Arslan	University of Luxembourg
Cristian Axenie	Huawei European Research Center
Hanane Azzag	Université Sorbonne Paris Nord
Pedro Baiz	Imperial College London
Idir Benouaret	CNRS, Université Grenoble Alpes
Laurent Besacier	Laboratoire d'Informatique de Grenoble
Antonio Bevilacqua	Insight Centre for Data Analytics
Adrien Bibal	University of Namur
Wu Bin	Zhengzhou University
Patrick Blöbaum	Amazon
Pavel Blinov	Sber Artificial Intelligence Laboratory
Ludovico Boratto	University of Cagliari
Stefano Bortoli	Huawei Technologies Duesseldorf
Zekun Cai	University of Tokyo
Nicolas Carrara	University of Toronto
John Cartlidge	University of Bristol
Oded Cats	Delft University of Technology
Tania Cerquitelli	Politecnico di Torino

Prithwish Chakraborty   IBM
Rita Chattopadhyay   Intel Corp.
Keru Chen   GrabTaxi Pte Ltd.
Liang Chen   Sun Yat-sen University
Zhiyong Cheng   Shandong Artificial Intelligence Institute
Silvia Chiusano   Politecnico di Torino
Minqi Chong   Citadel
Jeremie Clos   University of Nottingham
J. Albert Conejero Casares   Universitat Politécnica de Vaécia
Evan Crothers   University of Ottawa
Henggang Cui   Uber ATG
Tiago Cunha   University of Porto
Padraig Cunningham   University College Dublin
Eustache Diemert   CRITEO Research
Nat Dilokthanakul   Vidyasirimedhi Institute of Science and Technology
Daizong Ding   Fudan University
Kaize Ding   ASU
Michele Donini   Amazon
Lukas Ewecker   Porsche AG
Zipei Fan   University of Tokyo
Bojing Feng   National Laboratory of Pattern Recognition, Institute
                                    of Automation, Chinese Academy of Science
Flavio Figueiredo   Universidade Federal de Minas Gerais
Blaz Fortuna   Qlector d.o.o.
Zuohui Fu   Rutgers University
Fabio Fumarola   University of Bari "Aldo Moro"
Chen Gao   Tsinghua University
Luis Garcia   University of Brasília
Cinmayii   University of the Philippines Mindanao
   Garillos-Manliguez
Kiran Garimella   Aalto University
Etienne Goffinet   Laboratoire LIPN-UMR CNRS
Michael Granitzer   University of Passau
Xinyu Guan   Xi'an Jiaotong University
Thomas Guyet   Institut Agro, IRISA
Massinissa Hamidi   Laboratoire LIPN-UMR CNRS
Junheng Hao   University of California, Los Angeles
Martina Hasenjaeger   Honda Research Institute Europe GmbH
Lars Holdijk   University of Amsterdam
Chao Huang   University of Notre Dame
Guanjie Huang   Penn State University
Hong Huang   UGoe
Yiran Huang   TECO
Madiha Ijaz   IBM
Roberto Interdonato   CIRAD - UMR TETIS
Omid Isfahani Alamdari   University of Pisa

Guillaume Jacquet	JRC
Nathalie Japkowicz	American University
Shaoxiong Ji	Aalto University
Nan Jiang	Purdue University
Renhe Jiang	University of Tokyo
Song Jiang	University of California, Los Angeles
Adan Jose-Garcia	University of Exeter
Jihed Khiari	Johannes Kepler Universität
Hyunju Kim	KAIST
Tomas Kliegr	University of Economics
Yun Sing Koh	University of Auckland
Pawan Kumar	IIIT, Hyderabad
Chandresh Kumar Maurya	CSE, IIT Indore
Thach Le Nguyen	The Insight Centre for Data Analytics
Mustapha Lebbah	Université Paris 13, LIPN-CNRS
Dongman Lee	Korea Advanced Institute of Science and Technology
Rui Li	Sony
Xiaoting Li	Pennsylvania State University
Zeyu Li	University of California, Los Angeles
Defu Lian	University of Science and Technology of China
Jiayin Lin	University of Wollongong
Jason Lines	University of East Anglia
Bowen Liu	Stanford University
Pedro Henrique Luz de Araujo	University of Brasilia
Fenglong Ma	Pennsylvania State University
Brian Mac Namee	University College Dublin
Manchit Madan	Myntra
Ajay Mahimkar	AT&T Labs
Domenico Mandaglio	Università della Calabria
Koji Maruhashi	Fujitsu Laboratories Ltd.
Sarah Masud	LCS2, IIIT-D
Eric Meissner	University of Cambridge
João Mendes-Moreira	INESC TEC
Chuan Meng	Shandong University
Fabio Mercorio	University of Milano-Bicocca
Angela Meyer	Bern University of Applied Sciences
Congcong Miao	Tsinghua University
Stéphane Moreau	Université de Sherbrooke
Koyel Mukherjee	IBM Research India
Fabricio Murai	Universidade Federal de Minas Gerais
Taichi Murayama	NAIST
Philip Nadler	Imperial College London
Franco Maria Nardini	ISTI-CNR
Ngoc-Tri Ngo	The University of Danang - University of Science and Technology

Anna Nguyen	Karlsruhe Institute of Technology
Hao Niu	KDDI Research, Inc.
Inna Novalija	Jožef Stefan Institute
Tsuyosh Okita	Kyushu Institute of Technology
Aoma Osmani	LIPN-UMR CNRS 7030, Université Paris 13
Latifa Oukhellou	IFSTTAR
Andrei Paleyes	University of Cambridge
Chanyoung Park	KAIST
Juan Manuel Parrilla Gutierrez	University of Glasgow
Luca Pasa	Università degli Studi Di Padova
Pedro Pereira Rodrigues	University of Porto
Miquel Perelló-Nieto	University of Bristol
Beatrice Perez	Dartmouth College
Alan Perotti	ISI Foundation
Mirko Polato	University of Padua
Giovanni Ponti	ENEA
Nicolas Posocco	Eura Nova
Cedric Pradalier	GeorgiaTech Lorraine
Giulia Preti	ISI Foundation
A. A. A. Qahtan	Utrecht University
Chuan Qin	University of Science and Technology of China
Dimitrios Rafailidis	University of Thessaly
Cyril Ray	Arts et Metiers Institute of Technology, Ecole Navale, IRENav
Wolfgang Reif	University of Augsburg
Kit Rodolfa	Carnegie Mellon University
Christophe Rodrigues	Pôle Universitaire Léonard de Vinci
Natali Ruchansky	Netflix
Hajer Salem	AUDENSIEL
Parinya Sanguansat	Panyapiwat Institute of Management
Atul Saroop	Amazon
Alexander Schiendorfer	Technische Hochschule Ingolstadt
Peter Schlicht	Volkswagen
Jens Schreiber	University of Kassel
Alexander Schulz	Bielefeld University
Andrea Schwung	FH SWF
Edoardo Serra	Boise State University
Lorenzo Severini	UniCredit
Ammar Shaker	Paderborn University
Jiaming Shen	University of Illinois at Urbana-Champaign
Rongye Shi	Columbia University
Wang Siyu	Southwestern University of Finance and Economics
Hao Song	University of Bristol
Francesca Spezzano	Boise State University
Simon Stieber	University of Augsburg

Laurens Stoop	Utrecht University
Hongyang Su	Harbin Institute of Technology
David Sun	Apple
Weiwei Sun	Shandong University
Maryam Tabar	Pennsylvania State University
Anika Tabassum	Virginia Tech
Garth Tarr	University of Sydney
Dinh Van Tran	University of Padova
Sreekanth Vempati	Myntra
Herna Viktor	University of Ottawa
Daheng Wang	University of Notre Dame
Hongwei Wang	Stanford University
Wenjie Wang	National University of Singapore
Yue Wang	Microsoft Research
Zhaonan Wang	University of Tokyo and National Institute of Advanced Industrial Science and Technology
Michael Wilbur	Vanderbilt University
Roberto Wolfler Calvo	LIPN, Université Paris 13
Di Wu	Chongqing Institute of Green and Intelligent Technology
Gang Xiong	Chinese Academy of Sciences
Xiaoyu Xu	Chongqing Institute of Green and Intelligent Technology
Yexiang Xue	Purdue University
Sangeeta Yadav	Indian Institute of Science
Hao Yan	Washington University in St. Louis
Chuang Yang	University of Tokyo
Yang Yang	Northwestern University
You Yizhe	Institute of Information Engineering, Chinese Academy of Sciences
Alexander Ypma	ASML
Jun Yuan	The Boeing Company
Mingxuan Yue	University of Southern California
Danqing Zhang	Amazon
Jiangwei Zhang	Tencent
Xiaohan Zhang	Sony Interactive Entertainment
Xinyang Zhang	University of Illinois at Urbana-Champaign
Yongxin Zhang	Sun Yat-sen University
Mia Zhao	Airbnb
Tong Zhao	University of Notre Dame
Bin Zhou	National University of Defense Technology
Bo Zhou	Baidu
Louis Zigrand	Université Sorbonne Paris Nord

Sponsors

Contents – Part V

Automating Machine Learning, Optimization, and Feature Engineering

PuzzleShuffle: Undesirable Feature Learning for Semantic Shift Detection ... 3
Yusuke Kanebako and Kazuki Tsukamoto

Enabling Machine Learning on the Edge Using SRAM Conserving
Efficient Neural Networks Execution Approach 20
*Bharath Sudharsan, Pankesh Patel, John G. Breslin,
and Muhammad Intizar Ali*

AutoML Meets Time Series Regression Design and Analysis
of the AutoSeries Challenge 36
Zhen Xu, Wei-Wei Tu, and Isabelle Guyon

Methods for Automatic Machine-Learning Workflow Analysis 52
Lorenz Wendlinger, Emanuel Berndl, and Michael Granitzer

ConCAD: Contrastive Learning-Based Cross Attention for Sleep
Apnea Detection 68
Guanjie Huang and Fenglong Ma

Machine Learning Based Simulations and Knowledge Discovery

DeepPE: Emulating Parameterization in Numerical Weather Forecast
Model Through Bidirectional Network........................... 87
Fengyang Xu, Wencheng Shi, Yunfei Du, Zhiguang Chen, and Yutong Lu

Effects of Boundary Conditions in Fully Convolutional Networks
for Learning Spatio-Temporal Dynamics 102
*Antonio Alguacil, Wagner Gonçalves Pinto, Michael Bauerheim,
Marc C. Jacob, and Stéphane Moreau*

Physics Knowledge Discovery via Neural Differential
Equation Embedding 118
*Yexiang Xue, Md Nasim, Maosen Zhang, Cuncai Fan, Xinghang Zhang,
and Anter El-Azab*

A Bayesian Convolutional Neural Network for Robust Galaxy
Ellipticity Regression...................................... 135
*Claire Theobald, Bastien Arcelin, Frédéric Pennerath,
Brieuc Conan-Guez, Miguel Couceiro, and Amedeo Napoli*

Precise Weather Parameter Predictions for Target Regions
via Neural Networks   151
Yihe Zhang, Xu Yuan, Sytske K. Kimball, Eric Rappin, Li Chen,
Paul Darby, Tom Johnsten, Lu Peng, Boisy Pitre, David Bourrie,
and Nian-Feng Tzeng

Action Set Based Policy Optimization for Safe Power Grid Management   168
Bo Zhou, Hongsheng Zeng, Yuecheng Liu, Kejiao Li, Fan Wang,
and Hao Tian

Conditional Neural Relational Inference for Interacting Systems   182
Joao A. Candido Ramos, Lionel Blondé, Stéphane Armand,
and Alexandros Kalousis

Recommender Systems and Behavior Modeling

MMNet: Multi-granularity Multi-mode Network for Item-Level Share
Rate Prediction   201
Haomin Yu, Mingfei Liang, Ruobing Xie, Zhenlong Sun, Bo Zhang,
and Leyu Lin

The Joy of Dressing Is an Art: Outfit Generation Using
Self-attention Bi-LSTM   218
Manchit Madan, Ankur Chouragade, and Sreekanth Vempati

On Inferring a Meaningful Similarity Metric for Customer Behaviour   234
Sophie van den Berg and Marwan Hassani

Quantifying Explanations of Neural Networks in E-Commerce Based
on LRP   251
Anna Nguyen, Franz Krause, Daniel Hagenmayer, and Michael Färber

Natural Language Processing

Balancing Speed and Accuracy in Neural-Enhanced Phonetic
Name Matching   271
Philip Blair, Carmel Eliav, Fiona Hasanaj, and Kfir Bar

Robust Learning for Text Classification with Multi-source Noise
Simulation and Hard Example Mining   285
Guowei Xu, Wenbiao Ding, Weiping Fu, Zhongqin Wu, and Zitao Liu

Topic-to-Essay Generation with Comprehensive Knowledge Enhancement . . .   302
Zhiyue Liu, Jiahai Wang, and Zhenghong Li

Analyzing Research Trends in Inorganic Materials Literature Using NLP   319
Fusataka Kuniyoshi, Jun Ozawa, and Makoto Miwa

An Optimized NL2SQL System for Enterprise Data Mart   335
    Kaiwen Dong, Kai Lu, Xin Xia, David Cieslak, and Nitesh V. Chawla

Time Aspect in Making an Actionable Prediction
of a Conversation Breakdown.   351
    Piotr Janiszewski, Mateusz Lango, and Jerzy Stefanowski

Feature Enhanced Capsule Networks for Robust Automatic Essay Scoring . . .   365
    Arushi Sharma, Anubha Kabra, and Rajiv Kapoor

TagRec: Automated Tagging of Questions with Hierarchical
Learning Taxonomy   381
    V. Venktesh, Mukesh Mohania, and Vikram Goyal

Remote Sensing, Image and Video Processing

Checking Robustness of Representations Learned by Deep
Neural Networks.   399
    Kamil Szyc, Tomasz Walkowiak, and Henryk Maciejewski

CHECKER: Detecting Clickbait Thumbnails with Weak Supervision
and Co-teaching   415
    Tianyi Xie, Thai Le, and Dongwon Lee

Crowdsourcing Evaluation of Saliency-Based XAI Methods   431
    Xiaotian Lu, Arseny Tolmachev, Tatsuya Yamamoto, Koh Takeuchi,
    Seiji Okajima, Tomoyoshi Takebayashi, Koji Maruhashi,
    and Hisashi Kashima

Automated Machine Learning for Satellite Data: Integrating Remote
Sensing Pre-trained Models into AutoML Systems   447
    Nelly Rosaura Palacios Salinas, Mitra Baratchi, Jan N. van Rijn,
    and Andreas Vollrath

Multi-task Learning for User Engagement and Adoption in Live Video
Streaming Events   463
    Stefanos Antaris, Dimitrios Rafailidis, and Romina Arriaza

Social Media

Explainable Abusive Language Classification Leveraging User
and Network Data.   481
    Maximilian Wich, Edoardo Mosca, Adrian Gorniak, Johannes Hingerl,
    and Georg Groh

Calling to CNN-LSTM for Rumor Detection: A Deep Multi-channel Model
for Message Veracity Classification in Microblogs   497
  Abderrazek Azri, Cécile Favre, Nouria Harbi, Jérôme Darmont,
  and Camille Noûs

Correction to: Automated Machine Learning for Satellite Data:
Integrating Remote Sensing Pre-trained Models into AutoML Systems C1
  Nelly Rosaura Palacios Salinas, Mitra Baratchi, Jan N. van Rijn,
  and Andreas Vollrath

Author Index   515

Automating Machine Learning, Optimization, and Feature Engineering

PuzzleShuffle: Undesirable Feature Learning for Semantic Shift Detection

Yusuke Kanebako$^{(\boxtimes)}$ and Kazuki Tsukamoto

Ricoh Company, Ltd., Tokyo, Japan
{yuusuke.kanebako,kazuki.tsukamoto}@jp.ricoh.com

Abstract. When running a machine learning system, it is difficult to guarantee performance when the data distribution is different between training and production operations. Deep neural networks have attained remarkable performance in various tasks when the data distribution is consistent between training and operation phases, but performance significantly drops when they are not. The challenge of detecting Out-of-Distribution (OoD) data from a model that only trained In-Distribution (ID) data is important to ensure the robustness of the system and the model. In this paper, we have experimentally shown that conventional perturbation-based OoD detection methods can accurately detect non-semantic shift with different domain, but have difficulty detecting semantic shift in which objects different from ID are captured. Based on this experiment, we propose a simple and effective augmentation method for detecting semantic shift. The proposed method consists of the following two components: (1) PuzzleShuffle, which deliberately corrupts semantic information by dividing an image into multiple patches and randomly rearranging them to learn the image as OoD data. (2) Adaptive Label Smoothing, which changes labels adaptively according to the patch size in PuzzleShuffle. We show that our proposed method outperforms the conventional augmentation methods in both ID classification performance and OoD detection performance under semantic shift conditions.

Keywords: Semantic shift detection · Data augmentation · Out-of-distribution detection

1 Introduction

When running a machine learning system, it is difficult to guarantee performance when the data distribution is different between training and production operations. It is important to detect such data not included in the training data or build a model that can make predictions with low confidence for untrained data to ensure the reliability and safety of machine learning systems. Deep neural networks (DNNs) have attained remarkable performance in various tasks when

---

Y. Kanebako and K. Tsukamoto—Equal contribution.

© Springer Nature Switzerland AG 2021
Y. Dong et al. (Eds.): ECML PKDD 2021, LNAI 12979, pp. 3–19, 2021.
https://doi.org/10.1007/978-3-030-86517-7_1

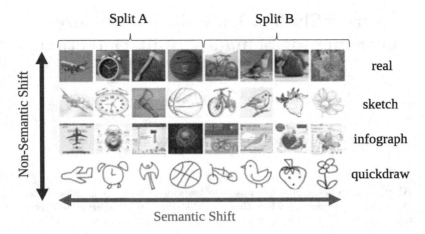

Fig. 1. Domainnet [17]

the data distribution is consistent between training and running phases. However, it is difficult to guarantee robustness when the domain changes between training and operation or when unexpected objects are captured. This challenge has been formulated as learning only In-Distribution (ID) data and detecting Out-of-Distribution (OoD) data [8], and many methods have been proposed in recent years [10,14–16,22].

The cause of the factor difference in distribution between ID and OoD does not distinguish in previous studies on OoD detection. As shown in Fig. 1, GenelizedODIN [10] uses the DomainNet Dataset [17] to separate the problem of OoD detection into two categories: Semantic Shift, in which the class of the object is different between ID and OoD in the same domain, and Non-Semantic Shift, in which the class of the object is the same, but the domain is different. The results showed that the previous OoD detection methods perform excellently to non-semantic shift detection but could not outperform the baseline MaxSoftmax-based method [8] in semantic shift detection.

Semantic shift detection is one of the most critical issues in the operation of machine learning systems. It is necessary to reject the prediction results or consider adding them as untrained data by lowering the confidence level of the prediction for unexpected objects. However, the prediction of DNNs is known to be high-confidence, and calibration by temperature scaling and adversarial training is reported to be effective for this problem [5,7]. In the framework of Bayesian DNNs, a learning method that theoretically guarantees the uncertainty of the prediction has been proposed [2,23]. Besides, some data augmentation methods show to improve the uncertainty and robustness of the DNNs model [9,24,27,28].

In this paper, we focus on the semantic shift in OoD detection. In the problem setting where the domain of image data is the same, but the classes of ID and OoD are different, the goal is to detect OoD data from a model that only

Fig. 2. A visual comparison of Cutout [1], AugMix [9], Mixup [28], CutMix [27], ResizeMix [19], Puzzle Mix [12], and our PuzzleShuffle

trained ID data. To address this problem, we propose a new data augmentation method named PuzzleShuffle. Our method was inspired by [14]. The key concept is to make the model explicitly train with data that has an undesirable feature. Figure 2 shows a comparison with conventional augmentation methods. PuzzleShuffle divides the image into some patches. And the patches are randomly rearranging to intentionally destroy the semantic information of the image, and then the models are trained with images that have undesirable features. The labels of the data to which PuzzleShuffle is applied are adaptively smoothed according to the patch size. For images with large patch size, we give labels close to one-hot distribution because we believe that there is still much semantic information, and for images with small patch size, we give labels close to uniform distribution because we believe that the semantic information is strongly corrupted. In this way, DNNs can learn to predict the semantic information with lower confidence as they move away from the ID. To verify our proposed method's effectiveness, we evaluated OoD detection's performance under the semantic shift using various datasets. As a result, we show that our proposed method outperforms the conventional augmentation methods in both the performance of ID classification accuracy and OoD detection performance under the semantic shift conditions.

In summary, our paper makes the following contributions:

- We show that the existing perturbation-based OoD detection methods cannot outperform the baseline method's OoD detection performance in semantic shift conditions.
- We show that adversarially trains the data with features not included in ID data and effectively improves OoD detection performance under the semantic shift conditions.
- We proposed a new simple and effective augmentation method to improve OoD detection accuracy under the semantic shift conditions.
- We show that the proposed method improves OoD detection performance in combination with any conventional augmentation methods.

2   Related Work

2.1   Out-of-Distribution Detection

The problem of OoD detection was formulated by [8], and the proposed method of separating ID and OoD using the maximum softmax value of DNNs is widely used as a baseline. ODIN [16] performs OoD detection by applying a perturbation to the input image such that the maximum softmax value increases. Similarly, Mahalanobis [15] also detect OoD using perturbation but assumes that the feature map's intermediate output follows a multivariate gaussian distribution and calculates the distance between the distributions during training and testing using the mahalanobis distance, and uses that value as the threshold for OoD detection. [10, 22] does not use perturbation and calculates the logit using cosine similarity instead of the linear transformation before the softmax function. [14] uses generative adversarial nets (GANs) [4] to generate boundary data ID and OoD and training generated data for confidence calibration and improvement of OoD detection. In any research, the problem of OoD detection under the semantic shift has not been solved. Besides, GeneralizedODIN [10] shows that the existing OoD methods cannot outperform the baseline method's [8] OoD detection performance in semantic shift conditions.

2.2   Data Augmentation

Data augmentation can improve the generalization performance of the model and uncertainty and robustness [9, 24, 27, 28]. CutMix [27] randomly cuts a portion of an image and pastes it at a position corresponding to the cut position in another image to improve performance, confidence calibration, and OoD detection. ResizeMix [19] pointed out that CutMix may not capture the intended object in the cropped image and clarified the importance of capturing the object and, ReizeMix outperforms CutMix by pasting a resized image instead of cropping the image. Mixup [28] proposes a method to compute convex combination of two images pixel by pixel, and Puzzle Mix [12] achieves effective mixup by using saliency information and graph cut. AugMix [9] improves the robustness and uncertainty evaluation by applying multiple augmentations to a single image and training the weighted combined image and the original image to be close in distribution. All the methods have improved test accuracy, OoD detection accuracy, robustness against distortion images, and uncertainty evaluation, but OoD detection under semantic shift has not been verified. Our method is novel in that it learns undesirable features as ID, and we propose that it can improve the OoD detection performance by giving appropriate soft labels to the data.

2.3   Uncertainty Calibration

In order to achieve high accuracy in the perturbation-based OoD detection described above, the confidence of the DNNs prediction must be properly calibrated. The confidence level of DNNs prediction is known to be high-confidence [5, 7], which means that the confidence level of DNNs is high even though the prediction results are wrong. Some Bayesian DNNs methods provide theoretical guarantees on uncertainty estimation [2,23]. In these methods, the estimation of uncertainty is theoretically guaranteed by using Dropout and Batch Normalization. However, although both of these methods achieve confidence calibration, they have not been reported as effective OoD detection methods.

In contrast to these works, we develop a new augmentation method for semantic shift detection. Inspired by [14], our method proposes a simple and effective augmentation method that can improve OoD detection performance under the Semantic Shift by explicitly training data with features not found in ID. First, we experimentally demonstrate the possibility of improving OoD detection performance by adversarial training data with uniformly distributed labels that have features not found in ID. Based on the results, we propose an augmentation method that intentionally corrupts the semantic information of ID data and learns the data as undesirable ID data.

3   Preliminaries

In this chapter, we conduct two preliminary verifications to propose a method to improve semantic shift detection performance. Subsection 3.1 discusses why the perturbation-based OoD detection method fails to detect OoD data under semantic shift conditions. In Subsect. 3.2, inspired by [14], we investigate adversarial training using explicitly semantic shift data that can improve the detection performance of semantic shift. The experimental settings are the same as those described in Sect. 5.

3.1   The Effects by Perturbation

We investigate the effectiveness of perturbation-based methods, which have been reported to be effective as OoD detection methods for semantic shift and non-semantic shift. We use MaxSoftmax [8], a baseline method, as the OoD detection method without perturbation, and ODIN [16] as the method with perturbation. To compare them, we choose four domains (real, sketch, infograph, and quickdraw) from the Domainnet [17] dataset and divide them into two classes: class labels 0-172 as A, and class labels 173-344 as B, for a total of eight datasets. The A group in the real domain use as ID, and the other groups evaluate as OoD. The results show in Table 1. The perturbation-based method detects OoD with higher accuracy than the method without perturbation in the non-semantic shift detection. However, the perturbation-based method is inferior to without perturbation in the semantic shift detection. The reason for this may be that the more similar the OoD features are to the ID features, the more they are embedded in the similar features by perturbation. ODIN tries to separate ID and OoD by adding perturbations to the image to increase the softmax value, so when similar features are obtained, OoD is also perturbed similarly to ID, making separation difficult. Figure 3 shows the result of plotting the intermediate features of semantic shift (real-B) and non-semantic shift (quickdraw-A) in two dimensions by tSNE. It can be seen that the non-semantic shift, which dynamically changes the trend of the image, produces features that are not similar to ID, while the semantic shift, which is in the same domain, produces features that are similar. Therefore, we hypothesize that it is important to explicitly learn undesirable features in order to improve the detection performance of semantic shift.

3.2   Adversarial Undesirable Feature Learning

In this section, we verify the hypothesis that semantic shift detection performance can be improved by explicitly learning undesirable features. We use the

Table 1. The OoD detection performance to semantic shift and non-semantic shift by perturbation

OoD	Shift		AUROC
	S	NS	Baseline/ODIN*
real-B	✓		**68.2**/65.0
sketch-A		✓	70.6/**75.0**
sketch-B	✓	✓	75.5/**78.8**
infograph-A		✓	75.6/**80.2**
infograph-B	✓	✓	76.65/**81.7**
quickdraw-A		✓	70.3/**96.0**
quickdraw-B	✓	✓	71.7/**96.7**

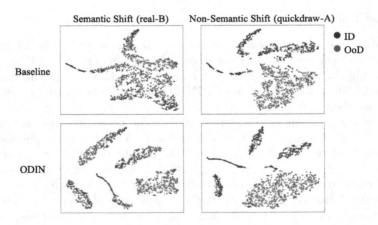

Fig. 3. Results of tSNE visualization of features from conv layer output in semantic shift and non-semantic shift data. The blue points indicate ID, and the red points indicate OoD. The semantic shift tends to extract features similar to ID, and the distribution of ID becomes closer to the semantic shift when the perturbation is applied. On the contrary, the non-semantic shift tends to extract features different from ID, and the ID distribution does not overlap with the non-semantic shift even after perturbation. (Color figure online)

CIFAR-10 dataset as ID and the CIFAR-100 dataset as OoD for adversarial training to verify this hypothesis. In adversarial training, the ID data is training with one-hot labels, and the OoD data is training with uniform distribution labels. We explicitly train the OoD data as undesirable features by training the OoD data with uniformly distributed labels. We split the 100 classes of CIFAR-100 into five types, from split1 to split5, based on 20 superclasses, and observe the effect of increasing the variation of OoD classes step by step. The ratio of the number of ID and OoD data included in a minibatch during training is 1:1. Table 2 shows the results. The results indicate that training the data that has undesirable features with uniform labels improves ID accuracy and OoD detection.

Figure 4 is the extracted features from the convolution layer of the model trained with OoD and without one. The results show that adversarial training of undesirable feature embeds the unobserved data to the non ID space, significantly improving the semantic shift detection performance. The hypothesis that

Table 2. ID accuracy and AUROC of each OoD split when adversarial learning with adding OoD step by step. The OoD split used for training has the AUROC value in bold for each testing OoD split.

train OoD	ID Acc.	AUROC				
		split1	split2	split3	split4	split5
None	86.65	82.41	80.40	79.87	81.61	78.85
split1	88.47	**96.94**	92.54	85.76	83.36	83.65
split1~2	88.94	**96.93**	**98.64**	89.41	83.48	86.32
split1~3	89.40	**97.39**	**98.30**	**97.20**	83.33	85.72
split1~4	89.65	**96.49**	**98.07**	**96.54**	**91.51**	84.10
split1~5	90.06	**96.27**	**97.83**	**96.54**	**92.77**	**92.86**

Test OoD: Split5

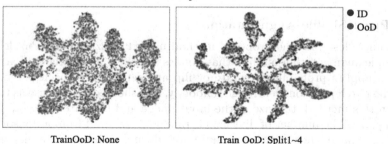

TrainOoD: None   Train OoD: Split1~4

Fig. 4. Results of tSNE visualization of features from conv layer output (Left: train OoD is none, Right: train OoD is split1-4). By explicitly training OoD data as undesirable features, we show that unobserved OoD data are embedded in places that are not ID regions.

adversarial training of undesirable features improves semantic shift detection by using OoD data is revealed. However, in a machine learning system operation, OoD data cannot be accessed in advance. Therefore, it is necessary to learn undesirable features using only ID data. To solve this problem, we propose a new augmentation method that destroys the semantic structure and intentionally induces semantic shift by shuffling the patches in the image like a puzzle.

4   Proposed Method

Figure 5 illustrates the proposed method. This method consists of two steps: (a) applying augmentation to the image and (b) adaptively changing the label according to the augmentation result. Details are described below.

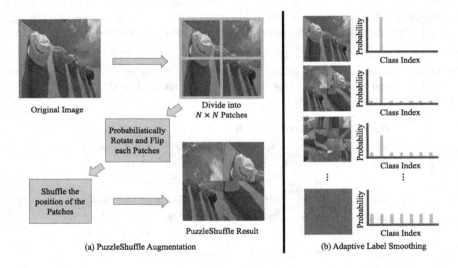

Fig. 5. Proposed method

4.1   PuzzleShuffle Augmentation

Algorithm 1 describes the proposed method named Puzzleshuffle. PuzzleShuffle is a simple augmentation method that divides an image into patches of arbitrary size and applies probabilistic rotate or flip augmentation to each patch. After then, the patch positions rearrange randomly, and shuffled images use as training data. In this method, the size of the input image and the patch size are limited to be square. The number of divisions is randomly selected from a divisor of the size of the image when creating a mini-batch during training. A similar method is PatchShuffle regularization [11] method, which randomly shuffles the pixels in a local patch in the image or feature map. Our method differs in that the patch size is variable and the shuffle is performed while preserving the global features, and the labels are changed adaptively according to the patch size as described below.

Algorithm 1: PuzzleShuffle Augmentation

Input: Dataset $\mathcal{D}$, probability p of applying PuzzleShuffle, Operations $\mathcal{O} = \{\text{rotate, flip}\}$

Output: A puzzle shuffled image $\tilde{x}$ and its adaptively changed label $\tilde{y}$

$divisors =$ CaluclateDivisor($image\_size$)

Sample $(x, y) \sim \mathcal{D}$

$\beta \sim$ Bernouli(p)

if $\beta = 1$ **then**

    Sample $patch\_size \sim$ RandomSelect(divisors)

    $divided\_images =$ DivideImage($x, patch\_size$)

    for $i = 1, ..., patch\_size \times patch\_size$ **do**

        $divided\_image_i \leftarrow \mathcal{O}(divided\_image_i)$

    ShufflePatchPosition($divided\_image$)

    $\tilde{x} \leftarrow divided\_image$

    Set $label$ according to Algorithm 2

    $\tilde{y} \leftarrow label$

else

    # Original data is returned.

    $\tilde{x} \leftarrow x$

    $\tilde{y} \leftarrow y$

4.2 Adaptive Label Smoothing

If the number of divisions in PuzzleShuffle is small, the structural information of the image remains, and if the number of divisions is large, the structural information is collapsed. Since images with many patches in PuzzleShuffle are like random noise, it is inappropriate to train them with one-hot labels. We propose a method to adaptively change the distribution of labels according to the number of divisions. When the number of divisions is large, we give label information with a distribution close to one-hot labels. When small, we give label information with a distribution close to the uniform distribution. Algorithm 2 has described Adaptive Label Smoothing. The basic idea is the same as that of Label Smoothing. The target class value discounts from the one-hot label and distributes the discounted value to other class labels. Label Smoothing has attained remarkable improvement of generalization performance as a regularization method for DNNs [18, 21]. We prepare a lookup table, a list of values from the inverse number of classes to 1.0, equally divided by image size and sorted in descending order. We select values from the lookup table using the selected divisor number at PuzzleShuffle Augmentation as an index and use the selected values as the target class values for Label Smoothing. In this way, the labels of PuzzleShuffle image assign according to the patch size.

Algorithm 2: Adaptive Label Smoothing

Input:
$C \cdots$ the number of classes
$LUT \cdots$ A lookup table of numbers from the inverse of C to 1.0, equally divided by image size and sorted in descending order.
$index \cdots$ Selected divisor by Altorithm 1
Output: Smoothed label $\hat{y}$
$score = LUT(index)$
$residual = (1 - score)/(C - 1)$

$$\hat{y}[i] = \begin{cases} score & \text{(if } i = y) \\ residual & \text{(otherwise)} \end{cases} \tag{1}$$

4.3 Motivation

The motivation for PuzzleShuffle is the effect of learning undesirable features, as shown in Sect. 3. In the semantic shift problem, the structural information of data is different between ID and OoD. Thus, we believe that it is important to learn structural information not available in ID data explicitly for semantic shift detection. In situations where OoD data is not available, it is necessary to create it from ID data. In [14], the boundary between ID and OoD is generated by GANs. However, GANs are generally expensive and difficult to learn stably, so we divided the image into patches and randomly rearranged the patches' positions. Convolutional neural networks tend to make decisions based on texture information rather than structural information of images [3]. Therefore, to detect semantic shift, we thought it is essential to give appropriate labels to images with broken structural information when learning structural information not present in ID data.

5 Experiments

5.1 Experimental Settings

Networks and Training Details: We use ResNet-34 [6] for all experiments. It is trained with batch size 128 for 200 epochs with and weight decay 0.0005. The optimizer is SGD with momentum 0.9, and the initial learning rate set to 0.1. The learning rate decreases by factor 0.1 at 50% and 75% of the training epochs.

Datasets: In the experiments, we use CIFAR-10/100 [13], Tiny ImageNet [20] (cropped and resized), LSUN [26] (cropped and resized), iSUN [25], Uniform noise, Gaussian noise and DomainNet [17]. If one of the CIFAR-10/100 use as ID, the other is evaluated as OoD. Tiny ImageNet, LSUN, iSUN, Uniform noise, and Gaussian noise are all used as OoD. The experiments using DomainNet follow the experimental method of GeneralizedODIN [10]. We divide the images in each domain into two groups: A for class labels 0-172 and B for class labels 173-344. The A group in the real domain use as ID, and the other groups evaluate as OoD.

Evaluation Metrics: Following previous OoD detection studies [8, 10, 14–16, 22], we use the area under the receiver operating characteristic curve (AUROC) and true negative rate at 95% true positive rate (TNR@TPR95) as the evaluation metrics. We also evaluate the classification performance of ID data. For all of these metrics, a higher value indicates better performance.

5.2 Compared Methods

We use Cutout [1], AugMix [9], Mixup [28], CutMix [27], ResizeMix [19], and Puzzle Mix [12] to compare augmentation methods. We evaluate these augmentation methods performance alone and in combination with standard augmentation (i.e., crop, horizontal flip) and the proposed methods. In comparison with the OoD detection method, we evaluate the performance of combining the proposed methods on the DomainNet dataset. We employed the MaxSoftmax-based method [8] as Baseline and compared it with ODIN [16] and Scaled Cosine [10].

5.3 Results

Comparison of Augmentation Method. Table 3 shows the results when each augmentation applies by itself and standard augmentations are not in use. In many experiments, our method has shown high OoD detection performance. In particular, we achieve high detection performance on datasets where the image is resized instead of cropped and Uniform and Gaussian noise datasets. This is

Table 3. Performance comparison of each augmentation methods.

ID	OoD	Method (AUROC/TNR@TPR95)							
		Baseline	Cutout	Mixup	CutMix	AugMix	ResizeMix	Pazzle Mix	Our
CIFAR-10	C100	80.5/19.8	82.6/23.0	81.1/24.3	81.4/22.8	82.7/21.1	68.8/27.3	80.7/26.3	**86.5/30.1**
	TINc	80.6/18.3	84.6/26.1	82.7/25.5	89.8/**39.5**	84.2/25.1	77.9/36.1	87.9/35.8	**89.9**/36.3
	TINr	76.0/16.6	80.9/21.7	80.6/23.7	89.0/39.2	84.2/24.4	75.4/42.3	92.5/49.8	**94.9/63.3**
	LSUNc	80.7/14.6	80.5/20.0	81.9/23.2	87.3/32.1	89.0/**38.9**	68.5/27.5	80.1/26.7	**90.1**/36.8
	LSUNr	80.8/19.3	86.3/29.5	84.6/30.2	92.3/49.1	86.8/28.1	88.5/60.1	94.0/55.8	**95.8/69.4**
	iSUN	79.8/20.1	85.4/27.6	83.5/28.7	91.8/47.6	86.3/28.0	85.6/54.9	94.1/56.8	**96.0/71.2**
	Uniform	86.0/19.3	88.8/31.0	81.0/11.3	83.8/19.3	97.7/83.7	92.7/51.9	96.2/72.4	**100.0/100.0**
	Gaussian	97.9/85.3	90.6/35.6	85.6/12.6	83.2/19.0	98.6/92.0	51.2/15.0	93.6/53.3	**100.0**/99.9
CIFAR-100	C10	66.6/9.7	69.4/12.2	70.3/11.7	70.0/11.3	69.6/12.1	**74.1/15.3**	71.3/71.3	73.6/14.7
	TINc	75.0/19.0	74.5/19.5	78.6/23.0	73.7/11.2	64.5/6.0	**82.1/26.7**	76.3/16.1	79.5/22.1
	TINr	69.2/12.5	62.9/9.1	67.5/9.2	45.4/1.7	73.1/14.3	78.7/21.5	54.3/4.7	**88.3/49.2**
	LSUNc	66.5/9.7	67.0/10.5	63.2/10.7	68.8/9.0	53.9/4.1	**77.4/21.2**	69.9/9.4	75.2/12.6
	LSUNr	71.5/13.8	63.8/8.2	69.2/8.7	45.3/1.3	73.2/14.6	80.4/22.6	54.6/3.6	**88.6/50.1**
	iSUN	69.4/12.1	62.1/7.1	67.6/7.7	44.4/1.1	70.0/12.2	78.2/20.0	54.2/3.6	**88.1/50.9**
	Uniform	57.5/0.9	35.4/0.1	29.4/0.0	60.0/0.7	33.5/0.0	20.0/0.0	91.0/49.3	**100.0/100.0**
	Gaussian	36.7/0.0	73.4/5.2	38.4/0.0	64.7/1.7	54.4/0.0	94.1/58.7	61.7/0.5	**100.0/100.0**

Table 4. The results of the combinations of augmentation methods. SA indicate using standard augmentation (i.e. crop and horizontal flip). The numbers in parentheses indicate the performance when the proposed method is combined, and the bold type indicates the improvement of performance by the proposed method.

ID	OoD	Method (AUROC/TNR@TPR95)					
		SA (+Our)	SA+Cutout (+Our)	SA+Mixup (+Our)	SA+CutMix (+Our)	SA+ResizeMix (+Our)	SA+Puzzle Mix (+Our)
CIFAR-10	C100	86.7/36.2 (**89.0/40.7**)	89.9/43.5 (**90.3/44.5**)	74.9/37.1 (**82.7/37.9**)	85.7/38.6 (**88.8/46.4**)	83.6/43.7 (**88.1/46.6**)	83.8/44.7 (**87.6/46.1**)
	TINc	91.7/49.5 (**92.2**/48.3)	94.1/58.6 (**94.3/60.0**)	83.2/58.1 (**86.3**/50.1)	96.1/73.1 (**96.7/77.2**)	91.2/54.4 (**93.1/56.6**)	97.1/84.0 (96.3/74.7)
	TINr	88.6/39.6 (**96.9/77.7**)	93.6/56.7 (**97.5/83.0**)	84.5/39.7 (**97.4/82.8**)	97.1/84.3 (**98.9/95.6**)	76.6/53.8 (**95.1/71.5**)	97.2/82.9 (**98.8/93.8**)
	LSUNc	93.6/57.5 (93.3/53.3)	93.9/57.7 (**95.4/66.6**)	84.8/68.4 (**89.0**/60.5)	94.4/61.1 (**97.1/81.0**)	91.2/53.5 (**93.8/58.4**)	95.3/77.1 (**96.1**/76.5)
	LSUNr	90.7/46.5 (**97.6/83.4**)	94.9/63.8 (**97.6/83.9**)	88.2/49.2 (**98.0/87.6**)	98.2/92.5 (**99.2/98.3**)	86.2/68.3 (**97.1/80.3**)	98.2/92.2 (**99.2/96.6**)
	iSUN	89.9/44.3 (**97.4/82.0**)	94.8/62.8 (**97.6/83.8**)	88.0/46.9 (**97.8/86.3**)	97.9/89.8 (**99.2/97.7**)	83.6/63.7 (**97.1/80.5**)	98.0/90.4 (**99.1/95.5**)
	Uniform	90.0/21.9 (**100.0/100.0**)	87.7/6.2 (**100.0/100.0**)	90.5/19.8 (**100.0/100.0**)	4.5/0.0 (**100.0/100.0**)	92.7/51.9 (**100.0/100.0**)	80.0/5.3 (**100.0/100.0**)
	Gaussian	98.1/89.1 (**100.0/100.0**)	97.2/84.9 (**100.0/100.0**)	98.0/92.9 (97.9/**95.1**)	78.0/3.2 (**97.0/99.8**)	84.5/7.0 (**100.0/100.0**)	59.0/0.0 (**100.0/100.0**)
CIFAR-100	C10	75.6/16.5 (**76.5/19.3**)	76.0/16.7 (75.2/16.4)	73.9/18.7 (**74.8/20.5**)	77.7/21.0 (75.9/**21.2**)	75.7/18.6 (**76.3/21.1**)	78.4/21.0 (77.2/20.5)
	TINc	82.3/26.5 (**83.8/32.0**)	80.5/22.7 (**82.6/29.0**)	84.7/39.6 (82.0/31.5)	86.6/36.5 (84.6/34.6)	82.5/30.4 (**85.0/33.9**)	88.9/40.8 (87.2/39.6)
	TINr	76.5/18.5 (**89.9/52.2**)	74.0/17.0 (**92.6/64.3**)	76.1/21.6 (**91.5/60.1**)	84.1/31.8 (**90.4/52.7**)	79.2/25.8 (**88.4/44.4**)	77.2/19.8 (**94.9/72.7**)
	LSUNc	79.5/21.3 (**82.7/29.3**)	75.7/16.4 (**83.5/29.9**)	83.5/38.1 (79.8/26.2)	84.1/31.2 (**84.6/34.4**)	79.7/26.5 (**82.1/27.7**)	85.4/32.1 (85.3/33.1)
	LSUNr	78.5/20.3 (**89.6/51.8**)	73.6/14.8 (**92.8/64.1**)	77.2/21.3 (**91.8/59.2**)	86.5/35.1 (**90.5/51.7**)	80.3/26.7 (**88.3/42.5**)	76.8/18.4 (**95.4/74.6**)
	iSUN	77.2/18.6 (**89.0/50.4**)	73.7/14.9 (**92.2/63.8**)	76.0/20.0 (**90.6/56.9**)	84.7/31.0 (**90.2/51.3**)	79.1/24.9 (**88.3/44.0**)	75.1/17.0 (**93.8/68.9**)
	Uniform	74.4/1.0 (**100.0/100.0**)	97.5/86.3 (**100.0/100.0**)	78.2/1.3 (**100.0/100.0**)	90.8/41.8 (**100.0/100.0**)	45.9/0.0 (**100.0/100.0**)	68.8/0.4 (**100.0/100.0**)
	Gaussian	52.8/0.0 (**98.8/97.7**)	80.7/0.0 (**100.0/100.0**)	60.1/0.0 (**100.0/100.0**)	89.6/24.3 (**100.0/100.0**)	35.9/0.0 (**99.7/98.9**)	61.0/0.0 (**99.8/100.0**)

because the proposed method can learn to focus on the image structure information and the minimum patch size is one pixel.

Combination of Augmentation Method. Table 4 shows the results of combining standard augmentation methods such as crop and horizontal flip with existing augmentation methods and our proposed method. The results show that for many augmentation methods, the combination of our proposed method can improve the performance of OoD detection. Table 5 shows the classification performance of ID data when using each augmentation method combined with our

Table 5. Comparison of ID classification accuracy. In all cases, we use the standard augmentation of crop and horizontal flip.

ID	Method	Classification accuracy
CIFAR-10	Baseline (+**Our**)	94.8(94.8)
	Cutout (+**Our**)	95.4(95.6)
	Mixup (+**Our**)	94.2(94.2)
	CutMix (+**Our**)	96.3(96.2)
	ResizeMix (+**Our**)	96.7(96.3)
	Puzzle Mix (+**Our**)	96.4(95.4)
CIFAR-100	Baseline (+ **Our**)	74.0(76.0)
	Cutout (+**Our**)	74.0(75.3)
	Mixup (+**Our**)	75.4(76.8)
	CutMix (+**Our**)	79.9(80.0)
	ResizeMix (+**Our**)	79.0(80.3)
	Puzzle Mix (+**Our**)	80.4(80.0)

Table 6. Results of combining the proposed method with the OoD detection method using DomainNet.

OoD	Shift		AUROC	TNR@TPR95
	S	NS	Baseline(+Our)/ODIN(+Our)/Cosine(+Our)	
real-B	✓		68.2(**71.5**)/65.0(**69.4**)/66.2(**69.9**)	9.7(**11.5**)/10.1(**11.8**)/8.6(**10.9**)
clipart-A		✓	67.6(**71.0**)/80.1(**81.5**)/70.2(**77.5**)	13.3(**15.5**)/30.5(**33.8**)/13.8(**21.3**)
clipart-B	✓	✓	74.8(**78.1**)/86.5(**87.5**)/77.0(**83.2**)	17.0(**19.4**)/38.2(**42.4**)/16.5(**24.6**)
infograph-A		✓	75.6(**77.6**)/80.2(**81.9**)/79.8(**85.4**)	16.9(**19.2**)/17.8(**23.0**)/20.7(**31.7**)
infograph-B	✓	✓	76.6(**79.2**)/81.7(**83.6**)/80.6(**86.6**)	16.9(**20.2**)/19.9(**25.9**)/20.5(**33.0**)
painting-A		✓	67.1(**71.1**)/55.7(**63.1**)/68.8(**75.4**)	11.0(**12.8**)/ 3.2(**4.2**)/11.9(**17.8**)
painting-B	✓	✓	73.3(**77.1**)/61.2(**69.4**)/74.3(**80.5**)	14.0(**16.7**)/ 4.6(**6.6**)/13.9(**20.7**)
quickdraw-A		✓	70.3(**77.2**)/96.0(**97.1**)/72.6(**78.9**)	12.1(**14.0**)/80.3(**84.4**)/11.0(10.8)
quickdraw-B	✓	✓	71.7(**78.7**)/96.7(**97.6**)/74.0(**80.2**)	12.2(**15.4**)/82.8(**87.4**)/11.8(11.3)
sketch-A		✓	70.6(**76.1**)/75.0(**80.8**)/72.9(**81.5**)	13.7(**18.2**)/22.1(**29.7**)/13.3(**22.7**)
sketch-B	✓	✓	75.5(**79.4**)/78.8(**83.8**)/77.4(**84.4**)	16.4(**21.1**)/24.0(**32.0**)/15.1(**25.3**)

proposed method. In all cases, the performance does not degrade significantly. Therefore, from Tables 4 and 5, we can see that our method can improve the OoD detection performance while maintaining the ID data classification performance.

Comparison of OoD Method. Table 6 shows the OoD detection results using the DomainNet dataset. The proposed method can improve the OoD detection performance in both cases of semantic shift and non-semantic shift. These results indicate that the proposed method learns features that only exist in ID data (i.e., real-A), thus improving the detection of semantic shifts and the detection

of non-semantic shifts. Besides, the proposed method improves the performance
of Baseline and existing OoD methods.

5.4   Analysis

Effect of Network Architecture. Table 7 shows the performance of the proposed method for different network architectures. It show that our proposed
method improves the performance of all network architectures.

Impact of Multiple Patch Sizes. Table 8 shows the results when PuzzleShuffle is performed with single patch size and with multiple sizes. On average, both
ID classification performance and OoD detection performance are higher when
multiple scales are combined. It shows that it is important to perform PuzzleShuffle with multiple sizes to learn more diverse undesirable features.

Impact of Labeling Method. Our proposed method changed the label according to the patch size, but a method to calculate the image similarity by images
before and after applying PuzzleShuffle is also possible. We use two image similarity metrics, SSIM [29] and the cosine similarity of feature vectors obtained
from the models trained by ImageNet [20]. The calculated image similarity is
applied to the score of Algorithm 2 to give a label. We also compare the results
with one-hot and uniform labels for all patch sizes. The results show in Table 9.
The image similarity obtained from the pre-trained model shows superior performance in ID classification and OoD detection. These results indicate that it
is important to appropriately reflect the similarity to the original image in the
label, which is a future challenge.

Table 7. Performance evaluation of various network architectures. We used standard
augmentation and combined our proposed method.

ID	OoD	Network	ID Acc	AUROC	TNR@TPR95
C-10	C-100	ResNet-34	94.8(**94.8**)	86.7(**89.0**)	36.2(**40.7**)
		WideResNet-28-10	95.3(**95.9**)	89.3(**89.7**)	43.1(**43.6**)
		DenseNet-100	94.4(**94.6**)	88.2(**89.0**)	35.2(**38.4**)
C-100	C-10	ResNet-34	74.0(**76.0**)	75.6(**76.5**)	16.5(**19.3**)
		WideResNet-28-10	79.8(79.3)	79.2(**80.1**)	20.8(**22.2**)
		DenseNet-100	75.1(**76.64**)	75.5(**75.7**)	17.8(**18.2**)

Table 8. Performance evaluation for various patch sizes.

ID	OoD	Num. of Div.	ID Acc.	AUROC	TNR@TPR95
C-10	C-100	1 × 1	94.8	89.0	36.2
		2 × 2	**95.3**	87.2	**41.0**
		4 × 4	94.5	86.7	37.1
		8 × 8	94.9	88.4	39.3
		16 × 16	93.5	84.8	35.2
		Multi-scale	94.8	**89.9**	40.7
C-100	C-10	1 × 1	74.0	75.6	16.5
		2 × 2	75.1	75.7	17.1
		4 × 4	73.5	74.8	16.3
		8 × 8	73.4	75.0	16.1
		16 × 16	74.4	75.5	16.7
		Multi-scale	**76.0**	**76.5**	**19.3**

Table 9. Performance evaluation using various labeling methods. * indicates using pre-trained model.

ID	OoD	Method	ID Acc	AUROC	TNR@TPR95
C-10	C-100	One-hot	94.8	**89.2**	39.0
		Uniform	94.5	85.6	44.2
		SSIM	94.5	85.4	42.7
		Cosine*	**95.5**	85.9	**49.8**
		Algorithm 2	94.8	89.0	40.7
C-100	C-10	One-hot	73.3	74.7	16.9
		Uniform	76.8	75.7	16.1
		SSIM	**77.7**	75.1	16.6
		Cosine*	77.4	**77.2**	**20.1**
		Algorithm 2	76.0	76.5	19.3

6   Conclusion

This paper focuses on OoD detection under semantic shift and shows that conventional OoD detection methods cannot detect semantic shift. Our proposed method improves the performance of OoD detection without degrading the performance of ID classification. In the future, we will study OoD detection that can detect not only the semantic shift but also the non-semantic shift and investigate more robust model construction and running machine learning systems.

References

1. DeVries, T., Taylor, G.W.: Improved regularization of convolutional neural networks with cutout. arXiv preprint arXiv:1708.04552 (2017)
2. Gal, Y., Ghahramani, Z.: Dropout as a Bayesian approximation: representing model uncertainty in deep learning. In: Proceedings of the 33rd International Conference on Machine Learning, pp. 1050–1059 (2016)
3. Geirhos, R., Rubisch, P., Michaelis, C., Bethge, M., Wichmann, F.A., Brendel, W.: ImageNet-trained CNNs are biased towards texture; increasing shape bias improves accuracy and robustness. In: International Conference on Learning Representations (2019)
4. Goodfellow, I., et al.: Generative adversarial nets. In: Advances in Neural Information Processing Systems, vol. 27 (2014)
5. Guo, C., Pleiss, G., Sun, Y., Weinberger, K.Q.: On calibration of modern neural networks. In: Proceedings of the 34th International Conference on Machine Learning, pp. 1321–1330 (2017)
6. He, K., Zhang, X., Ren, S., Sun, J.: Deep residual learning for image recognition. In: Proceedings of the IEEE Conference on Computer Vision and Pattern Recognition, pp. 770–778 (2016)
7. Hein, M., Andriushchenko, M., Bitterwolf, J.: Why ReLU networks yield high-confidence predictions far away from the training data and how to mitigate the problem. In: Proceedings of the IEEE/CVF Conference on Computer Vision and Pattern Recognition, pp. 41–50 (2019)

8. Hendrycks, D., Gimpel, K.: A baseline for detecting misclassified and out-of-distribution examples in neural networks. In: Proceedings of International Conference on Learning Representations (2017)
9. Hendrycks, D., Mu, N., Cubuk, E.D., Zoph, B., Gilmer, J., Lakshminarayanan, B.: AugMix: a simple method to improve robustness and uncertainty under data shift. In: International Conference on Learning Representations (2020)
10. Hsu, Y.C., Shen, Y., Jin, H., Kira, Z.: Generalized ODIN: detecting out-of-distribution image without learning from out-of-distribution data. In: Proceedings of the IEEE/CVF Conference on Computer Vision and Pattern Recognition, pp. 10951–10960 (2020)
11. Kang, G., Dong, X., Zheng, L., Yang, Y.: Patchshuffle regularization. arXiv preprint arXiv:1707.07103 (2017)
12. Kim, J.H., Choo, W., Song, H.O.: Puzzle mix: exploiting saliency and local statistics for optimal mixup. In: Proceedings of the 37th International Conference on Machine Learning, pp. 5275–5285 (2020)
13. Krizhevsky, A., Hinton, G.: Learning multiple layers of features from tiny images. Master's thesis, Department of Computer Science, University of Toronto (2009)
14. Lee, K., Lee, H., Lee, K., Shin, J.: Training confidence-calibrated classifiers for detecting out-of-distribution samples. In: International Conference on Learning Representations (2018)
15. Lee, K., Lee, K., Lee, H., Shin, J.: A simple unified framework for detecting out-of-distribution samples and adversarial attacks. In: Advances in Neural Information Processing Systems, vol. 31 (2018)
16. Liang, S., Li, Y., Srikant, R.: Enhancing the reliability of out-of-distribution image detection in neural networks. In: International Conference on Learning Representations (2018)
17. Peng, X., Bai, Q., Xia, X., Huang, Z., Saenko, K., Wang, B.: Moment matching for multi-source domain adaptation. In: Proceedings of the IEEE/CVF International Conference on Computer Vision, pp. 1406–1415 (2019)
18. Pereyra, G., Tucker, G., Chorowski, J., Kaiser, Ł., Hinton, G.: Regularizing neural networks by penalizing confident output distributions. arXiv preprint arXiv:1701.06548 (2017)
19. Qin, J., Fang, J., Zhang, Q., Liu, W., Wang, X., Wang, X.: ResizeMix: mixing data with preserved object information and true labels. arXiv preprint arXiv:2012.11101 (2020)
20. Russakovsky, O., et al.: ImageNet large scale visual recognition challenge. Int. J. Comput. Vis. **115**(3), 211–252 (2015)
21. Szegedy, C., Vanhoucke, V., Ioffe, S., Shlens, J., Wojna, Z.: Rethinking the inception architecture for computer vision. In: Proceedings of the IEEE Conference on Computer Vision and Pattern Recognition, pp. 2818–2826 (2016)
22. Techapanurak, E., Suganuma, M., Okatani, T.: Hyperparameter-free out-of-distribution detection using cosine similarity. In: Proceedings of the Asian Conference on Computer Vision (2020)
23. Teye, M., Azizpour, H., Smith, K.: Bayesian uncertainty estimation for batch normalized deep networks. In: Proceedings of the 35th International Conference on Machine Learning, pp. 4907–4916 (2018)
24. Verma, V., et al.: Manifold mixup: Better representations by interpolating hidden states. In: Proceedings of the 36th International Conference on Machine Learning, pp. 6438–6447 (2019)

25. Xu, P., Ehinger, K.A., Zhang, Y., Finkelstein, A., Kulkarni, S.R., Xiao, J.: TurkerGaze: crowdsourcing saliency with webcam based eye tracking. arXiv preprint arXiv:1504.06755 (2015)
26. Yu, F., Seff, A., Zhang, Y., Song, S., Funkhouser, T., Xiao, J.: LSUN: construction of a large-scale image dataset using deep learning with humans in the loop. arXiv preprint arXiv:1506.03365 (2015)
27. Yun, S., Han, D., Oh, S.J., Chun, S., Choe, J., Yoo, Y.: CutMix: regularization strategy to train strong classifiers with localizable features. In: Proceedings of the IEEE/CVF International Conference on Computer Vision, pp. 6023–6032 (2019)
28. Zhang, H., Cisse, M., Dauphin, Y.N., Lopez-Paz, D.: mixup: beyond empirical risk minimization. In: International Conference on Learning Representations (2018)
29. Wang, Z., Bovik, A.C., Sheikh, H.R., Simoncelli, E.P.: Image quality assessment: from error visibility to structural similarity. IEEE Trans. Image Process. **13**(4), 600–612 (2004)

Enabling Machine Learning on the Edge Using SRAM Conserving Efficient Neural Networks Execution Approach

Bharath Sudharsan[1]([✉]), Pankesh Patel[2], John G. Breslin[1], and Muhammad Intizar Ali[3]

[1] Confirm SFI Research Centre for Smart Manufacturing, Data Science Institute, NUI Galway, Galway, Ireland
{bharath.sudharsan,john.breslin}@insight-centre.org
[2] Artificial Intelligence Institute, University of South Carolina, Columbia, USA
ppankesh@mailbox.sc.edu
[3] School of Electronic Engineering, DCU, Dublin, Ireland
ali.intizar@dcu.ie

Abstract. Edge analytics refers to the application of data analytics and Machine Learning (ML) algorithms on IoT devices. The concept of edge analytics is gaining popularity due to its ability to perform AI-based analytics at the device level, enabling autonomous decision-making, without depending on the cloud. However, the majority of Internet of Things (IoT) devices are embedded systems with a low-cost microcontroller unit (MCU) or a small CPU as its brain, which often are incapable of handling complex ML algorithms.

In this paper, we propose an approach for the efficient execution of already deeply compressed, large neural networks (NNs) on tiny IoT devices. After optimizing NNs using state-of-the-art deep model compression methods, when the resultant models are executed by MCUs or small CPUs using the model execution sequence produced by our approach, higher levels of conserved SRAM can be achieved. During the evaluation for nine popular models, when comparing the default NN execution sequence with the sequence produced by our approach, we found that 1.61–38.06% less SRAM was used to produce inference results, the inference time was reduced by 0.28–4.9 ms, and energy consumption was reduced by 4–84 mJ. Despite achieving such high conserved levels of SRAM, our method 100% preserved the accuracy, F1 score, etc. (model performance).

Keywords: Edge AI · Resource-constrained devices · Intelligent microcontrollers · SRAM conservation · Offline inference

1 Introduction

Standalone execution of problem-solving AI on IoT devices produces a higher level of autonomy and also provides a great opportunity to avoid transmitting

© Springer Nature Switzerland AG 2021
Y. Dong et al. (Eds.): ECML PKDD 2021, LNAI 12979, pp. 20–35, 2021.
https://doi.org/10.1007/978-3-030-86517-7_2

data collected by the devices to the cloud for inference. However, at the core of a problem-solving AI is usually a Neural Network (NN) with complex and large architecture that demands a higher order of computational power and memory than what is available on most IoT edge devices. Majority of IoT devices like smartwatches, smart plugs, HVAC controllers, etc. are powered by MCUs and small CPUs that are highly resource-constrained. Hence, they lack multiple cores, parallel execution units, no hardware support for floating-point operations (FLOPS), low clock speed, etc.

The IoT devices are tiny in form factor (because FLASH, SRAM, and processor are contained in a single chip), magnitude power-efficient, and cheapest than the standard laptop CPUs and mobile phone processors. During the design phase of IoT devices, in order to conserve energy and to maintain high instruction execution speeds, no secondary/backing memory is added. For example, adding a high-capacity SD card or EEPROM can enable storing large models even without compression. But such an approach will highly affect the model execution speed since the memory outside the chipset is slow and also requires ≈100x more energy to read the thousands of outside-located model parameters.

The memory footprint (SRAM, Flash, and EEPROM) and computation power (clock speed and processor specification) of IoT devices are orders of magnitude less than the resources required for the standalone execution of a large, high-quality Neural Network (NN). Currently, to alleviate various critical issues caused by the poor hardware specifications of IoT devices, before deployment the NNs are optimized using various methods [12] such as pruning, quantization, sparsification, and model architecture tuning etc. Even after applying state-of-the-art optimization methods, there are numerous cases where the models after deep compression/optimization still exceed a device's memory capacity by a margin of just a few bytes, and users cannot optimize further since the model is already compressed to its maximum. In such scenarios, the users either have to change the model architecture and re-train to produce a smaller model (wasting GPU-days and electricity), or upgrade the device hardware (for a higher cost).

In this paper, we propose an efficient model execution approach to execute the deep compressed NNs. Our approach can comfortably accommodate a more complex/larger model on the tiny IoT devices which were unable to accommodate the same NNs without using our approach. The contributions of this paper can be summarised as follows:

- Our proposed approach shows high model execution efficiency since it can reduce the peak SRAM usage of a NN by making the onboard inference procedure follow a specific model execution sequence.
- Our approach is applicable to various NN architectures, and models trained using any datasets. Thus, users can apply it to make their IoT devices/products efficiently execute NNs that were designed and trained to solve problems in their use-case. We also implemented and made our approach freely available online.
- When the NNs optimized using state-of-the-art deep compression sequences exceed the device's memory capacity just by a few bytes margin, the users

cannot additionally apply any optimization approach since the model might be already maximum compressed or the users cannot find a study that contains methods compatible to the previous optimizations. In such scenarios, when our approach is used, the same NNs that couldn't fit on the user's device (due to SRAM overflow), can be comfortably accommodated due to the fact that our approach provides a model execution sequence that consumes less SRAM during execution.

- Orthogonal to the existing model memory optimization methods, our approach 100% preserves the deployed model's accuracy since it does not alter any properties and/or parameters of models, neither alter the standard inference software. Instead it instructs the device to just use the SRAM optimized execution sequence it provides.
- Many IoT devices running large NNs fail due to overheating, fast battery wear, and run-time stalling. The prime reason for such failure causing issues is the exhaustion of device memory (especially SRAM). To accurately estimate the memory consumed by models during execution on IoT devices, we provide a Tensor Memory Mapping (TMM) program that can load any pretrained models like ResNet, NASNet, Tiny-YOLO, etc., and can accurately compute and visualize the tensor memory requirement of each operator in the computation graph of any given model. A part of the approach proposed in this paper relies on the high-accuracy calculation results of TMM.

Outline. The rest of the paper is organized as follows; Sect. 2 briefs essential concepts and related studies. In Sect. 3, we present the complete proposed approach, and in Sect. 4, we perform an empirical evaluation that aims to justify the claims of our approach before concluding our paper in Sect. 5.

2   Background and Related Work

In Subsect. 2.1, we present the top deep model compression techniques that produce the smallest possible model, which can be executed on MCUs and small CPUs using our proposed approach. In Subsect. 2.2, we view the trained NN as a graph and explain its standard execution method, followed by the related studies comparable with our model execution approach.

2.1   Deep Model Compression

The approaches in this category employ various techniques to enable fitting large NNs on IoT devices. For instance, *Model design techniques* emphasize designing models with reduced parameters. *Model compression techniques* use quantization and pruning [12] based approaches. Quantization takes out the expensive floating-point operations by reducing it to a Q-bit fixed-point number, and pruning removes the unnecessary connections between the model layers. Other techniques such as layer decomposition [11], distillation [3], binarisation [5] is also applicable. Also, neural architecture search methods [15] can be used to design a network with only a certain floating-point operation count to fit within the memory budget of the MCUs. If users want to achieve a higher level of size reduction, let's assume when they aim to execute models like Tiny-YOLO and Inception v3 (23.9 MB after post-training quantization) on IoT devices, we recommend performing *Deep Model Compression*. Here the users, in a sequence, have to realize more than one of the briefed model optimization techniques.

After following the deep optimization sequence of their choice, the NNs become friendly enough to be executed on tiny devices. Additionally, when such deep optimized models are executed using our proposed approach, its peak on-device execution memory usage can be reduced.

2.2 Executing Neural Networks on Microcontrollers

A neural network is a graph with defined data flow patterns having an arrangement of nodes and edges, where nodes represent operators of a model, and graph edges represent the flow of data between nodes. The operator nodes in the model graph can be 2D convolutions (Conv2D), or Depthwise separable 2D convolution (DepthwiseConv2D), etc. These operator nodes can take more than one input to produce an output. Recently, a few ML frameworks have released tools to optimize model graphs in order to improve the execution efficiency of NNs. For example, the optimizer tool fuses adjacent operators and converts batch normalization layers into linear operations. In such model computation graphs, buffers are used to hold the input and output tensors before feeding them to the operators during the model execution. After execution, the items in the output buffer will be provided as input to the next operator, and the input buffers can be reclaimed by removing the stored data.

Structure of Computation Graphs. When executing a model, the graph nodes in both the regular graph and its optimized version are executed one by one in a topological fashion/order. For example, the VGG and AlexNet iteratively apply a linear sequence of layers to transform the input data. But, similar to the computation graph shown in Fig. 1, the newer networks like the Inception, NasNet, and MobileNet, etc. are non-linear as they contain branches. For these networks, the input data transformation is performed in divergent paths because the same input is accessible by numerous operators present in several layers i.e., the same input tensors are accessible for processing by several layers and operators. Hence when executing such branched models on MCUs, the execution method can have access to multiple operators.

Mapping Models on the MCU Memory. The typical small CPUs and MCUs based IoT devices have their on-chip memory partitioned into SRAM (read-write) and NOR-Flash (read-only). The complete memory requirement of a NN is mapped to these two partitions. Since SRAM is the only available read-write space, the intermediate tensors generated during model execution are stored here, increasing the peak SRAM usage on MCUs. The model parameters such as trainable weights, layers, constants, etc., do not change during the run-time (immutable in nature). Hence, they are converted into hex code and stored in the static Flash memory along with the application of the IoT use case.

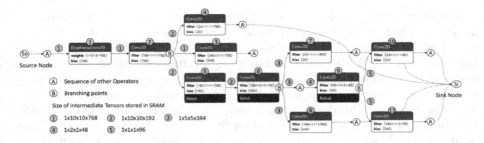

Fig. 1. A part of the COCO SSD MobileNet computation graph with its branched operators: When executing such graphs on IoT devices, our approach reduces the peak SRAM consumption by producing an optimized operators execution sequence. (Color figure online)

The most relevant work to ours are [10] and [9], where a NN execution runtime for MCUs is attached with their NAS. Next is the [1], which proposes a method for optimizing the execution of a given neural network by searching for efficient model layers. i.e., a search is performed to find efficient versions of kernels, convolution, matrix multiplication, etc., before the C code generation step for the target MCU. Both the methods aim to ease the deployment of NNs on MCUs, whereas our approach is to take any deep compressed model and during execution reduce its peak SRAM usage.

3   Efficient Neural Network Execution Approach Design

As discussed earlier, the trained model size and its peak SRAM need to be highly reduced due to the limited Flash and SRAM memory capacity of IoT devices. Here, we present our approach that can reduce the peak SRAM consumed by neural networks. We first describe our Tensor Memory Mapping (TMM) method in Subsect. 3.1. Then in Subsect. 3.2 and 3.3, we present the two parts of our proposed approach, followed by Subsect. 3.4 that combines both the parts and presents the complete approach in the form of an implementable algorithm.

3.1   Tensor Memory Mapping (TMM) Method Design

Before deployment, the memory requirement of models is often unknown or calculated with less accuracy. i.e., there will exist a few MB of deviations in the calculations. When the model is targeted to run on better-resourced devices like smartphones or edge GPUs, these few MB deviations do not cause any issues. But when users target the resource-constrained IoT devices (has only a few MB memory), then the low-accuracy calculation causes run-time memory overflows and/or restrict flashing model on IoT devices due to SRAM peaks. Based on our recent empirical study, we found that many IoT devices that are running large NNs fail due to overheating, fast battery wear, run-time stalling. The prime reason for such failure causing issues is the exhaustion of device memory (especially SRAM). Hence, this inaccurate calculation leads to a horrendous computing resource waste (especially the GPU days) and reduced development productivity. In this section, we thereby present our tensor memory mapping method, which can be realized to accurately compute and visualize the tensor memory requirement of each operator in any computation graph. We use this high-accuracy calculation method in the core algorithm design of our efficient neural network execution approach.

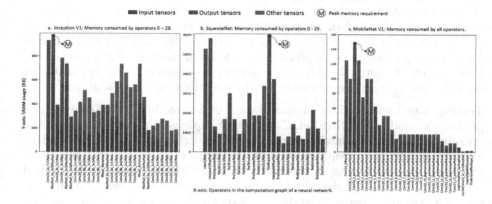

Fig. 2. Accurate computation and visualization of tensor memory requirement for each operator in NN computation graphs (performed using our TMM): The Algorithm 1 reduces the shown memory peaks by reordering operators to produce a new graph execution sequence.

Abstraction and Formalization. We treat the internal of neural networks as mathematical functions and formalize it as *tensor-oriented computation graphs* since the inputs and outputs of graph nodes/operators are a multi-dimensional array of numerical values (i.e., tensor variables). The shape of such a tensor is the element number in each dimension plus element data type. In the below equation, we formally represent a NN as a Directed Acyclic Graph (DAG), and we treat its execution as iterative forward and backward propagation via the graph branches.

$$NN_{DAG} = \langle \{op_i\}_{i=1}^n, \{(op_i, op_j)\}, \{p_k\}_{k=1}^m \rangle \qquad (1)$$

Here op_i are the graph operators, (op_i, op_j) is the connection to transmit output tensor from op_i as an input to op_j, and there are m hyperparameters p_k. Let the topological ordering of operators be $Seq = \langle op_{i_1}, op_{i_2}, \cdots, op_{i_n} \rangle$ that extends from the first graph edge such that $op_{i_i} <_{Seq} op_{i_k} \rightarrow (op_{i_k}, op_{i_j}) \notin NN_{DAG}$, where Seq is the operator execution sequence (we aim to find a memory friendly sequence in the later sections). In this graph, when visiting a node op, we need to calculate the memory it consumes to store (i) newly assigned tensors, (ii) previously assigned but still in-use tensors, (iii) reserved buffers. To calculate the memory consumption $M_{NN_{DAG}}$ of a graph NN_{DAG} we give the following formulae. We call the first two types of tensors as unreleased tensors.

$$M_{NN_{DAG}} = \max \{MF_{n_{init}}, MF_n(op_i) \mid op_i \in NN_{DAG}\} \tag{2}$$

Here, $MF_{n_{init}} = \sum MT_{sr}(t)$ is the function to compute the initial memory consumption, $MF_n(op) = MU_{res}(op) + MR(op)$ is the current memory consumption, $MU_{res}(op) = \sum_{t \in U_{res}T_{sr}(op)} MT_{sr}(t)$ is the function that computes memory requirement of unreleased tensors, $MR(op)$ function returns memory size of reserved buffers. The set of unreleased tensors are computed using $U_{res}T_{sr}$, and for a given tensor t, function MT_{sr} is used to find its allocated memory size. The Eq. 2 applies to models trained using any ML frameworks like TensorFLow, PyTorch, etc. to estimate the graph memory consumption, and applicable to calculate the memory requirements for any operators execution sequence.

Testing the Design. The implementation of our method is suitable for any pre-trained models like NASNet, Tiny-YOLO, SqueezeNet, etc. For each of the operators in any given model graph, our method computes the total required SRAM. i.e., the space required to store the input tensors + output tensors + other tensors, and then exports the detailed report in CSV format. Our method can also produce images that show the tensor memory requirement of each operator. For example, when we feed the Inception V1 that contains 84 graph nodes/-operators to our method, it produces Fig. 2a. (for brevity, we show only 0–29 operators) along with the detailed CSV report. Similarly, we test our method on SqueezeNet and MobileNet V1 and show the results in Fig. 2b–c. Thus by enabling visualization, our method helps users analyze multiple memory aspects of networks and obtain valuable insights that can guide them to customize their model graph for highly reduced memory. For example, we made the following observations; (i) Most of the Inception V1 nodes consume high memory to accommodate other tensors, whereas the MobileNet does not contain other tensors at all; (ii) Three nodes in SqueezeNet consume significantly higher memory than other nodes. Such nodes can be replaced with cheaper operators that perform the same tasks.

3.2   Loading Fewer Tensors and Tensors Re-usage

In the traditional model execution methods, multiple tensors of various sizes are loaded into the buffer (such bulk loading is the reason for causing peak memory usage) since the traditional methods execute operators requiring different size tensors. In contrast, our approach executes many operators by just loading a minimum number of tensors. This part of our approach also aims to achieve **SRAM conservation by tensors re-usage.** Here, our approach identifies and stores a particular set of tensors in the buffer (buffers are created within SRAM) and first executes the branch of the graph containing operators compatible with the stored tensors. Then in the next iteration, it loads tensors that are suitable as input for the set of operators belonging to the next branch, then performs the execution. After each iteration, the buffers are reclaimed.

For illustration purpose, in Fig. 1, the intermediate tensors of varying size are shown in blue circles ① to ⑤ which need to be stored in SRAM during the graph execution. Here, at the first branching point circled Ⓑ, when the default model execution software is utilized, the two tensors with blue circles ① and ② are loaded on the SRAM. Then it executes all the branched operators with rose circles ③ to ⑤. This method of loading many tensors and executing many operators leads to the most critical SRAM overflow issue, especially in the scenarios where multiple branches are emerging from one branching point.

3.3   Finding the Cheapest NN Graph Execution Sequence

The computational graphs of models perform the inference tasks in a collection of computational steps, where each step depends on the output from a few of the preceding steps. For example, in the graph of MobileNet shown in Fig. 1, these graph steps are the operators and rose circled ① → ② means the second operator depends on the output of the first. Since the computation graphs of most NNs are DAGs, we can enumerate orders/sequences to execute all the operators/computational steps.

As shown in Fig. 1, the modern NNs like MobileNet have divergent data flow paths. i.e., their computation graphs contain branches. As briefed in Subsect. 2.2, due to such a branched design, a given tensor can be accessed by operators in various branches. For example, in Fig. 1, the tensor with a blue circle ① of size $1 \times 10 \times 10 \times 768$ can be accessed by three Conv2D operators due to the presence of a branching point circled Ⓑ. Similarly, the tensor with blue circle ③ of size $1 \times 5 \times 5 \times 384$ is accessible by two Conv2D layers and by another sequence of operators circled Ⓐ. Such branched computation graphs provide freedom for the model execution software to alter the execution order/sequence of the operators. In the rest of this section, we show that any topological execution order of the graph nodes will result in a valid execution scheme; we then explain how our approach leverages this freedom to achieve its SRAM conservation goal.

Does Any Topological Execution Order of the NN Graph Nodes Result in a Valid Execution Scheme? DAGs of models have topological ordering and do not have cycles because the edge into the earliest vertex of a cycle would have to be oriented the wrong way. Therefore, every graph with a topological ordering is acyclic. But for a directed graph that is not acyclic, there can be more than one minimal subgraph with the same reachability relation. Where, for a complete DAG with N nodes, the search space contains $2^{N(N-1)/2}$ possible topological orders/structures. We consider the initial graph node as the *source node*, where the input data (i.e., sensor values that require predictions) is fed into the network, and the ending node as the *sink node*, where the inference results are transmitted to control the real world applications. In the graph of MobileNet from Fig. 1, let us assume the source node to be circle Ⓢ̲ₒ and the sink node to be circle Ⓢ̲ᵢ. Since NNs are DAGs, the topological execution numbers assigned to the nodes (operators) increases along the branched path (without forming any cycles) till the sink node. During this coverage, no graph vertices or nodes are skipped.

Formally, we define thus described topological process as $G_0 = (V, E)$, with operators $V = \{v_1, v_2, v_3, \ldots, v_{n-1}, v_n\}$ and E are the edges between operators. Here the operator execution order is a sequence containing all the operators $\in V, \{v_{k_1}, v_{k_2}, \cdots, v_{k_{n-1}}, v_{k_n}\}$ such that for all i, j $(0 \leqslant i, j \leqslant n)$, if there exists a path from v_{k_i} to v_{k_j}, then $i < j$. Briefly, if there is a path from operator v to operator w, then in the execution sequence, v should be set to be executed before w. Hence, the directed computation graph of NNs is a DAG *if and only if* it has a topological ordering. This explanation gives us two independent statements to prove; **First** we need to show if a directed graph follows a topological ordering of operator nodes, it is a DAG. **Second**, we need to show that all DAGs follows a topological ordering of operator nodes.

Proof One. Since a biconditional logical connective exists between the above two statements, either both statements are true or both are false. Hence, proving either the first or the second statement will suffice both. By contrapositive; if we prove that *if a NN graph is not a DAG, it can not have a topological ordering*, we can satisfy the first statement. In the following, we prove this.

When we assume the computation graph of a NN to not be a DAG, there will exist cyclic data flow between operators in the graph. For example, in $\{v_1, v_2, \cdots, v_k, v_1\}$, since there is a path from v_1 to v_2, the operator v_1 must appear before v_2 in the topological ordering scheme. But there is also a path from v_2 to v_1 via v_k making v_2 appear before v_1. If we implicate this scenario in Fig. 1, the execution sequence reaches the sink node (v_k) and then returns back to the source node (v_1), clearly voiding the main ordering principle of a DAG, hence proving the first statement. In the following, we also prove the second statement, but by induction.

Proof Two. We start to prove the second statement in *step one*. Here, we define the base case, which is a graph with just one operator. This graph is a DAG with topological ordering since the execution order starts from the source node, travels via the single operator, and ends at the sink node. In *step two*, we consider a topologically ordered DAG with multiple operators connected by n vertices as the induction hypothesis. In order to prove the second statement, for this induction step two, we need to show that the induction hypothesis implies that a DAG with $n+1$ vertices must have a topological ordering. To prove this, in *step three*, we take a NN graph with $n+1$ vertices/operators having one 0-degree vertex v_0. In *step four*, we remove the 0-degree vertex to obtain a computation graph with n vertices (similar to graph from step two). This resulting graph must be a DAG since the base graph from step two had no cycles, and also, in this step, we removed edges (not added).

According to the induction hypothesis from step two, since the resultant graph from step four is a DAG with n vertices, it also will have a topological ordering. Thus, a topological operators execution sequence can be constructed for the graph from step three that has $n+1$ operators, by prepending v_0 to the topological order of the n vertices DAG from step two.

Algorithm 1. Reducing the peak SRAM consumption by discovering an optimized operators execution sequence.

1: **Input**: Computation graph of the trained model.
2: **Output**: Cheapest graph execution order with reduced peak SRAM requirement.
3: $const_{tens}$ ▷ Constant tensors
4: $active_{tens}$ ▷ Active tensors that change during graph execution
5: set_{tens} ▷ Set of tensors
6: rem_{tens} ▷ Variable to store the remaining tensors
7: req_{tens} ▷ Tensors required to produce $tens$
8: **operator** $(tens)$ ▷ The operator that computes to produce tensors $tens$ and set_{tens}
9: $k \leftarrow \infty,\ s \leftarrow 0,\ k' \leftarrow 0$ ▷ Variables
10: **memory reduction**   ▷ Function to find the path that consumes minimum memory to compute all $tens \in set_{tens}$
11:     $const_{tens}, active_{tens} \leftarrow$ **Separate** $(set_{tens}, tens :$ **operator** $(tens)$ **is none**$)$
    ▷ Separate constant and active tensors
12:     **if** no $active_{tens}$ **then**
13:         **return** $\sum_{s \in const_{tens}} |c|$ ▷ No remaining operators to reorder. Send sizes of remaining $const_{tens}$
14:     **end if**
15:     **for** $tens$ in $active_{tens}$ **do**
16:         $rem_{tens} \leftarrow active_{tens}$ ▷ Remaining tensors need to be stored in memory
17:         $req_{tens} \leftarrow$ **operator** $(tens)\ .\ data$
18:         **if** any $(tens$ is used to produce rem where $rem \in rem_{tens})$ **then**
19:             $tens$ was used to produce rem. So in the future, the **operator** $(tens)$ will be executed   ▷ Result stored for re-use
20:         **end if**
21:                 ▷ At this stage, peak memory will be consumed either by; (i) the **operator** $(tens)$ that produced rem. In this case the peak is the memory of input tensors + output tensor + other tensors. (ii) other operators. i.e., recursive case **memory reduction** $(rem_{tens} \cup req_{tens})$
22:         $k' \leftarrow$ max $($**memory reduction** $(rem_{tens} \cup req_{tens}),$ $\sum_{t \in rem_{tens} \cup req_{tens} \cup \{tens\}} |t|)$
23:         $k \leftarrow$ min $(k,\ k')$
24:                 ▷ The cheapest graph execution order/path is decided here
25:     **end for**
26: **return** $\sum_{rem \in rem_{tens}} |rem|\ +\ k$

SRAM Conservation by Altering Operators Execution Sequence. Having proved that changing execution sequence of operators still produces a valid scheme; our approach achieves its memory conservation goal by intelligently selecting the execution branch that when executed consumes less SRAM (reduces the peak memory consumption) than the default sequence. For illustration purpose, in Fig. 1., if the model execution software follows the default operators execution order; the execution will start at the operator with a rose circle ① and follow the sequence till the operator with a rose circle ⑧, in the order of 1, 2, 3, 4, 5, 6, 7, 8. This unoptimized default order will consume a peak SRAM of 5900 Bytes. Whereas when our efficient execution approach is utilized, the operator execution order is altered to form a new sequence that will require a reduced SRAM of 5200 Bytes. This new order will be 1, 2, 5, 6, 7, 8, 3, 4. Here, the calculated SRAM consumption/requirement is the sum of the size of tensors stored in the operator's input and output buffers added with the tensor size of the output of previous or next operators. As explained in Subsect. 2.2, this third set of stored tensors are the input for the other operators that exist in the graph.

3.4   Core Algorithm

Discovering multiple topological orders of nodes in a computation graph belongs to the literature of graph optimization. The algorithm that we present in this section belongs here since we designed it considering the computation graph of a model as a DAG, and as proved in Subsect. 3.3, the execution of available nodes in any topological order will result in a valid execution sequence. When the computation graph of any given model is loaded into our algorithm, it analyzes the complete network by running through each branch of the network and finally discovering the cheapest graph execution path/sequence. The time consumed by the algorithm to produce the results depends on complexity $\mathcal{T}_|\left(|O|2^{|O|}\right)$, where $|O|$ is the total operators count. Since the latest network architectures contain hundreds of operators, our proposed algorithm is best-suited to run on better-to-high resource devices such as laptop CPUs. Our algorithm-generated optimized graph execution sequence should be used by the inference software when executing the target model on MCU-based IoT devices.

We present our complete approach in Algorithm 1. Here, Lines 10 to 25 is the core *memory reduction* function of our algorithm that performs the required tasks to reduce the peak SRAM usage by reordering operators to produce a new execution sequence. Before the core function, in Line 3 to 9, we declare all the function required variables. In Line 11, we remove the tensors that shall not be used as inputs by the operators in the graph. Also, the tensors that do not contain the operators that produced are taken out. Thus performed removal actions do not affect the model performance since the removed tensors are constant $const_{tens}$.

Next, in Line 18 to 20, we ensure that no operator nodes are executed twice. This is done by checking whether an operator node has produced any of the tensors (rem_{tens}) that are remaining after taking out $const_{tens}$. If such tensors are existing, in the future, the inference software might require to execute again the operators that produced those tensors. To conserve memory, in Line 19, the results of such operators that need to be re-executed are stored in the buffer for reuse. In fact, such re-execution can cause memory peaks. In Line 22, the *memory reduction* function is called multiple times in order to cover all the branches of the computation graph. Finally, in Line 26, the cheapest graph execution path is returned. When executing the thus produced reordered operators sequence on IoT devices, if the scope of loaded tensors is over, we recommend the inference software to reclaim the memory used by such tensors by removing them from the SRAM.

Table 1. Executing original models and its Algorithm 1 optimized versions: Comparing the peak SRAM usage, inference time, and the energy consumed for inference.

Model task/ category	Pre-trained model name	Quantized model without optimization			Quantized model with optimization using Algorithm 1		
		Peak SRAM usage (KB)	Inference time (ms)	Energy used (mJ)	Peak SRAM usage (KB)	Inference time (ms)	Energy used (mJ)
Image classification	MobileNetV1 [7]	98.304	1.6	27.59904	65.536(32 ↓)	0.96 (0.64 ↓)	16.55942 (11 ↓)
	SqueezeNet [6]	6195.200	12.4	213.8926	4816.896 (1378 ↓)	10.62 (1.78 ↓)	183.1886 (30.7 ↓)
	InceptionV1 [13]	1003.520	43.6	752.0738	802.816 (200 ↓)	38.9 (4.7 ↓)	671.0017 (81.0 ↓)
	MnasNet [14]	1605.632	7.4	127.6456	1204.224 (401 ↓)	5.7 (1.7 ↓)	98.32158 (29.3 ↓)
	NASNet mobile [17]	4511.660	63	1086.712	3834.284 (677 ↓)	61.2 (1.2 ↓)	1055.663 (31 ↓)
	DenseNet [4]	8429.568	246.3	4248.527	5221.264 (3208 ↓)	241.4 (4.9 ↓)	4164.005 (84 ↓)
Semantic segmentation	DeepLabv3 [2]	5639.592	38.2	658.927	5548.116 (91 ↓)	37.07 (1.13 ↓)	639.435 (19 ↓)
Pose estimation	PoseNet [8]	6575.904	22.3	384.661	4383.936 (2191 ↓)	19.4 (2.9 ↓)	334.638 (50 ↓)
Text detection	EAST [16]	5324.800	43.38	748.278	3686.400 (1638 ↓)	43.10 (0.28 ↓)	743.449 (4 ↓)

4   Experimental Evaluation

In this section, we perform an empirical evaluation to answer the following questions.

- To what levels can the proposed approach increase the model execution efficiency by reducing the peak SRAM usage of NNs?
- Is the approach suitable to diverse NN architectures and NNs trained using various datasets?
- Can the approach produce an optimized operators execution sequence for already optimized or deep compressed models?
- Does optimization using the proposed approach impact the accuracy or performance of the model?

We start the evaluation by downloading popular pre-trained TensorFlow Lite models (.tflite format) from TensorFlow Hub. For comprehensiveness, the models selected to evaluate our approach belong to various problem domains ranging from image classification to text detection and are listed in Table 1. As described in Subsect. 3.4, since the chosen models contain hundreds of operators, the complexity of our algorithm will be high. Hence, we conduct the evaluation on a standard NVIDIA GeForce GPU-based Ubuntu laptop with Intel (R) Core (TM) i7-5500 CPU @ 2.40 GHz. After the download, we first load and execute each model on the same laptop using the default execution sequence of operators and tabulate the corresponding peak SRAM usage, unit inference time, and the energy consumed to execute the model and perform inference.

In the same setup, we next apply the implementation of Algorithm 1 on each model and tabulate the obtained results in Table 1, next to the results obtained when executing models using their default execution sequence. During the evaluation, for statistical validation, the reported inference time and the consumed energy corresponds to the average of 5 runs. In order to perform analysis, in Table 1, we subtract the values reported under *Quantized Model with Optimization using Algorithm 1* with values under *Quantized Model without Optimization* and plot bar-graphs for each model in Fig. 3. Based on this, in the remainder subsections, we analyze and discuss the benefits achieved as a result of optimizing models using our proposed approach.

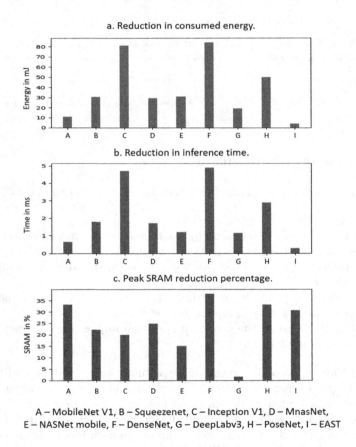

A – MobileNet V1, B – Squeezenet, C – Inception V1, D – MnasNet,
E – NASNet mobile, F – DenseNet, G – DeepLabv3, H – PoseNet, I – EAST

Fig. 3. Benefits achieved after optimization using our proposed approach.

4.1 SRAM Usage

In practice, there are many cases where ML models optimized using state-of-the-art deep compression sequences exceed the target device's SRAM capacity just by a few KB margin. In such cases, users cannot additionally apply any optimization approach since it might not match the previous optimizer components, or the model might already be maximum compressed. So they either have to alter the model architecture and re-train to produce a smaller model (waste of GPU days and electricity) or upgrade the IoT device hardware (loss of money). In the remainder of this section, we show how our approach can enable the accommodation and execution of memory overflow issues causing models on IoT devices.

We take the quantized DenseNet with its default execution sequence and feed it to our TMM program from Sect. 3.1. From the resultant computed memory requirement for each operator in the default graph, the 24th operator showed the peak SRAM consumption of 8429.568 KB. Next, after applying our Algorithm 1 on DenseNet, the resultant memory-friendly graph execution sequence, when evaluated by the TMM program, showed the peak memory of only 5221.264 KB (peak reduced by 38.06%).

Similarly for MobileNet V1, the peak SRAM usage reduced from 98.304 KB to 65.536 KB (see Table 1). Here our approach has reduced the memory peak by 32.76 KB (by 33%). In Fig. 3c, we plot thus calculated peak SRAM reduction percentage for MobileNet V1 (label A in x-axis) and the remaining 8 models selected for evaluation. The maximum peak SRAM reduction of 38.06% was achieved for the DenseNet and the least of 1.61% reduction for DeepLabv3. It is apparent from the results that the execution sequence produced by our approach is applicable for a wide range of ML models that have diverse network architectures. Also, since it reduces the SRAM peaks, the models that are still large after optimization can be accommodated on tiny IoT devices. Thus, our approach eliminates the re-training step that aims to produce small models, and also, the device hardware need not be upgraded to accommodate the models.

4.2 Model Performance

As a part of experimental results, we report that despite the SRAM conservation, the model executed using the SRAM optimized sequence provided by Algorithm 1 showed the same performance (accuracy, F1 score, etc.) as the models when executed with their default sequence. This is because, unlike existing methods, ours does not alter any properties/parameters of models, neither alter the standard inference software (just instructs to use a different model execution sequence). Also, as proved in Sect. 3.3, the SRAM optimized sequence produced by our approach is a valid model execution sequence. This 100% model performance preservation characteristics enable even tiny IoT devices to produce high accuracy offline analytics results.

4.3   Inference Time and Energy Consumption

Here in order to investigate the impact of our approach on inference/model execution performance, we execute each model first with their default execution sequence, then with the memory peak reduced sequence produced by our approach. We report the difference in inference time and consumed energy for both default and optimized sequence in Table 1 and show it in Fig. 3a–b. For the same tasks performed on the same device using the same datasets, the new graph execution sequence for DenseNet shows the maximum inference time reduction of 4.9 ms and the least of 0.28 ms reduction for EAST. We also achieved 4–84 mJ less energy to perform unit inference since executing the model using the SRAM optimized sequence produced by our approach is 0.28–4.9 ms faster than the default sequence.

In realistic scenarios, to infer using a stream of data input, the deployed model is executed in a loop. Here, even the minor inference speedups and energy conservation produced by our approach get multiplied, driving the IoT devices close to producing real-time edge analytics results at a lower power cost. Thus, even the autonomous tiny IoT devices can efficiently control real-world IoT applications by making timely predictions/decisions and also perform offline model inference without affecting the operating time of battery-powered devices.

5   Conclusion

In this paper, we presented an approach to efficiently execute (with reduced SRAM usage) deeply optimized (maximally compressed) ML models on resource-constrained devices. For nine popular models, when comparing the default model execution sequence with the sequence produced by our approach, we showed that 1.61–38.06% less SRAM was used to produce inference results, the inference time was reduced by 0.28–4.9 ms, and energy consumption was reduced by 4–84 mJ. As well as achieving highly conserved SRAM levels, our method 100% preserved the model performance. Thus, when users apply the approach presented in this paper, they can: (i) Execute large-high-quality models on their IoT devices/products without needing to upgrade the hardware or alter the model architecture and re-train to produce a smaller model; (ii) Devices can control real-world applications by making timely predictions/decisions; (iii) Devices can perform high accuracy offline analytics without affecting the operating time of battery-powered devices.

Acknowledgements. This publication has emanated from research supported in part by a research grant from Science Foundation Ireland (SFI) under Grant Number SFI/16/RC/3918 (Confirm) and also by a research grant from Science Foundation Ireland (SFI) under Grant Number SFI/12/RC/2289_P2 (Insight), with both grants co-funded by the European Regional Development Fund.

References

1. Tinyml - how TVM is taming tiny. https://tvm.apache.org/2020/06/04/tinyml-how-tvm-is-taming-tiny
2. Chen, L.C., Papandreou, G., Schroff, F., Adam, H.: Rethinking atrous convolution for semantic image segmentation. arXiv preprint arXiv:1706.05587 (2017)
3. Hinton, G., Vinyals, O., Dean, J.: Distilling the knowledge in a neural network. arXiv preprint arXiv:1503.02531 (2015)
4. Huang, G., Liu, Z., Van Der Maaten, L., Weinberger, K.Q.: Densely connected convolutional networks. In: Proceedings of the IEEE Conference on Computer Vision and Pattern Recognition (2017)
5. Hubara, I., Courbariaux, M., Soudry, D., El-Yaniv, R., Bengio, Y.: Binarized neural networks. In: Advances in Neural Information Processing Systems (2016)
6. Iandola, F.N., Han, S., Moskewicz, M.W., Ashraf, K., Dally, W.J., Keutzer, K.: SqueezeNet: AlexNet-level accuracy with 50x fewer parameters and <0.5 mb model size. arXiv preprint arXiv:1602.07360
7. Jacob, B., et al.: Quantization and training of neural networks for efficient integer-arithmetic-only inference. In: Proceedings of the IEEE Conference on Computer Vision and Pattern Recognition (2018)
8. Kendall, A., Grimes, M., Cipolla, R.: PoseNet: a convolutional network for real-time 6-DOF camera relocalization. In: Proceedings of the IEEE International Conference on Computer Vision (2015)
9. Liberis, E., Dudziak, Ł., Lane, N.D.: µNAS: constrained neural architecture search for microcontrollers. arXiv preprint arXiv:2010.14246
10. Lin, J., Chen, W.M., Lin, Y., Cohn, J., Gan, C., Han, S.: MCUNet: tiny deep learning on IoT devices. arXiv preprint arXiv:2007.10319 (2020)
11. Qiu, Q., Cheng, X., Calderbank, R., Sapiro, G.: DCFNet: deep neural network with decomposed convolutional filters. arXiv preprint arXiv:1802.04145 (2018)
12. Sudharsan, B., Breslin, J.G., Ali, M.I.: RCE-NN: a five-stage pipeline to execute neural networks (CNNs) on resource-constrained IoT edge devices. In: Proceedings of the 10th International Conference on the Internet of Things (2020)
13. Szegedy, C., et al.: Going deeper with convolutions. In: Proceedings of the IEEE Conference on Computer Vision and Pattern Recognition (2015)
14. Tan, M., e al.: MnasNet: platform-aware neural architecture search for mobile. In: Proceedings of the IEEE Conference on Computer Vision and Pattern Recognition (2019)
15. Tan, M., Le, Q.V.: EfficientNet: rethinking model scaling for convolutional neural networks. arXiv preprint arXiv:1905.11946 (2019)
16. Zhou, X., et al.: East: an efficient and accurate scene text detector. In: Proceedings of the IEEE Conference on Computer Vision and Pattern Recognition (2017)
17. Zoph, B., Vasudevan, V., Shlens, J., Le, Q.V.: Learning transferable architectures for scalable image recognition. In: Proceedings of the IEEE Conference on Computer Vision and Pattern Recognition (2018)

AutoML Meets Time Series Regression Design and Analysis of the AutoSeries Challenge

Zhen Xu[1]([✉]), Wei-Wei Tu[1], and Isabelle Guyon[2,3,4]

[1] 4Paradigm, Beijing, China
{xuzhen,tuweiwei}@4paradigm.com
[2] LISN CNRS/INRIA, Gif-sur-Yvette, France
guyon@chalearn.org
[3] University Paris-Saclay, Gif-sur-Yvette, France
[4] ChaLearn, California, USA

Abstract. Analyzing better time series with limited human effort is of interest to academia and industry. Driven by business scenarios, we organized the first Automated Time Series Regression challenge (AutoSeries) for the WSDM Cup 2020. We present its design, analysis, and post-hoc experiments. The code submission requirement precluded participants from any manual intervention, testing automated machine learning capabilities of solutions, across many datasets, under hardware and time limitations. We prepared 10 datasets from diverse application domains (sales, power consumption, air quality, traffic, and parking), featuring missing data, mixed continuous and categorical variables, and various sampling rates. Each dataset was split into a training and a test sequence (which was streamed, allowing models to continuously adapt). The setting of "time series regression", differs from classical forecasting in that covariates at the present time are known. Great strides were made by participants to tackle this AutoSeries problem, as demonstrated by the jump in performance from the sample submission, and post-hoc comparisons with AutoGluon. Simple yet effective methods were used, based on feature engineering, LightGBM, and random search hyper-parameter tuning, addressing all aspects of the challenge. Our post-hoc analyses revealed that providing additional time did not yield significant improvements. The winners' code was open-sourced (https://www.4paradigm.com/competition/autoseries2020).

1 Introduction

Machine Learning (ML) has made remarkable progress in the past few years in time series-related tasks, including time series classification, time series clustering, time series regression, and time series forecasting [9,14].To foster research in time series analysis, several competitions have been organized, since the onset of machine learning. These include the Santa Fe competition[1], the Sven Crone

---

[1] https://archive.physionet.org/physiobank/database/santa-fe/.

© Springer Nature Switzerland AG 2021
Y. Dong et al. (Eds.): ECML PKDD 2021, LNAI 12979, pp. 36–51, 2021.
https://doi.org/10.1007/978-3-030-86517-7_3

competitions[2], several Kaggle comptitions including M5 Forecasting[3], Web Traffic Time Series Forecasting[4], to name a few. While time series forecasting remains a very challenging problem for ML, successes have been reported on problems of time series regression and classification in practical applications [11,14].

Despite these advances, switching domain, or even analysing a new dataset from the same domain, still requires considerable human engineering effort. To address this problem, recent research has been directed to Automated Machine Learning (AutoML) frameworks [3,15], whose charter is to reduce human intervention in the process of rolling out machine learning solutions to specific tasks. AutoML approaches include designing (or meta-learning) generic reusable pipelines and/or learning machine architectures, fulfilling specific task requirements, and designing optimization methods devoid of (hyper-)parameter choices. To stimulate research in this area, we launched with our collaborators a series of challenges exploring various application settings[5], whose latest editions include the Automated Graph Representation Learning (AutoGraph) challenge at the KDD Cup AutoML track[6], Automated Weakly Supervised Learning (AutoWeakly) challenge at ACML 2019[7], Automated Computer Vision (AutoCV) [10]) challenges at IJCNN 2019 and ECML PKDD 2019, etc.

This paper presents the design and results of the Automated Time Series Regression (AutoSeries) challenge, one of the competitions of the WSDM Cup 2020 (Web Search and Data Mining conference) that we co-organized, in collaboration with 4Paradigm and ChaLearn.

This challenge addresses "time series regression" tasks [4]. In contrast with "strict" forecasting problems in which *forecast* variable(s) y_t should be predicted from **past** values **only** (often **y** values alone), **time series regression** seeks to predict y_t using **past** $\{t - t_{min}, \cdots, t - 1\}$ AND **present** t values of one (or several) "covariate" *feature* time series $\{\mathbf{x}_t\}$[8]. Typical scenarios in which $\mathbf{x}_t$ is known at the time of predicting y_t include cases in which $\mathbf{x}_t$ values are scheduled in advance or hypothesized for decision making purposes. Examples include: *scheduled events* like upcoming sales promotions, *recurring events* like holidays, or *forecasts* obtained by external accurate simulators, like weather forecasts. This challenge addresses in particular **multivariate** time series regression problems, in which $\mathbf{x}_t$ is a feature vector or a matrix of *covariate* information, and $\mathbf{y_t}$ is a vector. The domains considered include air quality, sales, parking, and city traffic forecasting. Data are feature-based and represented in a "tabular" manner. The challenge was run with **code submission** and the participants were evaluated on the Codalab challenge platform, without any human intervention, on five

---

[2] http://www.neural-forecasting-competition.com/.

[3] https://www.kaggle.com/c/m5-forecasting-accuracy.

[4] https://www.kaggle.com/c/web-traffic-time-series-forecasting.

[5] http://automl.chalearn.org, http://autodl.chalearn.org.

[6] https://www.automl.ai/competitions/3.

[7] https://autodl.lri.fr/competitions/64.

[8] In some application domains (not considered in this paper), even **future** $\{t + 1, \cdots, t + t_{max}\}$) values of the covariates may be considered. An example would be "simultaneous translation" with a small lag.

datasets in the feedback phase and five different datasets in the final "private" phased (with full blind testing of a single submission).

While future AutoSeries competitions might address other difficulties, this particular competition focused on the following 10 questions:

Q1: **Beyond autoregression: Time series regression.** Do participants exploit covariates/features $\{\mathbf{x}_t\}$ to predict y_t, as opposed to only past y?

Q2: **Explainability.** Do participants make an effort to provide an explainable model, e.g., by identifying the most predictive features in $\{\mathbf{x}_t\}$?

Q3: **Multivariate/multiple time series.** Do participants exploit the joint distribution/relationship of various time series in a dataset?

Q4: **Diversity of sampling rates.** Can methods developed handle different sampling rates (hourly, daily, etc.)?

Q5: **Heterogeneous series length.** Can methods developed handle truncated series either at the beginning or the end?

Q6: **Missing data.** Can methods developed handle (heavily) missing data?

Q7: **Data streaming.** Do models update themselves according to newly acquired streaming test data (to be explained in Subsect. 2.2)?

Q8: **Joint model and HP selection.** Can models select automatically learning machines and hyper-parameters?

Q9: **Transfer/Meta learning.** Are solutions provided generic and applicable to new domains or at least new datasets of the same domain?

Q10: **Hardware constraints.** Are computational/memory limitations observed?

2   Challenge Setting

2.1   Phases

The AutoSeries challenge had three phases: a **Feedback Phase**, a **Check Phase** and a **Private Phase**. In the Feedback Phase, five "feedback datasets" were provided to evaluate participants' AutoML models. The participants could read error messages in log files made available to them (e.g., if their model failed due to missing values) and obtain performance and ranking feedback on a leaderboard. When the Feedback Phase finished, five new "private datasets" were used in the Check Phase and the Private Phase. The Check Phase was a brief transition phase in which the participants submitted their models to the platform to verify whether the model ran properly. No performance information or log files were returned to them. Using a Check Phase is a particular feature of this challenge, to avoid disqualifying participants on the sole ground that their models timed out, used an excessive amount of memory, or raised another exception possible to correct without specific feedback on performance. Finally in the Private Phase, the participants submitted blindly their debugged models, to be evaluated by the same five datasets as in Check Phase.

As previously indicated, in addition to the five feedback datasets and five private datasets, two public datasets were provided for offline practice.

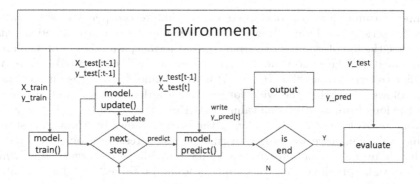

Fig. 1. Challenge protocol. `train`, `update`, and `predict` methods must be provided by participants. Such methods are under control of timers, omitted in the figure.

2.2 Protocol

The AutoSeries challenge was designed based on real business scenarios, emphasizing **automated machine learning** (AutoML) and **data streaming**. First, as in other AutoML challenges, algorithms were evaluated on various datasets entirely hidden to the particpants, **without any human intervention**. In other time series challenges, such as Kaggle's Web Traffic Time Series Forecasting[9]), participants downloaded and explored past training data, and manually tuned features or models. The AutoSeries challenge forced the participants to design **generic methods**, instead of developing *ad hoc* solutions. Secondly, test data were streamed such that at each time point t, historical information of past time steps $\mathbf{x}_{train}[: t-1]$, $\mathbf{y}_{train}[: t-1]$ and features of time t, X_test[t] were available for predicting $\mathbf{y}_t$. In addition to the usual **train** and **predict** methods, the participants had to prepare a method **update**, together with a strategy to update their model at an appropriate frequency, once fully trained on the training data. Updating too frequently might lead to run out of time; updating not frequently enough could result in missing recent useful information and performance degradation. The protocol is illustrated in Fig. 1.

2.3 Datasets

The datasets from the Feedback Phase and final Private Phase are listed in Table 1. We purposely chose datasets from various domains, having a diversity of types of variables (continuous/categorical), number of series, noise level, amount of missing values, and sampling frequency (hourly, daily, monthly), and level of nonstationarity. Still, we eased the difficultly by including in each of the two phases datasets having some resemblance.

Two types of tabular formats are commonly used: the "wide format" and "long format"[10]. The wide format facilitates visualization and direct use with

---

[9] https://www.kaggle.com/c/web-traffic-time-series-forecasting.

[10] https://doc.dataiku.com/dss/latest/time-series/data-formatting.html.

machine learning packages. It consists in one time record per line, with feature values (or series) in columns. However, for large number of features and/or missing values, the long format is preferred. In that format, a minimum of 3 columns are provided: (1) date and time (referred to as "Main_Timestamp"), (2) feature identifier (referred to as "ID_Key"), (3) feature value. Pivoting is an operation, which allows converting the wide format into the long format and vice-versa. From the long format, given one value of ID_Key (or a set of ID_Keys), a particular time series is obtained by ordering the feature values by Main_Timestamp. In this challenge, since we address a time series regression problem, we add a fourth column (4) "Label/Regression_Value" providing the target value, which must always be provided. A data sample in found in Table 2 and data visualizations in Fig. 2.

2.4   Metrics

The metric used to judge the participants is the RMSE. For each datasets, the participant's submissions are run in the same environment, and ranked according to the RMSE for each dataset. Then, an overall ranking is obtained from the average dataset rank, in a given phase. In post challenge analyses, we also used two other metrics: SMAPE and Correlation (CORR). The formulas are provided below. y means ground truth target. $\hat{y}$ is the prediction. $\bar{y}$ is the mean. N is total number of unique Id combinations (IdNum in Table 1) and T is number of timestamps. For evaluation, these metrics are run on the test sequences only.

$$\text{RMSE} = \sqrt{\frac{1}{NT}\sum_{n=1}^{N}\sum_{t=1}^{T}(y_{nt} - \hat{y}_{nt})^2} \tag{1}$$

$$\text{SMAPE} = \frac{1}{NT}\sum_{n=1}^{N}\sum_{t=1}^{T}\frac{|y_{nt} - \hat{y}_{nt}|}{(|y_{nt}| + |\hat{y}_{nt}| + \epsilon)/2} \tag{2}$$

$$\text{CORR} = \frac{\sum_{n=1}^{N}\sum_{t=1}^{T}(y_{nt} - \bar{y})(\hat{y}_{nt} - \bar{\hat{y}})}{\sqrt{\sum_{n=1}^{N}\sum_{t=1}^{T}(y_{nt} - \bar{y})^2}\sqrt{\sum_{n=1}^{N}\sum_{t=1}^{T}(\hat{y}_{nt} - \bar{\hat{y}})^2}} \tag{3}$$

2.5   Platform, Hardware and Limitations

The AutoSeries challenge is hosted on CodaLab[11], an open sourced challenge platform. We provide 4-core 30 GB memory CPU and no GPU is available. Participants may submit at most 5 times per day. A docker is provided[12] for executing submissions and for offline development. Participants can also install external packages if necessary.

---

[11] https://autodl.lri.fr/.
[12] https://hub.docker.com/r/vergilgxw/autotable.

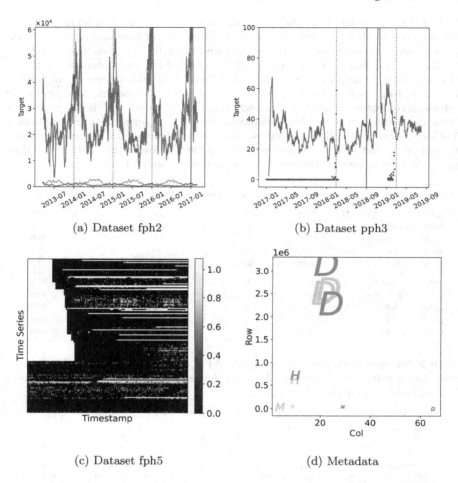

(a) Dataset fph2

(b) Dataset pph3

(c) Dataset fph5

(d) Metadata

Fig. 2. Dataset visualization. (a) Sample visualization of dataset fph2. Four time series (after smoothing and resampling). Training data is available until end of 2016 (red vertical solid line). Yearly prieriodicity is indicated by dashed vertical lines to highlight seasonalities. One notices large differences in series amplitudes and patterns of seasonality. **(b) Sample visualization of dataset pph3.** Two time series (after smoothing). Training data is available until end of 2018-08 (red vertical solid line). The purple time series suffers from missing values (certain items have zero sales most of the time). A clear trend exists in blue time series, unlike the example shown in (a). **(c) Heatmap visualization of dataset fph5.** White means missing value. Black means zero target value (sales). Several common issues can be observed: (1) many items don't sell most of the time; (2) presence of many missing values; (3) time series vary in lengths and are not aligned; (4) different time series have totally different scales. **(d) All dataset metadata visualization.** X axis is the number of columns. Y axis is the number of rows. The symbol letter shape represents the time period: Monthly, Daily, or Hourly. The symbol color represents the phase: green for "feedback" and orange for "private". The symbol size represents the number of lines in the dataset. (Color figure online)

Table 1. Statistics of all 10 datasets. Sampling "Period" is indicated in (M)inutes, (H)ours, (D)ays. "Row" and "Col" are the total number of lines and columns, in the long format. Columns includes: Timestamp, (multiple) Id_Keys, (multiple) Features, and Target. "KeyNum" is the number of Id_Keys (called Id_Key combination, e.g., in a sales problem Product_Id and Store_Id.) "FeatNum" indicates the number of features for each Id_Key combination (e.g., for a given Id_Key corresponding to a product in a given store, features include price, and promotion.) "ContNum" is the number of continuous features and "CatNum" is the number of categorical features; CatNum + ContNum = FeatNum. "IdNum" means the number of unique Id_Key combinations. One can verify that Col = 1 (timestamp) + KeyNum + FeatNum + 1 (target). "Budget" is the time in seconds that we allow participants' models to run.

Dataset	Domain	Period	Row	Col	KeyNum	FeatNum	ContNum	CatNum	IdNum	Budget
fph1	Power	M	39470	29	1	26	26	0	2	1300
fph2	AirQuality	H	716857	10	2	6	5	1	21	2000
fph3	Stock	D	1773	65	0	63	63	0	1	500
fph4	Sales	D	3147827	23	2	19	10	9	8904	3500
fph5	Sales	D	2290008	23	2	19	10	9	5209	2000
pph1	Traffic	H	40575	9	0	7	4	3	1	1600
pph2	AirQuality	H	721707	10	2	6	5	1	21	2000
pph3	Sales	D	2598365	23	2	19	10	9	6403	3500
pph4	Sales	D	2518172	23	2	19	10	9	6395	2000
pph5	Parking	M	35501	4	1	1	1	0	30	350

Table 2. Sample data for dataset fph2. A1 = timestamp. A2, A3, A4, A5, A7 = continuous features. A6 = categorical feature (hashed). A8, A9 = Id columns (hashed). Hashing is used for privacy. A10 = target.

A1	A2	A3	A4	A5	A6	A7	A8	A9	A10
2013-03-01 00:00:00	−2.3	1020.8	−19.7	0.0	−457...578	0.5	657...216	−731...089	13.0
2013-03-01 01:00:00	−2.5	1021.3	−19.0	0.0	511...667	0.7	657...216	−731...089	6.0
2013-03-01 02:00:00	−3.0	1021.3	−19.9	0.0	511...667	0.2	657...216	−731...089	22.0
...	...	...	...	...	...	...	...	...	...
2017-02-28 19:00:00	10.3	1014.2	−12.4	0.0	495...822	1.8	784...375	156...398	27.0
2017-02-28 20:00:00	9.8	1014.5	−9.9	0.0	−286...752	1.5	784...375	156...398	47.0
2017-02-28 21:00:00	9.1	1014.6	−12.7	0.0	−213...128	1.7	784...375	156...398	18.0

2.6   Baseline

To help participants get started, we provided a baseline method, which is simple but contains necessary modules in the processing pipeline. Many paticipants' submissions were derived from this baseline. In what follows, we decompose solutions (baseline and winning methods) into three modules: **feature engineering** (including time processing, numerical features, categorical features), **model training** (including models used, hyperparameter tuning, ensembling) and **update strategy** (including when and how to update models with the steaming test data). For the baseline, such modules include:

– **Feature engineering.** Multiple calendar features are extracted from the time stamp: *year, month, day, weekday,* and *hour.* Categorical variables (or

strings) are hashed to unique integers. No preprocessing is applied to numerical features.
- **Model training.** A single LightGBM [7] model is used. A LightGBM regressor is instantiated by predetermined hyperparameters and there is no hyperparameter tuning.
- **Update strategy.** Since the test data comes in a streaming way, we need an update strategy to incorporate new test data and adjust our model. However, due to time limit on update procedure, we can't update too frequently. The update strategy used in baseline is simple. We split all test timestamps by 5 segments and for every segment, we retrain the lightGBM with old training data and new segment of test data.

Table 3. Answers to the 10 challenge question. All of them are tackled to certain extent. Orange checkmark means the solution is trivial, though answers the question.

Question	Answered?	Comment
Q1 **Beyond autoregression**	✔	Features $\{x_t\}$ are leveraged
Q2 **Explainability**	✔	LightGBM outputs feature importance
Q3 **Multivariate/multiple time series**	✔	All training data is used to fit
Q4 **Diversity of sampling rates**	✔	Multiple calendar features are extracted
Q5 **Heterogeneous series length**	✔	Long format data facilitates the issue
Q6 **Missing data**	✔	Missing data is imputed by mean value
Q7 **Data streaming**	✔	Models are retrained every few steps
Q8 **Joint model and HP selection**	✔	Randomized grid search is applied
Q9 **Transfer/Meta Learning**	✔	Metadata (size, IdNum) is considered
Q10 **Hardware constraints**	✔	Model training time is recorded

2.7 Results

The AutoSeries challenge lasted one month and a half. We received over 700 submissions and more than 40 teams from both Academia (University of Washington, Nanjing University, etc.) and Industry (Oura, DeepBlue Technology, etc.), coming from various countries including China, United States, Singapore, Japan, Russia, Finland, etc. In the Feedback Phase[13], the top five participants are: **rekcahd, DeepBlueAI, DenisVorotyntsev, DeepWisdom, Kon** while in the Private Phase, the top five participants are: **DenisVorotyntsev, Deep-BlueAI, DeepWisdom, rekcahd, bingo**. It can be seen that team **rekcahd** seems to overfit on the Feedback Phase (additional experiments are provided in Subsect. 3.2). All winners use LightGBM [7] which is boosting ensemble of decision trees dominating most tabular challenges. Only 1st winner and 2nd winner implements hyperparameter tuning module which is really a key to successful generalisation in AutoSeries. We briefly summarize the solutions and provide a detailed account in Appendix.

---

[13] https://autodl.lri.fr/competitions/149#results.

- **Feature engineering.** Calendar features e.g., *year*, *month*, *day* were extracted from timestamp. Lag/shift and diff features added to original numerical features. Categorical features were encoded in various ways to integers.
- Model training. Only linear regression models and LightGBM were used. Most participants used default or fixed hyperparameters. Only the first winner made use of HPO. The second winner optimized only the learning rate. LightGBM provides built-in feature importance/selection. Model ensembling was obtained by weighting models based on their performance in the previous round.
- Update strategy. All participants updated their models. The update period was either hard coded, computed as a fixed fraction of the time budget, or re-estimating on-the-fly, given remaining time.

We verified (in Table 3) that the challenge successfully answered the ten questions we wanted addressed (see Sect. 1).

3   Post Challenge Experiments

This section presents systematic experiments, which consolidate some of our findings and extend them. We are particularly interested in verifying the generalisation ability of winning solutions on a larger number of tasks, and comparing them with open-sourced AutoSeries solutions. We also revisit some of our challenge design choices to provide guidelines for future challenges, including time budget limitations, and choice and number of datasets.

3.1   Reproducibility

First, we reproduce the solutions of the top four participants and the baseline methods, on the 10 datasets of the challenge (from both phases). In the AutoSeries challenge, we only used the RMSE (Eq. 1) for evaluation. For a more thorough comparison, we also include the SMAPE (Eq. 2) here for calculating the relative error (which is particualrly useful when the ground truth target is small, e.g., in the case of sales). The results are shown in the Table 4a and 4b, Fig. 3a and 3b. We ran each method on each dataset for 10 times with different random seed. For simplicity, we use *1st DV, 2nd DB, 3rd DW, 4th Rek* to denote solutions from top 4 winners.

We can observe that clear improvements have been made by the top winners, compared to the baseline, and both RMSE and SMAPE are significantly reduced. From Fig. 3a we can further visualize that, while sometimes the winners' solutions are close in RMSE, their SMAPE are totally different, which implies the necessity of using multiple metrics for evaluation.

3.2   Overfitting and Generalisation

Based on our reproduced results, we analyse potential overfitting which is visualized in Fig. 3c. Among each run (based on a different random seed), we rank solutions on feedback phase datasets and private phase datasets separately. Rankings are based on RMSE as in AutoSeries challenge. After 10 runs, we plot the mean and std of the ranking as a region. This shows that 4th Rek overfits to feedback datasets since it performs very well in feedback phase but poorly in private phase. But it is also interesting to visualize that 1st DV has a good generalisation: although it is not the best in feedback phase, it achieves great results in private phase. Including hyperparameter search may have provided the winner with a key advantage.

3.3   Comparison to Open Source AutoML Solutions

In this section, we turn our attention to comparing AutoSeries with similar open-source solutions. However, to the best of our knowledge, there is no publicly available AutoML framework dedicated to time series data. Current features (categorized by three modules of solutions as in Sec) of open source packages, which can be used to tackle the problems of the challenge with some engineering effort, are summarized in Table 5.

Packages like Featuretools, tsfresh focus on (tabular, temporal) feature engineering; they do not provide trainable models and should be used in conjuction with another package. Prophet and GluonTS are known for rapid prototyping with time series, but they are not AutoML packages (in the sense that they do not come with automated model selection and hyper-parameter selection). AutoKeras is an package focusing more on image and text, with KerasTuner[14] for neural architecture search. Google AutoTable meets most of our requirements, but is not open sourced, and is not dedicated to time series. Moreover, Google AutoTable costs around 19 dollars per hour in order to train on 92 computing instances at the same time, which is far more than our challenge settings.

At last, we selected AutoGluon for comparison, as being closest to our use case. AutoGluon provides end-to-end automated pipelines to handle tabular data without any human intervention (e.g., hyperparameter tuning, data preprocessing). AutoGluon includes many more candidate models and fancier ensemble methods than the wining solutions, but its feature engineering is not dedicated to multivariate time series. For example, it doesn't distinguish time series Id combinations to summarize statistics of one particular time series. We ran AutoGluon on all 10 datasets with default parameters except for using RMSE as evaluation metric and best_quality as presets parameter. The results are summarized in Table 6 column AutoGluon. Not surprisingly, vanilla AutoGluon can only beat the baseline, and it is significantly worse than the winning solutions. We further compile AutoGluon with 1st winner's time series feature engineering and update the models the same way as in baseline. The results are in Table 6 column

---

[14] https://keras-team.github.io/keras-tuner/.

Table 4. Post-challenge runs. We repeated 10 times all runs on all datasets for the top ranking submissions of the private phase. Each run is based on a different random seed. Error bar are indicated (one standard deviation) unless no variance was observed (algorithm with no stochastic component, such as 4th Rek).

Dataset	Phase	Baseline	1st DV	2nd DB	3rd DW	4th Rek
fph1	Feedback	100±10	40.7±0.2	**40.0±0.1**	40.1±0.2	40.7
fph2	Feedback	18000±2000	237±2	244±1	243.9±0.3	**230.7**
fph3	Feedback	3000	600±20	53.04±0.01	**52.4**	108.4
fph4	Feedback	6.9±0.3	3.66±0.02	2.760±0.007	NA	**2.632**
fph5	Feedback	8.6±0.7	5.76±0.01	5.760±0.005	5.780±0.003	**5.589**
pph1	Private	400±6	**200±2**	223.5±0.7	200±10	420.8
pph2	Private	17000±4000	**240±2**	253.7±0.5	260±2	246.7
pph3	Private	9±3	**6.20±0.02**	6.330±0.007	6.40±0.03	6.56
pph4	Private	12±2	4.0±0.2	3.80±0.04	3.700± 0.004	**3.32**
pph5	Private	300±30	**50±1**	100	60±20	167.8

(a) RMSE comparison.

Dataset	Phase	Baseline	1st DV	2nd DB	3rd DW	4th Rek
fph1	Feedback	140±20	100.0±0.5	104.00±0.07	104.00±0.04	**40.77**
fph2	Feedback	140±10	**30±1**	40.0±0.3	38.8±0.1	33.9
fph3	Feedback	38.49	5.0±0.1	0.770±0.001	**0.700±0.001**	1.674
fph4	Feedback	190±1	191.0±0.1	191.0±0.1	NA	**186.2**
fph5	Feedback	**170±1**	173.5±0.1	174.1± 0.1	172.9±0.1	170.6
pph1	Private	12.7±0.6	**6.1±0.2**	6.49±0.03	6.4±0.8	12.14
pph2	Private	140±10	**24.0±0.5**	35.0±0.3	31.0±0.1	32.75
pph3	Private	180±3	180.0±0.7	181.0±0.1	180±0.1	**174.7**
pph4	Private	170±3	170±1	167.8±0.1	170.0±0.2	**164**
pph5	Private	40±2	**6.0±0.5**	9	8±3	30.61

(b) SMAPE comparison.

Table 5. Supported features comparison between various open-source packages and the AutoSeries winning solution (also open-sourced).

Solutions	FeatureEngineering	ModelTraining	StreamingUpdate	TimeManagement
Featuretools [6]	Tabular	✗	✗	✗
tsfresh	Temporal	✗	✗	✗
Prophet [12]	✗	✓	✗	✗
GluonTS [1]	Temporal	✓	✗	✗
AutoKeras [5]	✗	✓	✗	✓
AutoGluon [2]	Tabulfar	✓	✗	✓
Google AutoTable	Tabular	✓	✓	✓
AutoSeries	**Temporal**	✓	✓	✓

[a] https://github.com/blue-yonder/tsfresh.
[b] https://cloud.google.com/automl-tables.

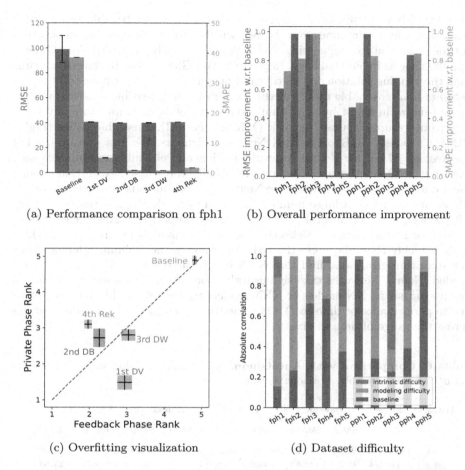

(a) Performance comparison on fph1

(b) Overall performance improvement

(c) Overfitting visualization

(d) Dataset difficulty

Fig. 3. Post challenge experiments. (a) Performance comparison on dataset fph1. We compare RMSE and SMAPE of all solutions on dataset fph1. Performances in RMSE are significantly better than the baseline for all winning teams, but all winners perform similarly. In contrast the SMAPE metric differentiates the winners, which focuses more on relative error. **(b) Performance improvement on all datasets.** Both RMSE and SMAPE errors from best methods are compared to the baseline performance. The improvement ratio is calculated by (baseline score - best score)/baseline score. **(c) Did the participants overfit the feed-back phase tasks?** Rankings are based on 10 runs: for each run, we rank separately in order to validate the stability of methods. Regions show the mean and std of rankings over multiple runs. Methods in the upper triangle are believed to overfit, e.g., 4th winner's solution. Methods in the lower triangle are believed to generalize well, e.g., 1st winner's solution. **(d) How well did we choose the 10 datasets?** As explained in Subsect. 3.5, we use absolute correlation as a bounded metric for calculating a notion of intrinsic difficulty (gray bar) and modeling difficulty (orange bar) of the 10 datasets. Datasets with high modeling difficulty and low intrinsic difficulty are better choices in a benchmark. (Color figure online)

FE+AutoGluon. AutoGluon can now indeed achieve comparable results with best winner and sometimes even better, which strongly implies the importance of time series feature engineering. Note that we didn't limit strictly AutoGluon's running time as in our challenge. In general, AutoGluon takes 10 times more time than the winning solution and it still can't output a valid performance on the four datasets in a reasonable time. For the six AutoGluon's feasible datasets, we further visualize in Fig. 4 by algorithm groups. AutoGluon contains mainly three algorithm groups: Neural Network (MXNet, FastAI), Ensemble Trees (Light-GBM, Catboost, XGBoost) and K-Nearest Neighbors. We first plot on the left the average RMSE for Neural Networks models and ensemble tree models each (we omit KNN methods since they are usually the worst). Note that among the 6 datasets, 3 datasets don't use Neural Network for final ensemble (so their RMSE are set to be a large number for visualization). On 2 datasets (bottom left corner), however, Neural Networks can be competitive. This encourages us to explore in the future the effectiveness of deep models on time series which evolve quickly these days. On the right, we average the training/inference time per algorithm group and find that KNN can be used for very fast prediction if needed. Neural Networks take significantly more time. Points above the dotted line mean that no NN models or KNN models are chosen for this dataset (either due to performance or time cost). Only the tree-based methods provide solutions across the range of dataset sizes.

Table 6. Comparison with AutoGluon. NA means a missing value: AutoGluon did not terminate within a reasonable time.

Dataset	Phase	Baseline		1st DV		AutoGluon		FE + AutoGluon	
		RMSE	SMAPE	RMSE	SMAPE	RMSE	SMAPE	RMSE	SMAPE
fph1	Feedback	99.04	142.59	40.69	102.19	90.19	**26.45**	**40.57**	105.31
fph2	Feedback	17563	142.64	**236.6**	26.63	14978	59.94	263.74	**25.51**
fph3	Feedback	3337	38.49	**623.32**	**4.99**	6365	116.14	3159	31.08
fph4	Feedback	6.91	**187.58**	**3.66**	190.94	NA	NA	NA	NA
fph5	Feedback	8.63	174.45	**5.76**	**173.54**	NA	NA	NA	NA
pph1	Private	422.37	12.65	218.83	6.11	2770.70	9.46	**212.68**	**5.85**
pph2	Private	16851	139.31	**242.41**	23.46	15028	57.04	269.85	**22.98**
pph3	Private	8.78	178.45	**6.21**	**177.08**	NA	NA	NA	NA
pph4	Private	11.54	174.94	**3.74**	**168.4**	NA	NA	NA	NA
pph5	Private	309.33	39.2	**50.37**	**5.91**	949.4	20.52	65.22	6.65

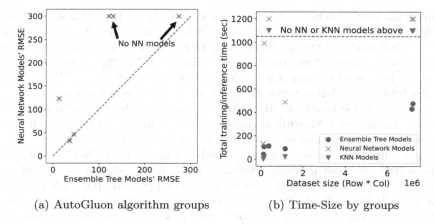

(a) AutoGluon algorithm groups   (b) Time-Size by groups

Fig. 4. AutoGluon experiments. (a) Average performance for two algorithm groups. Here we compare the average RMSE of Neural Network models and Ensemble Tree models. Among the six feasible datasets for AutoGluon with time series feature engineering, 3 of them don't choose Neural Networks as ensemble candidates. Ensemble trees have always significantly better performances. On 2 datasets, Neural Networks are quite competitive. **(b) Average time costs of candidate models.** When dataset is large, only ensemble tree models are chosen. When dataset is medium, KNN is fastest, followed by tree models. Neural Networks take significantly more time.

3.4   Impact of Time Budget

In the AutoSeries challenge, time management is an important aspect. Different time budgets are allowed for different datasets (as shown in Table 1). Ideally, AutoSeries solutions should take into account the allowed time budget and adapt all modules in the pipeline (i.e., different feature engineering, model training and updating strategy based on different allowed time budgets). We double the time budget and compare the performance in Appendix. In general, no obviously stable improvement can be observed. We also try to half the time budget and most solutions can't even produce valid predictions meaning that no single model training is finished. This could be because that we set the defaults budgets too tight but it also shows from another perspective that participants' solutions overfit to the challenge design (default time budget).

3.5   Dataset Difficulty

After a challenge finishes, another important issue for the organizers is to validate the choice of datasets. This is particularly interesting for AutoML challenges since the point is to generalize to a wide variety of tasks. Inspired by difficulty measurements in [10], we want to define intrinsic difficulty and modeling difficulty. By intrinsic difficulty we mean the irreducible error. As a surrogate to the intrinsic difficulty, we use the error of the best model. By modeling difficulty, we mean the range or spread of performances of candidate models. To separate well competition participants, we want to choose datasets of low intrinsic difficulty and high modeling difficulty. In [10], a notion of intrinsic difficulty and modeling difficulty is introduced for classification problems. Here we adapt such ideas and choose another bounded metric, the correlation (CORR) (Eq. 3). In fact, correlation has been used in many time series papers as a metric [8, 13]. We calculate the absolute correlation between the prediction sequence and ground truth test sequence. We define **Intrinsic difficulty** as 1 minus the best solution's absolute correlation score; and **Modeling difficulty** as the difference between the best solution's absolute correlation score and the provided baseline score.

These difficulty measures are visualized in Fig. 3d. It is obvious that both intrinsic difficulty and modeling difficulty differ from datasets to datasets. A posteriori, we can observe that some datasets like pph1 and pph5 are too easy, while pph3 is too difficult. In general, feedback datasets are of higher quality than private datasets, which is unfortunate. However, it is also possible that participants overfit the feedback datasets and thus, by using the best performing methods to estimate the intrinsic difficulty, we obtain a biased estimation.

4   Conclusion and Future Work

In this challenge, we introduce an AutoML setting with streaming test data, aiming at pushing forward research on Automated Time Series, and also having an impact on industry. Since there were no open sourced AutoML solutions dedicated to time series prior to our challenge, we believe the open sourced AutoSeries solutions fill this gap and provide a useful tool to researchers and practitioners. AutoSeries solutions don't need a GPU which facilitates their adoption.

The solutions of the winners are based on lightGBM. They addressed all challenge questions, demonstrating the feasibility of automating time series regression on datasets of the type considered. Significant improvements were made compared to the provided baseline. Our generalisation and overfitting experiments show that hyperparameter search is key to generalize. Still, some of the questions were addressed in a rather trivial way and deserve further attention. Explainabilty boils down to the feature importance delivered by lightGBM. In future challenge designs, we might want to quantitatively evaluate this aspect. Missing data were trivially imputed with the mean value. Hyper-parameters were not thoroughly optimized by most participants, and simple random search was used (if at all). Our experiments with the AutoGluon package demonstrate that much can be done in this direction to further improve results. Additionally, no sophisticated method of transfer learning or meta-learning was used. Knowledge transfer was limited to the choice of features and hyper-parameters performed on the feedback phase datasets. New challenge designs could include testing meta-learning capabilities on the platform, by letting the participant's code meta-train on the platform, e.g., not resetting the model instances when presented with each new dataset.

Other self criticisms of our design include that some datasets in the private phase may have been too easy or too difficult. Additionally, the RMSE alone could not separate well solutions, while a combination of metrics might be more revealing. Lastly, GPUs were not provided. On one hand this forced the participants to deliver practical rapid solutions; on the other hand, this precluded them from exploring neural time series models, which are rapidly progressing in this field.

Finally, winning solutions overfitted to the provided time budgets (no improvement with more time and fail with less time). An incentive to encourage participants to deliver "any-time-learning" solutions as opposed to "fixed-time-learning" solutions is to use the area under the learning curve as metric, as we did in other challenges. We will consider this for future designs.

References

1. Alexandrov, A., et al.: GluonTS: probabilistic and neural time series modeling in Python. J. Mach. Learn. Res. **21**(116), 1–6 (2020)
2. Erickson, N., et al.: AutoGluon-tabular: robust and accurate AutoML for structured data (2020)
3. Hutter, F., Kotthoff, L., Vanschoren, J. (eds.): Automated Machine Learning. Methods, Systems, Challenges. The Springer Series on Challenges in Machine Learning. Springer, Cham (2019). https://doi.org/10.1007/978-3-030-05318-5
4. Hyndman, R.J., Athanasopoulos, G. (eds.): Forecasting: principles and practice. OTexts (2021). https://otexts.com/fpp3/. Accessed 25 Mar 2021
5. Jin, H., Song, Q., Hu, X.: Auto-Keras: an efficient neural architecture search system. In: KDD (2019)
6. Kanter, J.M., Veeramachaneni, K.: Deep feature synthesis: towards automating data science endeavors. In: IEEE International Conference on Data Science and Advanced Analytics, DSAA (2015)
7. Ke, G., et al.: LightGBM: a highly efficient gradient boosting decision tree. In: Advances in Neural Information Processing Systems (2017)
8. Lai, G., Chang, W., Yang, Y., Liu, H.: Modeling long- and short-term temporal patterns with deep neural networks. In: SIGIR (2018)
9. Lim, B., Zohren, S.: Time series forecasting with deep learning: a survey (2020)
10. Liu, Z., et al.: Towards automated computer vision: analysis of the AutoCV challenges 2019. Pattern Recogn. Lett. **135**, 196–203 (2020)
11. Tan, C.W., Bergmeir, C., Petitjean, F., Webb, G.I.: Time series extrinsic regression. Data Min. Knowl. Disc. **35**(3), 1032–1060 (2021). https://doi.org/10.1007/s10618-021-00745-9
12. Taylor, S.J., Letham, B.: Forecasting at scale. PeerJ Prepr. **5**, e3190v2 (2017)
13. Wang, L., Chen, J., Marathe, M.: DEFSI: deep learning based epidemic forecasting with synthetic information. In: AAAI (2019)
14. Wang, Z., Yan, W., Oates, T.: Time series classification from scratch with deep neural networks: a strong baseline. In: International Joint Conference on Neural Networks (2017)
15. Yao, Q., et al.: Taking human out of learning applications: a survey on automated machine learning (2018)

Methods for Automatic Machine-Learning Workflow Analysis

Lorenz Wendlinger[1]([✉]) [ID], Emanuel Berndl[2], and Michael Granitzer[1] [ID]

[1] Chair of Data Science, University of Passau, Innstraße 31, 94032 Passau, Germany
{lorenz.wendlinger,michael.granitzer}@uni-passau.de
[2] ONE LOGIC GmbH, Kapuzinerstraße 2c, 94032 Passau, Germany
emanuel.berndl@onelogic.de

Abstract. Developing real-world Machine Learning-based Systems goes beyond algorithm development. ML algorithms are usually embedded in complex pre-processing steps and consider different stages like development, testing or deployment. Managing workflows poses several challenges, such as workflow versioning, sharing pipeline elements or optimizing individual workflow elements - tasks which are usually conducted manually by data scientists. A dataset containing 16 035 real-world Machine Learning and Data Science Workflows extracted from the ONE DATA platform (https://onelogic.de/en/one-data/) is explored and made available. Based on our analysis, we develop a representation learning algorithm using a graph-level Graph Convolutional Network with explicit residuals which exploits workflow versioning history. Moreover, this method can easily be adapted to supervised tasks and outperforms state-of-the-art approaches in NAS-bench-101 performance prediction. Another interesting application is the suggestion of component types, for which a classification baseline is presented. A slightly adapted GCN using both graph- and node-level information further improves upon this baseline. The used codebase as well as all experimental setups with results are available at https://github.com/wendli01/workflow_analysis.

Keywords: Graph neural networks · Structured prediction · Neural Architecture Search

1 Introduction

Using machine learning (ML) in the real world can require extensive data munging and pre-processing. Successful ML application thus needs to emphasize not only on the ML algorithm at hand, but also the context, i.e., the complete ML workflow. Practical ML worklows show a certain complexity, in the number of components (i.e., data aggregation, pre-processing, fitting and inference) and in terms of data flow, but also during their development in terms of versioning, testing and sharing. Consequently, ML workflows become an important asset that needs to be managed properly - comparable to software artifacts in software engineering [31]. A recently published case study from Amershi et al. [1]

© Springer Nature Switzerland AG 2021
Y. Dong et al. (Eds.): ECML PKDD 2021, LNAI 12979, pp. 52–67, 2021.
https://doi.org/10.1007/978-3-030-86517-7_4

showed the uptake of for example agile software engineering techniques for managing ML workflows and identified also several hurdles. One hurdle originates from knowledge sharing in a team developing ML workflows as well as the expertise of the people themselves while a second hurdle clearly identified the need of proper dataset management and a strict testing setup including hyper-parameter optimization within a workflow. Overall, workflow management has to support an highly iterative development process.

In this work, we start from the hypothesis that the development of ML Workflows requires techniques like code completion, coverage analysis and testing support, but focused on the particular properties of ML workflows. We therefore develop semi-automated workflow recommendation and composition techniques - based on Graph-Convolutional Neural Networks - for supporting development teams in knowledge sharing and efficient workflow testing. More precisely, we make the following contributions:

1. We analyze a large dataset of real-world data-science workflows consisting of 815 unique workflows in a total of 16035 versions from very diverse industrial data science scenarios. We analyze the workflows and show that a large portion of the components relate to data wrangling and pre-processing, rather than to algorithmic aspects.
2. We define three tasks for semi-automatically supporting the management of ML workflows, namely finding similar workflows, suggesting and refining components as well as structure-based performance prediction. While the former two support ML engineers in workflow creation and composition, the latter improves hyper-parameter tuning efficiency and reduces testing time.
3. We develop baseline graph-level feature set for representing ML workflows and develop a Graph-Convolutional Network dubbed P-GCN exploiting version history of workflows in order to represent workflows and enable component suggestion and refinement. Contrary to much of the existing work based on graph embeddings (c.f. the survey [32]), we consider heterogeneous node properties and edge directions in workflows.

We show that the P-GCN can produce high-quality dense representations that preserve the inherent structure of the dataset. Furthermore, we demonstrate that the P-GCN can learn complex mappings on DAG data by applying it to structural performance prediction on NAS-Bench-101. In this task, it outperforms state-of-the-art methods. Thirdly, it can be used to refine and suggest components using an internal hybrid node- and graph-level representation and thereby outperforms a strong baseline in both tasks.

In the following, we give a detailed motivation and definition for the supported tasks in Sect. 2 and go over related work in Sect. 3. Our P-GCN model is defined in Sect. 4 and Sect. 5 lists the used datasets as well as relevant qualities. We design experiments and present results for workflow similarity in Sect. 6, for structural performance prediction in Sect. 7 and, finally, for component refinement and suggestion in Sect. 8.

Source code including all experimental setups with results as well as datasets are made available for reproducibility.

2   Problem Definition

The creation, maintenance and management of ML workflows requires a powerful descriptive framework such as the ONE DATA platform. Versioning and tracking of results is especially important for efficient and reproducible work. Such a system, in turn, lends itself to the creation of a workflow library that can be a useful resource itself. To effectively leverage this resource, methods for the automatic processing of workflows are needed. In the following, we present concepts that can lead to improvements in three key areas.

Workflow Similarity. Considering similar workflows can help developers in reusing existing work and knowledge. Finding such workflows remains difficult. In contrast to explicit meta-information for describing a workflow, grouping based on structure alone does not require extra time on the user side and is more general. However, the space of graph definitions is very high-dimensional and sparse, making most distance measures defined over it meaningless and hard to interpret. Another challenge is that graphs are a variable length structure, while for most similarity calculations fixed length representations are required.

A common approach to solving this problem is the transformation to a dense lower-dimensional representation space. Between such representations, meaningful distances can be computed and used for grouping. Such representations can also be used as features for performance prediction or other meta-learning tasks.

Component Refinement and Suggestion. Another useful tool in the design of workflows is the automatic suggestion of components for a workflow. More specifically, a model is to predict the best fitting component type for a node in a workflow. This decision is based on patterns learned from a corpus of workflows created by experts. Therefore, it can be formulated as a many-class classification, a supervised learning task.

Two scenarios can be differentiated, depending on how much information about the rest of the workflow is available at prediction time. In *Component Suggestion*, only the nodes ancestral to the considered node are known and, consequently, at training time its decedents are artificially removed. For *Component Refinement*, the whole workflow is available, except for information about the considered node.

Structural Performance Prediction. Performance prediction on DAGs can be useful in both manual and automated search. It allows for focus on promising instances and thereby makes the search more efficient. A reliable performance predictor can reduce the number of costly executions for evaluation while keeping regret low. The most useful predictors use only structural information and therefore do not necessitate execution of the architecture.

This is especially useful for Neural Architecture Search (NAS) as each evaluation corresponds to full training with back propagation on a test dataset and is therefore computationally expensive.

3 Related Work

Workflow Management. There are many systematic approaches to the design and management of user-defined processing workflows [3,12,17]. However, despite the availability of workflow repositories and collection, they remain underused for most methods that automate parts of the workflow creation process. Friesen et al. propose the use of graph kernel, frequently occurring subgraphs and paths for recommendation and tagging of bio-informatics processes in [7].

Graph Representations. The main challenge in analyzing graph data is the high dimensionality and sparsity of the representation. This poses problems for manual analysis as well as for many automated methods designed for dense data.

Many algorithms for creating unsupervised node embeddings, a dense representation that preserves distance-based similarity, have been devised to solve this problem. Basic algorithms such as Adamic Adar [18] or Resource Allocation [34] use local node information only.

DeepWalk [23] is the first deep learning approach to network analysis and takes inspiration from methods for word embedding generation, such as Word2vec [20]. Representations are learned on random walks that preserve the context of a node and can be used for supervised learning tasks such as node classification.

Graph2vec [21] is a modification of document embedding models that produces whole graph embeddings by considering subgraph co-occurrence. However, it does not use edge direction which incurs significant data loss if applied to workflow DAGs.

Graph Classification. Graph Convolutional Networks were introduced by Kipf et al. in [15]. They capture the neighborhood of a node through convolutional filters, related to those known from Convolutional Neural Networks for images.

Shi et al. [26] construct a GCN based neural network assessor that uses a global node to obtain whole-graph representations.

Tang et al. construct a relational graph for similarities between graphs based on representations learned in an unsupervised manner through an auto-encoder in [29]. A GCN regressor is fed this information and produces performance predictions for each input graph.

Lukasik et al. propose smooth variational graph embeddings for neural architecture search in [19]. They are based on an autoencoder neural network in which both the decoder and encoder consider the backward pass and the forward pass of an architecture.

Ning et al. propose GATES in [22], a generic encoding scheme that uses knowledge of the underlying search space with an attention mechanism for structural performance prediction in Neural Architecture Search. It is suitable for both node-heterogeneous and edge-heterogeneous graph data.

4 Residual Graph-Level Graph Convolutional Networks

In this section, we introduce our graph convolutional model dubbed P-GCN that offers a robust aggregation method for whole-graph representation learning and related supervised tasks. We also go over the basics of graph convolutions and adjacent techniques used for P-GCN.

Graph Convolutional Networks [15] can be used to compute node-level functions. They take the graph structures and node features as input. These features may be one-hot encoded node types or any other type of feature such as more detailed node hyper-parameters. Similar to convolutions in image recognition, multiple learned filters are used to aggregate features from neighboring nodes via linear combination. Quite like CNNs, GCNs derive their expressive power from the stacking of multiple convolution layers that perform increasingly complex feature extraction based on the previous layers' output. Usually, a bottleneck is created by stacking multiple layers and adding a smaller last convolutional layer. This forces the model to compress information and create a denser and more meaningful representation of size F_L.

Formally, we consider ML workflows as heterogeneous directed acyclic graph (**DAG**) representing the data flow between different data processing components. Specifically, $\mathcal{G} = (\mathcal{V}, \mathcal{E}, \lambda_l)$ represents a graph with nodes (or vertices) $\mathcal{V}$ and edges $\mathcal{E} \subseteq \{(u,v) : u, v \in \mathcal{V} \wedge u \neq v\}$. A mapping $\lambda : \mathcal{V} \to \{0,1\}^{n_l}$ assigns a one-hot-encoded label, or node type, to each node. E expresses data flow between nodes while the node class $\lambda(v)$ is the kind of data processing component that v represents, of a total n_l possible component types.

A graph convolution in layer ℓ of L layers with filter size F_ℓ on node v of $\mathcal{G}$ is defined as

$$f_i^{(\ell+1)}(\mathcal{G}, v) = \sum_{u \in \Gamma_i(\mathcal{G}, v)} \Theta^{(\ell+1)} f^{(\ell)}(\mathcal{G}, u) z(v) \tag{1}$$

with a layer weight matrix $\Theta^{(\ell+1)} \in \mathbb{R}^{F^{(\ell)} \times F^{(\ell+1)}}$ and $i = 1$. $\Gamma_i(\mathcal{G}, v)$ is the ith neighborhood of v w.r.t. $\mathcal{G}$ and $z(u)$ is a normalization, usually the inverse square root of the node degrees. Self-loops are added artificially to $\mathcal{G}$ as $\mathcal{E} = \mathcal{E} \cup \{(u,v) : v \in \mathcal{V}\}$ so the representation $f_i^{(\ell+1)}(\mathcal{G}, v)$ also contains $f_i^{(\ell)}(\mathcal{G}, v)$. For the first layer, the input features are used as node representations, i.e., $f^{(0)}(\mathcal{G}, v) = \lambda_l(v)$ and $F^{(0)} = n_l$.

Topology adaptive GCNs [6] are an extension of the graph convolution that considers neighborhoods of hop sizes up to k. This changes the convolution in layer ℓ to

$$f^{(\ell+1)}(\mathcal{G}, v) = \sum_{i \in \{1..k\}} \Theta_i^{(\ell+1)} f_i^{(\ell)}(\mathcal{G}, v) z(v) \tag{2}$$

with learned weights $\Theta^{(\ell+1)} \in \mathbb{R}^{kF_\ell \times F^{\ell+1}}$ for each layer. We adopt this method with k set to 2 for its flexibility and improved expressive power.

In this way, a GCN can generate meaningful node-level representations, i.e., a F_L sized representation for each $v \in \mathcal{V}$. If we want graph-level outputs, i.e., one embedding that encodes the structure of a whole graph, pooling can be used. More specifically, we use a function $g_i : \mathbb{R}^{|V| \times F_L} \to \mathbb{R}^{F_L}$ to obtain a fixed-size representation regardless of graph size. We can use a set G of pooling functions such as mean, min, max or stdev for each embedding dimension for improved robustness. This produces an output of size $|G| \times F_L$ for each graph. These pooling results are then scaled via batch normalization [10] and aggregated via a weighted sum, resulting in an output of size F_L:

$$f^{(L+1)}(\mathcal{G}, g_i) = \sum_{v \in \mathcal{V}} g_i \left(f_i^{(L+1)}(\mathcal{G}, v) \right)$$

$$f^{(L+1)}(\mathcal{G}) = \sum_{i \in \{1...|G|\}} \Theta_i^{(L+1)} Z \left(f^{(L+1)}(\mathcal{G}, g_i) \right) \tag{3}$$

with learned weights $\Theta^{(L+1)} \in \mathbb{R}^{F^{L+1} \times |G| \ F^L}$ and normalization function Z.

For unsupervised tasks, $f^{(L+1)}(\mathcal{G})$ is the final model output. The model can also be adapted to supervised tasks by adding dense layers that function like an MLP estimator. For classification, a *softmax*-activated dense layer with the appropriate number of outputs for the predicted classes can be added. For regression, a dense layer with one output serves as the last layer.

To help convergence, batch normalization [10] is applied to each graph convolution's output to reduce the co-variate shift during training. Furthermore, Batch normalization after the pooling helps reduce the impact of different scales induced by the different pooling operations. According to the Ioffe et al., they also provide some regularization. This also means that convolutional layers that are followed by a batch normalization do not require a learned bias, as their output is scaled to zero mean anyway.

Skip connections as introduced by He et al. in [8] are automatically added between layers of matching size so residuals can be learned explicitly, which can help deeper architectures converge and generally improve performance, c.f. [5]. This changes the feature computation to

$$f_{\text{res}}^{(\ell+1)}(\mathcal{G}) = \begin{cases} \sigma \left(f^{(\ell+1)}(\mathcal{G}) + f_{\text{res}}^{(\ell-1)}(\mathcal{G}) \right) & \text{if } F^{(\ell-1)} = F^{(\ell+1)} \\ \sigma \left(f^{(\ell+1)}(\mathcal{G}) \right) & \text{otherwise} \end{cases} \tag{4}$$

with a non-linear activation function $\sigma : \mathbb{R} \to \mathbb{R}$, rectification in our case. In the same vein, a dropout layer [9] is added after the last graph convolution to obtain a model that generates more robust representations.

P-GCN is trained in mini-batches with adaptive momentum [14] and exponential learning rate decay. As over-fitting can be a problem in complex settings, weight decay is applied automatically with a factor of 0.01.

5 Datasets

This section introduces the datasets used to develop and validate our methods.

The ONE DATA data science workflow dataset ODDS-full[1] comprises 815 unique workflows in temporally ordered versions obtained from a broad range of real-world machine learning solutions realized using the ONE DATA platform. Consequently, the data set distinguishes itself from available academic datasets, especially when analyzing potential ML workflow support for real-world applications. A version of a workflow describes its evolution over time, so whenever a workflow is altered meaningfully, a new version of this respective workflow is persisted. Overall, 16 035 versions are available.

ODDS workflows represent machine learning workflows expressed as node-heterogeneous DAGs with 156 different node types. They can represent a wide array of data science and machine learning tasks with multiple data sources, model training, model inference and data munging. These node types represent various kinds of processing steps of a general machine learning workflow and are grouped into 5 broad categories, which are listed below.

Load Processors for loading or generating data (e.g. random number generator).
Save Processors for persisting data (possible in various data formats, via external connections or as a contained result within the ONE DATA platform) or for providing data to other places as a service.
Transformation Processors for altering and adapting data. This includes e.g. database-like operations such as renaming columns or joining tables as well as fully fledged dataset queries.
Quantitative Methods Various aggregation or correlation analysis, bucketing, and simple forecasting.
Advanced Methods Advanced machine learning algorithms such as BNN or Linear Regression. Also includes special meta processors that for example allow the execution of external workflows within the original workflow.

An example workflow is shown in Fig. 1. Any metadata beyond the structure and node types of a workflow has been removed for anonymization purposes.

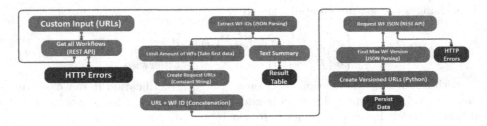

Fig. 1. Example workflow used in the ONE DATA platform.

ODDS, a filtered variant, which enforces weak connectedness and only contains workflows with at least 5 different versions and 5 nodes, is available as the default version for unsupervised and supervised learning (Table 1).

---

[1] Available at https://zenodo.org/record/4633704.

Table 1. Statistics for the full and filtered ONE DATA data science workflow datasets as well as NAS-Bench-101.

Statistic	ODDS-full	ODDS	NAS-Bench-101 [33]
Unique workflows	815	284	423k
Instances	16035	8639	1.27M
Node types	156	121	5
Mean graph size	$42.78_{\pm63.27}$	$57.21_{\pm69.34}$	$8.73_{\pm0.55}$

As a second data set we use NAS-bench-101 [33], which was published as a benchmark dataset for Neural-Architecture-Search (NAS) and NAS meta-learning. It consists of architectures sampled from a common search space focusing on standard machine learning tasks. These represent cells constructed of high-level CNN operations from which CNNs are generated by stacking them with a fixed strategy. 423k such architectures were trained with the same back-propagation schema on the image recognition task CIFAR-10 [16]. We use their accuracies in this task as our prediction target. Consequently, this can be seen as a sampling of generalization power for neural architectures and is therefore well suited for structural performance prediction.

6 Workflow Similarity

In the following, we will describe different approaches for creating dense representations from heterogeneous DAGs, starting with simple graph features and ending with deep-learning methods with the aim to detect similar workflows.

Evaluation Methodology. As learning such representations is an unsupervised task, quantitative evaluation is difficult. However, the structure imposed by the version groups of ODDS enables the definition of two informative criteria. One of those, dubbed the **G**roup **C**luster **S**core, indicates how well the embeddings are suited to clustering tasks. This is done by generating a clustering and evaluating how well it represents the workflow groups. Agglomerative clustering via Ward linkage [11] was chosen for this task due to its robustness and determinism. The V-Measure [24], defined as the harmonic mean of homogeneity and completeness, of this clustering is reported as the GCS.

Furthermore, the **T**riplet **R**atio **S**core indicates how closely instances of a workflow group are embedded together. It is defined as the mean of the distance to positive instances divided by the distance to negative instances for each sample. Consequently, lower triplet ratio scores are better.

Results. Simple graph features can be used to group workflows. Some of them are computed on the graph-level, such as the number of nodes or number of edges. Others, such as centrality measures, are extracted on the node-level and can be aggregated via their mean or other statistical moments. In the case of heterogeneous graphs, they can require significant manual feature engineering to respect the different node types. Furthermore, they do not produce a generally dense representation, as certain features can be sparse for some classes of DAGs.

As a compromise, we choose to use the feature set presented in [27] and extend them with the number of distinct node types in the graph and the count of nodes for the most frequent node type.

Graph Convolutional Networks have been shown to create meaningful embeddings on some data without training, c.f. [15]. However, we can use methods for learning on grouped data to generate useful embeddings. For this method, distinct workflows can be regarded as groups with their versions representing members of those groups. A P-GCN model can be trained to minimize the distance within groups while maximizing the distance to members of other groups. This can be achieved via triplet loss. Triplet loss is calculated on triplets of samples, where the current sample is the so-called *anchor*. Based on the group of this anchor, a *positive* instance from the same group as well as a *negative* instance from another group are sampled. Triplet loss can be computed either based on the ranking of these samples or on the ratio between their distances. *Triplet margin loss*, as described in [30], is a ranking loss that forces the model to embed anchor and positive closer together than anchor and negative.

Table 2. Representation quality for different methods on ODDS. For non-deterministic models, mean and standard deviation across 5 trials with different random states are given.

Approach	GCS	TRS
Graph2Vec [21]	$0.596_{\pm 0.0045}$	$0.453_{\pm 0.0039}$
FeatherGraph [25]	0.76	0.351
Basic graph-level features	0.701	0.339
Untrained P-GCN	$0.884_{\pm 0.0013}$	$0.4_{\pm 0.0069}$
Triplet margin loss P-GCN	$\mathbf{0.901}_{\pm 0.0038}$	$\mathbf{0.113}_{\pm 0.0032}$

As can be seen in Table 2, P-GCN trained with triplet margin loss produces high-quality representations for the high-dimensional data of ODDS. They have both significantly better GCS as well as TRS scores compared to traditional approaches that cannot natively use directed or heterogeneous graphs. Interestingly, embeddings generated with an untrained, fully random P-GCN achieve competitive GCS scores with relatively high repeatability.

Hyper-parameters are given in Table 3. P-GCN benefits from large minibatches and high exponential learning rate decay in this task to achieve smoother convergence behavior.

7   Structural Performance Prediction

For predicting workflow performance based on a workflow structure, we adapted the P-GCN model towards a regression task by adding fully connected layers

Table 3. P-GCN parameter setting for unsupervised learning on ODDS.

Parameter name	Default value
GCN layer sizes	(128, 128, 128, 128, 128, 64)
Pooling operations	(max, min, mean, stdev)
Epochs	50
Dropout probability	0.05
Batch size	1000
Learning rate	0.01
Learning rate decay	0.9

with non-linearities after the graph convolutions. These function like an MLP regressor after the GCN-based feature extraction, but are trained jointly. Layer Normalization as per Ba et al. [2] is applied to the output of each dense layer for improved convergence behavior.

We use a combined loss, a linear combination of MSE loss and hinge pairwise ranking loss as defined in [22]. For true accuracy y and prediction $\hat{y}$ of length N and margin $m = 0.05$:

$$L_c(y, \hat{y}) = w_1 \cdot MSE(y, \hat{y}) + w_2 \cdot L_r(y, \hat{y})$$

$$L_r(y, \hat{y}) = \sum_{j=1}^{N} \sum_{i:\ y_i > y_j} \max\left(0, m - (\hat{y}_i - \hat{y}_j)\right) \tag{5}$$

A focus on low squared error or high ranking correlation can be facilitated through the respective weights w_1 and w_2. This is important since we found that many low-error predictions have low correlation and vice-versa.

Table 4. Parameter setting for the P-GCN for supervised learning on NAS-Bench-101. All other hyper-parameters are set as before, c.f. Table 3.

N_l	Parameter	Parameter name	Default value
1000		Dense layer sizes	(64,)
		Training epochs	150
		Learning rate decay	0.95
	w_1	MSE loss weight	0.5
	w_2	Hinge ranking loss weight	0.5
		Batch size	100
381		Batch size	50
1906		Batch size	200

Analogous to the method of Lukasik et al. in [19], the back-propagation used in the training of ANNs can be considered by reversing the edge direction of an individual architecture. The predictor is presented both versions and produces a single prediction. P-GCN does this by jointly aggregating over the node-level representations of both passes.

Evaluation Methodology. The specific architectures in the training set can have a large impact on predictor performance. Therefore multiple trials with different dataset splits, 5 in our case, need to be performed for proper evaluation. The remaining instances are randomly partitioned into N_l training instances and a test set of size 50 000 for each fold. For each of these trials, the pseudo-random number generator used for initialization of network parameters is used with a different seed as well. This setup enables us to assess repeatability.

As this is a regression task, multiple metrics can be use to quantitatively evaluate predictions. Mean squared error alone is unsuitable as it is difficult to interpret and can be low for meaningless predictions. In most searches, performance predictions are only compared with other predictions. It is therefore not important that they exhibit low error with the target, but rather that they show high correlation with the target. Furthermore, as many search methods rank candidates by performance, ranking correlation can be considered the most important measure.

Concordant with [29], we choose mean squared error, Pearson correlation ρ_p and the Kendall Tau ranking coefficient τ_k [13] as evaluation criteria.

Results. Performance prediction was performed on NAS-bench-101 [33]. Results for 5 random folds are listed in Table 5. Our method offers improvements over state of the art methods with respect to the most important ranking correlation τ_k. This is despite the fact that P-GCN is a purely supervised method and does not need any information beyond the N_l training instances.

Table 5. Performance prediction results on NAS-Bench-101 with N_l training instances. Mean and standard deviation over 5 random trials for multiple evaluation criteria.

N_l	Criterion	SVGe [19]	GCN [26]	Tang et al. [29]	GATES [22]	P-GCN (Ours)
381	τ_k	-	-	-	0.7789	**0.7985**±0.008
1000	τ_k	-	-	0.6541±0.0078	-	**0.8291**±0.0.0067
	MSE	**0.0028**±0.00002	-	0.0031±0.0003	-	0.0038±0.0.00016
	ρ_p	-	**0.819**	0.5240±0.0068	-	0.589±0.024
1906	τ_k	-	-	-	0.8434	**0.8485**±0.0013

Figure 2 offers a more detailed look at the predictions for one fold. There is a strong linear relationship, but also a bias resulting from the used combined loss.

By altering the loss weights w_1 and w_2, the focus can be shifted towards one of the two prediction goals - low error or high correlation. The corresponding predictive performance can be observed in Fig. 3. The default configuration with equally weighted losses does not impair performance w.r.t. to τ_k. Optimizing P-GCN purely for MSE produces predictions with a MSE of $0.00206_{\pm 0.00009}$, which is an improvement over the state of the art, SVGe's $0.0028_{\pm 0.00002}$.

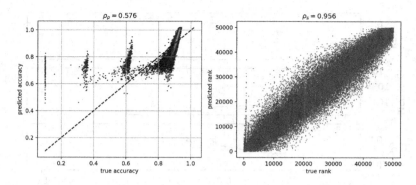

Fig. 2. P-GCN performance predictions for 50 000 test instances from a single fold with $N_l = 1\ 000$. Comparison of performance values (left) and rankings (right) with corresponding correlations, i.e., Pearson and Spearman coefficient, given.

8 Component Refinement and Suggestion

For component refinement, a set of basic numeric features can be extracted from the complex workflow structures to form a baseline. To this end, we use the same graph-level and node-level features employed for workflow similarity computation (c.f. Sect. 6), based on the set constructed in [27]. However, for this task, the node-level centrality measures are not aggregated. Furthermore, they are supplemented with harmonic centrality, pagerank, load centrality and katz centrality. For obvious reasons, node betweenness centrality is used instead of edge betweenness centrality. Additionally, the number of descendents and ancestors of each node as well as the longest shortest path to and from each node are added to provide explicit information about its position in the workflow.

The P-GCN model can be adapted to this task. To produce a joint representation of both the considered node and the workflow it is part of, their representations are concatenated. More precisely, the node representation is appended to the outputs of the pooling functions. This is fed into an MLP like the one used for structual performance prediction, c.f. Sect. 7. A softmax-activated output layer with a neuron for each class is added and the network's categorical cross-entropy loss is optimized via back-propagation. Due to the larger training set sizes, small alterations to the training schema were necessary, c.f. Table 6.

Table 6. Parameter setting for the hybrid P-GCN for node-level classification on ODDS. All other hyper-parameters are set as before, c.f. Table 4.

Parameter	Default value
Epochs	50
Loss	Categorical cross-entropy
Dropout	0.25
Batch size	5000

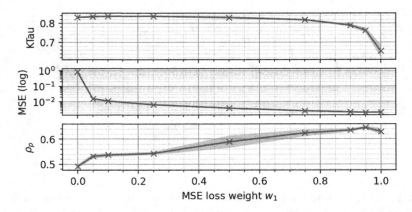

Fig. 3. P-GCN performance on NAS-bench-101 in response to varying MSE loss weight w_1 for the combined loss function. Shown with $\pm$ stdev confidence intervals across 5 folds. The hinge ranking loss weight w_2 is set to $1 - w_1$.

The hybrid P-GCN constructed such can utilize both information about the considered node as well as its workflow, extracted through the same graph convolutional functions. This removes the need for manually engineered and hard to generalize graph-level features that capture this node context.

Evaluation Methodology. In an application case, we can expect a component refinement model to be applied to unseen workflows only, i.e., such that differ from those in the training set. To obtain a realistic evaluation with respect to this use case, multiple grouped splits are used. The data set is split into a certain percentage of groups for training while the rest is withheld for testing. For this task, the groups are created by the distinct workflows. 5 splits with 80% of groups used for training and the rest withheld for testing are created in this way.

For component suggestion we only consider nodes with at least 5 ancestors to guarantee a minimum level of information available for the prediction.

As this prediction tasks requires inference for every single node of every graph, the computational load is very high. This can be remedied by removing very similar instances, i.e., by sub-sampling over the version history with a factor of 10, starting with the newest revision. As a result, only every 10th version of a workflow is present in the data-set used for training and testing.

Results. Various classifiers were tested on the basic feature set. Many of these are superior to the dummy classifier baseline, c.f. Table 7. The best performing method is a random forest classifier [4]. The hybrid P-GCN with slightly adapted hyper-parameters, c.f. Table 6, outperforms the basic methods in both tasks.

While graph-level P-GCN achieves competitive results in component refinement and node-level P-GCN performs well in component suggestion, neither approach excels in both tasks. The hybrid P-GCN however can deliver high quality prediction in both scenarios, showing that it is a best-of-both-worlds approach.

Table 7. Performance comparison for component refinement and component suggestion in 5 random grouped folds. Results for P-GCNs and a set of different classifiers using basic graph features.

Classifier	Component refinement		Component suggestion	
	Accuracy	Top 5 accuracy	Accuracy	Top 5 accuracy
Dummy classifier	$0.245_{\pm0.063}$	$0.452_{\pm0.025}$	$0.179_{\pm0.02}$	$0.527_{\pm0.051}$
Random Forest	$0.553_{\pm0.07}$	$0.744_{\pm0.047}$	$0.461_{\pm0.078}$	$0.715_{\pm0.067}$
Node-level P-GCN	$0.442_{\pm0.08}$	$0.758_{\pm0.061}$	$\mathbf{0.48}_{\pm0.059}$	$\mathbf{0.755}_{\pm0.06}$
Graph-level P-GCN	$0.578_{\pm0.041}$	$0.759_{\pm0.028}$	$0.27_{\pm0.043}$	$0.584_{\pm0.062}$
Hybrid P-GCN	$\mathbf{0.643}_{\pm0.074}$	$\mathbf{0.798}_{\pm0.046}$	$0.461_{\pm0.08}$	$0.748_{\pm0.06}$

In component refinement, it achieves a mean accuracy 0.643 ± 0.074 and a mean Top-5-accuracy 0.798 ± 0.046. This means that, on average, for 4 out of 5 nodes, the correct component type can be found in the top 5 predictions and for 5 out of 8 nodes the prediction is correct.

Despite the limitation to nodes with at least 5 ancestors, component suggestion is a distinctly more challenging task. However, P-GCN methods still outperform the strong random forest baseline by a significant margin.

9 Conclusion

The management and analysis of real-world ML workflows poses a number of interesting challenges, most prominently the generation of meaningful representations and component refinement. For both tasks, a baseline with adequate performance is presented and evaluated on the ODDS dataset.

P-GCN is a modification of node-level topology adaptive GCNs that shows promise for supervised tasks as well as unsupervised tasks with surrogate targets. It can generate meaningful representations for the highly complex data structures of ODDS and outperforms the state of the art in NAS-Bench-101 performance prediction. Additionally, it can be configured to create joint node- and graph-level representations and thereby outperforms a strong baseline in a node classification task on ODDS.

Since P-GCN is used in a purely supervised manner for regression tasks, it does not require a large-scale sampling or rule-based definition of the search space for generating unsupervised representations, as other methods do. The pooling method also makes it suitable for search spaces with varying graph sizes. Furthermore, input node features can easily be extended to cover node hyper-parameters or arbitrary numerical attributes. These properties make the adaptation of P-GCN to other performance prediction tasks trivial. Especially the predictive performance on supervised tasks with more complex and varied DAGs, such as those generated by CGP-CNN [28], would provide further insight into the capabilities of the model. P-GCN's generality also makes it an interesting candidate for transductive transfer as well. Multiple domains with information processing expressed in DAG architectures are worth considering.

Acknowledgments. This work has been partially funded by the Bavarian Ministry of Economic Affairs, Regional Develoment and Energy under the grant 'CrossAI' (IUK593/002) as well as by BMK, BMDW, and the Province of Upper Austria in the frame of the COMET Programme managed by FFG. It was also supported by the FFG BRIDGE project KnoP-2D (grant no. 871299).

References

1. Amershi, S., et al.: Software engineering for machine learning: a case study. In: 2019 IEEE/ACM 41st International Conference on Software Engineering: Software Engineering in Practice (ICSE-SEIP), pp. 291–300. IEEE (2019)
2. Ba, J.L., Kiros, J.R., Hinton, G.E.: Layer normalization. arXiv preprint arXiv:1607.06450 (2016)
3. Basu, A., Blanning, R.W.: A formal approach to workflow analysis. Inf. Syst. Res. **11**(1), 17–36 (2000)
4. Breiman, L.: Random forests. Mach. Learn. **45**(1), 5–32 (2001)
5. Bresson, X., Laurent, T.: Residual gated graph convNets. arXiv preprint arXiv:1711.07553 (2017)
6. Du, J., Zhang, S., Wu, G., Moura, J.M., Kar, S.: Topology adaptive graph convolutional networks. arXiv preprint arXiv:1710.10370 (2017)
7. Friesen, N., Rüping, S.: Workflow analysis using graph kernels. In: LWA, pp. 59–66. Citeseer (2010)
8. He, K., Zhang, X., Ren, S., Sun, J.: Deep residual learning for image recognition. In: Proceedings of the IEEE Conference on Computer Vision and Pattern Recognition, pp. 770–778 (2016)
9. Hinton, G.E., Srivastava, N., Krizhevsky, A., Sutskever, I., Salakhutdinov, R.R.: Improving neural networks by preventing co-adaptation of feature detectors. arXiv preprint arXiv:1207.0580 (2012)
10. Ioffe, S., Szegedy, C.: Batch normalization: accelerating deep network training by reducing internal covariate shift. arXiv preprint arXiv:1502.03167 (2015)
11. Ward, J.H., Jr.: Hierarchical grouping to optimize an objective function. J. Am. Stat. Assoc. **58**(301), 236–244 (1963). https://doi.org/10.1080/01621459.1963.10500845
12. Kaushik, G., Ivkovic, S., Simonovic, J., Tijanic, N., Davis-Dusenbery, B., Deniz, K.: Graph theory approaches for optimizing biomedical data analysis using reproducible workflows. bioRxiv, p. 074708 (2016)
13. Kendall, M.G.: A new measure of rank correlation. Biometrika **30**(1/2), 81–93 (1938)
14. Kingma, D.P., Ba, J.: Adam: a method for stochastic optimization. arXiv preprint arXiv:1412.6980 (2014)
15. Kipf, T.N., Welling, M.: Semi-supervised classification with graph convolutional networks. arXiv preprint arXiv:1609.02907 (2016)
16. Krizhevsky, A., Hinton, G., et al.: Learning multiple layers of features from tiny images (2009)
17. Li, J., Fan, Y., Zhou, M.: Timing constraint workflow nets for workflow analysis. IEEE Trans. Syst. Man Cybern. Part A Syst. Hum. **33**(2), 179–193 (2003)

18. Liben-Nowell, D., Kleinberg, J.: The link-prediction problem for social networks. J. Am. Soc. Inform. Sci. Technol. **58**(7), 1019–1031 (2007)
19. Lukasik, J., Friede, D., Zela, A., Stuckenschmidt, H., Hutter, F., Keuper, M.: Smooth variational graph embeddings for efficient neural architecture search. arXiv preprint arXiv:2010.04683 (2020)
20. Mikolov, T., Sutskever, I., Chen, K., Corrado, G.S., Dean, J.: Distributed representations of words and phrases and their compositionality. Adv. Neural. Inf. Process. Syst. **26**, 3111–3119 (2013)
21. Narayanan, A., Chandramohan, M., Venkatesan, R., Chen, L., Liu, Y., Jaiswal, S.: graph2vec: learning distributed representations of graphs. arXiv preprint arXiv:1707.05005 (2017)
22. Ning, X., Zheng, Y., Zhao, T., Wang, Y., Yang, H.: A generic graph-based neural architecture encoding scheme for predictor-based NAS (2020)
23. Perozzi, B., Al-Rfou, R., Skiena, S.: DeepWalk: online learning of social representations. In: Proceedings of the 20th ACM SIGKDD International Conference on Knowledge Discovery and Data Mining, pp. 701–710 (2014)
24. Rosenberg, A., Hirschberg, J.: V-measure: a conditional entropy-based external cluster evaluation measure. In: Proceedings of the 2007 Joint Conference on Empirical Methods in Natural Language Processing and Computational Natural Language Learning (EMNLP-CoNLL), pp. 410–420 (2007)
25. Rozemberczki, B., Sarkar, R.: Characteristic functions on graphs: birds of a feather, from statistical descriptors to parametric models (2020)
26. Shi, H., Pi, R., Xu, H., Li, Z., Kwok, J.T., Zhang, T.: Multi-objective neural architecture search via predictive network performance optimization (2019)
27. Stier, J., Granitzer, M.: Structural analysis of sparse neural networks. Procedia Comput. Sci. **159**, 107–116 (2019)
28. Suganuma, M., Shirakawa, S., Nagao, T.: A genetic programming approach to designing convolutional neural network architectures. In: Proceedings of the Genetic and Evolutionary Computation Conference, pp. 497–504 (2017)
29. Tang, Y., et al.: A semi-supervised assessor of neural architectures. In: Proceedings of the IEEE/CVF Conference on Computer Vision and Pattern Recognition, pp. 1810–1819 (2020)
30. Balntas, V., Riba, E., Ponsa, D., Mikolajczyk, K.: Learning local feature descriptors with triplets and shallow convolutional neural networks. In: Wilson, R.C., Hancock, E.R., Smith, W.A.P. (eds.) Proceedings of the British Machine Vision Conference (BMVC), pp. 119.1–119.11. BMVA Press (September 2016). https://doi.org/10.5244/C.30.119
31. Weißgerber, T., Granitzer, M.: Mapping platforms into a new open science model for machine learning. Inf. Technol. **61**(4), 197–208 (2019)
32. Wu, Z., Pan, S., Chen, F., Long, G., Zhang, C., Philip, S.Y.: A comprehensive survey on graph neural networks. IEEE Trans. Neural Netw. Learn. Syst. **32**, 4–24 (2020)
33. Ying, C., Klein, A., Christiansen, E., Real, E., Murphy, K., Hutter, F.: NAS-Bench-101: towards reproducible neural architecture search. In: International Conference on Machine Learning, pp. 7105–7114. PMLR (2019)
34. Zhou, T., Lü, L., Zhang, Y.C.: Predicting missing links via local information. Eur. Phys. J. B **71**(4), 623–630 (2009)

ConCAD: Contrastive Learning-Based Cross Attention for Sleep Apnea Detection

Guanjie Huang$^{(\boxtimes)}$ and Fenglong Ma$^{(\boxtimes)}$

College of Information Sciences and Technology, Pennsylvania State University,
State College, PA 16802, USA
{gzh8,fenglong}@psu.edu

Abstract. With recent advancements in deep learning methods, automatically learning deep features from the original data is becoming an effective and widespread approach. However, the hand-crafted expert knowledge-based features are still insightful. These expert-curated features can increase the model's generalization and remind the model of some data characteristics, such as the time interval between two patterns. It is particularly advantageous in tasks with the clinically-relevant data, where the data are usually limited and complex. To keep both implicit deep features and expert-curated explicit features together, an effective fusion strategy is becoming indispensable. In this work, we focus on a specific clinical application, i.e., sleep apnea detection. In this context, we propose a contrastive learning-based cross attention framework for sleep apnea detection (named ConCAD). The cross attention mechanism can fuse the deep and expert features by automatically assigning attention weights based on their importance. Contrastive learning can learn better representations by keeping the instances of each class closer and pushing away instances from different classes in the embedding space concurrently. Furthermore, a new hybrid loss is designed to simultaneously conduct contrastive learning and classification by integrating a supervised contrastive loss with a cross-entropy loss. Our proposed framework can be easily integrated into standard deep learning models to utilize expert knowledge and contrastive learning to boost performance. As demonstrated on two public ECG dataset with sleep apnea annotation, ConCAD significantly improves the detection performance and outperforms state-of-art benchmark methods.

Keywords: Contrastive learning · Cross attention · Sleep apnea detection

1 Introduction

According to the National Institutes of Health of USA, 50 to 70 million people have chronic sleep disorders [4], and the sleep disorders of sleep can increase the

© Springer Nature Switzerland AG 2021
Y. Dong et al. (Eds.): ECML PKDD 2021, LNAI 12979, pp. 68–84, 2021.
https://doi.org/10.1007/978-3-030-86517-7_5

risk of many related diseases, such as hypertension and cardiovascular patholo-
gies [41]. Sleep apnea is one of the most common sleep disorders, which is an
abnormal respiratory activity repeatedly occurring during sleep. The current
primary method for diagnosing sleep apnea requires the patient to record the
polysomnogram (PSG) in a clinic setup, which is very inconvenient and belated.
Thus, how to automatically and effectively detect sleep apnea is a challenge,
especially in the earlier stages.

Towards this end, automatic sleep apnea detection methods [3,10,18,30,35]
have been developed to simplify the diagnostic procedure, which including traditional machine learning methods and deep learning methods. Existing studies [29,35,38] have shown that deep learning models perform better than the
traditional machine learning ones, which require expert knowledge to manually
extract features. However, these hand-crafted expert features are still valuable
and insightful. While researching the most appropriate hand-crafted features is
time-consuming, there are a number of hand-crafted features that can be leveraged right away, as summarized by previous studies over centuries. In this context, we proposed a **cross-attention mechanism** to combine the deep features
and the hand-crafted features to take advantage of both of them appropriately.

On the other hand, the regular deep learning methods usually train with the
cross-entropy (CE) loss for a classification task. Since the CE loss only focuses
on learning the necessary features to solve the classification task over known
training data, it can be easily impaired by the mislabeled data [42], which further
hinders the quality of the learned representations [25]. To alleviate the problem, a
common solution is to collect more data so that the model can learn more general
features without excessive discrimination. However, this solution is particularly
impractical in clinically relevant data, such as electrocardiography (ECG), where
the data are always limited, and the labeling is prone to human errors. As a
remedy, we design a novel hybrid loss that integrates the cross-entropy loss with
a **contrastive loss**. The contrastive loss helps to learn more general and robust
features by clustering similar data and pushing apart dissimilar ones.

To sum up, in this work, we proposed a novel CONtrastive learning-based
Cross Attention for sleep apnea Detection (ConCAD) using single ECG data.
To the best of our knowledge, our work is the first to successfully integrate
contrastive learning for sleep apnea detection. Our major contributions of this
paper are as follows:

- We propose a cross attention mechanism to combine the deep features and
  expert knowledge-based features, which automatically fuses the features by
  emphasizing the important parts based on each other synergistically.
- We design a novel hybrid loss that encompasses both the cross-entropy (CE)
  loss and the supervised contrastive (SC) loss. The SC loss help to learn more
  general and robust by minimizing the ratio of intra-class to inter-class similarity while cross-entropy CE loss focus on discovering the useful features to
  solve the classification task.
- We demonstrate state-of-the-art classification performance on two public
  ECG datasets outperforming all benchmark methods.

- We show that our proposed framework of contrastive learning-based cross attention has better generalization ability, especially when the number of labeled training data is limited, comparing to a naive deep learning method without it.
- Both the cross attention mechanism and contrastive learning can be painlessly integrated into standard deep learning models.

2   Related Work

In this section, we review the studies related to the proposed ConCAD model, including the work on sleep apnea detection, cross attention mechanism for feature fusion, and contrastive learning.

2.1   Sleep Apnea Detection

The standard approach to diagnose sleep apnea requires the patient to sleep overnight at a clinic setup and record the polysomnography (PSG) by various physiological sensors, and then the outputs of PSG are visually inspected by a clinical expert to give a diagnosis [6,19,23]. This process is always inconvenient and uncomfortable. Thus, some studies have begun to simplify the procedure of diagnosing sleep apnea by only using a single physiological data, such as ECG [3,10], EEG [2], and the respiration signal [31]. Among these physiological data, ECG is a less intrusive option and also strongly related to sleep apnea.

To this end, several studies [18,32] manually extract hand-crafted features and feed them to classifiers (e.g., random forest, support vector machine) for sleep apnea detection. Recently, with the development of deep learning methods, some studies [3,35] extract RR interval (RRI) and the R-peak envelope (RPE) and build deep learning model to automatically learn representation and detect sleep apnea. Furthermore, several studies [10,30] develop deep learning models to directly learn features from the raw ECG data and detect sleep apnea in an end-to-end style.

2.2   Attention-Based Feature Fusion

Another line of related work is feature fusion, which aims at combining different features to obtain a more effective representation. The frequently-used feature fusion techniques are concatenation [29], summation [11], and multiplication [39]. However, these operations evenly combine all the features together without considering the importance of each feature. Some of the features gathered will help the model make the right decision, while others can lead to significant misjudgment [21].

Recently, the usage of attention learning mechanism has shown remarkable performance improvement for different tasks, such as natural language processing [37], image classification [33], and object tracking [8]. The attention mechanism highlights the effective discriminant parts of features while suppressing the redundant parts to a certain degree. To further take advantage of the features extracted from multi-modality inputs, a cross attention mechanism has been proposed to derive an attention mask from different inputs mutually. In [22], the authors use one modality (LiDAR) to generate an attention mask that controls the spatial features of a different modality (HSI). In [16], the authors derive cross attention maps for each pair of class features and query sample features to highlight specific regions and make the extracted features more discriminative.

2.3 Contrastive Learning

All of the deep learning methods mentioned above are trained by the cross-entropy loss. The cross-entropy loss is the most commonly-used one in the classification tasks, which calculates the difference between the actual probability distribution of the data and the predicted probability distribution of the model [28]. As we previously introduced, the cross-entropy loss has some limitations. Thus, a supervised contrastive loss is added as an auxiliary regularization to alleviate problems in our proposed framework.

The contrastive loss has recently been widely used in self-surprised learning [5,14,24], which aims at clustering the similar data and pushing apart the dissimilar data. A supervised version of the contrastive loss is proposed by [20] to leverage the label information. Their proposed supervised contrastive learning contains two steps: First, the supervised contrastive loss is used to learn a representation to cluster the data from the same class and separate the data from different classes; Second, they froze the model and add a multi-layer perceptron (MLP) as a classifier on its top for the classification task. Recently, supervised contrasting learning has been used for different applications, such as image classification [20], few-shot classification [25], and semantic segmentation [36].

3   Methodology

The goal of this work is to design an effective framework for leveraging the power of both the deep learning-based features and the expert knowledge-based features simultaneously to enhance the performance of sleep apnea detection. Towards this goal, we propose ConCAD as shown in Fig. 1, which is based on the contrastive learning framework to obtain better representations and utilizes a cross-attention mechanism to fuse different types of features.

Concretely, ConCAD is achieved by three steps as shown in Fig. 1. Firstly, the original raw data are passed through a feature extractor to learn the deep features. Simultaneously, the expert knowledge is passed through a feature extractor with a relatively shallow network to learn the expert features. Besides, data augmentation can be used before passing the data. Secondly, the deep features and the expert features are fed to a cross-attention module, which automatically fuses the features by emphasizing the important parts based on each other synergistically. Thirdly, the resultant attention-weighted features are mapped into a projection space for learning a representation with high intra-class similarity and low inter-class similarity to improve the classification accuracy by contrastive learning. Then, the learned representation is fed to the classification modules to output the probability of sleep apnea events.

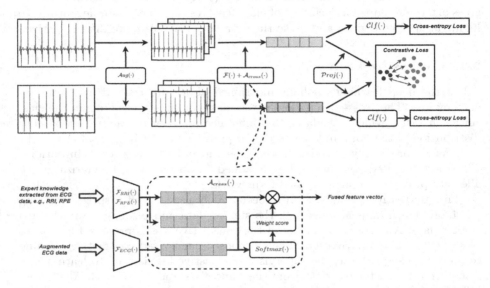

Fig. 1. Overall of our proposed ConCAD framework. The cross attention fuses the features by generating an attention weight mask. It can highlight the effective discriminant parts and suppress the irrelevant parts of features from ECG, RRI, and RPE collaboratively. Besides the cross-entropy, the supervised contrastive loss is also computed to optimize the intra-class to inter-class similarity ratio.

As we are going to demonstrate the proposed ConCAD on the datasets of ECG-based sleep apnea detection, the ECG data are certainly considered as the raw data input. Furthermore, the RRI and RPE manifest their effectiveness to detect sleep apnea by many research works [1,9]. Consequently, the RRI and RPE are chosen as the expert knowledge input for our proposed framework.

3.1   Expert Feature Extraction and Data Augmentation

The expert knowledge is summarized by previous researches over the last centuries. In the field of detecting sleep apnea using ECG data, several previous studies [1,9] have shown that the RR intervals (RRI) and R-peak envelope (RPE) are effective. To prepare the RRI and RPE data, we first detect the locations of the R-peaks by the Hamilton algorithm [13]. Then, we calculate the distance between R-peaks as the RRI and use the amplitudes of the R-peaks as the RPE. Since the RRI can be easily disturbed by unexpected ECG spikes, a median filter is used to eliminate the disturbance as suggested by [7]. Besides, since the number of the RRI or RPE is not always the same by giving a fixed time duration (e.g., 1 min), cubic interpolation was used to resample them to the same length [35].

We are augmenting the ECG, RRI and RPE by two simple approaches: random time shift and reversion. Given the data $\mathbf{x} = [x_0, x_1, x_2, \ldots, x_n]$, the random time shift will obtain $\mathbf{x}_{shift} = [x_t, x_{1+t}, x_{2+t}, \ldots, x_{n+t}]$, where t is a randomly-generated number and represents the number of data points to shift. The revision will generate $\mathbf{x}_{reverse} = [x_n, x_{n-1}, \ldots, x_2, x_1, x_0]$. The augmentation is conducted in each batch to provide more positives (i.e., instances with the same label) during batch training, which benefits a more robust clustering of the projection space. The augmentation process is presented as $\mathcal{Aug}(\mathbf{x})$.

3.2   Feature Extractor

The feature extractor should be designed case by case. In this study, we have three feature extractors, which are used for learning features from the ECG, RRI and RPE separately, and named $\mathcal{F}_{ECG}$, $\mathcal{F}_{RRI}$ and $\mathcal{F}_{RPE}$.

$\mathcal{F}$ consists of 4 convolution blocks. The first three blocks are made of one convolutional layer, one batch normalization layer, one ReLU activation layer, one maxpooling layer, and one dropout layer. The feature map size of the convolutional layer in the first block is chosen to cover data points of two contiguous beats in case that the patterns between beats get missed. The last convolution block does not contain the maxpooling and dropout layer. The module can be represented as $\mathbf{x}' = \mathcal{F}(\mathbf{x}; \theta_F)$, where $\mathbf{x}$ represents the input data and θ_F denotes the parameters of the module.

Since there are three kinds of data, we have three corresponding extractors, which are $\mathbf{x}'_{ECG} = \mathcal{F}_{ECG}(\mathbf{x}_{ECG}; \theta_{F_{ECG}})$ for ECG data, $\mathbf{x}'_{RRI} = \mathcal{F}_{RRI}(\mathbf{x}_{RRI}; \theta_{F_{RRI}})$ and $\mathbf{x}'_{RPE} = \mathcal{F}_{ECG}(\mathbf{x}_{RPE}; \theta_{F_{RPE}})$ for the expert knowledge-based features. More details of the extractors are described in Appendix A.

3.3   Cross Attention

Not all the deep features and expert features contribute equally to the classification task. Thus, we design a cross-attention module, $\mathcal{A}_{cross}$, to collaboratively learn their importance and concentrate more on the important ones. The cross-attention is designed to ask the model to concentrate on the particular features, which contribute more to distinguish the instances from different classes. Before computing the cross attention, since the outputs of feature extractors are likely to have different dimensions, we need to project them to the same space by a linear transformation. Given $\mathbf{x}' \in \mathbb{R}^{m \times n}$, the transformation is

$$\mathbf{x}'' = \mathbf{u}^{\top} \mathbf{x}' \mathbf{V} \tag{1}$$

where $\mathbf{u} \in \mathbb{R}^m$ and $\mathbf{V} \in \mathbb{R}^{n \times k}$ are trainable parameters.

After it, $\mathbf{x}''_{ECG}$, $\mathbf{x}''_{RRI}$ and $\mathbf{x}''_{RPE}$ have the same dimension k. Then, we are going to compute the attention weights. Specifically,

$$\boldsymbol{\alpha} = Softmax([\alpha_{ECG}, \alpha_{RRI}, \alpha_{RPE}])$$
$$\alpha_i = \mathbf{w}_i^\top \mathbf{x}''_i + b_i \tag{2}$$

where $i \in \mathcal{S} = \{ECG, RRI, RPE\}$ and $\boldsymbol{\alpha} \in \mathbb{R}^3$ is the attention weights. $\mathbf{w} \in \mathbb{R}^k$ and $b \in \mathbb{R}$ are trainable parameters. The transformed $\mathbf{x}''$ is passed through an one-layer MLP to learn the importance of different types of features synergistically, and the importance is normalized by a softmax function.

Lastly, we compute the context vector $\mathbf{c}$ by

$$\mathbf{c} = \sum_{i \in \mathcal{S}} \alpha_i \mathbf{x}''_i \tag{3}$$

The context vector is the fused feature vector that is the weighted sum of features from different inputs based on the learned importance. The cross-attention module can be represented as $\mathcal{A}_{cross}([\mathbf{x}'_{ECG}, \mathbf{x}'_{RRI}, \mathbf{x}'_{RPE}]; \theta_{A_{cross}})$.

3.4 Contrastive Learning.

For most of the conventional classification tasks, cross-entropy (CE) loss is commonly used to adjust model weights during training. However, CE loss may be impaired by noisy labels [42] and induce representations with excessive discrimination towards training data [25]. In order words, CE loss is likely to result in sub-optimal generalization.

As a remedy, contrastive learning is adopted to assist the model to learn more general and robust features by maximizing intra-class similarity while minimizing inter-class similarity. Concretely, we propose a novel hybrid loss, which utilizes the supervised contrastive (SC) loss [20] as an auxiliary regularization to the standard CE loss.

Contrastive Loss. The SC loss aims at simultaneously increasing the agreement among instances in positive pairs and encouraging the difference among instances in negative pairs. The instances with the same label form the positive pairs, and the instances with the different labels are considered as negative pairs. Specifically, the SC loss is computed in two steps. We first project the input, i.e., the fused feature vector, to a lower dimension space by a one-layer MLP, and the low dimension vector is normalized to the unit hypersphere by L2 norm, $\mathbf{z} = \mathcal{P}roj_{SC}(\mathcal{A}_{cross}([\mathbf{x}'_{ECG}, \mathbf{x}'_{RRI}, \mathbf{x}'_{RPE}]; \theta_{A_{cross}})$. Then, the SC loss can be computed by

$$\mathcal{L}_{SC} = -\sum_{i=1}^{N} \frac{1}{N_{y_i}} \log \frac{\sum_{j=1}^{N} \mathbb{1}_{[y_i = y_j]} \exp(\text{sim}(\mathbf{z}_i, \mathbf{z}_j)/\tau)}{\sum_{k=1}^{N} \mathbb{1}_{[k \neq i]} \exp(\text{sim}(\mathbf{z}_i, \mathbf{z}_k)/\tau)}, \tag{4}$$

where N is the batch size, and N_{y_i} is the number of samples with the same label in each batch. $\mathbb{1}_{[\cdot]}$ denotes an indicator function. $\text{sim}(\cdot)$ represents the measure of similarity, and here the cosine similarity is used, i.e., $\text{sim}(u, v) = u \cdot v / \|u\| \|v\|$. τ is a hyperparameter that controls the strength of penalties on negative pairs [34].

In the SC loss formula, the numerator represents the similarity of the positives, and the denominator represents the similarity of everything else in regard to z_i. The optimization of this formula pulls together the positives and pushes apart everything else. That is, instances from the same class will form a closer cluster while the distances between clusters are increased in the projected hypersphere. As a result, the model learns more general features instead of naively learning the features for the classification task over the known training data.

Hybrid Loss. As described in [20], the standard SC loss requires two separate steps for a classification task: first, they train the feature extractor with the SC loss to learn a representation vector; second, they freeze the feature extractor and train a classifier on the vector using the CE loss. However, the SC loss usually requires a very large batch size to achieve decent and stable performance. For example, [20] uses a batch size of 6,144. On the other hand, the CE loss only works in the second step and cannot update the model parameters in the feature extractor, which means that the CE loss does not make any contribution to learning the feature representation.

To alleviate these problems, we use the SC loss as an auxiliary regularization term and integrate it with the CE loss. Specifically, we propose a new hybrid loss, which is the summation of CE and SC losses with a scaling parameter λ to control the contribution of each loss:

$$\mathcal{L}_{hybrid} = \lambda \mathcal{L}_{CE} + (1 - \lambda) \mathcal{L}_{SC}. \tag{5}$$

With the proposed hybrid loss, the model can take advantage of both the CE and SC losses simultaneously. The CE loss can learn effective features for classification tasks with small batch sizes, and the SC loss helps to promote these features to be more general and robust by minimizing the intra-class to inter-class similarity ratio. To train the model with the proposed hybrid loss, we project the fused feature vector to lower dimension hypersphere by $\mathcal{P}roj_{SC}(\cdot)$ to calculate the SC loss. At the same time, the fused feature vector is sent to fully-connected layers $\mathcal{C}lf(\cdot)$ to calculate the CE loss as shown in the last step of Fig. 1. All the layers except the last one in $\mathcal{C}lf(\cdot)$ use the ReLU activation, while the last layer is operated on the softmax activation, and its unit number needs to be identical to the number of classes. In addition, we will discard $\mathcal{P}roj_{SC}(\cdot)$ during prediction so that the proposed model has the same number of parameters as a model with only the CE loss.

4 Experiments and Results

4.1 Datasets

In our experiments, two datasets, i.e., Apnea-ECG [26] and MIT-BIH Polysomnographic [17] obtained from Physionet [12], are used for performance evaluation and comparison. Both datasets are publicly available and have been used to study sleep apnea detection methods in previous researches.

- **Apnea-ECG**: The apnea-ECG database is provided by Philipps University, which is the most commonly-used dataset for ECG-based sleep apnea studies. It contains 70 single-lead ECG recordings of varying lengths between 7 h to 10 h, sampled at the rate 100 Hz. Each segment of 1 min ECG data is annotated by the expert as either apnea or normal event. The datasets are officially split into two sets by the provider: a released set of 35 recordings and a withheld set of 35 recordings. After removing the data with an unreasonable heartbeat rate, the released set contains 16,888 segments and the withheld set contains 17,120 segments.
- **MIT-BIH Polysomnographic**: The MIT-BIH Polysomnographic database is collected by Boston's Beth Israel Hospital Sleep Laboratory. It contains over 80 h of polysomnographic (PSG) recordings during sleep. Each recording includes a single channel of ECG annotated beat-by-beat and EEG and respiration signals. Each segment of 30 s of data is annotated with respect to sleep stages and apnea. After removing the data with an unreasonable heartbeat rate, the final dataset contains 9,717 segments.

4.2   Compared Methods

To valid the performance of our framework, we use several state-of-the-art methods as our benchmark methods:

- Support Vector Machine (**SVM**), Random Forest (**RF**), K-Nearest Neighbor (**KNN**), and Multi-Layer Perception (**MLP**) is adopted with 10 popular hand-crafted features from ECG data (e.g., RMSSD, NN50, etc.) as benchmark methods according to the work by [18].
- **LeNet-5**: In [35], a LeNet-5 convolutional neural network is used to learn features from RRI and RPE for sleep apnea detection.
- **CNN+LSTM**: In [3], three different deep learning architectures are proposed. We adopt their best performing architecture, i.e., CNN+LSTM, as one of our benchmark methods.
- **ResNet**: In [38], a strong baseline model with the ResNet structure is proposed for time series classification, including ECG classification. So, we also use it for comparison.
- **CNN-4**: In [10], a four-layer CNN-based model with a novel pooling layer is proposed to detect sleep apnea from ECG data directly, and we compare it with our proposed method as well.
- **CNN-6**: In [30], several models with a different number of convolutional layers are designed to predict sleep apnea with ECG data. We adopt their best performing one, which contains 6 convolutional layers for our comparison.

We also compare the following approaches to show the improvements of the proposed framework step by step:

- $\mathcal{F}_{ECG} + \mathcal{Clf}$: It employs a CNN-based feature extractor to learn the deep features from the raw ECG data and then classifies the targets by several fully-connected layers. It can be considered as a standard architecture of a naive deep learning model.
- $\mathcal{F}_{ECG} + \mathcal{F}_{RRI} + \mathcal{F}_{RPE} + \mathcal{Clf}$: Besides the raw ECG data, RRI and RPE are used as expert knowledge inputs. A simple concatenation is used to combine the features from ECG, RRI, and RPE. Then the concatenated features are sent for classification.
- $\mathcal{F}_{ECG} + \mathcal{F}_{RRI} + \mathcal{F}_{RPE} + \mathcal{A}_{cross} + \mathcal{Clf}$: Instead of using the simple concatenation, a cross-attention mechanism is proposed to collaboratively fuse the features from ECG, RRI, and RPE.
- $\mathcal{F}_{ECG} + \mathcal{F}_{RRI} + \mathcal{F}_{RPE} + \mathcal{A}_{cross} + \mathcal{Proj}_{SC} + \mathcal{Clf}$: The proposed hybrid loss is used to update the model's parameter by adding an auxiliary projection during training to learn more general and useful features.
- ConCAD ($\mathcal{Aug} + \mathcal{F}_{ECG} + \mathcal{F}_{RRI} + \mathcal{F}_{RPE} + \mathcal{A}_{cross} + \mathcal{Proj}_{SC} + \mathcal{Clf}$): Data augmentation is used with the architecture mentioned above to help learn more general and robust features to boost performance.

4.3 Experiment Setup

For the Apnea-ECG dataset, we train all the models, including the proposed ConCAD method and benchmark methods, on the released dataset and test them on the withheld dataset. For the MIT-BIH PSG dataset, 10-fold cross-validation is used to examine the performance as there is no predefined training and test set. Moreover, since some existing studies [29,35,40] have shown that adjacent segment information helps analyze the sleep-related problems, the labeled segment with its surrounding ±2 segments of the ECG data is also included in our study. Thus, we will examine segments of 1 and 5 min on the Apnea-ECG dataset and test segments of 0.5 and 2.5 min on the MIT-BIH PSG dataset. In addition, some of the deep learning-based benchmark methods (i.e., [10,30,38]) are modified by increasing the pooling size and replacing flatten layer with GlobalAveragePooling for the input of 5 min and 2.5 min as they do no have a version to handle data with adjacent segments, and processing very long vector with their original structures exceeds our hardware memory limitation.

The proposed ConCAD model is trained by the AMSGrad optimizer [27], and all its parameters are initialized using HeNormal initializer [15]. An initial learning rate of 0.005 is chosen and it decreases to 0.001 after certain epochs (e.g. 200 epochs). Moreover, the L2 regularization is added to the feature extractor $\mathcal{F}$ to prevent the model from overfitting into the noise or artifacts.

4.4 Results and Discussions

We compare the performance of ConCAD with other state-of-the-art benchmark methods, and the results are listed in Table 1. Our proposed framework achieves an accuracy of 88.75% with the 1 min segment input and 91.22% with the 5 min segment input on the Apnea-ECG dataset, and 82.50% with the 1 min segment input and 83.47% with the 5 min segment input on the MIT-BIH PSG dataset, which outperforms other benchmark methods. Besides, we can see deep learning methods can adaptively learn features from a different length of input while the machine learning methods with hand-crafted features are more sensitive to the change of the input length. With the adjacent segments, the deep learning model can learn more effective features for classification tasks.

We also examine the proposed framework step by step to show the effectiveness of each step in Table 2. We can see that the performance can be worse if we simply concatenate the deep features with the expert features as some of the features will help the model make the better judgment possible, while others are likely to act as noise and thereby lead to more errors. With the cross attention module $\mathcal{A}_{cross}$, we can see that the model learns a better-fused feature representation by learning an attention mask synergistically from each other. The new fused feature representation maintains the effective discriminant parts of features while suppressing the irrelevant parts. Specifically, the accuracy improves

Table 1. Accuracy of the proposed framework with other state-of-the-art methods on Apnea-ECG and MIT-BIH PSG datasets.

Methods		Ref	Apnea-ECG		MIT-BIH PSG	
			1 min	5 min	0.5 min	2.5 min
Feature Based Machine Learning (ML)	SVM	[18]	74.57	67.52	70.02	70.30
	RF		74.86	72.30	70.54	68.15
	KNN		71.81	67.80	69.51	68.12
	MLP		74.81	70.59	71.28	71.20
Deep Learning (DL)	LeNet-5	[35]	83.17	87.25	72.49	78.82
	CNN+LSTM	[3]	82.77	86.12	75.80	80.79
	ResNet	[38]	83.57	85.33	77.29	79.23
	CNN-4	[10]	81.65	84.42	73.56	76.92
	CNN-6	[30]	82.12	84.37	79.69	82.25
Proposed method	ConCAD		**88.75**	**91.22**	**82.50**	**83.47**

Table 2. Accuracy of different architectures of the proposed framework on Apnea-ECG and MIT-BIH PSG datasets

Architectures	Apnea-ECG		MIT-BIH PSG	
	1 min	5 min	0.5 min	2.5 min
$\mathcal{F}_{ECG} + \mathcal{C}lf$	83.41	85.48	78.83	80.11
$\mathcal{F}_{ECG} + \mathcal{F}_{RRI} + \mathcal{F}_{RPE} + \mathcal{C}lf$	83.23	87.64	79.42	80.60
$\mathcal{F}_{ECG} + \mathcal{F}_{RRI} + \mathcal{F}_{RPE} + \mathcal{A}_{cross} + \mathcal{C}lf$	85.35	89.43	80.22	81.77
$\mathcal{F}_{ECG} + \mathcal{F}_{RRI} + \mathcal{F}_{RPE} + \mathcal{A}_{cross} + \mathcal{P}roj_{SC} + \mathcal{C}lf$	87.16	90.85	81.83	82.83
ConCAD	**88.75**	**91.22**	**82.50**	**83.47**

from 83.41% to 85.35% with the 1 min segment input and from 85.48% to 89.43% with the 5 min segment input on the Apnea-ECG dataset. On the MIT-BIH PSG dataset, the accuracy improves from 78.83% to 80.22% with the 0.5 min segment input and from 80.11% to 81.77% with the 2.5 min segment input.

In Table 2, we can also see a further improvement by using the proposed hybrid loss, which takes advantage of both CE loss and SC loss. The accuracy increases to 87.16% with the 1 min segment input and 89.43% with the 5 min segment input on the Apnea-ECG dataset. On the MIT-BIH PSG dataset, the accuracy increases to 81.83% with the 1 min segment input and 81.77% with the 5 min segment input. The hybrid loss boosts the performance by promoting the model to learn more general and discriminant feature representations in case that the model overfits into the training data by learning features with excessive discrimination.

In Fig. 2, the t-SNE plots show the learned feature representation with the CE loss and the proposed hybrid loss. We can see that the hybrid loss promotes to more compact clustering of the instances from the same class while the representation with CE loss is more scattered. We also think the attention module benefits contrastive learning by focusing on parts of features when increasing the agreement among instances in positive pairs and encouraging the difference

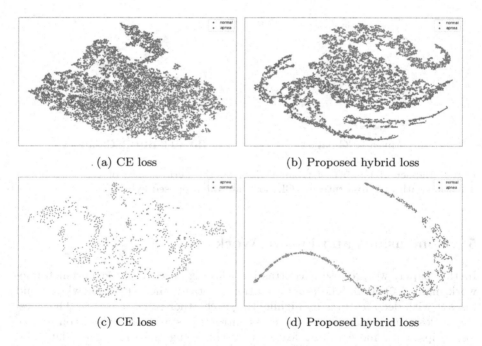

(a) CE loss   (b) Proposed hybrid loss

(c) CE loss   (d) Proposed hybrid loss

Fig. 2. The t-SNE plots of the fused feature vector on the withheld set of Apnea-ECG (a and b) and the validation set of MIT-BIH PSG (c and d), comparing the cross-entropy (CE) loss with the proposed hybrid loss.

among instances in negative pairs. It is similar to human behavior that human usually tends to recognize an unseen data by comparing the most relevant parts with known ones. By using data augmentation, the proposed model can learn more general feature representation with a more clear boundary and achieve better performance.

In addition, the hybrid loss enables the classification tasks with limited training labeled data. The limitation of labeled data is a prevalent and critical problem in the healthcare field. We train the model by using a fraction of the training set and test it on the entire test set. The results are shown in Fig. 3 in terms of the macro F1 score. F1 score can clearly show the quality of the model when the dataset is imbalanced. With only 1% of the training data, the proposed ConCAD model can still achieve an F1 score of 0.67 on the Apnea-ECG dataset and 0.59 on the MIT-BIH PSG dataset. However, the CE loss performs poorly and skews into the majority class. Furthermore, the proposed model only requires 10% of the training data to make a reasonable classification while the naive deep learning model needs more than 50% to get decent performance. Hence, the proposed model with hybrid loss largely outperforms a naive deep learning approach with CE loss on smaller datasets.

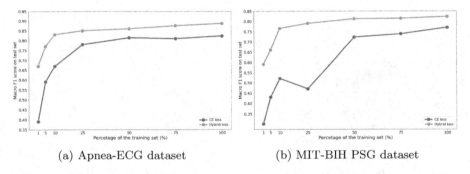

(a) Apnea-ECG dataset   (b) MIT-BIH PSG dataset

Fig. 3. Impact of number of training data on the performance of the sleep apnea detection with the cross entropy (CE) loss and the proposed hybrid loss.

5   Conclusions and Future Work

In this paper, we propose a contrastive learning-based cross attention framework, named ConCAD. The cross attention leverages the expert knowledge and fuses it with deep features by highlighting each other collaboratively. The novel hybrid loss that encompasses the cross-entropy loss and supervised contrastive loss helps learn more robust features by clustering the same class data and pushing apart data of different classes in projection space. Moreover, we show the proposed framework achieves state-of-the-art results on two public ECG datasets. Furthermore, we show that the proposed framework has better generalization ability with limited labeled training data. We conclude that the ECG data with adjacent segments helps to detect the sleep apnea occurrence through the experiment.

In future work, we plan to study more ECG data augmentation techniques that would help contrastive learning to generate better representations. We also plan to develop more interfaces to allow different formats of expert knowledge (e.g., electronic health records) to be integrated into our framework.

Appendix A

The feature extractor are different for different data and tasks. In this study, we design a CNN-based extractors for ECG, RRI and RPE separately. The structure of the extractor for two dataset are also different as their ECG data have different sampling frequency and noise. The details are shown in the table below. The ConvBlock(number of filters, kernel size, stride) is made of one convolutional layer, one batch normalization layers, one ReLU activation layer (Table 3).

Table 3. The details of the feature extractors used for ECG, RRI and RPE on Apnea-ECG and MIT-BIH PSG.

$\mathcal{F}_{ECG}$ (Apnea-ECG)	$\mathcal{F}_{ECG}$ (MIT-BIH PSG)	$\mathcal{F}_{RRI}, \mathcal{F}_{RPE}$
ConvBlock(64,100,20)- MaxPool(2)-Dropout(0.5)- ConvBlock(64,8,4)- MaxPool(2)-Dropout(0.5)- ConvBlock(128,4,2)- MaxPool(2)-Dropout(0.5)- ConvBlock(128,4,2)	ConvBlock(64,60,5)-MaxPool(2)- Dropout(0.5)-ConvBlock(128,8,3)- ConvBlock(128,8,3)-MaxPool(2)- Dropout(0.5)-ConvBlock(256,4,2)- ConvBlock(256,4,2)–MaxPool(2)- Dropout(0.5)-ConvBlock(128,4,1)- ConvBlock(128,4,1)	ConvBlock(64,8,4)- MaxPool(2)-Dropout(0.5)- ConvBlock(64,4,2)- MaxPool(2)-Dropout(0.5)- ConvBlock(128,2,1)- MaxPool(2)-Dropout(0.5)- ConvBlock(128,2,1)

References

1. Al-Abed, M.A., Manry, M., Burk, J.R., Lucas, E.A., Behbehani, K.: Sleep disordered breathing detection using heart rate variability and r-peak envelope spectrogram. In: 2009 Annual International Conference of the IEEE Engineering in Medicine and Biology Society, pp. 7106–7109. IEEE (2009)
2. Almuhammadi, W.S., Aboalayon, K.A., Faezipour, M.: Efficient obstructive sleep apnea classification based on eeg signals. In: 2015 Long Island Systems, Applications and Technology, pp. 1–6. IEEE (2015)
3. Almutairi, H., Hassan, G.M., Datta, A.: Detection of obstructive sleep apnoea by ecg signals using deep learning architectures. In: 2020 28th European Signal Processing Conference (EUSIPCO), pp. 1382–1386. IEEE (2021)
4. Altevogt, B.M., Colten, H.R., et al.: Sleep Disorders and Sleep Deprivation: An Unmet Public Health roblem. National Academies Press (2006)
5. Banville, H., Chehab, O., Hyvarinen, A., Engemann, D., Gramfort, A.: Uncovering the structure of clinical eeg signals with self-supervised learning. J. Neural Eng. **18**, 046020 (2020)

6. Bloch, K.E.: Polysomnography: a systematic review. Technol. Health Care **5**(4), 285–305 (1997)
7. Chen, L., Zhang, X., Song, C.: An automatic screening approach for obstructive sleep apnea diagnosis based on single-lead electrocardiogram. IEEE Trans. Autom. Sci. Eng. **12**(1), 106–115 (2014)
8. Chu, Q., Ouyang, W., Li, H., Wang, X., Liu, B., Yu, N.: Online multi-object tracking using cnn-based single object tracker with spatial-temporal attention mechanism. In: Proceedings of the IEEE International Conference on Computer Vision, pp. 4836–4845 (2017)
9. De Chazal, P., Heneghan, C., Sheridan, E., Reilly, R., Nolan, P., O'Malley, M.: Automatic classification of sleep apnea epochs using the electrocardiogram. In: Computers in Cardiology 2000, vol. 27 (Cat. 00CH37163), pp. 745–748. IEEE (2000)
10. Dey, D., Chaudhuri, S., Munshi, S.: Obstructive sleep apnoea detection using convolutional neural network based deep learning framework. Biomed. Eng. Lett. **8**(1), 95–100 (2017). https://doi.org/10.1007/s13534-017-0055-y
11. Feichtenhofer, C., Pinz, A., Zisserman, A.: Convolutional two-stream network fusion for video action recognition. In: Proceedings of the IEEE Conference on Computer Vision and Pattern Recognition, pp. 1933–1941 (2016)
12. Goldberger, A.L., et al.: Physiobank, physiotoolkit, and physionet: components of a new research resource for complex physiologic signals. Circulation **101**(23), e215–e220 (2000)
13. Hamilton, P.: Open source ecg analysis. In: Computers in Cardiology, pp. 101–104. IEEE (2002)
14. He, K., Fan, H., Wu, Y., Xie, S., Girshick, R.: Momentum contrast for unsupervised visual representation learning. In: Proceedings of the IEEE/CVF Conference on Computer Vision and Pattern Recognition, pp. 9729–9738 (2020)
15. He, K., Zhang, X., Ren, S., Sun, J.: Delving deep into rectifiers: surpassing human-level performance on imagenet classification. In: Proceedings of the IEEE International Conference on Computer Vision, pp. 1026–1034 (2015)
16. Hou, R., Chang, H., Ma, B., Shan, S., Chen, X.: Cross attention network for few-shot classification. arXiv preprint arXiv:1910.07677 (2019)
17. Ichimaru, Y., Moody, G.: Development of the polysomnographic database on cd-rom. Psychiatry Clin. Neurosci. **53**(2), 175–177 (1999)
18. Jezzini, A., Ayache, M., Elkhansa, L., Al Abidin Ibrahim, Z.: ECG classification for sleep apnea detection. In: 2015 International Conference on Advances in Biomedical Engineering (ICABME), pp. 301–304. IEEE (2015)
19. Kapur, V.K., et al.: Clinical practice guideline for diagnostic testing for adult obstructive sleep apnea: an American academy of sleep medicine clinical practice guideline. J. Clin. Sleep Med. **13**(3), 479–504 (2017)
20. Khosla, P., et al.: Supervised contrastive learning. arXiv preprint arXiv:2004.11362 (2020)
21. Lin, C.J., Lin, C.H., Jeng, S.Y.: Using feature fusion and parameter optimization of dual-input convolutional neural network for face gender recognition. Appl. Sci. **10**(9), 3166 (2020)
22. Mohla, S., Pande, S., Banerjee, B., Chaudhuri, S.: Fusatnet: dual attention based spectrospatial multimodal fusion network for hyperspectral and lidar classification. In: Proceedings of the IEEE/CVF Conference on Computer Vision and Pattern Recognition Workshops, pp. 92–93 (2020)

23. Nikolaidis, K., Kristiansen, S., Goebel, V., Plagemann, T., Liestøl, K., Kankan-halli, M.: Augmenting physiological time series data: a case study for sleep apnea detection. In: Brefeld, U., Fromont, E., Hotho, A., Knobbe, A., Maathuis, M., Robardet, C. (eds.) ECML PKDD 2019. LNCS (LNAI), vol. 11908, pp. 376–399. Springer, Cham (2020). https://doi.org/10.1007/978-3-030-46133-1_23
24. Oord, A.v.d., Li, Y., Vinyals, O.: Representation learning with contrastive predic-tive coding. arXiv preprint arXiv:1807.03748 (2018)
25. Ouali, Y., Hudelot, C., Tami, M.: Spatial contrastive learning for few-shot classi-fication. arXiv preprint arXiv:2012.13831 (2020)
26. Penzel, T., Moody, G.B., Mark, R.G., Goldberger, A.L., Peter, J.H.: The apnea-ecg database. In: Computers in Cardiology 2000, vol. 27 (Cat. 00CH37163), pp. 255–258. IEEE (2000)
27. Reddi, S.J., Kale, S., Kumar, S.: On the convergence of adam and beyond. arXiv preprint arXiv:1904.09237 (2019)
28. Rumelhart, D.E., Hinton, G.E., Williams, R.J.: Learning representations by back-propagating errors. Nature 323(6088), 533–536 (1986)
29. Supratak, A., Dong, H., Wu, C., Guo, Y.: Deepsleepnet: a model for automatic sleep stage scoring based on raw single-channel eeg. IEEE Trans. Neural Syst. Rehabil. Eng. 25(11), 1998–2008 (2017)
30. Urtnasan, E., Park, J.U., Joo, E.Y., Lee, K.J.: Automated detection of obstruc-tive sleep apnea events from a single-lead electrocardiogram using a convolutional neural network. J. Med. Syst. 42(6), 1–8 (2018)
31. Van Steenkiste, T., Groenendaal, W., Deschrijver, D., Dhaene, T.: Automated sleep apnea detection in raw respiratory signals using long short-term memory neural networks. IEEE J. Biomed. Health Inf. 23(6), 2354–2364 (2018)
32. Varon, C., Caicedo, A., Testelmans, D., Buyse, B., Van Huffel, S.: A novel algo-rithm for the automatic detection of sleep apnea from single-lead ecg. IEEE Trans. Biomed. Eng. 62(9), 2269–2278 (2015)
33. Wang, F., et al.: Residual attention network for image classification. In: Proceed-ings of the IEEE Conference on Computer Vision and Pattern Recognition, pp. 3156–3164 (2017)
34. Wang, F., Liu, H.: Understanding the behaviour of contrastive loss. arXiv preprint arXiv:2012.09740 (2020)
35. Wang, T., Lu, C., Shen, G., Hong, F.: Sleep apnea detection from a single-lead ecg signal with automatic feature-extraction through a modified lenet-5 convolutional neural network. PeerJ 7, e7731 (2019)
36. Wang, W., Zhou, T., Yu, F., Dai, J., Konukoglu, E., Van Gool, L.: Exploring cross-image pixel contrast for semantic segmentation. arXiv preprint arXiv:2101.11939 (2021)
37. Wang, Y., Huang, M., Zhu, X., Zhao, L.: Attention-based lstm for aspect-level sen-timent classification. In: Proceedings of the 2016 Conference on Empirical Methods in Natural Language Processing, pp. 606–615 (2016)
38. Wang, Z., Yan, W., Oates, T.: Time series classification from scratch with deep neural networks: a strong baseline. In: 2017 International Joint Conference on Neural Networks (IJCNN), pp. 1578–1585. IEEE (2017)
39. Wu, L., Wang, Y., Li, X., Gao, J.: What-and-where to match: deep spatially mul-tiplicative integration networks for person re-identification. Pattern Recogn. 76, 727–738 (2018)
40. Yadollahi, A., Moussavi, Z.: Acoustic obstructive sleep apnea detection. In: 2009 Annual International Conference of the IEEE Engineering in Medicine and Biology Society, pp. 7110–7113. IEEE (2009)

41. Young, T., Peppard, P.E., Gottlieb, D.J.: Epidemiology of obstructive sleep apnea: a population health perspective. Am. J. Resp. Crit. care Med. **165**(9), 1217–1239 (2002)
42. Zhang, Z., Sabuncu, M.R.: Generalized cross entropy loss for training deep neural networks with noisy labels. arXiv preprint arXiv:1805.07836 (2018)

Machine Learning Based Simulations and Knowledge Discovery

DeepPE: Emulating Parameterization in Numerical Weather Forecast Model Through Bidirectional Network

Fengyang Xu, Wencheng Shi, Yunfei Du, Zhiguang Chen[✉], and Yutong Lu

School of Computer Science and Engineering, Sun Yat-sen University,
Guangzhou, China
{xufy9,shiwch}@mail2.sysu.edu.cn
{yunfei.du,zhiguang.chen,yutong.lu}@nscc-gz.cn

Abstract. To make weather/climate modeling computationally afford-
able, subgrid-scale physical processes in the numerical models are usually
represented by semi-empirical parameterization schemes. For example,
planetary boundary layer (PBL) parameterizations are used in atmospheric models to represent the diurnal variation in the formation and
collapse of the atmospheric boundary layer—the lowest part of the atmosphere. We consider the problem of developing an accurate alternative
to physics-based PBL parameterizations for speeding up the operation
of atmosphere modeling. Our contributions are twofold. The first contribution is to propose a deep neural network emulator, called DeepPE,
that focuses on simulating nonlocal closures in the PBL to capture cross-
layer large eddies. We also explore a transfer method to maintain accuracy when applying a trained model to systems with different external
forcing. We provide a comparison with three data-driven approaches as
well as multi-task fine-tuning in predicting the PBL vertical profiles out-
putted by the Yonsei University (YSU) parameterization in the Weather
Research Forecast (WRF) climate model over 16 locations. The experiment results show that our method can better simulate the vertical profiles within the boundary layer of velocities, temperature, wind speed,
and water vapor over the entire cycle. And they also indicate that it
achieves a comparable generalization performance with less computational cost.

Keywords: Neural networks · Supervised learning · Environmental
sciences

1 Introduction

Scientists use numerical models to understand complex earth systems and make
predictions [29], which work by dividing the components of the system into large
boxes, known as grid boxes. Most of the current climate and weather models
have typical horizontal grid resolution of O(100) km and O(10) km , respectively.

© Springer Nature Switzerland AG 2021
Y. Dong et al. (Eds.): ECML PKDD 2021, LNAI 12979, pp. 87–101, 2021.
https://doi.org/10.1007/978-3-030-86517-7_6

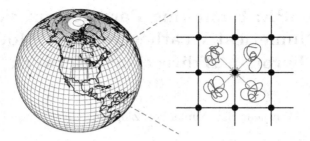

Fig. 1. Parameterization in atmospheric model. The numerical model based on grid boxes simulates atmospheric motion at the current spatial resolution (grid-scare), but it receives variables that reflect the influence of the subgrid-scale physical processes.

However, many important physical processes and mechanisms that take place on smaller spatial scales, such as atmospheric and oceanic turbulent circulations, are too small to be explicitly modelled, or some phenomena are not fully understood [30], and therefore model developers must resort to parameterizations (Fig. 1). That is, parameterization is a physical-based or semi-empirical approximation of small-scale, subgrid processes in large-scale resolved processes [21]. While these parameterizations are designed to be computationally efficient, calculation of a model *physics* (different from *dynamic core* with clear analytical solutions, which contains all the subgrid processes using parameterization) package still takes a good portion of the total computational time. For example, in the Community Atmospheric Model (CAM) developed by the National Center for Atmospheric Research (NCAR), with a spatial resolution of approximately 300 km and 26 vertical levels, the physical parameterizations account for about 70% of the total computational burden [16].

Moreover, an increasing need in the climate community is performing high spatial-resolution simulations, which calls for exponential growth of computing power [2], to assess risk and vulnerability due to climate variability at a local scale, while generating large-ensemble simulations in order to address uncertainty in the model projections. All these have caused great challenges to the time efficiency of numerical model operation. Thus, developing novel and computationally efficient emulators to parameterization [18], enabling researchers to generate finer resolution simulations and more ensemble members, are urgently needed and are at the forefront of research.

Since deep neural networks (DNNs) have excellent ability in nonlinear fitting, we have a strong intuition in favor of the potential of DNNs in parameterization emulation. In this work, we introduce an approach to emulate an existing planetary boundary layer (PBL) parameterization using 22-year-long output of 16 different locations created by the Weather Research Forecast (WRF) climate model through deep learning. The aim is to build an DNN-based algorithm to empirically understand the process in the numerical weather/climate models that could be used to replace the physics parameterizations that were derived from observational studies. This method would be computationally efficient and making the generation of large-ensemble simulations feasible at very high spatial/temporal resolutions with limited computational resources.

Developing a fast PBL parameterization emulator is challenging. First, PBL is the region adjacent to the earth's surface where small-scale turbulence, which is difficult to be observed, fully understood, let alone modeled, is induced by wind shear and/or thermal convection and occurs almost continuously in space and time [7]. Secondly, the emulator is expected to have certain extrapolation (i.e., generalization) ability, according to the First Law of Geography: "everything is related to everything else, but near things are more related than distant things" [27]. Nevertheless, the natural and anthropogenic variabilities cause non-stationarities, which can limit the applicability of a data-driven model that is trained with a dataset that contains a large amount of data but from a small part of the non-stationary distribution in climate modeling [23].

To tackle these challenges, our method has two components. The first is Deep parameterization emulator (DeepPE), taking account of ideas of traditional physical methods, that can be used to replace the PBL parameterization in the WRF model. The second component is a transfer mechanism to further reduce the cost of calculations. Specifically, we hope that a trained model can be applied to systems with different forcing (e.g., turbulence). This problem is closely associated with two (or three) widely-studied topics: multi-task learning (or transfer learning) and life-long learning, but has not been extensively addressed in any of them. In life-long learning, different tasks are trained over time, and the model accommodates new knowledge while retaining existing experience. [22]. However, data of all tasks is gained at the same time in our case. Our problem is more similar to multi-task learning, where model can generalize better on original task by sharing representations between related tasks [24], whereas in our case reducing total training time, rather than improving performance, is more concerned.

2   Related Work

Data-driven Parameterization Emulation. As far as we know from the literature available, experts have tried to use shallow neural networks to accelerate parameterization very early [4,15]. [17] utilized a single layer network to imitate an atmospheric longwave radiation parameterization for the National Center for Atmospheric Research (NCAR) Community Atmospheric Model (CAM) and obtained a 50–80 times acceleration. Compared with our approach, they are cruder and hardly competitive in accuracy when used for parameterization of more complex physical processes.

In recent years, because of the superior performance in modeling the underlying nonlinear functional relationship between inputs and outputs of systems, DNNs are particularly appealing for emulations of physical parameterization in the numerical weather and climate modeling. DNNs have been used to fit a simple chaotic dynamical system to prove their feasibility for atmosphere modeling. [8] used the two-time scale model proposed in [20], henceforth the L96 system,

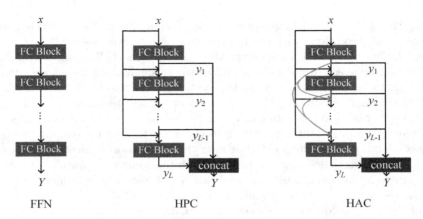

Fig. 2. Three existing DNNs' emulators: Fully connected feed-forward neural network (FFN), hierarchically connected network with previous layer only connection (HPC) and hierarchically connected network with all previous layers connected (HAC). FC Block is the full-connected feature extraction block.

which is a common baseline model for evaluating both parameterization and data assimilation techniques due to its transparency and computational cheapness, as a test bed to evaluate the performance of GAN in stochastic parameterization. Coincidentally, almost at the same time [3] also used a simple fully connected network to model the multi-scale L96 system [26] thus proving the performance of DNNs in sub-grid parameterization. Our work simulates a practical parameterization scheme, the Yonsei University (YSU) scheme [12], in commonly used WRF atmospheric model instead of a toy model, which is more applicable.

Fast emulation to PBL parameterization has been investigated for DNNs in the seminal work of [28]. It introduced three types of networks employed as our baseline methods (Fig. 2), and attempted to generalize the model trained by data from a single location to its neighbors. However, although they also tried to make the network capture the mixing between PBL vertical layers by using domain knowledge to guide the design, only the effects of higher altitude layers from previous layers are considered, which is inconsistent with the fact that in the scenario of nonlocal mixing, the vertical exchange between PBL is mutual. In addition, their experiments also show a serious decline in model performance in partial locations.

Multi-task Learning. Transfer learning [25], as well as multi-task learning [31] is widely studied, and it is out the scope of this paper to review all of them. We briefly review the most representative and related works. Our case is similar to multi-task learning in the sense that we also construct general representations which are task-agnostic [6], though we focus on reducing the total training time without decreasing the accuracy. Since all our datasets are labelled, *Fine-tuning*, which starts with a pre-trained model on the source task and trains it further

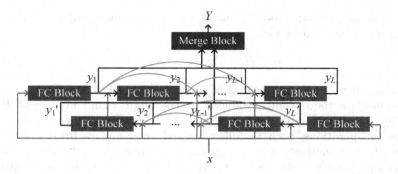

Fig. 3. Deep parameterization emulator (DeepPE) architecture.

on the target task, is arguably the most widely used approach [14]. In computer vision tasks, the fine-tuning methods have been studied fruitfully: models have been able to automatically determine which layers to fine-tune per target instance [10,11]. Yet because our model tries to simulate the correlation of the vertical profiles of PBL, it is difficult to freeze parameters of partial layers. And, in our problem, the sizes of the target dataset and source dataset have subtle differences, allowing fine-tuning the whole network with low possibility of overfitting.

3 Methods

3.1 Problem Definition

The PBL parametric emulator aims to create a mapping of near-surface characteristics (x) to vertical profiles of the model prognostic and diagnostic fields (y):

$$Y = \mathcal{F}(x, \Theta) \tag{1}$$

where $x \in \mathbb{R}^N$ and $Y = [y_1, y_2, \ldots, y_L]$, $y_l \in \mathbb{R}^M$ ($l = 1, 2, \ldots, L$). N, M and L are the number of input characteristics, output atmospheric variables and vertical layers. $\mathcal{F}$ is the parameterization method, here is DeepPE, and Θ denotes the parameters of $\mathcal{F}$. The observations of output variables are used as ground truth.

3.2 Deep Parameterization Emulator

As shown in Fig. 3, our DeepPE mainly consists two parts: bidirectional hierarchically all previous connected network (BiHAC) and merge block. BiHAC is an extension of HAC which akin to *Dense block* [13] but output layers are evenly lie in the network instead of only at the end. More specifically, each basic feature extraction module (FC block, Full Connected block) receives the output from all previous modules and the network input as input feature. This procedure can be expressed as:

$$y_l = H_{\text{FC},l}([x, y_1, \ldots, y_{l-1}]) \tag{2}$$

where $H_{\text{FC},l}$ denotes the operations of the l-th FC block. $[x, y_1, \ldots, y_{l-1}]$ refers to the concat of the outputs produced by the input x and FC blocks $H_{\text{FC},1}$, $\ldots$, $H_{\text{FC},(l-1)}$, resulting in $N + (l-1) \times M$ features. $H_{\text{FC},l}$ can be a composite function of operations, such as full-connected layer and rectified linear units (ReLU) [9]. The motivation for designing this skip-connected structure is not only back-propagation efficiency, but, from the physical process perspective, it takes the nonlocal mixture between vertical layers in PBL into account. As reported by [5], compared with solely local mixing processes, nonlocal mixing processes are shown to perform more accurately in simulating deeper mixing within an unstable PBL.

Although HAC consider the features transported from surface and all the points that below the given points, unsurprisingly it is not sufficient. Effects of currents or vertical eddies within the PBL are not unidirectional [5], hence when we use the network layers to simulate the vertical exchange of PBL, we should explore the upward and downward information transmission. We add the reverse structure:

$$y_l' = H_{\text{FC},l}'([x, y_L', \ldots, y_{l+1}']) \tag{3}$$

where $H_{\text{FC},l}'$ denotes the l-th FC block of the reverse structure and $y_l' \in \mathbb{R}^M$ is another output of l-th PBL, which contain information of top-down vertical convective transfer.

At this stage, we have two output predictions, then the Merge block is utilized to combine them up. The Merge operation can be simple addition, here we use attention mechanism [1] to choose the final outputs:

$$Y_{\text{DtoT}} = [y_1, \ldots, y_L] \tag{4}$$

$$Y_{\text{TtoD}} = [y_1', \ldots, y_L'] \tag{5}$$

$$Y = f(Y_{\text{TtoD}}^T W_Q^T W_K Y_{\text{DtoT}}) Y_{\text{DtoT}}^T W_V^T \tag{6}$$

where $W_Q, W_K, W_V \in \mathbb{R}^{m \times m}$ are three weight matrices to be learned, and f is *softmax* function. The attention mechanism queries each reverse output layer for each down-top output layer, and integrates the information of the bidirectional chain into the final output with a weighted sum based on their relevance, which is more flexible than direct addition.

3.3 Transfer Scheme

Unlike the traditional developing process of parameterization, most of the extrapolation strategy for parameterization DNNs is training the model by data from one location (source dataset) before applying it to other locations (target datasets), or continue to fine-tuning using the data from target locations.

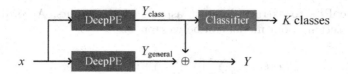

Fig. 4. Transfer Framework of DeepPE. $\oplus$ denotes element-wise sum.

According to the development paradigm of non-data-driven approach [21], simulations of all small-scale processes are built upon the modeling of basic atmospheric motion. In other words, our network should first summarize the general mapping across locations, and then extract the specific features of individual location. Therefore, we mix the data from all locations to train a general model, and then use the data for different tasks to fine-tune it. We have K classes, and classifier was added to distinguish task classes (Fig. 4), while a unique DeepPE and a general DeepPE are respectively used to fit process in different locations and extract common information among grids.

$$Y_{\text{class}} = \boldsymbol{DeepPE}(x, \theta_{\text{class}}) \tag{7}$$
$$Y_{\text{general}} = \boldsymbol{DeepPE}(x, \theta_{\text{general}}) \tag{8}$$
$$Y = Y_{\text{class}} + Y_{\text{general}} \tag{9}$$

Here, the function $\boldsymbol{DeepPE}(\cdot, \cdot)$ is a shorthand for Eq.(2)–(6) and θ represents all the parameters of DeepPE. The final outputs are concatenation of the outputs from unique space and shared space. The linear classifier can estimate what kinds of locations the data comes from:

$$\boldsymbol{Classifier}(Y_{\text{class}}, \theta_{\text{classifier}}) = softmax(b_{\text{classifier}} + W_{\text{classifier}}Y_{\text{class}}) \tag{10}$$

3.4   Training

Given a training set $\left\{x^{(i)}, \hat{Y}^{(i)}\right\}_{i=1}^{d}$, where d is the number of training samples (batch size) and $\hat{Y}^{(i)}$ is the ground truth observation of $x^{(i)}$, the optimization objective is defined as:

$$\arg\min_{\Theta}(\mathcal{L}(\Theta)) \tag{11}$$

We use MSE loss function to optimize DeepPE, while the parameters of the network are trained to minimise the cross-entropy of the predicted and true distributions on the task numbers. And inspired by [19], orthogonality constraints can penalize redundant latent representations and encourages the unique and general DeepPEs to extract different aspects of the inputs. Accordingly, the final loss function of our model can be written as:

$$
\begin{aligned}
\mathcal{L}(\Theta) = & \frac{1}{d} \sum_{i=1}^{d} \left\| \hat{Y}^{(i)} - Y^{(i)} \right\|_2^2 \\
& - \lambda \frac{1}{d} \sum_{i=1}^{d} \sum_{j=1}^{K} \hat{y}_{\text{class}}^j \log(y_{\text{class}}^j) \\
& + \gamma \frac{1}{d} \sum_{i=1}^{d} \left\| Y_{\text{class}}^T Y_{\text{general}} \right\|_F^2
\end{aligned}
\tag{12}
$$

where λ and γ are hyper-parameter. $\hat{y}_{\text{class}}^j$ and y_{class}^j are the ground-truth label as well as prediction probabilities of task number. $\|\cdot\|_2^2$ and $\|\cdot\|_F^2$ are the Euclidean norm and the squared Frobenius norm.

4   Experiments

This section describes the experiments performed to demonstrate the effectiveness of DeepPE when applied to one location data and data from multiple locations.

4.1   Datasets

The data[1] we used in this study is the PBL parameterization (YSU scheme, in which the vertical diffusion equation term includes the nonlocal mixing by convective eddies) dataset published by [28], which is a 22-year output from the regional climate model WRF version 3.3.1, driven by NCEP-R2 for the period 1984–2005. The input is 16 near surface characteristics, including 2m water vapor($Q2$) and air temperature($T2$), 10 m zonal and meridional wind ($U10$, $V10$), ground heat flux (GRDFLX), incoming shortwave radiation (SWDOWN), incoming longwave radiation (GLW), PBL height (PBLH), sensible heat flux (HFX), latent heat flux (LH), surface friction velocity (UST), ground temperature (TSK), soil temperature at 2m below the ground (TSLB), soil moisture at 0–0.3 cm below the ground (SMOIS), and a geostrophic wind component at 700 hPa (Ug, Vg). The results of WRF model simulations referred to as observations, which contain 17 vertical profiles of the following five model prognostic and diagnostic fields: temperature (tK), water vapor mixing ratio(QVAPOR), zonal and meridional wind (U, V), as well as vertical motions (W).

The 22-year data was partitioned into three parts: a training set consisting of 20 years of 3-hourly data to train the model; a validation set consisting of 1 year data used for tuning algorithm's hyper-parameters and to control over-fitting; and a test set consisting of 1-year records for prediction and evaluations. We use

---

[1] Retrieved from https://github.com/pbalapra/dl-pbl.

data from a site in the midwestern United States (Logan, Kansas; 38.8701°N, 100.9627°W) to test the performance of DeepPE, and use it and data from 15 sites nearby (within a $\sim 1100\,km \times 1100\,km$ area) for transfer experiments.

4.2 Experimental Setup

For the performance test of DeepPE, we apply a chronological 11-fold crossover experiment, that is, the first experiment uses 1984 data to verify, 1985 data to test, and 1986–2005 data to train; at the second run, 1985 data are used as verification set, 1986 data as test set and left data as training set; etc. For the transfer experiments, only the last fold setting is chosen. Our baselines are FFN, HPC, HAC (Sect. 2), all of them as well as DeepPE have 16 hidden units, and 2 full-connected layers in the FC Block.

The networks are trained using Adam optimizer on a 4-GPU machine and each GPU has 16/64 (single/transfer) clips in a mini-batch (so in total with a mini-batch size of 64/256 clips). We train models for 100 epochs in total, starting with a learning rate of 0.001 and reducing it to 10e–5 follow the cosine annealing strategy. For preprocessing, we applied StandardScaler (removes the mean and scales each variable to unit variance) and MinMaxScaler (scales each variable between 0 and 1) transformations before training, and we applied the inverse transformation after prediction so that the evaluation metrics are computed on the original scale. For the implementation, we use Pytorch (version 1.2.0) and our code is available at https://github.com/shiwch/DeepPE_Model.

We use mean absolute error (MAE), root mean squared error (RMSE), r2 score (r2) and Pearson correlation coefficient (PCC) to evaluate the performance of various algorithms, where MAE and RMSE range from 0 to positive infinity, the smaller the better; r2 and PCC range from negative infinity to 1 and negative 1 to positive 1, respectively, 0 indicates no skill and 1 is the perfect score.

5 Results

5.1 DeepPE Performance Analysis

Table 1 and 2 shows the average results of 11-fold crossover experiment and the first four lines indicates that DeepPE achieves the best performance in terms of all evaluation metrics on five predicted variables. Compared to the best performance gained by three baseline approaches, our model respectively shows $2.9\% \sim 9.2\%$, $0.8\% \sim 11.1\%$, $4.0\% \sim 19.2\%$, $4.1\% \sim 21.7\%$ and $0.8\% \sim 18.1\%$ improvement in terms of MAE, RMSE, PCC and r2 on U, V, W, tK and QVA-POR. Among them the smallest improvement occurs in the PCC of V and QVA-POR (both 0.8%), this is because the baseline methods show greate results on all predicted variables except vertical motions(W) in the PCC, of which our model improves 19.2%. It is also worth noting that the increase in vertical wind direction and wind speed of our model relative to HAC is more obvious than that of HAC relative to HPC. All these appeals show that DeepPE has a better ability to fit the vertical mixing.

Table 1. Comparison of predictive ability between four emulators in terms of U, V, W.

	U				V				W			
	MAE	RMSE	PCC	r2	MAE	RMSE	PCC	r2	MAE	RMSE	PCC	r2
FFN	3.638	4.563	0.000	−0.003	5.885	7.224	0.000	−0.004	3.003	5.245	0.000	−0.002
HPC	1.324	1.805	0.901	0.817	1.573	2.143	0.951	0.905	0.027	0.049	0.449	0.276
HAC	1.331	1.810	0.902	0.818	1.577	2.148	0.951	0.905	0.025	0.048	0.453	0.294
DeepPE	**1.208**	**1.676**	**0.916**	**0.842**	**1.402**	**1.960**	**0.959**	**0.920**	**0.024**	**0.046**	**0.540**	**0.351**
DeepPE-d	1.255	1.729	0.910	0.832	1.475	2.039	0.956	0.914	0.025	0.048	0.473	0.304
HAC-32	1.286	1.762	0.907	0.827	1.521	2.086	0.954	0.911	0.025	0.048	0.468	0.301
DeepPE-8	1.231	1.706	0.911	0.836	1.433	1.994	0.957	0.917	0.024	0.047	0.515	0.333

Table 2. Comparison of predictive ability between four emulators in terms of tK, QVAPOR.

	tK				QVAPOR			
	MAE	RMSE	PCC	r2	MAE ($\times 10^{-3}$)	RMSE ($\times 10^{-3}$)	PCC	r2
FFN	8.521	1.017	0.000	−0.007	3.109	3.641	0.000	−0.004
HPC	1.152	1.625	0.984	0.967	0.439	0.647	0.972	0.946
HAC	1.173	1.615	0.984	0.968	0.452	0.660	0.972	0.945
DeepPE	**0.922**	**1.327**	**0.988**	**0.977**	**0.372**	**0.559**	**0.980**	**0.961**
DeepPE-d	1.032	1.445	0.987	0.973	0.413	0.614	0.975	0.951
HAC-32	1.091	1.514	0.986	0.972	0.437	0.645	0.974	0.948
DeepPE-8	0.956	1.370	0.988	0.975	0.382	0.573	0.979	0.958

In addition, we evaluate the effectiveness of merge component of DeepPE with an ablation study. DeepPE-d represent only the merge block in DeepPE is replaced by a simple element-wise addition operation. It can be seen that its performance is between the DeepPE and HAC, which implies that the new reverse path is beneficial to the PBL variables fitting. However, compared to the layer-by-layer accumulation with adaptive weights based on attention mechanism, addition turns up to be rigid to merge bidirectional paths.

Table 3. Training details of emulators.

	Training time per fold ($s/100epoch$)	Test time (s)	Total parameters
FFN	3 103.18	0.008 3	4 307
HPC	3 147.28	0.013 7	6 533
HAC	3 228.60	0.009 0	17 493
DeepPE	4 580.97	0.011 8	41 827
HAC-32	3 223.96	0.009 2	35 925
DeepPE-8	4 590.12	0.011 5	24 563

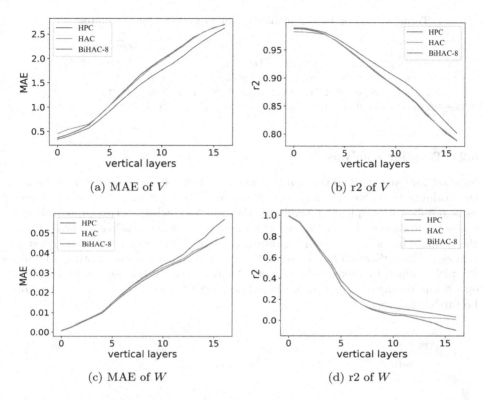

Fig. 5. Performance of HAC, HPC and BiHAC-8 in each vertical layer on vertical wind.

In order to verify that the advancement of DeepPE in modeling physical processes is attributed to the design of network structure rather than the increased number of neurons, we provide comparison models DeepPE-8 and HAC-32, having 8 as well as 32 units in each hidden layer respectively, and both have 2 full-connected layers in the FC Block which is the same as other experimental methods. Our network maintains performance even with fewer parameters (last two lines of Table 1, 2 and 3). Nonetheless, it has no advantage in training time, even with fewer parameters. Fortunately, the model is mainly doing interference rather than training in application, and the interference time of DeepPE does not increase much.

Figure 5 further shows how DeepPE benefits from the added reverse chain. We find that it has a prominent contribution to the high layers which are difficult to predict. In addition, the bidirectional propagating features also make its performance slides more smooth between two adjacent layer, rather like HPC and HAC (more obvious in a specific location experiment, please refer to Fig. 6 for details)

Table 4. Comparison of predictive ability between four emulators.

	U		V		W		tK		QVAPOR	
	MAE	r2	MAE	r2	MAE	r2	MAE	r2	MAE ($\times 10^{-4}$)	r2
DeepPE-tran	1.409	0.835	1.544	0.877	0.028	0.263	1.033	0.971	4.286	0.946
DeepPE-test	1.428	0.830	1.547	0.873	0.028	0.291	1.036	0.971	4.397	0.938
DeepPE-tune	1.291	0.842	1.386	0.884	0.027	0.310	0.827	0.978	3.555	0.957

5.2   Transfer Analysis

Similarly, we also verified the transfer method (Table 4). In order not to increase the training time, we use a subset of mixed data for training, which has the same number of samples as a single site dataset. Meanwhile, because our transfer scheme contains two DeepPEs, we halve the number of neurons in each layer, thus the training has a similar training duration (4395.87 s). We see that our transfer strategy (DeepPE-tran) has achieved better results than directly testing original DeepPE trained by mixed sub-dataset (DeepPE-test). When we put another 10 epoch fine-tuning (DeepPE-tune) based on each site data, the performance will be further improved.

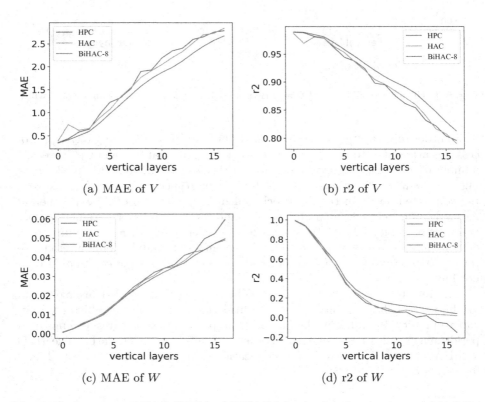

(a) MAE of V (b) r2 of V

(c) MAE of W (d) r2 of W

Fig. 6. Performance of HAC, HPC and BiHAC-8 in the fourth experiment of 11-fold crossover.

6 Conclusion

In this work, we introduce an approach to emulate existing physical parameterizations in atmospheric models through deep learning. A bidirectional network takes into account the domain-specific features: there are different local and nonlocal closure approximations in use in non-data-driven method for the small, locally generated turbulent eddies as well as cross-layer large eddies. This computationally efficient PBL parameterization emulator can quickly predict the variables with decent accuracy. Experiments show that our emulator is superior to others. In addition, we propose a transfer scheme to enable the emulator to have better generalization capabilities without increasing the calculation.

Acknowledgements. This research was supported by the National Key R&D Program of China 2018YFB0204303, and was also supported in part by the Natural Science Foundation of China under Grant No. U1811464, No.U1811463, in part by the Projects for Guangdong Introducing Innovative and Enterpreneurial Teams under Grant NO. 2016ZT06D211, in part by the Major Program of Guangdong Basic and Applied Research under Grant 2019B030302002, in part by the Guangdong Natural Science Foundation under Grant 2018B030312002. It was also sponsored by Zhejiang Lab (NO. 2021KC0AB04). We also appreciate the insightful comments and feedbacks from anonymous reviewers.

References

1. Bahdanau, D., Cho, K., Bengio, Y.: Neural machine translation by jointly learning to align and translate. arXiv preprint arXiv:1409.0473 (2014)
2. Bernstein, L., Bosch, P., Canziani, O., Chen, Z., Christ, R., Riahi, K.: Ipcc, 2007: climate change 2007: synthesis report (2008)
3. Chattopadhyay, A., Subel, A., Hassanzadeh, P.: Data-driven super-parameterization using deep learning: experimentation with multiscale lorenz 96 systems and transfer learning. J. Adv. Model. Earth Syst. **12**(11), e2020MS002084 (2020)
4. Chevallier, F., Chéruy, F., Scott, N., Chédin, A.: A neural network approach for a fast and accurate computation of a longwave radiative budget. J. Appl. Meteorol. **37**(11), 1385–1397 (1998)
5. Cohen, A.E., Cavallo, S.M., Coniglio, M.C., Brooks, H.E.: A review of planetary boundary layer parameterization schemes and their sensitivity in simulating southeastern us cold season severe weather environments. Weather Forecast. **30**(3), 591–612 (2015)
6. Crawshaw, M.: Multi-task learning with deep neural networks: a survey. arXiv preprint arXiv:2009.09796 (2020)
7. Deardorff, J.W.: Parameterization of the planetary boundary layer for use in general circulation models. Mon. Weather Rev. **100**(2), 93–106 (1972)
8. Gagne, D.J., Christensen, H.M., Subramanian, A.C., Monahan, A.H.: Machine learning for stochastic parameterization: generative adversarial networks in the lorenz'96 model. J. Adv. Model. Earth Syst. **12**(3), e2019MS001896 (2020)
9. Glorot, X., Bordes, A., Bengio, Y.: Deep sparse rectifier neural networks. In: Proceedings of the Fourteenth International Conference on Artificial Intelligence and Statistics, pp. 315–323 (2011)

10. Guo, Y., Li, Y., Wang, L., Rosing, T.: Adafilter: adaptive filter fine-tuning for deep transfer learning. In: Proceedings of the AAAI Conference on Artificial Intelligence, vol. 34, pp. 4060–4066 (2020)
11. Guo, Y., Shi, H., Kumar, A., Grauman, K., Rosing, T., Feris, R.: Spottune: transfer learning through adaptive fine-tuning. In: Proceedings of the IEEE Conference on Computer Vision and Pattern Recognition, pp. 4805–4814 (2019)
12. Hong, S.Y., Noh, Y., Dudhia, J.: A new vertical diffusion package with an explicit treatment of entrainment processes. Mon. Weather Rev. **134**(9), 2318–2341 (2006)
13. Huang, G., Liu, Z., Van Der Maaten, L., Weinberger, K.Q.: Densely connected convolutional networks. In: Proceedings of the IEEE Conference on Computer Vision and Pattern Recognition, pp. 4700–4708 (2017)
14. Kornblith, S., Shlens, J., Le, Q.V.: Do better imagenet models transfer better? In: Proceedings of the IEEE/CVF Conference on Computer Vision and Pattern Recognition (CVPR) (2019)
15. Krasnopolsky, V.M.: Neural network-based forward model for direct assimilation of SSM/I brightness temperatures. NASA (19980004608) (1997)
16. Krasnopolsky, V.M., Fox-Rabinovitz, M.S.: Complex hybrid models combining deterministic and machine learning components for numerical climate modeling and weather prediction. Neural Netw. **19**(2), 122–134 (2006)
17. Krasnopolsky, V.M., Fox-Rabinovitz, M.S., Chalikov, D.V.: New approach to calculation of atmospheric model physics: Accurate and fast neural network emulation of longwave radiation in a climate model. Mon. Weather Rev. **133**(5), 1370–1383 (2005)
18. Leeds, W., Wikle, C., Fiechter, J., Brown, J., Milliff, R.: Modeling 3-D spatio-temporal biogeochemical processes with a forest of 1-D statistical emulators. Environmetrics **24**(1), 1–12 (2013)
19. Liu, P., Qiu, X., Huang, X.: Adversarial multi-task learning for text classification. arXiv preprint arXiv:1704.05742 (2017)
20. Lorenz, E.N.: Predictability: a problem partly solved. In: Proceedings of Seminar on predictability, vol. 1 (1996)
21. McFarlane, N.: Parameterizations: representing key processes in climate models without resolving them. Wiley Interdisc. Rev. Clim. Change **2**(4), 482–497 (2011)
22. Parisi, G.I., Kemker, R., Part, J.L., Kanan, C., Wermter, S.: Continual lifelong learning with neural networks: a review. Neural Netw. **113**, 54–71 (2019)
23. Rasp, S., Pritchard, M.S., Gentine, P.: Deep learning to represent subgrid processes in climate models. Proc. Nat. Acad. Sci. **115**(39), 9684–9689 (2018)
24. Ruder, S.: An overview of multi-task learning in deep neural networks. arXiv preprint arXiv:1706.05098 (2017)
25. Tan, C., Sun, F., Kong, T., Zhang, W., Yang, C., Liu, C.: A survey on deep transfer learning. In: Kůrková, V., Manolopoulos, Y., Hammer, B., Iliadis, L., Maglogiannis, I. (eds.) ICANN 2018. LNCS, vol. 11141, pp. 270–279. Springer, Cham (2018). https://doi.org/10.1007/978-3-030-01424-7_27
26. Thornes, T., Düben, P., Palmer, T.: On the use of scale-dependent precision in earth system modelling. Q. J. R. Meteorol. Soc. **143**(703), 897–908 (2017)
27. Tobler, W.R.: A computer movie simulating urban growth in the detroit region. Econ. Geogr. **46**(sup1), 234–240 (1970)
28. Wang, J., Balaprakash, P., Kotamarthi, R.: Fast domain-aware neural network emulation of a planetary boundary layer parameterization in a numerical weather forecast model. Geoscientific Model Dev. (Online) **12**(10), 4261–4274 (2019)
29. Warner, T.T.: Numerical Weather and Climate Prediction. Cambridge University Press (2010)

30. Williams, P.D.: Modelling climate change: the role of unresolved processes. Philos. Trans. R. Soc. A Math. Phys. Eng. Sci. **363**(1837), 2931–2946 (2005)
31. Zhang, Y., Yang, Q.: An overview of multi-task learning. Nat. Sci. Rev. **5**(1), 30–43 (2018)

Effects of Boundary Conditions in Fully Convolutional Networks for Learning Spatio-Temporal Dynamics

Antonio Alguacil[1,2]([✉]) [iD], Wagner Gonçalves Pinto[2] [iD], Michael Bauerheim[1,2] [iD], Marc C. Jacob[3] [iD], and Stéphane Moreau[1] [iD]

[1] Department of Mechanical Engineering, University of Sherbrooke, 2500, boul. de l'Université, Sherbrooke, QC J1K 2R1, Canada
[2] Department of Aerodynamics, Energetics and Propulsion, ISAE-Supaero, 10 Avenue Edouard Belin, 31055 Toulouse, France
{antonio.alguacil-cabrerizo,michael.bauerheim}@isae-supaero.fr
[3] Université de Lyon, École Centrale de Lyon, INSA Lyon, Université Claude Bernard Lyon 1, CNRS, LMFA, 69134 Écully, France

Abstract. Accurate modeling of boundary conditions is crucial in computational physics. The ever increasing use of neural networks as surrogates for physics-related problems calls for an improved understanding of boundary condition treatment, and its influence on the network accuracy. In this paper, several strategies to impose boundary conditions (namely padding, improved spatial context, and explicit encoding of physical boundaries) are investigated in the context of fully convolutional networks applied to recurrent tasks. These strategies are evaluated on two spatio-temporal evolving problems modeled by partial differential equations: the 2D propagation of acoustic waves (hyperbolic PDE) and the heat equation (parabolic PDE). Results reveal a high sensitivity of both accuracy and stability on the boundary implementation in such recurrent tasks. It is then demonstrated that the choice of the optimal padding strategy is directly linked to the data semantics. Furthermore, the inclusion of additional input spatial context or explicit physics-based rules allows a better handling of boundaries in particular for large number of recurrences, resulting in more robust and stable neural networks, while facilitating the design and versatility of such types of networks. (Datasets, code and supplementary material are available at https://gitlab.isae-supaero.fr/a.alguacil/boundary_conditions_fcn_dyn).

Keywords: Boundary conditions · Fully convolutional neural network · Padding · Heat equation · Wave equation

Supported by the French "Programme d'Investissements d'avenir" ANR-17-EURE-0005 and the Natural Sciences and Engineering Research Council of Canada (NSERC). W.G.P. and M.B. are supportted by the French Direction Générale de l'Armement (DGA) through the AID POLA3 project.

Y. Dong et al. (Eds.): ECML PKDD 2021, LNAI 12979, pp. 102–117, 2021.
https://doi.org/10.1007/978-3-030-86517-7_7

1   Introduction

Recent advances in deep learning have shown an increased use of neural networks to create surrogate models for physics-related problems. In particular, Convolutional Neural Networks (CNN) have been employed in a wide variety of applications, leveraging their efficient parameter sharing property and their ability to capture long-range spatial correlations. However, most of the existing works limit themselves to one particular problem setup, keeping the same boundary conditions (BCs) throughout the entire training data [11], thus being unable to be generalized to other types of boundary conditions. Ideally, a flexible neural network framework should be able to work with several types of boundary conditions, without the need of retraining the network for each new problem setup. It is thus crucial to understand how boundary conditions are treated by data-driven CNNs in order to improve their generalization capabilities. The general theme of border effect for CNNs has been broadly studied in the image processing community [6,13]. Still today, such border effects can have a strong influence in state-of-the-art architectures employed in image segmentation [2]. The usual zero-padding strategy leads to border pixel artifacts and blind spots where the network accuracy drops. Solutions have been proposed to treat borders through separate filters [7] for the edges, corners and inner pixels or to consider the padded pixels as missing information, through the use of Partial Convolution strategies [12]. Other works demonstrate how CNNs implicitly learn spatial position [8,9], using the padded pixels to serve as anchors, i.e., as a reference for filter activation in border regions. The use of circular convolution [19] eliminates such border effects, but can only be employed on "panoramic" datasets.

Yet, the previous works have been devised for image segmentation/classification tasks and it is still unclear whether such studies can be directly transposed for regression and recurrent tasks, as usually encountered when modeling physics. In such contexts, a small error in the boundary prediction may lead to large errors elsewhere in the computational domain. A typical example of such a phenomenon is the development of non-reflecting boundary conditions in computational acoustics [17], which avoids the unphysical reflection of waves back into the computational domain. If left untreated, undesired reflections can pollute the calculated solution. To the author's knowledge, there is a lack of clear results regarding what is the optimal strategy for the treatment of such boundary conditions when employing CNNs for spatio-temporal evolving problems, for which BC errors can propagate and contaminate the whole computational domain. Indeed, only few works specifically focus on the boundary treatment problem. Some employ explicit rules in relatively simple cases, such as in periodic domains, as in the case of turbulence modeling [16]. For more complex boundary treatments, previous works are found in the context of Physical-Informed Neural Networks (PINN) [18] employed in combination with CNN [4], where hard constraints are imposed through the padding mechanism. However, only simple Dirichlet or Neumann conditions on static problems are considered, leaving dynamical conditions out of the study. In the case of spatio-temporal modeling, CNNs and Recurrent Neural Networks (RNN) architectures have been employed indistinctly. Mathieu et al. [15] designed a Multi-Scale CNN for video prediction, which was later employed in physics-related applications [1,11]. Fotiadis et al. [3] compared recurrent and convolutional approaches in physics-based applications and found that CNNs can be successfully employed as spatio-temporal predictors, with greater accuracy than RNNs and lesser training costs. In all previous cases, the treatment of boundary conditions was not explicitly studied, leaving the effect of BC treatments on such spatio-temporal evolving systems as an open question.

In this paper, the effects of several boundary condition treatments are characterized when modeling a space-time evolving problem using Convolutional Neural Networks. Three strategies are employed for handling the boundary conditions: (i) an implicit treatment through padding only, (ii) adding some extra spatial context to the network input and finally, (iii) explicitly encoding the boundary condition rules into the network output. These strategies are tested on a series of datasets with varying boundary conditions, modeling the two dimensional wave and heat equations.

The main contributions are the following:

(i) For problems with simple Neumann boundary conditions, padding allows CNNs to efficiently model the physics; yet, mimicking the actual semantics of the dataset makes it only possible for problems with fixed BCs;
(ii) The addition of an extra spatial context makes the neural network less sensitive to the choice of padding;
(iii) Explicit coding of neural networks gives the best accuracy in the more challenging test cases where dynamical effects are needed to model BCs, allowing more robust predictions by first enforcing physics.

2 Method

This section describes the methodology to predict the spatio-temporal dynamics of physics-related quantities using a convolutional neural network. The trained network follows a typical auto-regressive strategy [5] to produce time series of high-dimensional state vectors. The focus is put on the treatment of boundary conditions in the context of convolutional networks, in order to reproduce the desired physics. Several algorithms for the treatment of such BCs are presented and later evaluated in Sect. 4.

2.1 Learning an Auto-Regressive Model

Dynamical systems can be modeled through a discrete time-invariant model f acting on a delayed state vector $X^t = \{s^t, s^{t-1}, ..., s^{t-k-1}\}$ composed of k discrete temporal states s^i which may lie on a high dimensional space. Formally, the discretized time-dynamics read:

$$Y^{t+1} = f(X^t),\tag{1}$$

where $Y^{t+1} = \{s^{t+1}\}$ corresponds to the next state in the time series.

In order to generate an approximate model for f, a neural network $\hat{f}_\theta$, parametrized by its weights and biases θ, is trained on a dataset composed of input-target tuples $\{X^t, Y^{t+1}\}_i$ through a supervised optimisation problem, based on an error metric $\mathcal{L}$, such that

$$\hat{f}_\theta = \arg \min_{f_\theta} \sum_i \left\{ \mathcal{L} \left[f_\theta(X^t), Y^{t+1} \right]_i \right\}. \tag{2}$$

Once an approximate solution is obtained, any time state s^T can be reached by employing an auto-regressive iterative strategy on the learned model, namely:

$$Y^T = \underbrace{\hat{f}_\theta \circ \hat{f}_\theta \circ ... \circ \hat{f}_\theta}_{T \text{ times}}(X^0) \tag{3}$$

where $\circ$ is the function composition operator.

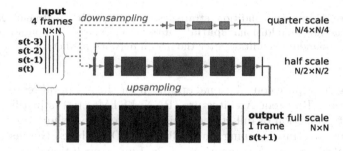

Fig. 1. Multi-scale fully convolutional neural network with 4 consecutive input states of size $N \times N$ and 3 resolution banks ($N/4$, $N/2$ and N). Grey arrows represent 2D convolutions and width of boxes the number of features. (Color figure online)

2.2   Neural Network Convolutional Architecture

The auto-regressive strategy can be employed to create surrogate models for physics-based quantities. In traditional fluid solvers, it is common to discretize both the time and space dimensions of physical quantities, such as pressure or velocity fields. This results in high dimensional state where the modeled equations are solved for each degree of freedom. To reduce the computational costs associated with training a neural network surrogate on such high-dimensional states, a convolutional neural network is employed due to its weight-sharing capabilities. In order to efficiently treat the intrinsic multi-scale features of fluid flows, a Multi-Scale fully convolutional neural network [11,15] is employed, as shown in Fig. 1. The input state is composed of four consecutive vectors $X^t = \{s^{t-3}, s^{t-2}, s^{t-1}, s^t\}$, in order to provide the network with additional temporal context.

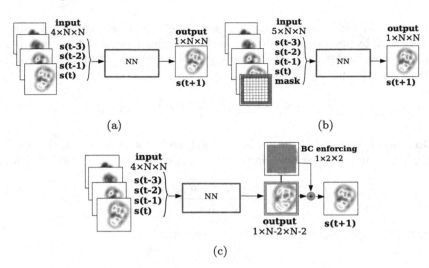

Fig. 2. Three boundary condition strategies: (a) **implicit** treatment using only padding, (b) adding an additional **spatial context** to the network input or (c) **explicitly** encoding boundary condition after the neural network prediction.

The usage of a pure convolutional approach for the temporal regression problem instead of a Recurrent Architecture (RNN, LSTMs etc.) is justified because of the direct prediction of full states, which only require a short temporal span for their accurate time-stepping (in traditional PDE solvers, the discretization of the time-derivatives). A comparison of the fully convolutional approach with LSTM approaches performed in [3] confirms this observation, as the convolutional architecture achieves better accuracies on such types of problems.

2.3 Boundary Condition Treatment

In traditional solvers, the proper modeling of boundary conditions is key to accurate numerical resolution of the partial differential equations. Therefore, understanding the boundary condition treatment is fundamental if convolution neural networks are to be employed as surrogates for physics-based models. Boundary conditions are intrinsically linked to the concept of padding in fully convolutional networks: additional information must be created at the borders before each convolution in order to keep the same image input resolution at the output. However, the value of the additional pixel information is not known *a priori*, and several padding strategies are available to encode this information. **Zero padding**, where the additional information is filled with zero values; **replication padding**, where the values at border pixels are replicated multiple times into the padding area; **reflection padding**, where an axial symmetry is performed along boundary edges; and **circular padding** (or **periodic**) which wraps values from the opposite boundary in the same spatial direction.

While padding is the most straightforward strategy to impose BCs in CNNs, it is unclear how the padding type affects the predictions, neither how an optimal choice is connected to the physical BC type (Dirichlet, Neumann, etc.). Moreover, padding is fixed for the entire database considered, which lacks of versatility when targeting physical systems with multiple possible BCs. Consequently, the objective of this work is to study if an optimal boundary treatment strategy can be found when employing CNNs for physics-based regression. Three types of strategies are considered:

Implicit: It consists in applying exclusively padding to the input and successive feature maps. This is the most common approach in convolutional networks and forces the network to implicitly learn the boundary physics. It is up to the network designer to chose an adequate padding strategy which usually results in a long trial-and-error process until finding an optimal solution. It constitutes the baseline method of this work (Fig. 2a) and the four padding strategies presented above are investigated (**zeros**, **replication**, **reflection** and **circular**).

Spatial Context: The second strategy consists in concatenating an additional channel to the network input, which is consisting in a Boolean mask indicating the position of border pixels. This is formalized as follows:

$$\mathbb{I}(x) = \begin{cases} 1, & \text{if } x \in \partial\mathcal{D} \\ 0, & \text{otherwise} \end{cases} \tag{4}$$

where $\mathcal{D}$ represents the domain of interest, $\partial\mathcal{D}$ its boundaries and x are the spatial coordinate. The motivation for such a strategy is to provide the network with an increased spatial context. Figure 2b depicts such a strategy. Note that padding is still employed in order to maintain the size of feature maps. This strategy is inspired by other works such as Liu et al. [14], where giving explicit spatial information to CNNs is shown to crucially improve their generalization capability. Yet, the correlation between the spatial extended context, the padding type, and the actual physical BC is still unclear, thus being studied here.

Explicit Encoding: Finally, the third strategy (Fig. 2c) consists in explicitly encoding the boundary condition. In practice, the boundary pixel values are imposed after the network output [4], before stepping into the optimization step during the training phase. The way the value is imposed depends on the mathematical modeling of the boundary (Dirichlet, Neumann condition, etc.).

All three strategies are compared in this work, by combining them with the four types of padding previously mentioned.

2.4   Loss Function

The loss function for training the aforementioned Multi-Scale network is defined as:

$$\mathcal{L} = \frac{1}{N} \sum_{k=1}^{N} \{\lambda_{L2}\mathcal{L}_2 + \lambda_{GDL}\mathcal{L}_{GDL}\} \tag{5}$$

where $\mathcal{L}_2 = ||Y^{t+1} - \hat{Y}^{t+1}||_2^2$ and $\mathcal{L}_{GDL} = ||\partial_x Y^{t+1} - \partial_x \hat{Y}^{t+1}||^2 + ||\partial_y Y^{t+1} - \partial_y \hat{Y}^{t+1}||^2$, where a classical mean square metric is employed, both for the state vector and its spatial derivatives, denoted as Gradient Difference Loss (GDL) as in [15]. This loss drives the optimization towards achieving sharper predictions, and compensates for the smoothing of the predicted signal in the long term prediction of spatio-temporal series.

Note that the training focuses only on the next time-step prediction. Therefore, the auto-regressive prediction of a long time series of state vectors is a generalization problem.

3   Applications: Time-Evolving PDEs

The studied modeled is applied to create data-driven surrogates of spatio-temporal evolving partial differential equations. In practice, two applications are investigated: a hyperbolic PDE (acoustic wave propagation) and parabolic PDE (heat equation). The emphasis is put on the influence of the boundary conditions on the ability of surrogate data-driven models to reproduce accurately the underlying dynamics. Thus, several types of boundary conditions are studied for each case, which are detailed next.

3.1   Acoustic Propagation of Gaussian Pulses

The first application corresponds to the surrogate modelling of a 2D acoustic wave equation in a quiescent medium with speed of sound c_0, written in terms of the acoustic density $\rho = \rho(x, y, t)$, with p Gaussian density pulses as initial conditions:

$$\frac{\partial^2 \rho}{\partial^2 t} + c_0 \nabla^2 \rho = 0 \qquad (6a)$$

$$\rho(x, y, t = 0) = \sum_i^p \varepsilon^i \exp\left\{ -\frac{\log 2}{(b^i)^2} \left[(x - x_0^i)^2 + (y - y_0^i)^2 \right] \right\} \qquad (6b)$$

where ∇^2 is the Laplacian operator, (x_0^i, y_0^i), ε^i and b^i represent respectively the spatial positions of the center, the amplitude and the half-width of the ith initial pulse.

Boundary Conditions: Three cases of boundary conditions are considered, representative of typical configurations found in acoustics. Each one of the BC constitutes a dataset to be employed for training a surrogate model:

– *Reflecting walls (Dataset 1 - D1)*: Hard-reflecting walls, representing interior acoustics, which is modeled with a Neumann boundary condition: $\nabla \rho \cdot \boldsymbol{n} = 0$.
– *Periodic walls (Dataset 2 - D2)*: Periodic conditions to model infinitely repeating domains.
– *Absorbing walls (Dataset 3 - D3)*: Radiation boundary conditions, modeling propagation of waves into the far-field (exterior acoustics). The challenge is to avoid spurious reflections that can pollute the computational domain.

3.2   Diffusion of Temperature Spots

Second, the diffusion of temperature spots is studied, modeled by the following heat equation on the temperature $T = T(x, y, t)$ with p Gaussian density pulses as initial conditions:

$$\frac{\partial T}{\partial t} + \alpha \nabla^2 T = 0 \qquad (7)$$

where α denotes the thermal diffusivity of the medium. The intial conditions are identical as those employed in Eq. (6b) (Gaussian temperature spots).

Boundary Conditions: Here, an additional dataset is generated, called Adiabatic walls (Dataset 4 - D4): Zero-flux adiabatic walls, modeled as Neumann boundary conditions $\nabla T \cdot \boldsymbol{n} = 0$.

3.3   Datasets Generation and Parameters

The datasets of input-target fields is generated offline with the multi-physics open-source Palabos Lattice-Boltzmann Method (LBM) [10] numerical solver. We solve equations (6) and (7) for a duration T with a time-step of Δt. A two-dimensional square domain is considered.

Acoustic Datasets (D1 to D3): Each set is composed of 600 LBM simulations, each with $T = 231$ discrete time snapshots. Only the acoustic density fields ρ are recorded in square domains of physical length $L \times L$, discretized with $N = 200$ cells per spatial direction and a spatial step of $\Delta x = L/N = 0.5$. The LB time step is set to $\Delta t_{LBM} = 0.0029 D/c_0$, with $c_0 = 1/\sqrt{3}\Delta_x/\Delta t_{LBM}$ is the speed of sound. The four input density fields fed into the Neural network are equally spaced in time with $\Delta t_{NN} = \Delta t_{LBM}$. A random number of Gaussian pulses in the range $p \in [1, 5]$ are used as initial conditions, with fixed amplitude and half-width, $\varepsilon = 0.001$ and $b/\Delta x = 12$. The initial location of the pulses (x_0, y_0) is also randomly sampled inside the domain following an uniform distribution. A 500/100 training/validation random split is employed for the simulations.

Temperature Datasets (D4) : Each set is composed of 550 LBM simulations, each with $T = 160$ discrete time snapshots. The temperature fields T are recorded in square domains of physical length $L \times L$, discretized with $N = 200$ cells per spatial direction and a spatial step of $\Delta x = L/N = 0.005$. The heat diffusivity is set to $\alpha = 8\Delta_x^2/\Delta t_{LBM}$, where the LB time step is set to $\Delta t_{LBM} = 1$. The four input temperature fields fed into the Neural network are sampled so that $\Delta t_{NN} = 4\Delta t_{LBM}$. The same initial conditions as in D1-D3 are employed. A 400/150 training/validation random split is employed for the simulations.

4   Results

To perform the evaluation of the presented boundary condition treatment, a Multi-Scale network with 0.4 million parameters is trained on the four proposed

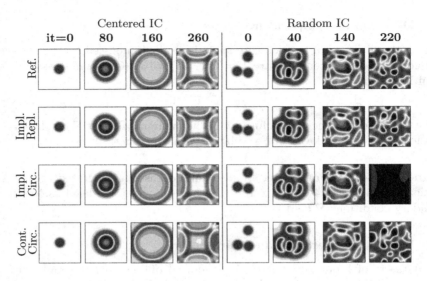

Fig. 3. Snapshots of propagating density waves for CNN trained on Dataset 1, for two different initial conditions ($it = 0$): centered Gaussian pulse (4 left columns) and randomly sampled Gaussian pulses (4 right columns). Different boundary condition treatments are compared to the LBM reference (top row). For the implicit strategy, the best (resp. worse) results regarding the employed padding are shown in the second row (replication) and third row (circular). For the spatial context strategy, results with circular padding are shown in the last row.

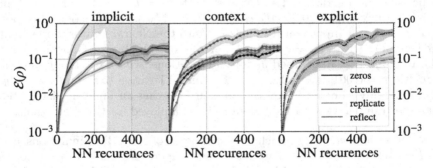

Fig. 4. Relative error evolution for case D1 (reflecting walls). 3 boundary treatments are employed: implicit (full lines), spatial context (lines and star marker), and explicit BC treatment (dashed-dotted lines). Results are averaged over 25 different initial conditions and shaded area represent the standard deviation. (Color figure online)

datasets. As discussed in Sect. 2.4 the network is trained to minimize Eq. 5 for a single-step prediction. The Adam optimizer is employed, with a learning rate initially set to 10^{-4}, and decaying by 20% each time the loss reaches a plateau. The loss weights are set to $\lambda_{L2} = 0.02$ and $\lambda_{GDL} = 0.98$. Data augmentation is employed through the random rotation of input-target tuples, and input fields are normalized by their mean and standard deviation. Batch size is kept at 32. Trainings are performed on a NVIDIA V100 GPU and convergence is achieved at about 1000 epochs for each run.

Wave Equation with Reflecting Boundary: The first case of study corresponds to Dataset 1. Density waves are fully reflected back into the domain after the interaction with boundaries. Therefore, the waves stay in the computational domain for infinitely long times, as no viscous dissipation is present. To test the presented approaches, 24 initial conditions with 1 to 5 Gaussian pulses randomly located in the initial domain are used as inputs for the auto-regressive model. A 25th initial condition is also tested, with the particular case of a Gaussian pulse initially centered in the domain, whose solution leads to strong symmetric solutions and is thus challenging for the neural network. For each initial condition (generated with the LBM solver), the auto-regressive strategy can recursively predict the density fields over a time horizon of T iterations, by using the previous prediction as a new input. In order to improve the neural network robustness versus the error accumulation over time, an *a posteriori* correction is employed to improve the predictions after each time step [1], based on the conservation of acoustic energy over time in this particular application.

Table 1. Averaged relative error for Dataset 1 (hard reflecting walls) at iteration $it = 600$ for 25 random initial conditions. Bolded results represent best padding for each strategy.

METHOD	PADDING											
	ZEROS			CIRCULAR			REPLICATE			REFLECT		
	MIN	MAX	AVG	MIN	MAX	AVG	MIN	MAX	AVG	MIN	MAX	AVG
IMPLICIT	0.118	0.271	0.189	0.306	∞	∞	0.098	0.145	**0.119**	0.133	0.319	0.234
CONTEXT	0.114	0.213	**0.176**	0.134	0.453	0.670	0.161	0.235	0.204	0.138	0.297	0.226
EXPLICIT	0.768	0.580	1.392	0.291	0.780	0.545	0.079	0.341	0.148	0.064	0.157	**0.094**

The error is evaluated in terms of relative root mean square error at each neural network iteration, namely $\mathcal{E}(\rho) = \sqrt{||\rho_t - \hat{\rho}_t||_2}/\sqrt{||\rho_t||_2}$, where $\hat{\rho}_t$ is the high-dimensional density prediction at iteration t. Here, the time horizon is set at $T = 600$ iterations. For the explicit method, Neumann boundary conditions on the density fields are used to model such conditions. A first-order finite difference discretization is employed.

Results are qualitatively evaluated in Fig. 3. For two different initial conditions (centered pulse and 3 randomly sampled pulses), the LBM reference (top row) is compared to several of the proposed approaches. For the implicit case (i.e., only padding) strategies, the best model regarding the employed padding is shown in the second row, corresponding to the replication padding. For both initial conditions, the auto-regressive strategy follows closely the ground truth data. In the third row, the implicit strategy employs circular padding and shows a good agreement in the case of the initially centered Gaussian pulse. However, for the other initial condition, the prediction diverges after some iterations. At iteration $it = 80$, the pulse arriving on the left wall is re-injected at the right wall, mimicking the behavior of periodic boundaries instead of the reflecting ones, on which the network has been trained.

Figure 4 shows the time-evolution of the error averaged over 25 initial conditions for the different evaluated methods and Table 1 presents the error values for the last prediction at $it = 600$. For the implicit strategy, results show that choosing the optimal padding crucially depends on the data physics. With circular padding the network is incapable of reproducing the desired physics except for some particular initial conditions, illustrated by the increased variance area signaling the presence of outliers. This is due to the artifacts discussed previously. For the rest of available padding (zeros, replication and reflection), the replication padding solution performs better than the other two even if error levels remain acceptable (below 2% relative error).

Such observations agree with other studies performed in image segmentation [2]: circular padding limits the CNNs ability to encode position information and can only be used with spatially periodic data. To further investigate this claim, an additional spatial context channel is employed, while maintaining circular padding. As observed in Fig. 4 (middle plot, red curve), the error is significantly lower than the one obtained with the implicit strategy. This suggests that the additional input serves as an spatial anchor for the CNN to encode the hard-reflecting wall, which was not possible using only circular padding. The last row in Fig. 3 demonstrates this improvement, even though the prediction error is higher than the one obtained with other padding methods. Furthermore, the addition of the spatial context channel reduces the overall variability of the chosen padding effects. While the error slightly increases for the replication case versus the implicit strategy, all three padding methods converge to similar error evolutions. This suggests that the additional context channel may force the network to explicitly learn similar convolutional kernels for boundary treatment, while this is not guaranteed by the implicit case.

For the explicit case, results in Fig. 4 (right) show that the replicate and reflection padding cases achieve the lowest errors, while zero and circular paddings have larger errors. Note that the network is only trained for a one-step prediction and the explicit enforcing of the boundary is performed after each prediction. Thus, the explicit enforcing is only processed by the network in the auto-regressive context. Results show whether the employed padding strategy is compatible with the enforced boundary values: reflecting and replication padding can be though as first-order finite difference approximations of spatial derivatives, for 1-pixel padding. Zero and circular paddings are on the other hand not compatible with the enforced boundary values, performing worse in both cases.

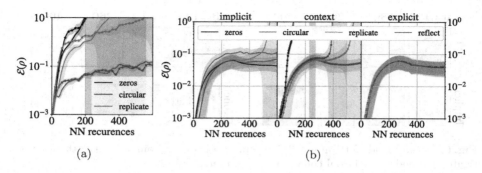

Fig. 5. Relative errors for (a) D2 (periodic) and (b) D3 (absorbing) cases. Full lines represent the implicit strategy and lines and star markers the spatial context strategy.

This behavior highlights the possible benefits of employing explicitly boundary rules as long as the subsequent padding follows the same logic. It also calls to directly enforce such explicit rules in the input padding mechanism, which is left for future work.

Wave Equation with Periodic BCs: The second case of study corresponds to Dataset 2, where all four wall boundaries are set as periodic walls in the training data. The objective is that the neural network reproduces the wave propagation in an infinitely-repeating domain.

The relative error evolution over time is depicted in Fig. 5a, for the implicit and spatial context methods. The explicit method is not employed here as it is equivalent to the implicit one: physical solvers employ additional "ghost cells" to wrap values from one boundary to another [16]. Also, reflect padding is not shown as it behaves very similarly to the replicate strategy. Results show that for both implicit and spatial context strategies, only circular padding is able to achieve acceptable error levels, with an average relative error of 15% for the implicit case and 11% at $it = 600$ for the spatial context. The use of a padding strategy other than circular produces unphysical behavior at the boundaries, as the network is incapable of copying by itself the values arriving at one border to the opposite one.

Wave Equation with Absorbing BCs: The third test case corresponds to the neural network trained on dataset 3, with non-reflecting (absorbing) boundary conditions. This case is significantly more challenging than the two previous cases: the initial Gaussian pulse is expected to propagate into the far field and completely leave the computational domain. Thus, the underlying data distribution is changing over time: the initial acoustic energy $||\rho||_2$ tends towards 0 as $t \to \infty$. The challenge for the network is to correctly propagate the initial pulses outside the domain without spurious reflections at boundaries. As the signal energy tends towards zero, the error is now calculated relatively to the *initial*

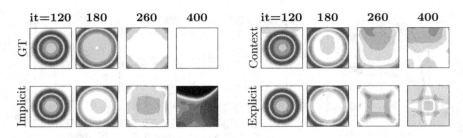

Fig. 6. Density fields for Dataset 3 and replicate padding, comparing the three investigated methods. The initial conditions is a centered pulse.

density, i.e., $\mathcal{E}(\rho) = \sqrt{||\rho_t - \hat{\rho}_t||_2}/|\rho_{t=0}|$. Similarly to previous experiments, 25 different initial conditions are employed for the auto-regressive tests.

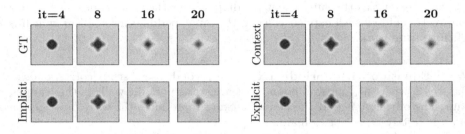

Fig. 7. Temperature fields for Dataset 4 and replicate padding, comparing the three investigated methods. The initial condition is a centered pulse.

For the explicit method, a local one-dimensional (LODI) non-reflecting equation is used to impose the values a the boundaries, which reads $\partial_t\rho + c_0 \boldsymbol{\nabla}\rho\cdot\boldsymbol{n} = 0$ [17]. First order finite differences are employed to discretize both the spatial and temporal derivatives.

The evolution of the error for 600 auto-regressive iterations is shown in Fig. 5b. For both the implicit and spatial context cases, the results show a high variability between the different cases. While the implicit strategy with zero and reflect padding manage to produce low-error results, the other two padding strategies lead to diverging simulations. In contrast, when the spatial context is employed, the circular padding performs better, while the other three methods diverge. This unstable network behavior shows the complexity of the non-reflecting case in comparison to the previous ones. Instabilities can be directly related to the appearance of artifacts when the pulse impinges the walls, as can be seen in Fig. 6: before the pulse arrival at the wall ($it = 120$), all methods show stable and accurate predictions, while larger errors are shown after the first interaction with the BCs ($it > 160$). Such artifacts lead in some cases to the unbounded growth of the density amplitude, leading to instabilities.

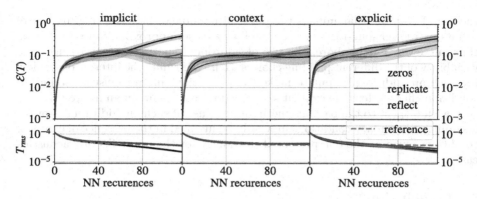

Fig. 8. Results for case D4. (Top) Relative MSE on temperature fields, (bottom) temporal evolution of the spatially averaged temperature T_{rms}.

Interestingly, the explicit encoding of boundaries has an important stabilizing effect: all four padding methods show a very similar error evolution. Figures 6 show the main differences between implicit, spatial context and explicit approaches for the same padding (replicate): the first two cases initially damp the outgoing waves more efficiently. However, some unphysical reflections (shown by the asymmetry of density the fields) lead to artifacts remaining in the computational domain, eventually leading to instabilities in the implicit case. In the explicit case, even though reflections also exist, they follow a symmetric pattern, which corresponds very closely to the one found for reflecting walls, as seen in Fig. 3. This suggests that the explicit encoding adds physical consistency to the network BC treatment. The error in that approach arises mainly from the lack of proper BC modeling, as the aforementioned LODI hypothesis cannot handle efficiently two-dimensional effects, typical of Gaussian pulses, for example at corners. Future research should reach out towards the treatment of such transverse effects [17]. This demonstrates that blending prior physics knowledge when handling boundary conditions can improve CNNs accuracy and robustness.

Heat Equation with Adiabatic Boundary: The last studied application corresponds to the diffusion of temperature pulses, modeled by the heat Eq. (7), when employing Neumann boundary conditions in the temperature (adiabatic walls). The same testing strategy is employed, by simulating 25 initial conditions for $T = 120$ neural network recurrences. Two metrics are employed to analyze the network, the relative mean square error $\mathcal{E}(T) = \sqrt{||T_t - \hat{T}_t||_2}/\sqrt{||T_t||_2}$ and the evolution of the average spatial temperature T_{rms} that tends to a constant value in time, as no heat flux exists at the adiabatic walls.

Results in Figs. 7 and 8 confirm some of the previously made observations for Neumann BCs (reflecting acoustic case): the implicit padding strategy seems appropriate to correctly reproduce the dynamics, when those remain simple, such as in the diffusion of temperature fields. However, the error evolution shows that the use of padding that mimics the physical boundary condition (e.g. replicate) results in a lower long-term error, as well as a closer approximation to the T_{rms} constant level. Furthermore, the use of additional spatial context reduces the variability of the different padding choices, thus confirming the interest of employing such an additional input to make the predictions more robust to boundary condition effects. Here the explicit strategy does not results in an improved accuracy with respect to the baseline methods. The use of such an additional a-priori encoding of boundary conditions may only be justified in the presence of complex conditions, such as the one presented previously for the non-reflecting acoustics case.

5   Conclusion

This paper presents an exhaustive comparison between several available methods to treat boundary conditions in fully convolutional neural networks for spatio-temporal regression, in the context of hyperbolic and parabolic PDEs. Such temporal regression tasks are highly sensitive to the well-posedness of boundary conditions, as small localized errors can propagate in time and space, producing instabilities in some cases. The characterization of such boundaries is crucial to improve the neural network accuracy.

The main outcomes are summarized next: employing padding alone yields accurate results only when the chosen padding is compatible with the underlying data. The addition of a spatial context channel seems to increase the robustness of the network in simple cases (Neumann boundaries), but fails for the more complex non-reflecting boundary case. Finally, the explicit encoding of boundaries, which enforces some physics constraints on border pixels, clearly demonstrates its superiority in such cases, allowing to design more robust neural networks. Such an approach should be further investigated in order to understand its coupling with the neural network behavior, and its extension to problems with several types of boundary conditions.

References

1. Alguacil, A., Bauerheim, M., Jacob, M.C., Moreau, S.: Predicting the propagation of acoustic waves using deep convolutional neural networks. In: AIAA Aviation Forum, Reston, VA, p. 2513 (2020)
2. Alsallakh, B., Kokhlikyan, N., Miglani, V., Yuan, J., Reblitz-Richardson, O.: Mind the pad - CNNs can develop blind spots. In: 9th International Conference on Learning Representations (ICLR), Vienna, Austria (2021)
3. Fotiadis, S., Pignatelli, E., Bharath, A.A., Lino Valencia, M., Cantwell, C.D., Storkey, A.: Comparing recurrent and convolutional neural networks for predicting wave propagation. In: ICLR 2020 Workshop on Deep Learning and Differential Equations (2020)
4. Gao, H., Sun, L., Wang, J.X.: PhyGeoNet: physics-informed geometry-adaptive convolutional neural networks for solving parametric PDEs on irregular domain. J. Comput. Phys. **428**, 110079 (2021)

5. Geneva, N., Zabaras, N.: Modeling the dynamics of PDE systems with physics-constrained deep auto-regressive networks. J. Comput. Phys. **403**, 109056 (2019)
6. Hamey, L.G.: A functional approach to border handling in image processing. In: International Conference on Digital Image Computing: Techniques and Applications, (DICTA), Adelaide, Australia (2015)
7. Innamorati, C., Ritschel, T., Weyrich, T., Mitra, N.J.: Learning on the edge: investigating boundary filters in CNNs. Int. J. Comput. Vis. **128**(4), 773–782 (2019). https://doi.org/10.1007/s11263-019-01223-y
8. Islam, A., Jia, S., Bruce, N.D.B.: How much position information do convolutional neural networks encode? In: 8th International Conference on Learning Representations (ICLR), Addis Ababa, Ethiopia (2020)
9. Kayhan, O.S., Van Gemert, J.C.: On translation invariance in CNNs: convolutional layers can exploit absolute spatial location. In: Proceedings of the IEEE/CVF Conference on Computer Vision and Pattern Recognition (CVPR), pp. 14274–14285. Virtual Event (2020)
10. Latt, J., et al.: Palabos: parallel lattice boltzmann solver. Comput. Math. Appl. **81**, 334–350 (2020)
11. Lee, S., You, D.: Data-driven prediction of unsteady flow over a circular cylinder using deep learning. J. Fluid Mech. **879**(1), 217–254 (2019)
12. Liu, G., et al.: Partial convolution based padding. arXiv preprint arXiv:1811.11718 (2018)
13. Liu, R., Jia, J.: Reducing boundary artifacts in image deconvolution. In: 15th IEEE International Conference on Image Processing, San Diego, CA, pp. 505–508 (2008)
14. Liu, R., et al.: An intriguing failing of convolutional neural networks and the Coord-Conv solution. In: Advances in Neural Information Processing Systems, Montréal, Canada, vol. 31, pp. 9605–9616 (2018)
15. Mathieu, M., Couprie, C., LeCun, Y.: Deep multi-scale video prediction beyond mean square error. In: 4th International Conference on Learning Representtions, ICLR (2016)
16. Mohan, A.T., Lubbers, N., Livescu, D., Chertkov, M.: Embedding hard physical constraints in neural network coarse-graining of 3D turbulence. In: ICLR 2020 Workshop Tackling Climate Change with Machine Learning. arXiv (2020)
17. Poinsot, T.J., Lele, S.K.: Boundary conditions for direct simulations of compressible viscous flows. J. Comput. Phys. **101**(1), 104–129 (1992)
18. Raissi, M., Perdikaris, P., Karniadakis, G.E.: Physics-informed neural networks: a deep learning framework for solving forward and inverse problems involving nonlinear partial differential equations. J. Comput. Phys. **378**, 686–707 (2019)
19. Schubert, S., Neubert, P., Poschmann, J., Pretzel, P.: Circular convolutional neural networks for panoramic images and laser data. In: 2019 IEEE Intelligent Vehicles Symposium (IV), Paris, France, pp. 653–660 (2019)

Physics Knowledge Discovery via Neural Differential Equation Embedding

Yexiang Xue[1(✉)], Md Nasim[1], Maosen Zhang[1], Cuncai Fan[2],
Xinghang Zhang[3], and Anter El-Azab[3]

[1] Department of Computer Science, Purdue University, West Lafayette, IN, USA
{yexiang,mnasim,maosen}@purdue.edu
[2] The University of Hong Kong, Hong Kong SAR, China
cuncai@hku.hk
[3] School of Materials Engineering, Purdue University, West Lafayette, IN, USA
{xzhang98,aelazab}@purdue.edu

Abstract. Despite much interest, physics knowledge discovery from experiment data remains largely a manual trial-and-error process. This paper proposes neural differential equation embedding (`NeuraDiff`), an end-to-end approach to learn a physics model characterized by a set of partial differential equations directly from experiment data. The key idea is the integration of two neural networks – one recognition net extracting the values of physics model variables from experimental data, and the other neural differential equation net simulating the temporal evolution of the physics model. Learning is completed by matching the outcomes of the two neural networks. We apply `NeuraDiff` to the real-world application of tracking and learning the physics model of nano-scale defects in crystalline materials under irradiation and high temperature. Experimental results demonstrate that `NeuraDiff` produces highly accurate tracking results while capturing the correct dynamics of nano-scale defects.

Keywords: Physics knowledge discovery · Neural differential equation embedding · Nano-scale materials science

1 Introduction

The advancement and application of machine learning in the last decade has been crucial in many domains. In spite of its wide outreach, the potential to leverage machine learning for scientific discovery in a closed loop has not been fully realized. Real-world experimentation and physics-based simulation provide a *forward* approach to *validate* a given physics model. The accuracy of a hypothetical model can be verified by testing if the simulated results match actual experiments. Nonetheless, the more important *backward* learning task, namely, knowledge discovery and refinement of physics models from experimental data, remains largely a manual trial-and-error process relying on the intuitions and inspirations from the physicists (upper panel Fig. 1). Recently, a series of research

© Springer Nature Switzerland AG 2021
Y. Dong et al. (Eds.): ECML PKDD 2021, LNAI 12979, pp. 118–134, 2021.
https://doi.org/10.1007/978-3-030-86517-7_8

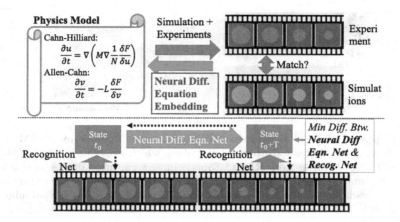

Fig. 1. (Upper) Physics experiments and simulation provides a forward approach to validate a physics model. Our Neural Differential Equation Embedding (`NeuraDiff`) is a backward approach to learn a physics model directly from experimental data. **(Lower)** The high-level idea of `NeuraDiff`. A recognition network extracts model parameters at time t_0, which are fed to a neural differential equation net to simulate evolution for T steps, and are compared with the recognized results at $t_0 + T$. Back-propagation is utilized to match the output of the recognition and the neural differential equation net.

[9,16,29,36,39,40] aim at learning partial differential equations from data. However, they did not achieve fully automatic physics model identification from experiment data because the input of these models are the trajectories of differential equations, which may be unavailable from experiment data and need to be extracted as a separate step. Unfortunately this is the case in the application domain considered in this paper.

We develop neural differential equation embedding (`NeuraDiff`), an end-to-end approach to learn a physics model characterized by a set of partial differential equations directly from experiment data. The key idea is the integration of two neural networks, one neural differential equation net simulating the temporal dynamics, and the second recognition net extracting the values of physics model variables from experimental data. The high level idea is shown in the lower panel of Fig. 1. Here, the recognition net extracts physics model variables at time t_0 and feed it to the differential equation net to simulate the temporal evolution for T steps. Then, the predicted model parameters are compared with the recognized values at time $t_0 + T$ and with additional annotations. Back-propagation is utilized to minimize the difference among the predictions of the recognition net, the differential equation net, and the annotations. The three predictions converge when the training is complete.

The development of `NeuraDiff` was motivated by the real-world application of tracking and learning the physics model of nano-scale crystalline defects in materials. These materials and alloys are critical for current nuclear fission reactors and future fusion devices. Nano-scale crystalline defects can appear in differ-

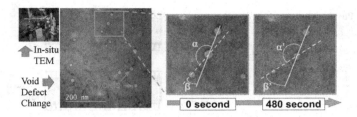

Fig. 2. (Upper Left) The Intermediate Voltage Electron Microscopy (IVEM) – Tandem Facility at the Argonne National Laboratory which provides in-situ TEM data. Source: anl.gov. **(Middle and Right)** Sample images captured during in-situ radiation experiments. The middle image shows void defects embedded in a Cu specimen at 350 °C and irradiation dose 0.25 − 1.00 dpa (dose increases with time). Void migration is illustrated by the change in sizes and in the angles of yellow lines in the right images. (Color figure online).

ent forms in these materials. Extreme environments of heat and irradiation can cause these defects to evolve in size and position. As shown in Fig. 2, void shaped defects are captured by transmission electron microscope (TEM) cameras during in-situ radiation experiments. These defects appear in round shapes, and drift in position as demonstrated by the change of angles α, β to α', β' respectively, as time progresses. They also change size. These changes can affect the physical and mechanical properties of the material in undesirable ways as discussed in [31]. For this reason, characterizing these defects is essential in designing new materials that can resist adverse environments.

In-situ radiation experiments are carried out to analyze the evolution of crystalline defects in materials. During these experiments, changes in a material specimen, subjected to high temperature and irradiation, is recorded through a TEM camera and stored in high-resolution high frame rate videos. The huge amount of data calls for a data-driven approach to expedite the video analysis, which can bring in new scientific knowledge and insights for alloy designs. However, manual video analysis requires huge effort. According to our calculation, it takes a graduate student 3.75 months to fully annotate the defects in a 10-minute in-situ video if he spends 5 min per frame and devotes 40 h a week. Phase-field modeling is a simulation tool commonly used to study the evolution of point defects. In this model, the evolution of the void shaped defects is characterized by a number of field variables. These field variables are continuous, vary rapidly at the interface of the void defects, and are governed by a number of differential equations. Data assimilation is often used to estimate the model parameters of a phase-field model from data. However, tuning phase-field models relies heavily on expert knowledge, and the results are often qualitative.

Our proposed `NeuraDiff` learns the phase field model automatically as described in [30] that governs the void nucleation and growth in irradiated materials, while provides accurate tracking of void clusters. `NeuraDiff` connects phase-field simulation and physics experiments, enabling an automatic pipeline to discover correct physics models from data. Our experimental results show that

`NeuraDiff` produces highly accurate tracking results while learning the correct physics. Our model's accuracy is close to 100% on both the synthetic dataset and a real-world in-situ dataset of Cu under $350\,^{\circ}C$ and an irradiation dose of 0.25–1.00 dpa (dose increases as time goes by). Moreover, our model learns the correct physics. The simulation based on the phase-field parameters learned by our model demonstrate similar dynamics as the ground truth, while a neural model without embedding physics cannot discover the correct dynamics. We also tested our model for transfer learning. Our `NeuraDiff` model correctly predicts the evolution of nano-structures from an unseen start condition while competing approaches cannot.

In summary, our contribution is as follows: 1) we propose `NeuraDiff`, an end-to-end approach integrating the recognition and the neural differential equation net to learn a physics model characterized by a set of partial differential equations directly from experiment data. 2) We apply `NeuraDiff` in tracking and learning the physics model of nano-scale crystalline defects in materials from in-situ experiments. Our approach enables detailed analysis of nano-structures at scale, which otherwise is beyond reach of manual efforts by materials scientists. 3) Our experimental results show that `NeuraDiff` produces close to 100% accuracy in tracking void defects. 4) Our `NeuraDiff` learns the correct physics while neural networks without embedding physics cannot. 5) Our `NeuraDiff` performs well in a transfer learning setting.

2   Phase-Field Model

Micro-structures in nano-scale physics are spatial arrangements of the phases that have different compositional and/or structural characters; e.g., the regions composed of different crystal structures and/or having different chemical compositions, grains of different orientations, domains of different structural variants, and domains of different electric or magnetic polarizations. The size, shape, volume fraction, and spatial arrangement of these micro-structural features determine the overall properties of multi-phase and/or multi-component materials.

In a phase-field model, micro-structures are defined by a set of field variables. Field variables are assumed to be continuous and changing rapidly across the interfacial regions. For example, in the phase field model of irradiated metals, 3 different phase-filed variables c_v, c_i and η together represent the system state. $c_v(\mathbf{r}, t)$ represents the voids concentration, $c_i(\mathbf{r}, t)$ represent interstitial concentration and $\eta(\mathbf{r}, t)$ differentiates between the two phases - solid phase and void phase (details discussed later). Here, $\mathbf{r} = (x, y)$ represents the spatial coordinates and t represents time. We work with 2-dimensional case in this paper, but high dimensional cases can be handled similarly.

c_v and c_i represents two types of defects in irradiated metals – voids and interstitials. Voids result from the missing of atoms in certain crystal lattice locations, as shown in Fig. 2. c_v is zero in the region consisting of 0% of voids and is one in regions of 100% voids. c_v changes continuously albeit rapidly at the interfaces of void and non-void regions. The interstitials, represented by c_i, are another variety of crystallographic defects, where atoms assume a normally unoccupied site in the crystal structure. c_i is defined similarly to c_v. The void cluster variable η is an order parameter that spatially differentiates the 2 phases. η takes a constant value $\eta = 0$ in the solid phase and $\eta = 1$ in the void phase.

Phase-field modeling leverages a set of differential equations of these field variables to model the microstructure evolution. The temporal evolution of a conserved field variable $u(\mathbf{r}, t)$ is governed by the Cahn–Hilliard [7] equation:

$$\frac{\partial u}{\partial t} = \nabla \cdot \left(M \nabla \frac{1}{N} \frac{\delta F}{\delta u} \right). \tag{1}$$

Here, F is the free energy. M is the diffusivities of the material species and N is the number of lattice sites per unit volume of the material. $\nabla = \left(\frac{\partial}{\partial x}, \frac{\partial}{\partial y} \right)$ is the diffusion operator. $\nabla \cdot \nabla$ is the laplacian i.e., $\nabla^2 f = \frac{\partial^2 f}{\partial x^2} + \frac{\partial^2 f}{\partial y^2}$. $\frac{\delta F}{\delta u}$ is the functional derivative. A non-conserved field variable v evolves according to the Allen–Cahn [1] equation:

$$\frac{\partial v}{\partial t} = -L \frac{\delta F}{\delta v}. \tag{2}$$

Here L is the mobility constant. In the phase-field model for irradiated metals, c_v, c_i are conserved field variables and η is a non-conserved field variable. Allen-Cahn and Cahn-Hilliard equations are the cornerstones of phase-field modeling. They offer good descriptions of the basic physics of many multi-phase systems.

Finite Difference Approach.   Finite difference is a useful tool to obtain numerical solutions to differential equations. Let $(x_1, \ldots, x_{N_x})$ and $(y_1, \ldots, y_{N_y})$ be a finite discretization of the x-axis and the y-axis covering the region of interest. We use uniform step sizes, i.e., $x_i - x_{i-1} = y_j - y_{j-1} = $ ds for all $i \in \{2, \ldots, N_x\}$ and $j \in \{2, \ldots, N_y\}$. As a result, the region is covered by a finite mesh of the size $N_x \times N_y$. We also assume the time is discretized into $(t_1, \ldots, t_{N_t})$ and $t_k - t_{k-1} = $ dt for $k \in \{2, \ldots, N_t\}$. Let $u(\mathbf{r}, t)$ be a function that depends on location $\mathbf{r} = (x, y)$ and time t. We discretize u onto this mesh by denoting $u_{i,j,k}$ as a shorthand for $u(x_i, y_j, t_k)$. The finite difference algorithm uses the finite difference to approximate derivatives. For example, the value of $\frac{\partial u}{\partial x}(x_i, y_j, t_k)$ can be approximated by:

$$(u(x_{i+1}, y_j, t_k) - u(x_i, y_j, t_k))/(x_{i+1} - x_i) = (u_{i+1,j,k} - u_{i,j,k})/\text{ds}.$$

Similarly, $\nabla^2 f$, the second order laplacian ∇^2 of a 2D function f, can be approximated by five point stencil centered second-order difference:

$$\nabla^2 f_{i,j,k} = \frac{1}{\text{ds}^2} \left(f_{i+1,j,k} + f_{i-1,j,k} + f_{i,j+1,k} + f_{i,j-1,k} - 4 f_{i,j,k} \right)$$

Using this idea, both the Cahn-Hilliard and the Allen-Cahn equations can be discretized. A finite approximate solution can be obtained by simulating the evolvement of field variables from a given starting state.

3   Problem Statement

Our phase field model of irradiated metals follow largely from the work of [30]. This model incorporates a coupled set of Cahn–Hilliard and Allen–Cahn equations to capture the processes of point defect generation and recombination, annihilation of defects at sinks. The phase-field model includes 3 field variables, c_v, c_i, and η, which vary both spatially and temporally. All of the variables are continuous, yet vary rapidly across interfaces. The total free energy F of the heterogeneous material is expressed in terms of the free energy of each constituent phases and interfaces:

$$F = N \int_V \left[h(\eta) f^s(c_v, c_i) + j(\eta) f^v(c_v, c_i) + \frac{\kappa_v}{2} |\nabla c_v| + \frac{\kappa_i}{2} |\nabla c_i| + \frac{\kappa_\eta}{2} |\nabla \eta| \right] dV.$$

Here, $f^s(c_v, c_i)$ is the contribution term from the solid phase. $h(\eta) = (\eta - 1)^2$ makes sure that f^s contributes 0 when $\eta = 1$. Similarly, $f^v(c_v, c_i)$ is the contribution term from the void phase, and $j(\eta) = \eta^2$. We use the formulation from [30] for f^s and f^v:

$$f^s(c_v, c_i) = E_v^f c_v + E_i^f c_i + k_B T[c_v \ln c_v + c_i \ln c_i + (1 - c_v - c_i) \ln(1 - c_v - c_i)],$$
$$f^v(c_v, c_i) = (c_v - 1)^2 + c_i^2.$$

According to the phase-field model, the dynamics of the field variables c_v, c_i and η should follow the Cahn-Hilliard and the Allen-Cahn equations. Nevertheless, new voids and interstitials can form due to irradiation and thermal fluctuation. Therefore, the standard equations need to be updated to the form:

$$\frac{\partial c_v}{\partial t} = \nabla \cdot (M_v \nabla \frac{1}{N} \frac{\delta F}{\delta c_v}) + \xi(\mathbf{r}, t) + P_v(\mathbf{r}, t) - R_{iv}(\mathbf{r}, t),$$
$$\frac{\partial c_i}{\partial t} = \nabla \cdot (M_i \nabla \frac{1}{N} \frac{\delta F}{\delta c_i}) + \zeta(\mathbf{r}, t) + P_i(\mathbf{r}, t) - R_{iv}(\mathbf{r}, t),$$
$$\frac{\partial \eta}{\partial t} = - L \frac{\delta F}{\delta \eta} + \iota(\mathbf{r}, t) + P_v(\mathbf{r}, t).$$

Here, ξ, ζ and ι are thermal fluctuation terms, modeling the fact that voids and interstitials can appear randomly in the environments of high temperature and irradiation. P_v and P_i reflect the voids (and interstitials) introduced during the irradiation process. Irradiation hits the surface of the materials and both voids and interstitials can form as a result. R_{iv} models the cancellation of voids and interstitials. We refer the details of these terms to the original publication [30]. In this model, the following set of parameters $P = \{E_v^f, E_i^f, k_B T, \kappa_v, \kappa_i, \kappa_\eta, M_v, M_i, L\}$ determine the evolution of nanovoids. Our physics learning task is to identify the values of these parameters from a partially annotated video of void dynamics.

We assume access to partial video annotations, in which part of regions in a subset of frames are annotated. For simplicity, we assume one pixel in one frame

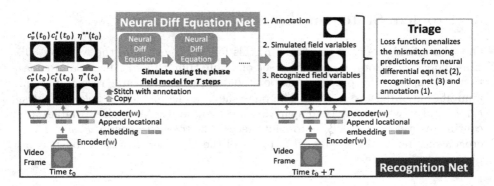

Fig. 3. The architecture of `NeuraDiff`. Our architecture consists of a recognition net, which predicts field variable values (c_v, c_i and η) based on video frames. A second neural differential equation net simulates phase field evolution for T steps. Finally, a loss function is applied which penalizes the difference among the predictions from the neural differential equation net, the recognition net and the annotations. Backpropagation is then used to train the two neural networks to minimize the loss function.

V can be in three states: 0 means the pixel is annotated to be in a solid state; i.e., $\eta = 0$; 1 means the pixel is annotated to be part of a void cluster; i.e., $\eta = 1$; * means the pixel is not annotated or the annotator is not sure of its state. We denote A as a matrix of these annotations, each entry of which is one of the three states for the corresponding pixel. The **physics-aware micro-structure tracking problem** is defined as:

- **Given**: $\{(t_1, V_1, A_1), (t_2, V_2, A_2), \ldots, (t_N, V_N, A_N)\}$ as a partially annotated video of nano-structural evolution, where $t_1, \ldots, t_N$ are time stamps, V_i is the video frame for the time stamp t_i and A_i is the partial annotation for V_i, in which each pixel is annotated to one of the three states.
- **Find**: (i) *track microstructures*: for each frame V_i, find matrix η_i, which contains the predicted η value for each pixel. (ii) *learn physics*: find the set of phase-field parameters P, along with the values of the unobserved variables c_v and c_i, which best fit the micro-structure evolution.

4   Neural Differential Equation Embedding

Our `NeuraDiff` model learns a physics model directly from experiment data via a tight integration of a neural differential equation net and a recognition net, embedding phase-field simulation into neural network learning. The high level idea is shown in Fig. 3. A recognition net extracts the values of the three field variables, c_v, c_i and η from noisy video frames. Taking these field variables as initial condition, the neural differential equation net uses the finite difference method to simulate a phase-field model. We implement the finite difference method as a convolutional neural net (details discussed later), the parameters of which can be updated via back-propagation. This architecture is related to the recurrent neural networks (RNN), where the same operational step is repeatedly applied during the forward pass. Contrary to RNNs, each step in our neural differential equation net represents a simulation step of the phase-field model.

`NeuraDiff` works through a triage process. First the recognition net extracts the three field variables from the video frame at time stamp t_0. The predicted field values are partially replaced by the groundtruth annotations (if they are present at t_0) and are sent to the neural phase-field net. The neural differential equation net then simulates the phase-field model for T steps and outputs the simulated field variable values at time $t_0 + T$. We also have partial annotations at the time $t_0 + T$ and the predictions of these field variable values from the recognition net. Ideally, if the recognition net is trained to predict the three field variables accurately and the neural differential equation net has the ground-truth parameters, then the three outcomes, namely, the simulated, the recognized field variables, and the partial annotations at the time $t_0 + T$ should match. Therefore, we enforce a loss function which penalizes the differences among the three outcomes. Back-propagation is then used to minimize this loss function. At the end of training, when the predictions from both neural nets and the partial annotations all match, the recognition net is able to extract phase field values from video frames and the neural differential equation net captures the correct phase-field parameters.

Recognition Net. The recognition net predicts the three field variables c_v, c_i and η from in-situ experiment video frames. Under a transmission electron microscope (TEM), void clusters, or the η variable, can be reliably observed (see Fig. 2 for void clusters in the actual TEM pictures). The void and interstitial defect percentages (c_v and c_i) are depicted as black shades but cannot be reliably observed due to noise caused by small perturbations, e.g., slight bending of the material samples. The bending of samples is in the scale of nanometers, which cannot be eliminated experimentally, even given the best effort. Therefore, we treat c_v and c_i as unobserved variables.

The η variable can be predicted mainly from the video frames by the recognition net, i.e., $\eta(.,t) = RN_\eta(V_t)$. As a way to estimate hidden variables c_v and c_i, we introduce location embedding vectors into our recognition net model. Let $l_1, \ldots, l_N$ be N vectors, where l_t is the location embedding vector for time t. The value of these vectors vary continuously and slowly with time t. Our first idea was to build the recognition net for c_v and c_i as $c_v(.,t) = RN_v(l_t)$ $c_i(.,t) = RN_i(l_t)$. Here, RN_v and RN_i are two neural nets which translate the location embedding vector l_t into the field variables c_v and c_i at time t, which are both matrices of the size $N_x \times N_y$. As a second idea, we also include the video frame at time t, V_t, as the input, since it offers partial information (the black shades). As the final result, the three field variables are predicted from an uniform architecture $c_v(.,t) = RN_v(l_t, V_t)$, $c_i(.,t) = RN_i(l_t, V_t)$, and $\eta(.,t) = RN_\eta(l_t, V_t)$.

In practice, the three recognition nets, $RN_v(l_t, v_t)$, $RN_i(l_t, v_t)$, $RN_\eta(l_t, v_t)$ are all implemented using the UNet architecture [35]. UNet follows a contracting then expanding neural path. Our motivation for using UNet as the recognition net stems from its wide use in scientific community, although in principle any pixelwise pattern recognition network can be used in this case. In our implementation, the input of the UNet are the video frames V_t. The location embedding vectors l_t are appended to the bottleneck vector in UNet.

Neural Phase-field Net.   One of our key contributions is to encode a finite difference phase field model as a differential equation network. As a result, the neural differential equation net can be embedded in the overall neural network architecture, allowing end-to-end training. The high-level idea is to use finite difference to approximate the Cahn-Hilliard and Allen-Cahn equations. We present here the details of embedding the Cahn-Hilliard equation. Similar process applies for the Allen-Cahn equation. Recall the Cahn-Hilliard equation is as follows:

$$\frac{\partial u}{\partial t} = \nabla \cdot \left(M\nabla \frac{1}{N} \frac{\delta F}{\delta u} \right) = \frac{M}{N} \nabla^2 \left(\frac{\delta F}{\delta u} \right).$$

Using finite difference approach as described previously, especially noting $\nabla^2 f$ can be approximated by the five point stencil centered second-order difference, $\nabla^2 f_{i,j,k} = \frac{1}{ds^2} \left(f_{i+1,j,k} + f_{i-1,j,k} + f_{i,j+1,k} + f_{i,j-1,k} - 4f_{i,j,k} \right)$, the Cahn-Hilliard equation can be written as:

$$u_{k+1} = u_k + \frac{M}{N} \frac{dt}{ds^2} Conv(\frac{\delta F}{\delta u}, K). \tag{3}$$

Here, u_k is a discretized matrix of field variable u in which the i,j-th entry of u_k is $u(x_i, y_j, t_k)$. The functional derivative $\frac{\delta F}{\delta u}$ is also a matrix, whose i,j-th entry is $\frac{\delta F}{\delta u}(x_i, y_j, t_k)$. $\frac{\delta F}{\delta u}$ can be derived by hand. $Conv$ means to convolve $\frac{\delta F}{\delta u}$ with kernel K, where

$$K = \begin{bmatrix} 0 & 1 & 0 \\ 1 & -4 & 1 \\ 0 & 1 & 0 \end{bmatrix}.$$

Equation 3 gives out a finite difference form to obtain the value of the field variable u_{k+1} in the next time stamp from the current value u_k. Interestingly, the temporal dynamics of u_k can be calculated via a convolutional operator, which can be implemented as a neural network layer relatively easily and subsequently embedded into `NeuraDiff`. Notice that a key difference between our neural differential equation net and a common convolutional layer is that, the convolution kernel is learned through training in a classical convolutional net. However, in our neural differential equation net, we keep the convolution kernel fixed, and learn the parameters associated with the variables in free energy F.

Overall Architecture and Training.   The overall architecture of `NeuraDiff` combines the recognition net with the neural differential equation net. We arrange the dataset into pairs of frames which are T time stamps apart: $\mathcal{D} = \{(t_i, V_{t_i}, A_{t_i}, V_{t_i+T}, A_{t_i+T}) \mid i = 1, \ldots, M\}$. Here, V_{t_i} is the video frame at the time stamp t_i. A_{t_i} is the annotation for V_{t_i}. V_{t_i+T} and A_{t_i+T} are the video frame and its annotations at the time stamp $t_i + T$. First, V_{t_i} are fed into the recognition net together with the location embedding l_{t_i} to produce the predicted field variables c_v^*, c_i^*, and η^* at time stamp t_i. We replace the portion of η^* with the ground-truth annotation if the annotation is available. The updated value of η^* is denoted by η^{**}. After this update, c_v^*, c_i^*, and η^{**} values are sent to the neural differential equation net to simulate for T steps. The results of the simulation are $c_v^*(t_i + T)$, $c_i^*(t_i + T)$, and $\eta^*(t_i + T)$. At $t_i + T$, the recognition net produces the recognized field variables $\hat{c}_v(t_i + T)$, $\hat{c}_i(t_i + T)$ and $\hat{\eta}(t_i + T)$. Along with the annotations, the triage loss function that the neural network model optimizes, penalizes three types of mismatches:

$$\mathcal{L} = L_{sim} + \lambda_1 L_{rec} + \lambda_2 L_{sim-rec}.$$

Here, L_{sim} denotes the loss function for the mismatch between simulated η^* and the annotations A: $L_{sim} = \|\mathbf{1}_A(\eta^* - A)\|^2$. $\mathbf{1}_A$ is the indicator matrix of annotations, the entry of which is 1 if the corresponding entry in A is not $*$. L_{rec} denotes the loss penalizing the mismatch between recognized $\hat{\eta}$ and the annotations A, $L_{rec} = \|\mathbf{1}_A(\hat{\eta} - A)\|^2$. $L_{sim-rec}$ denotes the penalties between the simulated and recognized phase-field variables $L_{sim-rec} = \left(\|\eta^* - \hat{\eta}\|^2 + \|c_v^* - \hat{c}_v\|^2 + \|c_i^* - \hat{c}_i\|^2\right)$ In these equations, all phase-field variables are at time $t_i + T$. λ_1 and λ_2 are hyper-parameters that balance the relative importance of terms. The entire neural network structure is trained via stochastic gradient descent. A minibatch of frames are sampled for the back-propagation algorithm in each iteration.

5 Related Work

AI Driven Scientific Discovery. There has been a recent trend to leverage AI for scientific discovery. In materials science, CRYSTAL is a multi-agent AI system to solve the phase-map identification problem in high-throughput materials discovery [15]. Neural models have been proposed to generate optimized molecule designs; see, e.g., the Attentive Multi-view Graph Auto-Encoders [28], the Junction Tree Variational Autoencoder [19,20], message passing neural networks [33]. Attia et al. demonstrate a machine learning methodology to efficiently optimize the parameter space for fast-charging protocols in electric-vehicles [3]. Bayesian optimization and reinforcement learning have also been used in budgeted experimental designs [4]. Embedding physics knowledge in machine learning has also attracted attention. The work by [27] adds to variational autoencoders constraints as regularization terms to improve the validity of the molecules generated. Grammar Variational Autoencoders [23] provided generative modeling of molecular structures by encoding and decoding directly to and from these parse trees, ensuring their validity. Constraint driven approaches such as satisfiability modulo theory (SMT) have also been used to ensure physically meaningful results [12]. In addition, the work of [38] proposed a new supervising approach to learn from physical constraints.

Embedding Optimization in Neural Architectures. Amos et al. proposed to embed quadratic program as a layer in an end-to-end deep neural network [2].

Recently, Ferber et al. proposes to embed a mixed integer program as neural network layers [13]. Devulapalli et al. proposed a neural network capable of back-propagating gradients through the matrix inverse in an end-to-end approach for learning a random walk model [11]. Dai et al. proposed learning good heuristics or approximation algorithms for NP-hard combinatorial optimization problems to replace specialized knowledge and trial-and-error [21]. The work by [18] proposes a new programming language for differentiable physical simulation. [37] proposes a graph network based simulator, where a stack of embedded graph networks in an encoder-decoder architecture is used to learn the dynamics of particles interacting in a 3D environment.

Learning PDEs. Previous work has discovered approaches that include physics information in machine learning, where the physics models are represented by differential equations; see, e.g., in turbulence prediction [32]. Bezenac et al. used a convolutional-deconvolutional (CDNN) module to predict the the motion field from a sequence of past images for sea surface temperature forecasting, motivated by the solution of a general class of partial differential equations [6]. Lutter et al. proposed Deep Lagrangian Networks that can learn the equations of motion of a mechanical system with a deep network efficiently while ensuring physical plausibility. Their approach incorporates the structure introduced by the ODE of the Lagrangian mechanics into the learning problem and learns the parameters in an end-to-end fashion [26]. Time-aware RNNs [10] utilized the similarities between a set of discretized differential equations and the RNN network to model the system equations from a physics system. PDE-Net [24] was proposed to accurately predict dynamics of complex systems by representing the PDEs with convolutional networks where all filters are properly constrained. Neural ODE was introduced in [8,25], where the output of a neural network is treated as the continuous-time derivative of input, thus providing an interface to incorporate differential equation modeling into machine learning models. Their work have inspired a number of other ideas. For example [22] uses a natural spline to handle irregularly observed time series data with Neural CDE model. While the original Neural ODE model was designed for continuous time modeling, discrete time modeling have been proposed as well by [29]. Hamiltonian neural network (HNN) as proposed in [16] uses partial derivatives of the final output instead of the actual output value, to approximate an energy function and build a Hamiltonian system with a neural network. To make learning easier in HNN, [14] propose a change in system representation along with explicit constraints. A separate line of work [5,17] exploit this connection to solve PDEs. Most of these works learn PDEs from the observed trajectories, which in many applications need to be extracted in separate steps. For example, the TEM videos in our application only provide partial information on the actual trajectories of the phase-field variables. Our NeuraDiff integrates a computer vision neural network with a PDE neural net in the discovery of physics models directly from experiment data.

Image Analysis for In-situ Data. Automated image segmentation models are being developed to identify defects and other nanostructures in TEM images [34]. However, they do not learn any physics models.

6 Experiments

Our experiments on both the synthetic and real-world datasets demonstrate that our `NeuraDiff` provides highly accurate tracking while at the same time learns the phase-field model that governs nanostructure dynamics.

Training. In the experiments, we first pre-train the recognition net before the entire architecture, to predict the 3 phase-field variables using only video frames. The details of this pre-training step is provided in the supplementary materials. The stochastic optimization algorithm we used for both pre-training and the actual training is Adam, with the initial learning rate set to be 0.01. Additional details on the experiment setup, train-test split and hyperparameter tuning are in the supplementary materials.

Table 1. Our `NeuraDiff` obtains similar and near perfect tracking accuracy as a UNet baseline in both the synthetic and the real-world datasets.

Accuracy	NeuraDiff	UNet baseline
Synthetic data	98.5%	99.9%
Real data	96.2%	96.4%

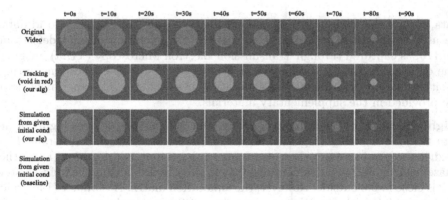

Fig. 4. Our `NeuraDiff` provides reliable tracking (2nd row, the tracking result of the voids shown in red) while learning the correct physics on synthetic data. In the third row, we simulate our `NeuraDiff` with the learned parameters from a given initial condition. The learned model simulates a similar dynamics as the original video. Nevertheless, a neural network baseline without embedding the phase-field model produces unsatisfying result (4th row). (Color figure online).

Dataset Description. We use both synthetic data and real-world in-situ experiment data to evaluate our model. Both the synthetic and real-world data are in high frame rate high resolution video format. For generating synthetic video

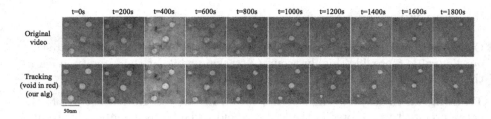

Fig. 5. Our `NeuraDiff` provides accurate tracking (in red) on real in-situ Kr ion experiment of Cu at the temperature of $350\,^\circ$C and $0.25 - 1.00$ displacements-per-atom (dpa) of irradiation. (Color figure online).

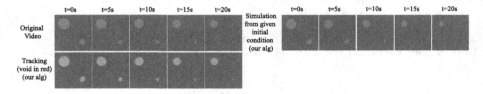

Fig. 6. Transfer learning result for `NeuraDiff`. It provides reliable tracking (second row, in red) and predict correct void progression on unseen data. `NeuraDiff` was trained on the dataset used in Fig. 4 including a single nanovoid, and was not fine tuned when evaluated on this dataset of several nanovoids. (Color figure online).

data, we use the void evolution model as described in [30]. For real-world data, we use the in-situ experiment video showing the evolution of void defects in Cu 110, as captured through Transmission Electron Microscopy (TEM) imaging. The details of synthetic data generation process, the testbed conditions during in-situ radiation experiments and annotation process for in-situ experiment data are provided in the supplementary materials.

Highly Accurate Tracking Accuracy. Our `NeuraDiff` provides highly accurate tracking accuracy, together with a UNet baseline, which were trained to predict the η values from the video frames using supervised learning. From the phase-field model, η varies continuously, is close to 1.0 within the void cluster and is close to 0.0 outside. However, the annotation matrix A is binary (1 for void cluster and 0 for others). We cut off η values at 0.5 and evaluate the accuracy in the following way: $\frac{1}{N_x N_y} \sum_{x,y} \mathbb{1}(\eta_{x,y} \geq 0.5, A_{x,y} = 1) + \mathbb{1}(\eta_{x,y} < 0.5, A_{x,y} = 0)$.

We can see from Table 1 that both our `NeuraDiff` and the UNet baseline produce close to optimum tracking results. The second row of Fig. 4 and Fig. 5 depict the actual tracking of void clusters on synthetic data as well as on real experimental data. The region of void clusters is highlighted with the red color. We can visually inspect that the tracking is close to optimum.

Capture the Physics. Aside from providing accurate and reliable tracking, our `NeuraDiff` also learns the phase-field model that correctly predicts void cluster evolution. Our key contribution lies within the fact that our model can learn the dynamics of void evolution without compromising the tracking task, and needs less data for the tracking purpose. With the learned parameters, we simulate the evolution of the phase-field variables from the initial condition of the synthetic dataset using finite difference. The initial condition is given as the first frame of the video, instead of the values of the three field variables. Our model has to infer their values from the recoginition net. The result is shown in the third row of Fig. 4. We can see that the dynamics closely resembles that of the original video, suggesting that our approach identifies the correct phase-field model. We point out that the learned parameters as well as the predicted c_v and c_i unobserved field variables are different from the original values used to synthesize the dataset. This suggests that there are multiple parameter values which lead to similar dynamics. We also evaluated our model performance in transfer learning. Here, the model was trained on the synthetic dataset involving one void, but was tested for both tracking and void evolution in an unseen dataset involving multiple void of different sizes (Fig. 6). Our model produces reasonable tracking results and simulates the correct dynamics. See more details in the supplementary materials.

We tried hard to use a neural network model to predict void evolution without embedding the phase-field model. However, the result is not satisfying. For example, in the fourth row of Fig. 4, we used the UNet to predict the next frame given the current frame. Then we use the UNet to synthesize the entire video via repeated predictions of the next frame. However, the performance is not satisfying as the noise quickly dominates the signals. We even tried to feed the neural network with the correct values of the three field variables and ask it to predict the next frame. Note the field variable values are not available in real-world experiments. The baseline neural network cannot predict the dynamics even with these additional inputs.

7   Conclusion

We present NeuraDiff, an end-to-end model to learn a physics model characterized by a set of partial differential equations directly from data. Our key idea is to embed the physics model as a multi-layer convolutional neural net into the overall neural architecture for end-to-end training. We apply NeuraDiff in the task of tracking and characterizing the dynamics of point defect clusters in materials under high temperatures and heavy irradiations. Our approach produces near perfect tracking and is able to capture a physics model that predicts future nanostructure dynamics correctly, which are not possible for pure data-driven machine learning models. Our model is validated on both synthetic and real experimental data. Future work include to scale up the computation for high dimensional, high frame rate videos, and to validate the physics models learned from our framework with more real-world irradiation experiments.

Acknowledgments. This research was supported by NSF grants IIS-1850243, CCF-1918327. We thank anonymous reviewers for their comments and suggestions.

References

1. Allen, S.M., Cahn, J.W.: Ground state structures in ordered binary alloys with second neighbor interactions. Acta Metallurgica **20**(3), 423–433 (1972)
2. Amos, B., Kolter, J.Z.: Optnet: differentiable optimization as a layer in neural networks. In: International Conference on Machine Learning, pp. 136–145 (2017)
3. Attia, P.M., et al.: Closed-loop optimization of fast-charging protocols for batteries with machine learning. Nature **578**(7795), 397–402 (2020)
4. Azimi, J., Fern, X.Z., Fern, A.: Budgeted optimization with constrained experiments. J. Artif. Int. Res. **56**(1), 119–152 (2016)
5. Beck, C., Weinan, E., Jentzen, A.: Machine learning approximation algorithms for high-dimensional fully nonlinear partial differential equations and second-order backward stochastic differential equations. J. Nonlinear Sci. **29**(4), 1563–1619 (2019). https://doi.org/10.1007/s00332-018-9525-3
6. de Bezenac, E., Pajot, A., Gallinari, P.: Deep learning for physical processes: incorporating prior scientific knowledge. In: International Conference on Learning Representations (2018)
7. Cahn, J.W., Hilliard, J.E.: Free energy of a nonuniform system. I. Interfacial free energy. J. Chem. Phys. **28**(2), 258–267 (1958)
8. Chen, R.T.Q., Rubanova, Y., Bettencourt, J., Duvenaud, D.K.: Neural ordinary differential equations. In: Bengio, S., Wallach, H., Larochelle, H., Grauman, K., Cesa-Bianchi, N., Garnett, R. (eds.) Adv. Neural Inf. Process. Syst. **31**, 6571–6583 (2018)
9. Chen, Z., Zhang, J., Arjovsky, M., Bottou, L.: Symplectic recurrent neural networks. In: 8th International Conference on Learning Representations, ICLR (2020)
10. Demeester, T.: System identification with time-aware neural sequence models. arXiv preprint arXiv:1911.09431 (2019)
11. Devulapalli, P., Dilkina, B., Xue, Y.: Embedding conjugate gradient in learning random walks for landscape connectivity modeling in conservation. In: Bessiere, C. (ed.) Proceedings of the Twenty-Ninth International Joint Conference on Artificial Intelligence, IJCAI-20, pp. 4338–4344. International Joint Conferences on Artificial Intelligence Organization (2020)
12. Ermon, S., Le Bras, R., Gomes, C.P., Selman, B., van Dover, R.B.: SMT-aided combinatorial materials discovery. In: Cimatti, A., Sebastiani, R. (eds.) SAT 2012. LNCS, vol. 7317, pp. 172–185. Springer, Heidelberg (2012). https://doi.org/10.1007/978-3-642-31612-8_14
13. Ferber, A., Wilder, B., Dilkina, B., Tambe, M.: Mipaal: mixed integer program as a layer. In: AAAI, pp. 1504–1511 (2020)
14. Finzi, M., Wang, K.A., Wilson, A.G.: Simplifying Hamiltonian and Lagrangian neural networks via explicit constraints. Adv. Neural Inf. Process. Syst. **33**, 13581 (2020)
15. Gomes, C.P., et al.: Crystal: a multi-agent AI system for automated mapping of materials' crystal structures. MRS Commun. **9**(2), 600–608 (2019)
16. Greydanus, S., Dzamba, M., Yosinski, J.: Hamiltonian neural networks. Adv. Neural Inf. Process. Syst. **32**, 15379–15389 (2019)
17. Han, J., Jentzen, A., Weinan, E.: Solving high-dimensional partial differential equations using deep learning. Proc. Nat. Acad. Sci. **115**(34), 8505–8510 (2018)
18. Hu, Y., et al.: Difftaichi: differentiable programming for physical simulation. In: 8th International Conference on Learning Representations, ICLR (2020)

19. Jin, W., Barzilay, R., Jaakkola, T.: Junction tree variational autoencoder for molecular graph generation. In: International Conference on Machine Learning (2018)

20. Jin, W., Yang, K., Barzilay, R., Jaakkola, T.: Learning multimodal graph-to-graph translation for molecule optimization. In: International Conference on Learning Representations (2018)

21. Khalil, E., Dai, H., Zhang, Y., Dilkina, B., Song, L.: Learning combinatorial optimization algorithms over graphs. In: Guyon, I., et al. (eds.) Adv. Neural Inf. Process. Syst. **30**, 6348–6358 (2017)

22. Kidger, P., Morrill, J., Foster, J., Lyons, T.: Neural controlled differential equations for irregular time series. arXiv:2005.08926 (2020)

23. Kusner, M.J., Paige, B., Hernández-Lobato, J.M.: Grammar variational autoencoder. In: International Conference on Machine Learning, pp. 1945–1954 (2017)

24. Long, Z., Lu, Y., Ma, X., Dong, B.: Pde-net: Learning pdes from data. In: International Conference on Machine Learning, pp. 3208–3216 (2018)

25. Lu, Y., Zhong, A., Li, Q., Dong, B.: Beyond finite layer neural networks: bridging deep architectures and numerical differential equations. In: International Conference on Machine Learning, pp. 3276–3285 (2018)

26. Lutter, M., Ritter, C., Peters, J.: Deep Lagrangian networks: using physics as model prior for deep learning. In: International Conference on Learning Representations (2018)

27. Ma, T., Chen, J., Xiao, C.: Constrained generation of semantically valid graphs via regularizing variational autoencoders. In: Advances in Neural Information Processing Systems, pp. 7113–7124 (2018)

28. Ma, T., Xiao, C., Zhou, J., Wang, F.: Drug similarity integration through attentive multi-view graph auto-encoders. In: Proceedings of the Twenty-Seventh International Joint Conference on Artificial Intelligence, IJCAI, pp. 3477–3483 (7 2018)

29. Matsubara, T., Ishikawa, A., Yaguchi, T.: Deep energy-based modeling of discrete-time physics. In: Advances in Neural Information Processing Systems 33 (NeurIPS2020) (2020)

30. Millett, P.C., El-Azab, A., Rokkam, S., Tonks, M., Wolf, D.: Phase-field simulation of irradiated metals: part i: void kinetics. Comput. Mater. Sci. **50**(3), 949–959 (2011)

31. Niu, T., et al.: Recent studies on void shrinkage in metallic materials subjected to in situ heavy ion irradiations. JOM **72**(11), 4008–4016 (2020). https://doi.org/10.1007/s11837-020-04358-3

32. Portwood, G.D., et al.: Turbulence forecasting via neural ode. arXiv preprint arXiv:1911.05180 (2019)

33. Raza, A., Sturluson, A., Simon, C.M., Fern, X.: Message passing neural networks for partial charge assignment to metal-organic frameworks. J. Phys. Chem. C **124**(35), 19070–19082 (2020)

34. Roberts, G., Haile, S.Y., Sainju, R., Edwards, D.J., Hutchinson, B., Zhu, Y.: Deep learning for semantic segmentation of defects in advanced stem images of steels. Sci. Rep. **9**(1), 1–12 (2019)

35. Ronneberger, O., Fischer, P., Brox, T.: U-net: convolutional networks for biomedical image segmentation. In: Navab, N., Hornegger, J., Wells, W.M., Frangi, A.F. (eds.) MICCAI 2015. LNCS, vol. 9351, pp. 234–241. Springer, Cham (2015). https://doi.org/10.1007/978-3-319-24574-4_28

36. Sæmundsson, S., Terenin, A., Hofmann, K., Deisenroth, M.P.: Variational integrator networks for physically structured embeddings. In: The 23rd International Conference on Artificial Intelligence and Statistics, vol. 108, pp. 3078–3087 (2020)

37. Sanchez-Gonzalez, A., Godwin, J., Pfaff, T., Ying, R., Leskovec, J., Battaglia, P.W.: Learning to simulate complex physics with graph networks. In: International Conference on Machine Learning (2020)
38. Stewart, R., Ermon, S.: Label-free supervision of neural networks with physics and domain knowledge. In: 31 AAAI Conference on Artificial Intelligence (2017)
39. Tong, Y., Xiong, S., He, X., Pan, G., Zhu, B.: Symplectic neural networks in Taylor series form for Hamiltonian systems. ArXiv abs/2005.04986 (2020)
40. Zhong, Y.D., Dey, B., Chakraborty, A.: Symplectic ode-net: learning hamiltonian dynamics with control. In: 8th International Conference on Learning Representations, ICLR (2020)

A Bayesian Convolutional Neural Network for Robust Galaxy Ellipticity Regression

Claire Theobald[1]([✉]), Bastien Arcelin[3], Frédéric Pennerath[2],
Brieuc Conan-Guez[2], Miguel Couceiro[1], and Amedeo Napoli[1]

[1] Université de Lorraine, CNRS, LORIA, 54000 Nancy, France
{claire.theobald,miguel.couceiro,amedeo.napoli}@loria.fr
[2] Université de Lorraine, CentraleSupélec, CNRS, LORIA, 57000 Metz, France
frederic.pennerath@centralesupelec.fr, brieuc.conan-guez@univ-lorraine.fr
[3] Université de Paris, CNRS, Astroparticule et Cosmologie, 75013 Paris, France
arcelin@apc.in2p3.fr

Abstract. Cosmic shear estimation is an essential scientific goal for large galaxy surveys. It refers to the coherent distortion of distant galaxy images due to weak gravitational lensing along the line of sight. It can be used as a tracer of the matter distribution in the Universe. The unbiased estimation of the local value of the cosmic shear can be obtained via Bayesian analysis which relies on robust estimation of the galaxies ellipticity (shape) posterior distribution. This is not a simple problem as, among other things, the images may be corrupted with strong background noise. For current and coming surveys, another central issue in galaxy shape determination is the treatment of statistically dominant overlapping (blended) objects. We propose a Bayesian Convolutional Neural Network based on Monte-Carlo Dropout to reliably estimate the ellipticity of galaxies and the corresponding measurement uncertainties. We show that while a convolutional network can be trained to correctly estimate well calibrated aleatoric uncertainty, -the uncertainty due to the presence of noise in the images- it is unable to generate a trustworthy ellipticity distribution when exposed to previously unseen data (i.e. here, blended scenes). By introducing a Bayesian Neural Network, we show how to reliably estimate the posterior predictive distribution of ellipticities along with robust estimation of epistemic uncertainties. Experiments also show that epistemic uncertainty can detect inconsistent predictions due to unknown blended scenes.

Keywords: Bayesian neural networks · Convolutional neural networks · Epistemic uncertainty · Uncertainty calibration · Cosmology.

The first author is preparing a PhD thesis at the LORIA Lab in the context of the AstroDeep Research Project (https://astrodeep.pages.in2p3.fr/website/projects/) funded by ANR under the grant ANR-19-CE23-0024.

Y. Dong et al. (Eds.): ECML PKDD 2021, LNAI 12979, pp. 135–150, 2021.
https://doi.org/10.1007/978-3-030-86517-7_9

1   Introduction

One of the goals of large galaxy surveys such as the *Legacy Survey of Space and Time* (LSST, [16]) conducted at the Vera C. Rubin Observatory is to study *dark energy*. This component of unknown nature was introduced in the current cosmological standard model to explain the acceleration of the Universe expansion. One way to probe dark energy is to study the mass distribution across the Universe. This distribution mostly follows the dark matter distribution, which does not interact with baryonic matter (i.e. visible matter) except through gravitation, as dark matter represents around 85% of the matter in the Universe. Consequently, cosmologists need to use indirect measurement techniques such as *cosmic shear*, which measures the coherent distortion of background galaxies images by foreground matter due to *weak gravitational lensing* [17]. In astrophysics, gravitational lensing is the distortion of the image of an observed source, induced by the bending of space-time, thus of the light path, generated by the presence of mass along the line of sight. The mass acts like a lens, in partial analogy with optical lenses, as illustrated in Fig. 1a. The weak gravitational lensing effect is faint (1% of galaxy shape measurement) and only statistical tools provide a way to detect a local correlation in the observed galaxies orientations. This correlation yields a local value at every point of the observable Universe, defining the cosmic shear field. As pictured in Fig. 1b, in an isotropic and uniform Universe orientations of galaxies are expected to follow a uniform distribution (left panel). The statistical average of their oriented elongations, hereafter called *complex ellipticities*, is expected to be null. In presence of a lens, a smooth spatial deformation field modifies coherently the complex ellipticities of neighboring galaxies so that their mean is no longer zero (right panel).

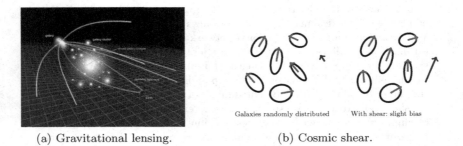

(a) Gravitational lensing.   (b) Cosmic shear.

Fig. 1. (a) Effect of gravitational lensing: the mass bends the light and deforms the images of the galaxies. (b) Weak lensing: the correlation between orientations and shapes of neighbour galaxies defines the cosmic shear. In blue: average ellipticity. Left: the expected ellipticity distribution. Right: the observed ellipticity distribution. Image: (a) NASA/ESA (Color figure online)

The unbiased measurement of cosmic shear is a major ambition of nowadays cosmology [21]. One avenue to estimate the cosmic shear locally is to combine individual galaxy ellipticity measurements. By looking deeper into the sky, that is to older objects, the next generation of telescopes will allow for the detection of a very large number of galaxies, potentially leading to very precise shear measurement and resulting in tight constraints on dark energy parameters.

Methods already exist to estimate galaxy ellipticities through direct measurement on images recorded by telescope cameras ([13] for example). This is a complex problem as, among other things, the shear signal is carried by faint galaxies which makes it very sensitive to background noise. Another central issue for current and coming surveys in galaxy shape determination, is the treatment of statistically dominant overlapping objects, an effect called *blending*. A current survey projects that 58% of the detected objects will appear blended [18] and this value is expected to reach around 62% for LSST [19]. To overcome this issue, solutions exist such as deblending [22–24]: the separation of overlapping objects. Yet, they are not perfect and rely on an accurate detection of blended scenes which is also a complex problem. As such, in addition to a precise estimation of the complex ellipticities, a reliable measurement of the uncertainties is crucial in order to discard, or at least decrease the impact of, unreliable and inaccurate measurements avoiding as much as possible the introduction of a bias into the shear estimation.

Classical ellipticity measurement methods usually adopt assumptions about the shape of the galaxies (for example via the shape of the window function in [13]) potentially resulting in model bias. In contrast, *convolutional neural networks* or CNNs [2] make it possible to learn and recognize complex and diverse galaxy shapes directly from data without making any other hypothesis than the representativeness of the training sample. They consequently are appropriate tools to learn the regression of galaxy ellipticities, even in the presence of noise and complex distortions. Yet standard CNNs can only measure the *aleatoric uncertainty*: the one due to the presence of noise in the data. They are unable to estimate the *epistemic uncertainty*, the one due to the limited number of samples a CNN has been trained with and to the model [1,9]. This second type of uncertainty is essential to detect outliers from the training samples, or formulated accordingly to our problem, to distinguish between reliable or unreliable galaxy ellipticity estimation. It is only accessible by considering neural network weights as random variables instead of constants, that is, by adopting a Bayesian approach. Consequently, we have focused our work on *Bayesian Deep Learning* [11] using Monte Carlo dropout (MC dropout) [1] as the mean to apply Bayesian inference to Deep Learning models.

Foreseeing a Bayesian estimation of the cosmic shear, combining galaxy ellipticity posteriors estimated directly from images (with blends or not) in different filters (or bands), this paper focuses on estimating reliable galaxy ellipticity posteriors from single band images. This is a necessary step to check that the proposed method efficiently estimates a calibrated aleatoric uncertainty and is able to minimize the impact of wrongly estimated ellipticity values due to outliers in the computation of the shear. We compare two networks trained on isolated galaxy images with or without noise in order to test for the calibration of aleatoric uncertainty. Regarding outliers, blended scenes are perfect examples. Note that these are illustrations of aleatoric or epistemic uncertainty sources. Most of cosmic shear bias sources such as detection, Point-Spread-Function (PSF) treatment, or selection for example [17,21], can fall in one or the other category. The estimation of galaxy ellipticity posterior from blended scenes in different bands is a harder problem that we will investigate in further work.

The contributions of this article are 1) to propose a Bayesian Deep Learning model that solves a complex multivariate regression problem of estimating the galaxy shape parameters while accurately estimating aleatoric and epistemic uncertainties; 2) to establish an operational protocol to train such a model based on multiple incremental learning steps; and 3) to provide experimental evidences that the proposed method is able to assess whether an ellipticity measurement is reliable. This is illustrated, in this paper, by the accurate differentiation between isolated galaxy or blended scenes, considered here as outliers, and the relationship between epistemic uncertainty and predictive ellipticity error. We also show that this last result could not be obtained with a classical, non Bayesian network.

The rest of the paper is organized as follows. In Sect. 2 we briefly describe the problem to be solved and comment on some of its peculiarities. We detail our proposed solution in Sect. 3. We analyse the results obtained on the various experiments we performed in Sect. 4, and we conclude and give the directions of further research in Sect. 5.

2   Estimating Galaxy Ellipticity from Images

As mentioned previously, it is possible to estimate cosmic shear combining individual measurements of galaxy shape. This shape information can be quantified by the complex ellipticity, which can be defined in cosmology as in Definition 1.

Definition 1. *Let E be an ellipse with major axis a, minor axis b, and with θ as its position angle. The complex ellipticity of E is defined as:*

$$\epsilon = \epsilon_1 + \epsilon_2\, i = \frac{1 - q^2}{1 + q^2}\, e^{2i\theta}, \tag{1}$$

where $q = \frac{b}{a}$ is the axis ratio of the ellipse.

An illustration of the ellipticity parameters is shown in Fig. 2a. The complex ellipticity defines a bijection between the orientation and the elongation of the ellipse on one side, and the unit disk on the other side, see Fig. 2b.

However, the process to achieve an unbiased measurement of cosmic shear, starting with the estimation of ellipticities, is going to be challenging for several reasons. We test the reliability of our networks prediction on noise and blending, two of the many possible bias sources in the cosmic shear estimation. Both of these issues result from the fact that the shear signal is mostly carried by faint galaxies. By definition, these objects have a low signal-to-noise ratio. The noise corrupts the galaxy images, making the shape estimation much harder (see Fig. 3b), and can introduce a bias in shear measurement [17].

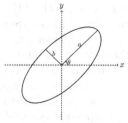
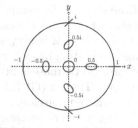

(a) Ellipticity parameters: major axis a, minor axis b, position angle θ.

(b) Bijective mapping between ellipse shapes and complex ellipticities.

Fig. 2. Geometric representation of the complex ellipticity. (a) The ellipse parameters. (b) The complex ellipticity defines a bijection between ellipse shapes and the unit disk. An ellipticity with low magnitude is close to a circle, while one with a high magnitude is closer to a straight line. The argument defines the orientation of the ellipse

(a) Isolated noiseless galaxy   (b) Isolated noisy galaxy   (c) Blended noisy galaxies

Fig. 3. Three different types of image complexity for the same galaxy: isolated without noise, isolated with noise, blended with noise. Notice how the noise slightly deforms the galaxy (b) and how the blended galaxies makes the ellipticity estimation very difficult (c) when compared to a simple isolated galaxy without noise (a)

Also, a large part of these faint objects will appear blended with foreground galaxies. Even in scenes where objects are only slightly overlapped, the apparent shape of the detected object does not correspond to a single galaxy model and an ellipticity measurement on this image could give a completely wrong result. Again, this work is the first step of a longer-term goal. Here, we target a reliable estimation of galaxy ellipticity posterior from single band images. This includes obtaining a well calibrated aleatoric uncertainty, tested here with and without the addition of Poisson noise on images, and an epistemic uncertainty allowing for minimization of the impact of untrustworthy measurement due to outliers (here, blended scenes).

We simulate LSST-like images, allowing us to control the parameters of the scene, e.g., the number of galaxies, their location on the image, and the level and type of noise applied. We consider four categories of simulated data: isolated centered galaxies without noise, isolated centered galaxies with noise, and blended scene with and without noise. Images are 64 × 64 pixels stamps simulated in the

brightest of the six bands corresponding to the LSST filters, each of them selecting a different part of the electromagnetic spectrum. These images are simulated
placing, in their center, a galaxy whose ellipticity is to be measured.

The image generating process relies on the GalSim library [14] and is based on
a catalog of parametric models fitted to real galaxies for the third Gravitational
Lensing Accuracy Testing (GREAT3) Challenge [20]. It consists in 1) producing
an image of a centered noiseless isolated galaxy from a model sampled randomly
from the catalog, with its corresponding physical properties (size, shape, orientation, PSF, brightness, redshift, etc.) 2) measuring the complex ellipticity
of the galaxy with the KSB algorithm [13] on the image and record it as the
image label, 3) possibly adding on random image location other galaxy images
(from 0 to 5) to generate blended scenes 4) possibly adding Poisson noise (as in
[24]). In this study and for sake of interpretability, we only provide as input to
our CNN the reference band (the brightest) which we use to define the target
ellipticity, making our images two-dimensional. Once again, while using multiple
bands is useful for blended galaxies [23,24], here we focus only on predicting the
ellipticity of a single centered galaxy with a correctly estimated uncertainty.

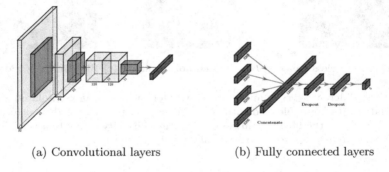

(a) Convolutional layers   (b) Fully connected layers

Fig. 4. Convolutional neural network architecture. (a) The input after augmentation
has dimensions $45 \times 45 \times 1$. Each convolutional block starts with a batch normalization
layer and has a PReLU activation. The first convolutional layer is of dimension $45 \times
45 \times 32$ with a 5×5 kernel size (in yellow), followed by a 2×2 Max-Pooling operation
(in orange). The second convolutional layer is $22 \times 22 \times 64$ with kernel size $3 \times
3$, followed by a 2×2 Max-Pooling operation. Then, we add two $11 \times 11 \times 128$
convolutional layers with a 3×3 kernel that ends with a final 2×2 Max-Pooling
operation, and the resulting feature maps are flattened into a 3200 fully connected layer
(in purple). (b) Each augmented image gives a 3200 fully connected layer convolutional
output (all augmented images share the same convolutional layers and filters), which
are then concatenated into a 12800 fully connected layer. The two final layers have
4096 neurons in the case of an MVN regression, 2048 else; with Maxout activation [15]
and dropout with a rate of 0.5. The output layer has 5 neurons in the case of an MVN
regression (Color figure online)

3 A Method to Assess Uncertainty in Ellipticity Estimation

3.1 Estimation of Noise Related Uncertainty

Our first goal is to reliably estimate the first layer of complexity in the galaxy images, the noise. Given the nature of the data, we will be using a CNN [2]. However, training a CNN to solve a standard regression problem with an L2 loss does not allow us to estimate the uncertainty due to the noise. Therefore, in place of a complex scalar output, we predict a 2D *multivariate normal distribution* (MVN) as an output: given an input image X, whose complex ellipticity is denoted Y and given weight parameters w, the network outputs an MVN $Y \sim \mathcal{N}(\mu(X, w), \Sigma(X, w))$. As such, the model is no longer trained on a simple L2 loss but rather on the log-likelihood of the MVN. The mean of the distribution $\mu(X, w)$, which is also the mode, serves as the predicted output, and the covariance matrix $\Sigma(X, w)$, which is also an output of the network, represents the so-called aleatoric uncertainty on the input data X. The model is therefore heteroscedastic, as $\Sigma(X, w)$ depends on the input X [9]. This allows our model to estimate the aleatoric uncertainty for each image individually. The determinant of $\Sigma(X, w)$, denoted $|\Sigma(X, w)|$, is a scalar measure of uncertainty, as it is directly related to the differential entropy, $\ln\left(\sqrt{(2\pi e)^2 |\Sigma(X, w)|}\right)$, of an MVN.

The architecture of our network is inspired by the work of Dielman, who proposed a simple model specifically tuned for the Galaxy Zoo challenge, therefore adapted to our data [12]. Each image is augmented in four different parts by cropping thumbnails from high resolution images, centered on spatial modes of light profile. Then each augmented image is fed to the CNN. The complete architecture is explained in Fig. 4. More details on the training process are explained in Sect. 3.3. Results obtained with this model are given in Sect. 4.1.

3.2 Estimation of Blend Related Uncertainty

As seen in Sect. 2, estimating the uncertainty due to the noise in the data is only one part of the problem. An estimated 60% of the images represent blended scenes, for which a direct estimation of ellipticity does not make sense in the context of this work. The uncertainty related to the blended images cannot be estimated simply with the variance of the MVN distribution. Indeed, in the case of a blended scene image, the network is not uncertain because of the noise but rather because this kind of images is not part of the training sample. This can be characterized by the epistemic uncertainty.

This uncertainty can be estimated using a Bayesian Neural Network (BNN), which assumes a probability distribution on the weights W of the network instead of a single point estimate [11]. Given a prior $p(w)$ on W and a set $\mathcal{D} = \{(X_i, Y_i)\}_i$ of observations, the resulting posterior distribution $p(w|\mathcal{D}) \propto p(\mathcal{D}|w) p(w)$ is analytically impossible to compute. A variational Bayes optimization method is necessary to derive an approximate posterior $q_\theta(w)$ parameterized by hyperparameters θ. In MC dropout [1,3], the considered search space includes all approximate posteriors resulting from applying *dropout* [7], i.e. multiplying every neuron output (of selected layers) by an independent Bernoulli variable. The dropout rate is set to the conventional value of 0.5, as this leads to an approximate posterior that can achieve well calibrated uncertainty estimates [1]. However, there are other ways to define the posteriors such as dropout rate tuning [5], or ensemble methods by training many networks [6]. During training, standard stochastic gradient descent techniques can be used, thanks to the reparameterization trick, to search for an approximate posterior maximizing locally the ELBO [1]. During testing, the *posterior predictive distribution* $p(Y|X, \mathcal{D})$ for some input X can be estimated using Monte Carlo sampling:

$$p(Y|X, \mathcal{D}) \approx \int p(Y|X, w) q_\theta(w) dw \approx \frac{1}{K} \sum_{k=1}^{K} p(Y|X, w_k), \tag{2}$$

where $(w_k)_{k=1}^{K} \sim q_\theta(w)$ refer to weights of K independent dropout samples.

In the case of a multivariate regression problem like ours, every distribution $p(Y|X, w_k)$ is a MVN so that the resulting posterior predictive distribution in Eq. 2 is a Gaussian mixture of order K. The uncertainty underlying this mixture can be summarized by its covariance matrix $\Sigma_{pred.}(X, \mathcal{D}) = \text{Cov}(Y|X, \mathcal{D})$. This matrix accounts for both aleatoric and epistemic uncertainties, whose respective contributions can actually be separated in a way that generalizes the variance decomposition described in Depeweg [10]:

$$\Sigma_{pred.}(X, \mathcal{D}) = \Sigma_{aleat.}(X, \mathcal{D}) + \Sigma_{epist.}(X, \mathcal{D}), \tag{3}$$

where the first term represents the aleatoric uncertainty and can be computed as the mean of the covariance matrices for each of the K output samples:

$$\Sigma_{aleat.}(X, \mathcal{D}) = \mathbb{E}_{W|\mathcal{D}}(\text{Cov}(Y|X, W)),$$

$$\approx \frac{1}{K} \sum_{k=1}^{K} \Sigma(X, w_k). \tag{4}$$

while the second represents the epistemic uncertainty and is estimated as the empirical covariance matrix of the K mean vectors produced as outputs:

$$\Sigma_{epist.}(X, \mathcal{D}) = \text{Cov}_{W|\mathcal{D}}(\mathbb{E}(Y|X, W)),$$

$$\approx \frac{1}{K} \sum_{k=1}^{K} (\mu(X, w_k) - \mu(X))(\mu(X, w_k) - \mu(X))^T. \tag{5}$$

where $\mu(X) = \frac{1}{K} \sum_{k=1}^{K} \mu(X, w_k)$.

Matrix $\Sigma_{epist.}(X, \mathcal{D})$ defines the epistemic uncertainty as the covariance matrix of the mean vectors over the posterior. This uncertainty will be high if the sampled predictions from each model vary considerably with respect to W. This would mean that no consistent answer can be deduced from the model and therefore it would be highly uncertain.

Finally when the context requires to reduce these uncertainty matrices to uncertainty levels so that they can be compared, their determinants are used to define two corresponding scalar quantities:

$$\mathcal{U}_{aleat.}(X, \mathcal{D}) = |\Sigma_{aleat.}(X, \mathcal{D})| \quad \text{and} \quad \mathcal{U}_{epist.}(X, \mathcal{D}) = |\Sigma_{epist.}(X, \mathcal{D})|.$$

3.3 Training Protocol

In order to train a BNN with an MVN output, the model needs to learn both the mean and the covariance matrix. The network's training diverges when trying to learn both at the same time, forcing us to separate the training into two steps. First, we train a simple neural network without an MVN output - we use only two output neurons representing the mean - using a L2 loss. Then, we transfer the filters of the convolutional layers into the model with an MVN output, but reinitialize the fully connected layers. This allows the model to converge smoothly as the mean of the MVN distribution has already been learned, allowing the covariance matrix to be calibrated accordingly.

This protocol works well when training on noiseless images of isolated galaxies but fails when training on noisy images. Indeed, overfitting occurs during the training of the network without MVN. When transferring the filters to the MVN model, the mean of the MVN is not well calibrated enough and the training of the BNN diverges. To fix this, we adjust the protocol for the model without MVN, adding noise incrementally during training: we first submit noiseless images, and modify 5% of the sample, switching from noiseless to noisy images, every 50 epochs for 1000 epochs. This prevents overfitting and allows the MVN model to converge after the transfer.

4 Experiments

4.1 Estimation of Uncertainty Related to Noise

In this section we show that using an MVN as an output allows for a reliable and well calibrated estimation of the aleatoric uncertainty, i.e. uncertainty related to the noise in the data.

In order to show that estimating the ellipticity of galaxies in the presence of background noise is complex and can induce incorrect predicted ellipticity values, we first train two simple CNNs without an MVN output: one on noiseless images and one on noisy images, accordingly tested on noiseless and noisy images respectively. Figure 5 shows the images of galaxy with their target complex ellipticity superimposed, as well as the predicted one.

The ellipse represents the estimated shape - with a fixed scale adapted for visualization - and the arrow is the corresponding complex ellipticity - modified with half its argument in order to be aligned with the main axis of the ellipse. On this example, we can qualitatively see that the galaxy ellipticity on the noisy image is harder to estimate as the noise deforms the shape of the galaxy. Figure 6 generalizes this observation as it shows a sample of the predicted ellipticities on

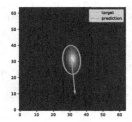 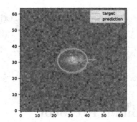

(a) Predicted ellipticity without noise   (b) Predicted ellipticity with noise

Fig. 5. Galaxy images with the predicted ellipticity superimposed on them. The arrow and the corresponding elliptic shape are rendered in an arbitrary scale for visualization purposes. In orange: the true ellipticity. In green: the predicted ellipticity (Color figure online)

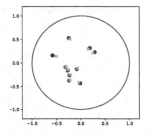 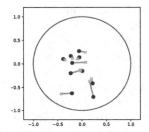

(a) Predicted ellipticities without noise   (b) Predicted ellipticities with noise

Fig. 6. Predicted ellipticities on the complex plane. In red: unit circle. In yellow: predicted ellipticities. In blue: target ellipticites. In green: difference between true and predicted values (Color figure online)

the complex plane within the unit circle, with the target ellipticity and the difference between predicted and targeted values.

While the model trained on noiseless data performs really well (Fig. 6a), it cannot achieve the same level of performance when trained on noisy data, losing part of its reliability (Fig. 6b). As such, using a simple CNN without any estimation of aleatoric uncertainty is not satisfying for our application.

We now train two Bayesian Convolutional Neural Networks with an MVN distribution to estimate both epistemic and aleatoric uncertainties, as seen in Sect. 3.2. Like the simple CNN models, we show in Fig. 7, the ellipticities estimated from the BNNs on the complex plane. We also add the 90% confidence ellipses of both epistemic, aleatoric and predictive uncertainties. We observe that in both cases, the epistemic uncertainty is low if not negligible, meaning that the model is confident in its predictions. Put another way, all K pairs of outputs $\mu(X, w_k)$ and $\Sigma(X, w_k)$ are roughly equal to their mean, respectively $\mu(X, w)$ and $\Sigma_{aleat.}(X, w)$, so that, according to Eq. 3 and Eq. 5, $\Sigma_{epist.}(X, w) \approx 0$ and $\Sigma_{pred.}(X, w) \approx \Sigma_{aleat.}(X, w)$. The aleatoric uncertainty is low for noiseless

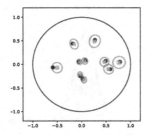
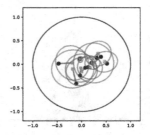

(a) Predicted ellipticities without noise   (b) Predicted ellipticities with noise

Fig. 7. Predicted ellipticities on the complex plane. In red: unit circle. In yellow: predicted ellipticities. In blue: target ellipticites. In light green: difference between true and predicted values. In pink: 90% epistemic confidence ellipse. In dark green: 90% aleatoric confidence ellipse. In grey: 90% predictive confidence ellipse (Color figure online)

images but higher for noisy ones, confirming that the noise corrupting galaxy images makes it more difficult for the model to consistently give an accurate ellipticity estimation.

Finally, in order to see if the MVN distribution is well calibrated, we standardize the output and check if the resulting distribution follows the standard distribution. More precisely, if we define:

$$Z(X,w) = \Sigma_{pred.}(X,w)^{-\frac{1}{2}}(Y - \mu(X,w)), \tag{6}$$

then the distributions of its two independent components $z_1 \sim Z(X,w)_1$ and $z_2 \sim Z(X,w)_2$ should be equivalent to the standard distribution $\mathcal{N}(0,1)$. Note that this is true only because all K output MVNs are confounded. Figure 8 shows that the standardized distributions for the model trained on noisy images are indeed well calibrated and therefore the model is neither overestimating nor underestimating the predictive uncertainty.

4.2   Estimation of Uncertainty Related to Blending

In the previous part we showed that our BNNs are well calibrated. Here we submit outliers to the networks in order to study the impact on epistemic uncertainty and whether it can be used to detect them. Our models have only been trained on images of isolated galaxies, but astrophysical images can contain multiple overlapped galaxies. In that case, asking the model to measure a single ellipticity does not make sense. If the epistemic uncertainty behaves as expected, then its measurement would allow us to detect when a predicted ellipticity is incorrect due to the presence of multiple galaxies in the image. We fed images of blended scenes to the two models trained on isolated galaxies (with or without noise), adding noise to the blended scenes only for the model trained on noisy images.

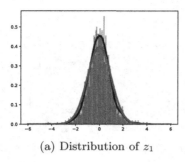
(a) Distribution of z_1

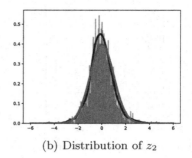
(b) Distribution of z_2

Fig. 8. Histogram of the standardized distributions on the model trained with noisy images. In red: standard bell curve. In blue: histogram of the standardized distribution with the smoothed curve. (Color figure online)

Results shown in Fig. 9 demonstrate that in both cases the predictions are particularly inexact when compared to the target ellipticity of the central galaxy. Also, and as expected, the epistemic uncertainty is much higher for these blended scenes than for isolated galaxy images. However, the aleatoric uncertainty gives incoherent values as the model has not been trained to evaluate it on blended images: notice how the aleatoric ellipses are more flattened with a lower area. Figure 10 permits to visualise the behavior of the epistemic uncertainty. It shows how the ellipticities sampled with dropout slightly diverge compared to the mean prediction. Here the model cannot give a consistent answer and therefore its prediction should be deemed untrustworthy.

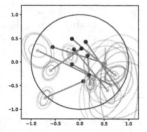
(a) Predicted ellipticities, without noise

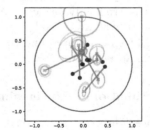
(b) Predicted ellipticities, with noise

Fig. 9. Predicted ellipticities on the complex plane for blended galaxies images. In red: unit circle. In yellow: predicted ellipticities. In blue: target ellipticites (label of the centered galaxy). In light green: difference between true and predicted values. In pink: 90% epistemic confidence ellipse. In dark green: 90% aleatoric confidence ellipse. In grey: 90% predictive confidence ellipse (Color figure online)

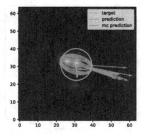
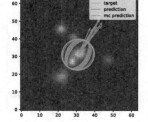

(a) Blended galaxies without noise   (b) Blended galaxies with noise

Fig. 10. Blended galaxies images with the predicted ellipticity superimposed on them. The arrow and the corresponding elliptic shape are rendered in an arbitrary scale for visualization purposes. In orange: the true ellipticity (label for the galaxy in the center). In green: the predicted ellipticity. In pink: the individual MC dropout predicted ellipses. The green ellipticity is therefore the mean of the pink ones. On both images the prediction is uncertain as the individual MC samples slightly diverge from the mean (Color figure online)

To quantify the quality of the epistemic uncertainty when it comes to detecting incoherent predictions due to outliers, we computed the ROC curves for each uncertainty type. More precisely, we reduce each covariance matrix (aleatoric, epistemic and predictive) to a scalar by computing its determinant. We interpret these estimates as a scoring function to assess whether an image is an outlier, i.e. a blended image: the higher the score, the more likely the image contains a blend. Finally, we compute for each of these scoring functions its ROC curve. We repeat that process for both networks trained with noisy and noiseless data. The results are shown in Fig. 11.

These ROC curves are also summarized by their associated Area Under Curve (AUC) on Fig. 11c. The epistemic uncertainty clearly appears as the most consistent "metric" to detect outliers and therefore to give useful information about the confidence in the model predictions. Even the predictive uncertainty performs worse than the epistemic one. This is especially true in the presence of noise since the aleatoric uncertainty then occupies a more important part of the predictive one compared to the noiseless case. Notice that the aleatoric ROC curve is mostly below the diagonal with an AUC below 0.5, meaning it performs worse than a random classifier. This is due to the fact that the model has not been trained to evaluate aleatoric uncertainty on blended scenes. As seen in Fig. 9, the aleatoric ellipses are more flattened in the blended cases, meaning its determinant is lower. Thus the aleatoric uncertainty is on average lower on blended scenes when compared to isolated ones.

To compensate, results of the complementary classifier for the aleatoric uncertainty are shown. It is still not as satisfying as the epistemic uncertainty. While using epistemic uncertainty to identify inconsistent predictions due to a lack of knowledge is highly effective, we note that few blended images still have low epistemic uncertainty due, for instance, to a large galaxy that obstructs all of the other ones, making the image actually closer to an isolated galaxy image.

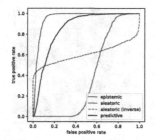
(a) ROC curve, model without noise

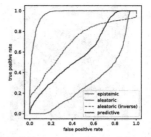
(b) ROC curve, model with noise

Uncertainty	AUC noiseless	AUC noise
Epistemic	**0.956**	**0.969**
Aleatoric	0.394	0.306
Aleatoric (inverse)	0.606	0.694
Predictive	0.856	0.594

(c) AUC values

Fig. 11. ROC curves for detecting outliers for aleatoric, epistemic and predictive uncertainty. (a) ROC curve, model without noise. (b) ROC curve model with noise. Since the aleatoric ROC curve gives incoherent answers on outliers (see Fig. 9), we also plot the complementary classifier as a dashed line. (c) AUC values for all uncertainties, for the model with and without noise. Here, the epistemic uncertainty is clearly the best to detect outliers, as its AUC value is close to 1 in both noisy and noiseless datasets

Finally, we evaluate how each type of uncertainty is a reliable representation of the risk of error in ellipticity prediction. Unfortunately, in the presence of blended images, the predictive distribution is no longer a simple MVN but a mixture of K well separated Gaussian distributions. The normalization process that allowed us to obtain the results presented in Fig. 8 is no longer applicable here. It is still possible to study the relationship between the uncertainty and the ellipticity prediction error testing a trivial rule: the higher the uncertainty, the more important we expect the error to be. To do so, we do three sorting of the images according to each uncertainty type, from the lowest uncertainty to the highest, on a scale from 0 to 0.4 for isolated objects, and from 0.4 to 1 for blended scenes. We then compute the mean ellipticity error considering the proportion of the sorted data from 0 to 1. For blended scenes the ellipticity prediction error is computed w.r.t. the ellipticity of the centered galaxy. We repeat this experiment twice, for networks trained on noiseless and noisy data. Finally we add an "oracle" curve where the data is sorted directly according to the ellipticity prediction error which represents a perfect sorting. Results are shown in Fig. 12.

Once again, epistemic uncertainty proves to be best suited to anticipate ellipticity predictive error. The samples with the lowest epistemic uncertainty have the lowest mean ellipticity error and conversely, while samples with low aleatoric uncertainty can already have high mean error. Consequently, on real astrophysical data, when the predictive ellipticity error is obviously unknown, relying on the epistemic uncertainty to reject, or minimize the impact of, a sample because of its probable predictive error is the best way to go.

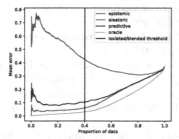 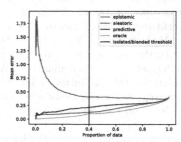

(a) Mean error curve, model without noise   (b) Mean error curve, model with noise

Fig. 12. Mean error curves w.r.t. data proportion for aleatoric, epistemic and predictive uncertainty. (a) Mean error curve without noise. (b) Mean error curve with noise. In black the threshold between the proportion of isolated galaxies: $[0, 0.4]$ and blended galaxies: $[0.4, 1]$. In pink the oracle curve, where the data is sorted by the predictive error. The closest a curve is to the oracle the better (Color figure online)

5 Conclusion

We developed a Bayesian approach to estimate the posterior distribution of galaxy shape parameters using convolutional neural networks and MC-Dropout. In addition to a precise measurement of the ellipticities, this approach provides a calibrated estimation of the aleatoric uncertainty as well as an estimation of the epistemic uncertainty. We showed that the latter is behaving according to expectations when applied to different kind of galaxy images, and is well-suited to identify outliers and to anticipate high predictive ellipticity error. These results confirm the suitability of Bayesian neural networks for galaxy shape estimation and incite us to continue exploring their use to go from ellipticity posterior distributions, estimated from multi-band galaxy images, to cosmic shear estimation.

References

1. Gal, Y., Ghahramani, Z.: Dropout as a Bayesian approximation: representing model uncertainty in deep learning. In: ICML Proceedings 2016, JMLR Workshop and Conference Proceedings, vol. 48, pp. 1050–1059 (2016)
2. LeCun, Y., Bengio, Y., Hinton, G.: Deep learning. Nature **521**(7553), 436–444 (2015)
3. Gal, Y.: Uncertainty in deep learning, PhD thesis, University of Cambridge (2016)
4. Gal, Y., Hron, J., Kendall, A.: Concrete dropout. In: NIPS Proceedings 2017, pp. 3581–3590 (2017)

5. Lakshminarayanan, B., Pritzel, A., Blundell, A.: Simple and scalable predictive uncertainty estimation using deep ensembles. In: NIPS Proceedings 2017, pp. 6402–6413 (2017)

6. Kendall, A., Gal, Y.: What uncertainties do we need in Bayesian deep learning for computer vision? In: NIPS Proceedings 2017, pp. 5574–5584 (2017)

7. Srivastava, N., et al.: Dropout: a simple way to prevent neural networks from overfitting. J. Mach. Learn. Res. **15**, 1929–1958 (2014)

8. Le, V.Q., Smola, J.A., Canu, S.: Heteroscedastic Gaussian process regression. In: ICML Proceedings 2005, vol. 119, pp. 489–496 (2005)

9. Hüllermeier, E., Waegeman, W.: Aleatoric and epistemic uncertainty in machine learning: a tutorial introduction. CoRR. http://arxiv.org/abs/1910.09457 (2019)

10. Depeweg, S., et al.: Decomposition of uncertainty in Bayesian deep learning for efficient and risk-sensitive learning. In: ICML Proceedings 2018, vol. 80, pp. 1192–1201 (2018)

11. Hinton, G.E., van Camp, D.: Keeping the neural networks simple by minimizing the description length of the weights. In: Proceedings of the Sixth Annual ACM Conference on Computational Learning Theory, COLT 1993, pp. 5–13 (1993)

12. Dieleman, S., Willett, W.K., Dambre, J.: Rotation-invariant convolutional neural networks for galaxy morphology prediction. Mon. Not. R. Astron. Soc. **450**(2), 1441–1459 (2015)

13. Kaiser, N., Squires, G., Broadhurst, T.: A method for weak lensing observations. Astrophys. J. **449**, 460–475 (1995)

14. Rowe B., et al.: GalSim: the modular galaxy image simulation toolkit. https://github.com/GalSim-developers/GalSim (2015)

15. Goodfellow, I.J., et al.: Maxout networks. In: Proceedings of the 30th International Conference on Machine Learning, vol. 28, pp. 1319–1327 (2013)

16. Abell, P.A., et al.: LSST science collaboration. LSST science book. arXiv, arXiv:0912.0201 (2009)

17. Kilbinger, M.: Cosmology with cosmic shear observations: a review. Rep. Prog. Phys. **78**, 086901 (2015)

18. Bosch, J., et al.: The hyper suprime-cam software pipeline. Publications of the Astronomical Society of Japan 70, S5 (2018)

19. Sanchez, J., Mendoza, I., Kirkby, D.P., Burchat, P.R.: Effects of overlapping sources on cosmic shear estimation: statistical sensitivity and pixel-noise bias. arXiv e-prints. arXiv:2103.02078 (2021)

20. Mandelbaum, R., et al.: The third gravitational lensing accuracy testing (GREAT3) challenge handbook. Astrophys. J. Suppl. Ser. **212**, 5 (2014)

21. Mandelbaum, R.: Weak lensing for precision cosmology. Annu. Rev. Astron. Astrophys. **56**, 393 (2018)

22. Bertin, E., Arnouts, S.: SExtractor: software for source extraction. Astron. Astrophys. Suppl. Ser. **117**, 393 (1996)

23. Melchior, P., et al.: SCARLET: source separation in multi-band images by constrained matrix factorization. Astron. Comput. **24**, 129 (2018)

24. Arcelin, B., Doux, C., Aubourg, E., Roucelle, C.: LSST dark energy science collaboration: deblending galaxies with variational autoencoders: a joint multi-band, multi-instrument approach. Mon. Not. R. Astron. Soc. **500**, 531 (2021)

Precise Weather Parameter Predictions for Target Regions via Neural Networks

Yihe Zhang[1], Xu Yuan[1(✉)], Sytske K. Kimball[2], Eric Rappin[3], Li Chen[1],
Paul Darby[1], Tom Johnsten[2], Lu Peng[4], Boisy Pitre[1], David Bourrie[2],
and Nian-Feng Tzeng[1]

[1] University of Louisiana at Lafayette, Lafayette, LA 70508, USA
xu.yuan@louisiana.edu
[2] University of South Alabama, Mobile, AL 36688, USA
[3] Western Kentucky University, Bowling Green, KY 42101, USA
[4] Louisiana State University, Baton Rouge, LA 70803, USA

Abstract. Flexible fine-grained weather forecasting is a problem of
national importance due to its stark impacts on economic development and human livelihoods. It remains challenging for such forecasting, given the limitation of currently employed statistical models, that
usually involve the complex simulation governed by atmosphere physical equations. To address such a challenge, we develop a deep learning-
based prediction model, called Micro-Macro, aiming to precisely forecast
weather conditions in the fine temporal resolution (i.e., multiple consecutive short time horizons) based on both the atmospheric numerical
output of WRF-HRRR (the weather research and forecasting model with
high-resolution rapid refresh) and the ground observation of Mesonet stations. It includes: 1) an Encoder which leverages a set of LSTM units
to process the past measurements sequentially in the temporal domain,
arriving at a final dense vector that can capture the sequential temporal patterns; 2) a Periodical Mapper which is designed to extract the
periodical patterns from past measurements; and 3) a Decoder which
employs multiple LSTM units sequentially to forecast a set of weather
parameters in the next few short time horizons. Our solution permits
temporal scaling in weather parameter predictions flexibly, yielding precise weather forecasting in desirable temporal resolutions. It resorts to a
number of Micro-Macro model instances, called modelets, one for each
weather parameter per Mesonet station site, to collectively predict a
target region precisely. Extensive experiments are conducted to forecast
four important weather parameters at two Mesonet station sites. The
results exhibit that our Micro-Macro model can achieve high prediction accuracy, outperforming almost all compared counterparts on four
parameters of interest.

1 Introduction

Weather forecasting in the temporal domain is a critical problem of national
importance, closely tied to the economic development and human livelihoods.

© Springer Nature Switzerland AG 2021
Y. Dong et al. (Eds.): ECML PKDD 2021, LNAI 12979, pp. 151–167, 2021.
https://doi.org/10.1007/978-3-030-86517-7_10

However, accurate forecasting remains open and quite challenging, especially in the context of precise and fine-grained prediction over multiple temporal resolutions. Such a short-term and fine temporal resolution prediction relates tightly to agriculture, transportation, water resource management, human health, emergency responses, and urban planning, essential for taking such timely actions as generating society-level emergency alerts on convection initiation, producing real-time weather guidance for highways and airports, among others.

To date, the most prominent and widely used national forecasting model is called Weather Research and Forecasting (WRF) with HRRR (High Resolution Rapid Refresh) [3]. It provides prediction for the weather parameters that cover the United States continent. However, it is an hourly prediction model, which can only coarsely forecast the weather parameters in the resolution of one hour, failing to capture finer time granularity needs (say, the 5- or 10-min time horizon) in weather forecasting. This is largely due to the high computation requirement and voluminous data outputs associated with this model, involving complex simulation of physical governing atmospheric flows [16]. Its prediction accuracy is far from satisfaction, as a result of the employed statistical models, whose capability of extracting fine-grained weather patterns is limited. Meanwhile, more than three dozen of regional Mesonet networks exist under the U.S. National Mesonet Program, with each network involving tens or hundreds of observational stations for gathering near-surface weather measurements periodically. Mesonet Stations provide site-specific real datasets in finer temporal granularity (typically in minutes). For example, our experimental evaluation makes use of datasets gathered by the SA Mesonet, which covers South Alabama by 26 observational stations to gather data once in every minute [2].

Recent advances in machine learning technologies have promoted weather forecasting into a new era. Many studies have attempted to leverage the neural network-centric techniques in weather forecasting, producing promising results. These techniques include, but are not limited to, the deep neural network (DNN), convolutional neural network (CNN), long short-term memory network (LSTM), generative adversarial network (GAN), and autoEncoder, for predicting such weather parameters as precipitation [14,18,27], wind direction and speed [4,10, 15,19], solar radiation [5,12], air quality [28], weather changes [13,26], and many others. However, known parameter forecasting models developed so far cannot yield accurate enough predictions in fine-grained temporal resolution over flexible time horizons.

This paper aims to develop a new forecasting model, termed Micro-Macro, for effective and precise prediction on weather parameters in the fine-gained temporal resolution, by taking both micro inputs from Mesonet Stations [1] and macro inputs from Weather Research and Forecasting (WRF) with HRRR (High Resolution Rapid Refresh) [3] computation outputs, for the first time. We leverage the prominent deep learning technologies that take the existing massive atmospheric data sets (resulting from WRF-HRRR numerical prediction) and surface observation data (gathered via existing Mesonet networks) as the input to produce fine-grained weather forecasting in the temporal domain for target regions of interest. Specifically, the developed model includes three components: 1) an Encoder which processes the time sequence data to capture the temporal domain variation of weather conditions, 2) a Periodical Mapper which extracts the periodical pattern of the time sequence data, and 3) a Decoder which predicts a sequence of values corresponding to different time points. Specifically, each LSTM unit in the Encoder can learn the key features from inputs and then outputs its hidden state to the next LSTM unit, which can continue to learn the key features from both the previous input and the current input, in terms of time sequence characteristics. This results in a dense vector, including rich information for the weather condition's variation in the temporal domain out of the atmospheric output and surface observation. Meanwhile, a Periodical Mapper can capture the periodical pattern of the data and generate a dense vector for enhancing the learning of temporal data patterns. Both dense vectors are used by the Decoder to forecast the weather parameters in the next few continued time horizons. This model incorporates the near surface observation and the atmospheric numerical output, which are complementary with each other to let our model better use relevant past measurements for forecasting, significantly improving prediction accuracy.

We conduct experiments to predict a set of weather parameters. Our experimental results show that the developed Micro-Macro model instances, dubbed modelets, outperform almost all the compared solutions in forecasting temperature, humidity, pressure, and wind speed.

2   Related Work

Abundant applications of machine learning techniques to weather forecasting exist. This section reviews the recent advances in such applications, which mostly follow two lines of work.

The first line aims to explore whether the neural network is capable of simulating the physical principles of atmosphere systems. In particular, Dueben et al. [11] employed two neural networks, i.e., Global NN and Local NN, to simulate the dynamics of a simple global atmosphere model at 500 hPa geopotential. The results concluded that prediction outcomes by the neural network models can be better than those of the coarse-resolution atmosphere models for a short duration under the 1-h time scale. Scher [21] applied the CNN structure with autoEncoder setup to learn the simplified general circulation models (GCMs), which can predict the weather parameters up to 14 days. Weyn et al. [25] leveraged the CNN with LSTM structure to achieve a 14-day lead time forecasting as well. Vlachas et al. [22] employed the LSTM model to reduce the order space of a chaotic system. However, known proposed solutions along this line all just focused on developing prediction models for simulated or simplified climate environments, without taking into account the real-world conditions, which tend to be rather complex. Their applicability and effectiveness on real environments are still questionable, given their complex conditions in practice. For example, the actual measurements from Mesonet stations are highly dependent on local conditions. In addition, their solutions cannot be applied to fine-grained predictions with flexible time horizons in the desirable temporal resolution.

The other line of work aims to leverage the neural networks to develop new models for the real-world weather parameters prediction. For example, [19] leveraged the LSTM and fully connected neural networks to predict the wind speed at an offshore site, by capturing its rapidly changing features. Grover et al. [13] combined the discriminatively trained predictive models with a deep neural network to predict the atmospheric pressure, temperature, dew point, and winds. [27] proposed a convolutional LSTM model to predict precipitation. Pan et al. [18] employed the CNN with delicately selected stacked frames for precipitation forecasting. [14] proposed a model with the autoEncoder structure to predict rainfalls. [4] forecasted the hurricane trajectories via an RNN structure. [12] and [5] employed the LSTM structures to predict the solar radiation and photovoltaic energy, respectively. [28] proposed a deep fusion network to predict air quality. [26] developed a deep-CNN model on a cubed sphere for predicting several basic atmospheric variables on a global grid. However, all aforementioned work still cannot predict weather parameters accurately in fine granularity over flexible time horizons, for a desirable temporal resolution. Hence, accurate weather prediction and fine-grained temporal resolution across flexible time horizons remains an open and challenging problem.

3   Pertinent Background

In this section, we describe Mesonet near surface observation and WRF-HRRR (Weather Research and Forecasting with High Resolution Rapid Refresh model) prediction model to illustrate their limitations in precise weather forecasting.

Mesonet [1] is a national supported program that comprises a set of automated weather stations located at some specific areas in the USA. Its towers aim to gather meteorological- and soil- measurements relevant to local weather phenomena. Each station monitors tens of atmospheric measurements, including temperature, rainfall, wind speed, and others, once per minute for every day since its establishment.

WRF with HRRR Prediction: The WRF model takes actual atmospheric conditions (i.e., from observations and analyses) as its input to produce outputs that serve a wide range of meteorological applications across national scales. WRF with HRRR weather forecast modeling system is nested in the Rapid Refresh model for predicting weather parameters that cover the United States continent with a resolution of 3 km for a total of 1059×1799 geo-grids. The prediction outputs are produced hourly, over the next consecutive 18 h. In each geo-grid, there are up to 148 parameters, representing the temperature, pressure, among many others, to signify the predicted weather condition. A 1059×1799 matrix is employed to keep each parameter's outputs, with each entry mapping to one geolocation of the United States map.

However, both Mesonet and WRF-HRRR have their respective limitations. For Mesonet, the involved stations are only for gathering the current near-surface measurements, unable to predict future values. For WRF-HRRR, its prediction accuracy is far from satisfaction, besides its hourly scale prediction to limit its suitability for meteorological applications that requires high temporal resolutions (say, 5 min, 15 min, or 30 min).

4 Learning-Based Modelets for Weather Forecasting

This paper aims to develop learning-based meteorology (abbreviated as Meteo) modelets, for correctly and concurrently predicting multiple weather parameters in a flexible and fine-temporal resolution, based on the inputs of both minute-level near-surface observations from Mesonet and WRF hourly atmospheric numerical outputs, referring respectively as the Micro and the Macro datasets. We take the Micro dataset as the main input and screen a set of relevant parameters in Macro dataset for incorporation to predict target weather parameters correctly. Our goal is to extract the temporal variation features from the previous measurements to precisely predict the weather condition in the next few time horizons (e.g., next T min, $2T$ mins, etc.). It is challenging as the two data sources have different scales in the temporal domains. To address such a challenge, the prominent machine learning technology is leveraged to learn the temporal sequence patterns from both datasets that can capture variation of weather conditions to predict specific parameters. A new Meteo modelet, named Micro-Macro, is developed to permit temporal downscaling and upscaling in weather parameter predictions flexibly, arriving at precise weather forecasting in desirable temporal resolutions. We will first outline a Micro model by just relying on the Micro dataset as the input for prediction. Then, we describe our Micro-Macro model which takes both Micro and Macro datasets as the input for precisely forecasting weather parameters via separate modelets (i.e., model instances) in the temporal domain.

4.1 Micro Model

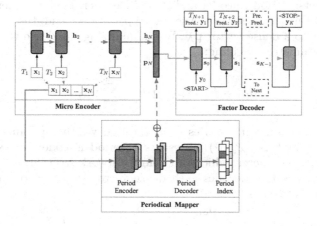

Fig. 1. Structure of micro model.

Most atmospheric data has the noticeable temporal sequence patterns and periodical patterns, whereas weather conditions (i.e., parameters) change continuously with time. To capture such patterns for forecasting in continuous T-minute horizons, we leverage a structure with an Encoder, a Decoder, and a Periodical Mapper, with the first two both include the (LSTM) networks and the last one is in the neural network structure, to capture the time sequence patterns and periodical patterns, respectively. The structure is shown in Fig. 1. Notably, although the encoder-decoder LSTM model has been widely applied to sequence tasks, e.g., language translation [9] and question answering [7], the physical meaning in each entry for the input vectors is not well explored. This results in the loss of affluent element-wise features, only to encode all features into a dense vector, which cannot work effectively here. The customized design is desired under our application context. The details of three components are illustrated as follows.

Micro Encoder. It comprises one LSTM network, to encode the temporal sequence data in a certain period into one single dense vector, representing the temporal feature variation. To forecast weather condition in next continuous T-*min* horizons, we consider the past $N \times T$ minutes surface observation from Mesonet as a sequence of data frames, with each one including T-*min* observed weather condition to serve as the input. Here, N represents the number of selected T-*min* intervals. The LSTM unit will learn the key features and update its corresponding hidden state vector (denoted as $\mathbf{h}_{t-1}$). Such a vector together with the next data frame is input to the next LSTM unit to produce a new hidden state vector $\mathbf{h}_t$, which can be logically modeled as follows:

$$\mathbf{h}_t = LSTM_h(\mathbf{h}_{t-1}, \mathbf{x}_t) \,, \tag{1}$$

where $LSTM_h$ represents a series of steps to generate the next hidden states and $\mathbf{x}_t$ denotes the data frame in time slot t. In the end, a dense vector $\mathbf{h}_N$ is generated, including the aggregated temporal patterns variation from N inputs.

Periodical Mapper. This design is used to process the input data sequence $\mathbf{x} = \{\mathbf{x}_1, \mathbf{x}_2, ..., \mathbf{x}_t, ..., \mathbf{x}_N\}$ for extracting the periodical patterns, comprising two core components: Period Encoder and Period Decoder. Each weather parameter i has a Period Encoder, with its dense vector $\mathbf{p}_{(i)}$. In the end, the sequence data $\mathbf{x}$ is encoded into a dense vector $\mathbf{p}_N$, by summarizing the dense vector from all M weather parameters, yielding:

$$\mathbf{p}_{(i)} = P_{e,i}(\bar{\mathbf{x}}_{(i)}), \ \mathbf{p}_N = \sum_{i=1}^{M_i} \mathbf{p}_{(i)} \,. \tag{2}$$

where $\bar{\mathbf{x}}_{(i)}$ is a vector with entries from the i-th weather parameter value of $\mathbf{x}_1, \mathbf{x}_2, ..., \mathbf{x}_t, ...,$ and $\mathbf{x}_N$, $P_{e,j}(\cdot)$ represents a Period Encoder corresponding to the i-th weather parameter, which is a neural network structure.

The Period Decoder decodes each dense vector $\mathbf{p}_{(i)}$ to a periodical index vector $\mathbf{p}_{\mathbf{o}(i)}$, expressed as

$$\mathbf{p}_{\mathbf{o}(i)} = P_{d,i}(\mathbf{p}_{(i)}) \,, \tag{3}$$

where $P_{d,i}$ is also a neural network structure. If the input temporal sequence $\bar{\mathbf{x}}_{(i)}$ matches a periodical pattern, the corresponding entry will be 1 and all other entries will be 0.

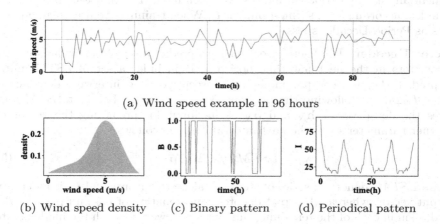

(a) Wind speed example in 96 hours

(b) Wind speed density   (c) Binary pattern   (d) Periodical pattern

Fig. 2. Example on periodical pattern discovery.

In the training phase, we derive the periodical index vector from the historical weather records. We run a toy example to explain this step. For example, Fig. 2(a) shows the wind speed within 96 h, taken from the Mesonet observation dataset. We first need to find a reference point, which shall help discover the periodical pattern of the weather records. Since the data distribution is unknown, we leverage Kernel Density Estimation [8] to find the density of observation values, with the density likelihood to yield:

$$\hat{f}_h(X_i) = \frac{1}{nh} \sum_{j=1}^{n} \Phi(\frac{X_i - X_{ij}}{h}) , \qquad (4)$$

where $\hat{f}_h(X_i)$ is the density function of measurement X_i. X_{ij} is the j-th observed value of X_i corresponding to a weather parameter. n is the total number of data points and h is an empirical parameter which is set to 0.85 in our experimental evaluation. Φ denotes the normal distribution. By maximizing Eqn. (4), we get the density distribution as shown in Fig. 2(b) and pick up the largest density point of $\hat{f}_h(X_i)$ as the reference point, i.e., 5.03. We then consider the area that covers top-15% density values as the reference area R_i. A binary sequence B_i of measurement X_i is then calculated. That is, if the observed value $X_{ij} \in R_i, 0 < j \leq n$, we set $B_{ij} = 1$, otherwise $B_{ij} = 0$, as shown in Fig. 2(c). Afterwards, we conduct the Discrete Fourier Transform (DFT) [24] on the sequence B_i to transform them to n complex numbers, denoted as $D_i : [D_{i1}, D_{i2}, ..., D_{ij}, ..., D_{in}]$. Then we calculate the periodogram $F_{ij} = \|D_{ij}\|^2$ for each complex number to get F_i. By taking Inverse Discrete Fourier Transform (IDFT) [17] on F_i, we derive the Periodic Correlation I_i [6], as shown in Fig. 2(d).

In the curve of I_i, we identify all peak values. Each interval between two neighboring peak values is denoted as one period. We equally divide each time period into $P = 24 * 60/T$ time slots and label the time slots from 1 to $24 * 60/T$ sequentially as the periodical indices. We then label the input sequence $\mathbf{x}$ with an index, according to $\mathbf{x}$'s timestamp on I_i. When training, we use Mean Square Loss as Period Decoder's loss.

Factor Decoder. The Factor Decoder is to predict a set of particular weather parameters in the next few time horizons. It includes a set of LSTM units, to predict the weather parameter at consecutive time intervals, denoted as $T_{N+1}, T_{N+2}, \cdots$, following the previous $N \times T$ minutes. The first LSTM unit takes the dense vector $\mathbf{h}_N$ and $\mathbf{p}_N$ as its input for predicting the vector of weather parameter $\mathbf{y}_1$ in the next interval T_{N+1} as follows,

$$\mathbf{y}_1 = LSTM_o(\mathbf{y}_0, \langle \mathbf{h}_N, \mathbf{p}_N \rangle) , \tag{5}$$

where $LSTM_o$ denotes a series of steps to calculate outputs and $\mathbf{y}_0$ is an empty output vector, whereas $\langle \mathbf{h}_N, \mathbf{p}_N \rangle$ denotes concatenation of $\mathbf{h}_N$ and $\mathbf{p}_N$. For the prediction in each of the remaining time intervals, we take both the hidden state vector $\mathbf{s}_k$ and the previous predicted vector $\mathbf{y}_k$ as inputs to update the current LSTM state. The new hidden state $\mathbf{s}_{k+1}$ can be logically expressed as: $\mathbf{s}_{k+1} = LSTM_s(\mathbf{s}_k, \mathbf{y}_k)$. Note that, we retake the pervious output as new input to update the new hidden state. The next output is given by $y_{k+1} = LSTM_o(\mathbf{y}_k, \mathbf{s}_{k+1})$.

In the training phase, each $(N \times T)$-minute data will be used as inputs and the data from subsequent M time intervals will be used for labeling. Here, M represents the number of time horizons that we aim to predict. For example, to predict a weather parameter, say temperature, we consider a set of relevant parameters in $N \times T$ minutes as the features and label the temperature values in the following time intervals of $T_{N+1}, T_{N+2}, \cdots, T_{N+M}$. As the surface observation data are generated once in every minute, we average the values of each parameter over T minutes as the features. Similarly, for labeling, we take the averaged temperature value within each T minutes. The N data frames (corresponding to the $(N \times T)$-minute past measurements) and the labeled temperature values (in M subsequent intervals) are inputted to Micro Encoder. At the Decoder, we start from the first LSTM unit and predict a set of weather parameters at the time interval of T_{N+1}. Both the hidden state from this LSTM network and the predicted value of T_{N+1} are then input to the second LSTM for predicting T_{N+2}. This step continues until all values for the next M time horizons are predicted.

4.2 Micro-Macro Model

As the number of observed parameters at Mesonet is limited, it is insufficient for forecasting just based on the Micro datset. Hence, we incorporate the Macro dataset as a complementary input to the model for better forecasting. Given the Macro dataset is hourly generated and surface observation is updated in each

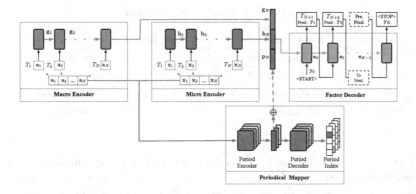

Fig. 3. The structure of micro-macro model.

minute, how to integrate such two data sources is still a challenging problem, as it requires downscaling the atmospheric output.

The structure of Micro-Macro model is shown in Fig. 3, which is similar to that of the Micro model, with a difference in the input that includes an additional Macro Encoder. In Macro Encoder, we divide each hour into $60/T$ time frames and use this hourly output from Macro dataset to represent the first time frame's value. The values of all remaining time frames are indicated as "Empty". All hourly datasets are processed in the same way. When inputting to the Encoder, if a frame has an empty value, the corresponding LSTM unit in the Macro Encoder takes only the hidden state vector from the previous unit as the input to self-update its hidden state vector; otherwise, it executes in the same way as in Micro Encoder. The Macro Encoder outputs a dense vector, denoted by $\mathbf{g}_N$, as depicted in Fig. 3. To extract the time sequence features from both Micro and Macro datasets, we concatenate the dense vectors ($\mathbf{h}_N$, $\mathbf{g}_N$, and $\mathbf{p}_N$) from the Micro Encoder, Macro Encoder and Periodic Selector, i.e., $\mathbf{h} = \langle \mathbf{h}_N, \mathbf{g}_N, \mathbf{p}_N \rangle$. The decoder in the Micro-Macro model is similar to that in the Micro model. It takes the concatenated dense vector $\mathbf{h}$ as its input to perform forecasting for subsequent time horizons sequentially. Notably, in both training and prediction phases, the Micro-Macro model takes the data of the same geo-grid from Micro and Macro datasets at an identical time interval.

5   Experiment

We conduct experiments to evaluate the performance of Macro-Micro model for precise weather parameters (i.e., temperature, humidity, pressure, and wind speed) prediction regionally.

5.1   Setting

Datasets. We take the near surface observation from SA Mesonet [2] and the WRF-HRRR [3] atmospheric numerical output as our experimental datasets,

Table 1. Parameter information

Parameter	Measurement	Mounting height	Measuring range
TEMP	Air Temperature	2 m	−40 to 60 °C
HUMI	Relative humidity	2 m	0 to 100%
PRES	Atmospheric pressure	1.5 m	600 to 1060 mb
WSPD	Wind speed	2 m	0 to 100 m/s

which are called as Micro and Macro datasets, respectively. The Micro dataset includes 26 automated weather stations for monitoring the real-time meteorological phenomena. The monitored weather conditions include temperature, rainfall, wind speed and direction, soil temperature and humidity, once in every minute. SA Mesonet stations Elberta and Atmore are selected for our experiments, with the former located closer to the Gulf Shore and the latter one away from the shore. In total, eight Micro-Macro model instances (called modelets) are involved, one for a weather parameter at each station site. We take the ground observation from years 2017 and 2018 as the training dataset, while taking the observation from 2019 as the test dataset. Macro dataset is the predicted output from WRF-HRRR model. The numerical output in the years 2017, 2018, and 2019, corresponding to the stations of Atmore and Elberta, are taken to conduct our experiments. To forecast temperature, humidity, pressure, and wind speed (see details in Table 1), we select their respective most relevant parameters from Micro dataset and ten most important parameters from the Macro dataset. Table 2 lists the most relevant parameters selected from Micro dataset for training the weather measurements of temperature, humidity, pressure, and wind speed, respectively. Table 3 lists 10 most important parameters that are selected from Macro dataset.

Table 2. Relevant parameters from micro dataset

Predictions	Measurement parameters
TEMP	Vitel_100cm_d, IRTS_Body, SoilCond, SoilWaCond_tc,
	Vitel_100cm_b, eR, wfv, Vitel_100cm_a, SoilCond_tc, RH_10m
HUMI	Temp_C, Vitel_100cm_d, Vitel_100cm_a, Vitel_100cm_b, AirT_2m, AirT_10m
	WndSpd_Vert_Min, SoilT_5cm, Pressure_1, PTemp, IRTS
PRES	RH_10m, SoilCond, Temp_C, Vitel_100cm_d,
	AirT_1pt5m, IRTS_Trgt, PTemp, Vitel_100cm_b, SoilSfcT, AirT_10m
WSPD	WndSpd_2m_WVc_1, WndSpd_10m, WndSpd_2m_Max,
	WndSpd_Vert_Tot, WndSpd_2m_Std, QuantRadn,
	WndSpd_2m_WVc_2, WndSpd_Vert, WndSpd_10m_Max, WndDir_2m

Table 3. Relevant parameters from macro dataset

Feature ID	Description
9	250hpa U-component of wind (m/s)
10	250hpa V-component of wind (m/s)
55	80 m U-component of wind (m/s)
56	80 m V-component of wind (m/s)
61	Ground moisture (%)
71	10 m U-component of wind (m/s)
72	10 m V-component of wind (m/s)
102	Cloud base pressure (Pa)
105	Cloud top pressure (Pa)
116	1000m storm relative helicity (%)

Compared Solutions. We compare our results with the following ones: *1) Observation:* We take the ground observation monitored in 2019 from Mesonet at stations Atmore and Elberta, respectively, to inspect our results; *2) WRF-HRRR:* The predicted atmospheric numerical output in 2019 from WRF-HRRR model; *3) SVR* [20]: A regression model based on support vector machine; *4) SNN-Micro* [11]: A neural network model which takes the Micro dataset for training; *5) SNN-both* [11]: A neural network model that takes the aligned data from both Micro and Macro datasets for training; *6) DUQ_{512}* [23]: A deep uncertainty quantification model which has one GRU layer with 512 hidden nodes; and *7) $DUQ_{512-512}$* [23]: A deep uncertainty quantification model which has two GRU layers with 512 hidden nodes in each layer.

Experiment Setup. We take data from the first season in 2017 and 2018 for training, and predict the weather conditions (i.e., temperature, humidity, pressure, and wind speed) in the same season in 2019. The time is divided with a sequence of $T = 5$-min intervals. We take each set of 60 min' (i.e., $N = 12$) data as the features, and label the weather parameter values in the following 30 min, with each 5 min as one time interval and the averaged value as the label. For prediction, we also take past 60 min' measurements as the input to forecast the next 6 continuous time intervals' values. As SNN-Micro and SNN-both cannot conduct the sequence of prediction, we only let it predict the next time interval immediately after every 60 min' measurement. Both of them employ the 3-layer neural network, with three hidden layers including 200, 100, and 20 neurons, respectively. The input sizes are 10 and 20 respectively.

Each LSTM in the Micro model includes 256 hidden states, whereas every Encoder and Decoder of the Micro-Macro model has 256 and 512 hidden states, respectively. Root Mean Squared Error (RMSE) is employed to gauge the prediction error: RMSE $= \sqrt{\frac{1}{n}\sum_{i=1}^{n}(Y_i - \hat{Y}_i)^2}$, where $\hat{\mathbf{Y}}$ and $\mathbf{Y}$ denote the vectors of predicted and observed values, respectively. n is the number of data values.

5.2 Overall Performance

Table 4. RMSE values of our modelets at atmore and elberta stations

		0 to 5 min	5 to 10 min	10 to 15 min	15 to 20 min	20 to 25 min	25 to 30 min
Atmore	TEMP	0.502	0.531	0.564	0.601	0.632	0.670
	HUMI	4.431	4.507	4.552	4.707	5.122	5.802
	PRES	1.087	1.133	1.139	1.156	1.184	1.235
	WSPD	0.396	0.552	0.572	0.658	0.709	0.833
Elberta	TEMP	0.424	0.468	0.471	0.475	0.479	0.485
	HUMI	1.852	1.873	1.893	1.905	1.933	2.015
	PRES	1.075	1.213	1.245	1.309	1.452	1.607
	WSPD	0.492	0.528	0.556	0.584	0.614	0.656

We conduct multiple experiments to forecast the values of various weather parameters of interest at different time points in the first season of 2019. Table 4 shows the averaged RMSE of our Micro-Macro model for forecasting the next 30-min weather conditions on temperature (TEMP), humidity (HUMI), pressure (PRES), and wind speed (WSPD) at the two representative SA Mesonet stations of Atmore and Elberta, when comparing to ground observations. Clearly, our modelets achieve very small RMSE values for predicting temperature, pressure, and wind speed. Although the RMSE values appear relatively larger for humility prediction at both stations, but when comparing to its wide measurement range (of 1 to 100%), these errors are negligible.

5.3 Comparing to Other Methods

Table 5. RMSE values of different methods for 5-min prediction

	Atmore				Elberta			
	TEMP	HUMI	PRES	WSPD	TEMP	HUMI	PRES	WSPD
WRF-HRRR	2.412	20.471	1.648	1.112	1.633	14.296	1.554	1.412
SVR	3.581	20.507	5.209	1.306	1.734	22.953	6.752	1.887
SNN-Micro	0.668	9.137	5.373	0.354	1.381	4.387	4.927	0.265
SNN-both	0.619	7.611	4.959	**0.330**	0.804	4.250	4.337	**0.264**
DUQ_{512}	0.812	5.668	2.714	0.592	0.645	3.524	3.513	0.541
$DUQ_{512-512}$	0.657	5.354	2.667	0.585	0.632	3.326	3.225	0.489
Micro-Macro	**0.502**	**4.431**	**1.087**	0.396	**0.424**	**1.852**	**1.075**	0.492

We next compare the results from our Micro-Macro model to those from other methods on forecasting temperature, humidity, pressure, and wind speed. Table 5

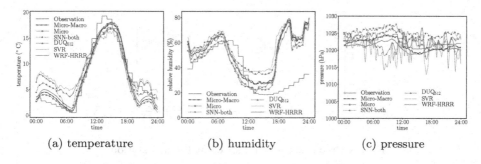

(a) temperature   (b) humidity   (c) pressure

Fig. 4. Prediction of temperature, humidity, and pressure at elberta station.

shows the prediction results of RMSE (in comparsion to ground observation) obtained from different methods for 5-min prediction. We can see our model outperforms all other models, with RMSE values of only 0.502, 4.431, 1.087 at Atmore, and with RMSE values of only 0.424, 1.852, 1.075 at Elberta, on the forecasting of temperature, humidity and pressure, respectively. On predicting wind speed Micro-Macro model beats WRF-HRRR, SVR, DUQ_{512}, and $DUQ_{512-512}$. SNN-Micro and SNN-both have similar prediction performance as our Micro-Macro model on predicting the wind speed parameter, but notably, they cannot conduct a sequence prediction for subsequent multiple time intervals. SVR performs the worst on predicting all parameters at both stations. WRF-HRRR also performs poorly on all parameters but pressure, which has better accuracy than all other models except for our Micro-Macro model. This demonstrates the necessity and importance of developing new meteorological modelets for nationwide use in lieu of WRF-HRRR.

For prediction result illustration, we randomly select one day in the first season of 2019 for forecasting its weather conditions, starting from 00:00 am to 11:59 pm. Figures 4(a), (b), and (c) exhibit the comparative results from our modelets versus those from the ground observation, WRF-HRRR output, Micro, SNN-both, DUQ_{512} and SVR, respectively for forecasting temperature, humidity, and pressure at Elberta station. We observe the curves of our modelets are most close to those from ground observation. This demonstrates that our modelets can continuously provide the best prediction results for the examined duration (of 24 h), in comparison to other methods. Figure 5 shows the results of forecasting wind speed by Micro-Macro model, SNN-both, DUQ_{512}, and WRF-HRRR output for the same day. Micro-Macro, SNN-both, and DUQ_{512} models exhibit similar forecasting performance, being far better than the WRF-HRRR output.

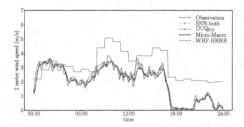

Fig. 5. Prediction of wind speed at Elberta station.

5.4   Ablation Study

The ablation study is next conducted to signify the necessity and importance of the *Periodical Mapper* component in our design. We denote Micro$^-$ and Micro-Macro$^-$ as the models precluding the *Periodical Mapper* from the Micro model and Micro-Macro model, respectively, for comparison. The RMSEs of different variants for 5-min prediction are listed in Table 6.

From this table, we observe that both Micro and Micro-Macro models significantly outperform their respective variants (i.e., Micro$^-$ and Micro-Macro$^-$ respectively) on predicting all four weather parameters at both stations, except that the Micro-Macro model is slightly inferior to the Micro-Macro$^-$ model on predicting wind speed at Elberta station. These results demonstrate that the inclusion of *Periodical Mapper* is important to help elevate the overall prediction performance. In addition, we also observe that our Micro-Macro model greatly outperforms the Micro model, demonstrating the necessity of incorporating both ground observation and the atmospheric numerical output for precise prediction.

Table 6. Results of ablation study

	Atmore				Elberta			
	TEMP	HUMI	PRES	WSPD	TEMP	HUMI	PRES	WSPD
Micro$^-$	0.620	7.892	2.845	5.220	1.289	6.034	3.022	0.682
Micro	0.583	7.279	2.653	5.122	1.064	5.756	2.985	0.467
Micro-Macro$^-$	0.526	4.494	1.114	4.970	0.467	1.860	1.088	**0.447**
Micro-Macro	**0.502**	**4.431**	**1.087**	**4.426**	**0.424**	**1.852**	**1.075**	0.492

5.5   Abnormal Weather Forecasting

We next validate the ability of our proposed Micro-Macro model for forecasting abnormal weather conditions. Four abnormal weather conditions are considered, i.e., chill, torridity, storm and rainstorm, which are assumed to associate with the lowest temperature, highest temperature, highest wind speed, and highest precipitation, respectively. We take the set of 5-min intervals in the first season of 2019 that have the lowest 5% temperature, highest 5% temperature, highest 5% wind speed and highest 5% precipitation from the Mesonet ground measurements. Our experiment is conducted to predict each respective weather parameter in 5-min intervals, with the one hour input.

Table 7. RMSE for abnormal weather prediction

	Chill	Torridity	Storm	Rainstorm
WRF-HRRR	3.098	1.534	5.269	1.694
SVR	3.711	1.715	6.311	4.219
$DUQ_{512-512}$	1.322	0.864	2.695	2.907
Micro	0.452	0.779	2.231	2.301
Micro-Macro	**0.311**	**0.642**	**2.045**	**1.637**

Table 7 lists the averaged RMSE values for different methods for forecasting chill, torridity, storm, and rainstorm, corresponding to lowest temperature, highest temperature, highest wind speed, and highest precipitation, respectively. Our Micro-Macro model clearly outperforms all other methods, with its RMSE values of 0.311, 0.642, 2.045, and 1.637, respectively, in forecasting chill, torridity, storm, and rainstorm. SVR is the poorest performer. WRF-HRRR performs worse than Micro, $DUQ_{512-512}$, and Micro-Macro, in forecasting chill, torridity, and storm. For rainstorm forecasting, it performs better than all other models except our Micro-Macro model. $DUQ_{512-512}$ performs worse than both Micro and Micro-Macro models. These results demonstrate the effectiveness of our Micro-Macro model for forecasting abnormal weather conditions.

6 Conclusion

This paper has dealt with a novel deep learning model which takes both the atmospheric numerical output and the ground measurements taken as inputs for the very first time, dubbed as the Micro-Macro model for precise regional weather forecasting in multiple short-term time horizons. Our model employs the LSTM structure to capture the temporal variation of weather conditions and incorporates two data sources that include most relevant parameters for individual weather parameter forecasting per Mesonet station site via one model instance, called a modelet. A Periodical Mapper is also designed based on the neural network and Fourier Transform to capture the periodical patterns of temporal data. Experimental results demonstrated that our modelets can achieve much better meteorological forecasting with finer time granularity than almost all examined counterparts, to address an urgent need of national importance.

Acknowledgement. This work was supported in part by NSF under Grants 1763620, 1948374, and 2019511. Any opinion and findings expressed in the paper are those of the authors and do not necessarily reflect the view of funding agency.

References

1. Mesonet. https://www.mesonet.org/ (2021)
2. South Alabama Mesonet. http://chiliweb.southalabama.edu/ (2021)
3. WRF Resources. http://home.chpc.utah.edu/~u0553130/Brian_Blaylock/wrf.html (2021)
4. Alemany, S., Beltran, J., Perez, A., Ganzfried, S.: Predicting hurricane trajectories using a recurrent neural network. In: Proceedings of AAAI Conference on Artificial Intelligence, vol. 33, pp. 468–475 (2019)
5. Arshi, S., Zhang, L., Strachan, B.: Weather based photovoltaic energy generation prediction using lstm networks. In: Proceedings of International Joint Conference on Neural Networks (IJCNN) (2019)
6. Bloomfield, P., Hurd, H.L., Lund, R.B.: Periodic correlation in stratospheric ozone data. J. Time Ser. Anal. **15**(2), 127–150 (1994)
7. Bordes, A., Chopra, S., Weston, J.: Question answering with subgraph embeddings. In: Proceedings of Conference on Empirical Methods in Natural Language Processing (EMNLP), pp. 615–620 (2014)
8. Botev, Z.I., Grotowski, J.F., Kroese, D.P., et al.: Kernel density estimation via diffusion. Ann. Stat. **38**(5), 2916–2957 (2010)
9. Cho, K., et al.: Learning phrase representations using RNN encoder-decoder for statistical machine translation. In: Proceedings of Conference on Empirical Methods in Natural Language Processing (EMNLP), pp. 1724–1734 (2014)
10. Dalto, M., Matuško, J., Vašak, M.: Deep neural networks for ultra-short-term wind forecasting. In: Proceedings of IEEE International Conference on Industrial Technology (ICIT), pp. 1657–1663 (2015)
11. Dueben, P.D., Bauer, P.: Challenges and design choices for global weather and climate models based on machine learning. Geoscientific Model Dev. **11**(10), 3999–4009 (2018)
12. Gensler, A., Henze, J., Sick, B., Raabe, N.: Deep learning for solar power forecasting-an approach using autoencoder and lstm neural networks. In: Proceedings of International Conference on Systems, Man, and Cybernetics (SMC), pp. 2858–2865 (2016)
13. Grover, A., Kapoor, A., Horvitz, E.: A deep hybrid model for weather forecasting. In: Proceedings of 21th ACM SIGKDD International Conference on Knowledge Discovery and Data Mining, pp. 379–386 (2015)
14. Hernández, E., Sanchez-Anguix, V., Julian, V., Palanca, J., Duque, N.: Rainfall prediction: a deep learning approach. In: Martínez-Álvarez, F., Troncoso, A., Quintián, H., Corchado, E. (eds.) HAIS 2016. LNCS (LNAI), vol. 9648, pp. 151–162. Springer, Cham (2016). https://doi.org/10.1007/978-3-319-32034-2_13
15. Qinghua, H., Zhang, R., Zhou, Y.: Transfer learning for short-term wind speed prediction with deep neural networks. Renewable Energy **85**, 83–95 (2016)
16. National Center for Atmospheric Research. Understanding the earth system (2019)
17. Oppenheim, A.V., Buck, J.R., Schafer, R.W.: Discrete-Time Signal Processing, vol. 2, Prentice Hall, Upper Saddle River (2001)

18. Pan, B., Hsu, K., AghaKouchak, A., Sorooshian, S.: Improving precipitation estimation using vonvolutional neural network. Water Resour. Res. **55**(3), 2301–2321 (2019)
19. Pandit, R.K., Kolios, A., Infield, D.: Data-driven weather forecasting models performance comparison for improving offshore wind turbine availability and maintenance. IET Renew. Power Gener. **14**(13), 2386–2394 (2020)
20. Radhika, Y., Shashi, M.: Atmospheric temperature prediction using support vector machines. Int. J. Comput. Theor. Eng. **1**(1), 55 (2009)
21. Scher, S.: Toward data-driven weather and climate forecasting: approximating a simple general circulation model with deep learning. Geophys. Res. Lett. **45**(22), 12–616 (2018)
22. Vlachas, P.R., Byeon, W., Wan, Z.Y., Sapsis, T.P., Koumoutsakos, P.: Data-driven forecasting of high-dimensional chaotic systems with long short-term memory networks. Proc. Royal Soc. Math. Phys. Eng. Sci. **474**(2213), 20170844 (2018)
23. Wang, B., et al.: Deep uncertainty quantification: A machine learning approach for weather forecasting. In: Proceedings of 25th ACM SIGKDD International Conference on Knowledge Discovery & Data Mining, pp. 2087–2095 (2019)
24. Weinstein, S., Ebert, P.: Data transmission by frequency-division multiplexing using the discrete fourier transform. IEEE Trans. Commun. Technol. **19**(5), 628–634 (1971)
25. Weyn, J.A., Durran, D.R., Caruana, R.: Can machines learn to predict weather? using deep learning to predict gridded 500-hpa geopotential height from historical weather data. J. Adv. Model. Earth Syst. **11**(8), 2680–2693 (2019)
26. Weyn, J.A., Durran, D.R., Caruana, R.: Improving data-driven global weather prediction using deep convolutional neural networks on a cubed sphere. J. Adv. Model. Earth Syst. **12**(9), e2020MS002109 (2020)
27. Xingjian, S.H.I., Chen, Z., Wang, H., Yeung, D.Y., Wong, W.K., Woo, W.C.: Convolutional lstm network: A machine learning approach for precipitation nowcasting. In: Proceedings of Advances in Neural Information Processing Systems, pp. 802–810 (2015)
28. Yi, X., Zhang, J., Wang, Z., Li, T., Zheng, Y.: Deep distributed fusion network for air quality prediction. In: Proceedings of 24th ACM SIGKDD International Conference on Knowledge Discovery & Data Mining, pp. 965–973 (2018)

Action Set Based Policy Optimization for Safe Power Grid Management

Bo Zhou, Hongsheng Zeng, Yuecheng Liu, Kejiao Li, Fan Wang[✉],
and Hao Tian

Baidu Inc., Beijing, China
{zhoubo01,zenghongsheng,liuyuecheng,likejiao,wang.fan,tianhao}@baidu.com

Abstract. Maintaining the stability of the modern power grid is becoming increasingly difficult due to fluctuating power consumption, unstable power supply coming from renewable energies, and unpredictable accidents such as man-made and natural disasters. As the operation on the power grid must consider its impact on future stability, reinforcement learning (RL) has been employed to provide sequential decision-making in power grid management. However, existing methods have not considered the environmental constraints. As a result, the learned policy has risk of selecting actions that violate the constraints in emergencies, which will escalate the issue of overloaded power lines and lead to large-scale blackouts. In this work, we propose a novel method for this problem, which builds on top of the search-based planning algorithm. At the planning stage, the search space is limited to the action set produced by the policy. The selected action strictly follows the constraints by testing its outcome with the simulation function provided by the system. At the learning stage, to address the problem that gradients cannot be propagated to the policy, we introduce Evolutionary Strategies (ES) with black-box policy optimization to improve the policy directly, maximizing the returns of the long run. In NeurIPS 2020 Learning to Run Power Network (L2RPN) competition, our solution safely managed the power grid and ranked first in both tracks.

Keywords: Power grid management · Reinforcement learning · Planning

1 Introduction

Electrical grid plays a central role in modern society, supplying electricity across cities or even countries. However, managing the well functioning of the power network not only suffer from the fluctuating power consumption and unexpected accidents in the network, but also faces challenges from the unprecedented utilization of renewable energy [17,30]. The control system has lower operational flexibility as more renewable energy power plant connects to the power grid.

B. Zhou, H. Zeng—Equal Contribution.

© Springer Nature Switzerland AG 2021
Y. Dong et al. (Eds.): ECML PKDD 2021, LNAI 12979, pp. 168–181, 2021.
https://doi.org/10.1007/978-3-030-86517-7_11

For example, wind power stations that rely on seasonal wind cannot provide such stable electricity throughout the year like traditional thermal power station. Other issues such as rapidly growing electric car deployment that increases fluctuations in electricity consumption across regions also pose new challenges.

There are many efforts on applying deep reinforcement learning (RL) in power grid management, the recent technique emerged as a powerful approach for sequential decision-making tasks [16,19,22]. Taking the grid states as input, the policy adjusts the power generation of each power plant to feed the loads safely [6,8,30]. To further improve operational flexibility, recent works also study on managing the power grid through topological actions (e.g., reconfiguring bus assignments and disconnecting power lines) [9,14,17].

While RL-based approaches for grid management have achieved impressive results, existing methods ignore the environmental constraints. In power grid management, there are a number of complex rules that the selected actions must follow. For example, the system should avoid operations that lead to the disconnection of some residents to the power grid. Based on such constraints, an intuitive solution is to discourage actions that violate the rules by adding penalties on the feedback. However, this approach does not guarantee that all the actions produced by the policy strictly satisfy the constraints [24].

In this work, we propose an action set based method to manage the power grid safely through topological operations while strictly meeting the environmental constraints. The algorithm builds on top of the search-based planning approach, which has recently shown performance that exceeds human in challenging tasks [22,25]. At the planning stage, rather than use the entire action space as the search space, we limit the search space to the action set produced by the policy model. We then test the outcome of each candidate action in the action set with the simulation function provided by the system and select the action that strictly meet the constraints. However, such module blocks the gradient route of back-propagation from the selected action to the policy. To address the problem, we introduce Evolutionary Strategies (ES) [21] to directly optimize the policy towards maximizing long-term returns, by regarding the planning stage as block-box. Our agent participated in NeurIPS 2020 L2RPN competition, which provides two challenging power grid environments: Track 1 with unexpected power line attacks and Track 2 with increasing proportion of renewable energy plants. We ranked 1st place in both tracks.[1]

2   Related Work

Traditional system protection scheme (SPS) [1,20,27,28] builds an expert system to maintain the electrical flow and voltage magnitudes within safe range. The system relies on the network state such as voltage and electricity load level to make the decision. If overloaded power lines are detected, the system is triggered

---

[1] Our code is available open-source at: https://github.com/PaddlePaddle/PARL/tree/develop/examples/NeurIPS2020-Learning-to-Run-a-Power-Network-Challenge.

to take actions following the expert rules. SPS has less computation complexity and can provide real-time operations. The limitation of SPS is that not all possible issues can be foreseen at the stage of designing the system, which may result in instability and eventually lead to large-scale collapse [15].

A number of approaches formulate the power network management as a control problem and apply control and optimization theory to solve it. [13] uses model predictive control (MPC) [10] to select actions by minimizing the cost of operations and voltage-deviations under the security constraints of the power network, with a linear approximation model for state prediction. [15] predicts the future states based on a simulation model with nonlinear differential-algebraic equations and adopts tree search for optimization. However, The performance of MPC-based often heavily relies on the accuracy of the dynamics model [12].

Prior work has also modeled the grid management as a Markov decision process (MDP) and adopted reinforcement learning for sequential decision making. [6,8,12] operates the power network by adjusting generator outputs or reducing the load of electricity. [5] proposes a hierarchical architecture to consider long-term reliability and provide real-time decision making, where the policy updates at a fast time-scale and the value function updates at a slow time-scale. To further improve the operation flexibility, recent research [9,14,17] studies on reconfiguring the topology for power grid management (e.g., switching the bus assignments of the loads and generators in a substation). [31] employs the after-state representation [26] to reduce the difficulty of modeling the large observation and action space, with a hierarchical policy to determine and adjust the network topology.

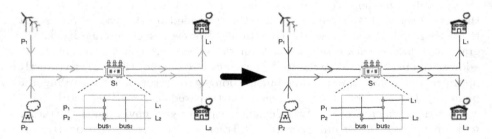

Fig. 1. Illustration of topology reconfiguration through the two buses in the substation. (Left) The two power stations provide electricity to two loads simultaneously, with the solid blue dots representing the connection between the object and the bus. (Right) Each power plant only transmits power to one load after the reconfiguration. (Color figure online)

3   Preliminary

In this section, we first formulate the objective of power grid management. Then we consider the task as a Markov decision process (MDP) and introduce the basic idea of search-based planning algorithm [4,25] proposed for solving sequential decision tasks. Our algorithm builds on top of the search-based planning approach.

3.1 Power Grid Management

There are three basic and important elements in a power grid: power plants P (indexed from $P_1...P_N$), substations S (indexed from $S_1...S_M$) and loads L (indexed from $L_1...L_K$). All the elements are connected together with the power lines that transmit power from one end to the other. There are two bus bars in each substation, and every element connected to the substations must be connected to one of them. We provide a toy example in Fig. 1 to demonstrate how to reconfigure the topology by switching the connected bus bar of the elements. There is a small grid with two power stations, where two stations provide power for two loads through the substation, and all the elements are connected to bus 1. If we reconnect P_1 and L_1 to the second bus, then each plant only provides power for only a load.

One of the most distinguished features of grid management is the topological graph that describes how electrical elements construct the power network. The grid can be represented as an undirected graph $G(V, E)$, where $V = (\boldsymbol{P}, \boldsymbol{S}, \boldsymbol{L})$ is the node-set composed of all the basic elements, and E is the edge set representing the connections of elements. Each edge $e_i(u, v, t_u, t_v)$ in E represents that a power line connects $u \in V$ with $v \in V$, and $t_u, t_v \in \{0, 1, 2\}$ indicate the bus to which the power line connects. 0 means that the line is not connecting with a substation, while 1 and 2 represent that the power line connects to the first and second bus, respectively. Denote the number of elements connected to each substation S_i by $|S_i|, i = 1, 2, .., M$. The number of possible topologies generated by bus switching in a substation is $2^{|S_i|}$, and for the whole power grid, there are $2^{\sum_{i=1}^{i=M} |S_i|}$ possible topology.

The task of operating the power grid is to maintain electrical flows of power lines within the acceptable range. If the current exceeds the maximum limit and maintains a high level, it will damage the power line and increase the transmission burden of other power lines, which may cause a large-scale power outage. For each power line $e_i \in E$, the ratio of current flow over maximum flow should stay at a safe level less than 1. The controller/operator can take three types of actions to avoid or address the overflow issue: (1) reconfiguring the bus connection in the substations (2) disconnecting or connecting the power lines (3) adjusting the power generation of the plants.

There are mainly two metrics for evaluating the controllers: the time horizon that power grid runs safely and the cost of operations decided by the controller (e.g., reconfiguring the grid topology and adjusting the power generation has different cost). The total reward of a controller can be defined as: $R = \sum_{t=0}^{T} r_t - c_t$, where $r_t > 0$ is the positive feedback and c_t is the operation cost. If any load or generator is disconnected from the power grid, the controller will not receive any reward since it takes dangerous actions that collapse the grid.

To encourage the study in power grid management, Coreso (European RSC) and RTE (French TSO) built a simulation environment named Grid2Op that runs the power system with real-time operations [17]. Grid2Op provides a training data set of multiple hundreds of yearly scenarios at 5 min resolution. More details about the environment can be found in Appendix A.

3.2   Search-Based Planning

We now consider power grid management as a MDP, defined by the tuple $(\mathcal{S}, \mathcal{A}, P, r, \gamma, \rho_0)$. $\mathcal{S}$ and $\mathcal{A}$ represent the state and action spaces, respectively. Note that $\mathcal{S}$ includes not only the topology graph mentioned in Sect. 3.1 but also other grid information such as power generation of the plants. We denote the distribution of initial states as ρ_0, the environment dynamics as $P(s_{t+1}|s_t, a_t)$. The reward function $r(s_t, a_t)$ relies on the state and action, and the discount factor $\gamma \in (0, 1)$ is used for accumulative reward computation.

At each time step t, the controller selects an action $a_t \in \mathcal{A}$ following the policy π, then the environment transits into the next state according to P. The optimal policy to obtain the maximum average reward can be defined:

$$\pi^* = \operatorname*{argmax}_{\pi} \mathbb{E}_{a_t \sim \pi, s_0 \sim \rho_0, s_{t+1} \sim P} \sum_{t=0}^{\infty} \gamma^t r(s_t, a_t) \tag{1}$$

Note that the selected actions must meet the power network constraints.

To obtain the optimal policy, we define the state-action value function $Q(s, a)$ to estimate the discounted future reward at state s after taking action a. Once the optimal $Q*(s, a)$ is learned, the optimal policy can be obtained by taking the action with the maximum estimated Q value at each step: $a^* = \operatorname{argmax}_{a \in \mathcal{A}} Q^*(s, a)$.

To learn the optimal Q function, we can adopt Monte Carlo sampling methods [2] and bootstrapping approaches [26] such as Q-learning [29]. The search-based planning algorithm combines the Monte Carlo method with tree search to gradually approximate the optimal Q function. At the search stage, the tree is traversed by simulation, starting from the root node of state s_t. At each simulation step, an action a_t is sampled from the search policy until a leaf node S_L is reached:

$$a_t = \operatorname*{argmax}_{a}(Q(s_t, a) + U(s_t, a)) \tag{2}$$

where $U(s_t, a) \propto \frac{1}{N(s_t, a)}$ is a bonus that decays with increasing visit count $N(s_t, a)$ to encourage exploration [25]. The bonus term can also be combined with a prior policy that matches the search policy: $U(s, a) \propto \frac{P(s,a)}{N(s_t, a)}$ [22, 25]. The future return of the leaf node is evaluated by a value function $V(s)$.

At the learning stage, The state action function $Q(s, a)$ and visit count $N(s, a)$ are updated along the traversed nodes. For the simulation trajectory $\tau(s_0, a_0, r_0, ..., s_{l-1}, a_{l-1}, r_{l-1})$ with length l, we can estimate the discounted

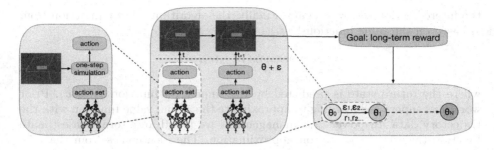

Fig. 2. Searching with the action set produced by the policy. (Left) At the search stage, the simulation rolls out the trajectory by searching the action candidates produced by the policy. (Middle) An exploration policy with parameter noise interacts with the environment and collects the feedback from the environment. (Right) The policy is updated towards maximizing the average future reward over the exploration and search trajectories.

future reward with the value function $V(s)$: $G(s_0, a_0) = \sum_{t=0}^{t=l-1} r_0 \gamma^t r_t + \gamma^l V(s_{l-1})$ [22]. For each edge (s, a) in the simulation path, we can perform the following update:

$$Q(s,a) = \frac{N(s,a) * Q(s,a) + G(s,a)}{N(s,a) + 1} \tag{3}$$

$$N(s,a) = N(s,a) + 1. \tag{4}$$

The value function $V(s)$ is updated through supervised learning that fits the average return starting from the state s.

4   Methodology

We now introduce a novel search-based planning algorithm that performs Search with the Action Set (SAS). The goal of the algorithm is to maximize the average long-term reward. We will detail how to optimize the policy towards this goal while strictly meeting the environment constraints. The overview of SAS is summarized in Fig. 2.

4.1   Search with the Action Set

At each simulation step, the policy $\pi_\theta(a_t|s_t)$ parameterized by θ outputs a vector of probabilities that the actions should be selected, and the top K actions with higher probabilities form the action-set A. We then leverage the simulation function $f_s(s_t, a)$ for action selection to ensure that the action strictly meets the constraints and rules, by simulating the outcome of each action $a \in A$ and filtering actions that violate the constraints. For notational simplicity, we denote the filtered action set by A again. Finally, the algorithm selects an action from A based on the value function $V(s_t)$:

$$a_t = \underset{a \in A}{\mathrm{argmax}}(V(f_s(s_t, a)). \tag{5}$$

where the future state is predicted by the simulation function $f_s(s_t, a)$. Prior work uses supervised learning to approximate the actual value function with the trajectory data. In power grid management, we found an alternative estimate function that does not rely on approximation. The idea comes from that the unsolved overloaded power line can induce more overloaded power lines and even lead to large-scale blackouts (i.e., large penalty). We thus define a risk function to monitor the overloaded power lines:

$$R_{isk} = max \frac{I_i}{I_{max_i}}, i = 1, 2, ..., |L|, \tag{6}$$

where I_i and I_{max_i} represent the current flow and the flow capacity of line L_i, respectively. The ratio $\frac{I_i}{I_{max_i}} > 1$ means that the power line i is overloaded. We replace the value function in Eq. 5 with the risk function and have:

$$a_t = \underset{a \in A}{\mathrm{argmin}}(R_{isk}(f_s(s_t, a)), \tag{7}$$

4.2   Policy Optimization

As shown in Eq. 7, action selection relies on the simulation function $f_s(s, a)$. If the $f_s(s, a)$ is known and differentiable, we can compute the backpropagating gradients and optimize the policy directly by maximizing the average reward in Eq. 1. However, it is often difficult to acquire the exact dynamics function $f_s(s, a)$ and it is unnecessarily differentiable in real-world applications. Though previous work uses an differentiable linear or nonlinear approximation function as an alternative [13,15], it introduces additional noise into optimization, and the performance highly relies on the accuracy of the approximation function [12].

To address the problem of obtaining the actual $f_s(s, a)$, we apply the black-box optimization of evolution strategies (ES) [7,21,23] to update the policy, which does not require backpropagating gradients for optimization. ES is an optimization algorithm inspired by natural evolution: A population of parameter vectors is derived from current parameters and evaluated in the environment, after which the parameter vectors with the highest scores will be recombined and form the next generation. In SAS, we repeatedly inject Gaussian noise ϵ into the parameter vector of the original policy and obtain a bunch of policies for exploration.

Overall, The optimization process repeatedly runs the two following phrases until the policy converges: (1) obtain the exploratory policy by perturbing the policy parameters with $\theta + \epsilon$ and evaluating its performance in power grid management (2) collect the sampled noise parameters ϵ and the related rewards for the computation of combined gradient, and update the policy parameter.

4.3 Discussion on Action Set Size

We now discuss the selection of action set size $K \in [1, N]$, where N is the number of actions. We first discuss the boundary values. If the size K is equal to 1, the algorithm reduces to the traditional RL method, since the function $f_s(s_t, a)$ can be omitted in Eq. 7 and we can perform backpropagation optimization to maximize the average return. When K is equal to N, the algorithm looks like an inefficient SPS system. It tries all the possible actions and selects the one with the lowest risk level, which is unacceptably time-consuming in the real-world and ignores the long-term return. The policy can also not be improved as the action selection always tests the whole action space, and policy is of no use.

Intuitively, a larger action set allows the algorithm to search more times at each step, which can improve the search result. We will further discuss the set size in the experiment section.

Algorithm 1. Action Set based Optimization

Require: Initialize the policy : π_θ
   Input learning rate α, noise stand deviation σ, and
   action set size K.
 1: **repeat**
 2:   **for** i in $\{1, 2..., n\}$ **do**
 3:      Sample Gaussian noise vector: $\epsilon \in N(0, I)$
 4:      Perturb the policy with $\epsilon_i * \sigma$: $\pi_{\theta + \epsilon_i * \sigma}(s, a)$
 5:      **while** not the end of the episode **do**
 6:         Top-K actions with higher probabilities forms a set
 7:         Select the action a according to Equation (7)
 8:      **end while**
 9:      Compute the total return r_i
10:      Record the exploration result (ϵ_i, r_i)
11:   **end for**
12:   Summarize the gradient $g = \frac{1}{n\sigma} \sum_{i=1}^n r_i \sigma_i$
13:   Update the policy $\theta \leftarrow \theta + \alpha g$
14: **until** π_θ converges

4.4 Algorithm Summary

Traditional search-based planning algorithms select the action mainly based on the reward estimation and exploration bonus, as shown in Eq. 2. In order to consider the environmental constraints, we split the action selection into two steps. The policy first outputs a number of action candidates as the limited search space, and then the algorithm starts to search with the set and filters actions that violate the constraints. The policy will not get a positive feedback if all the action candidates fail to meet the constraints. We further introduce ES with black-box optimization to maximize the average return such that the policy learns to output actions with a higher return while strictly meeting the constraints. The SAS algorithm is summarized at Algorithm 1.

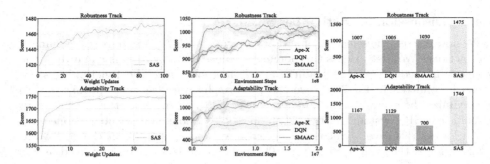

Fig. 3. Evaluation of SAS and baseline methods on robustness and adaptability tracks. (Left) Average return of SAS over weight update times. (Mid) Average return of RL methods over environment steps. (Right) Performance comparison at convergence. We plot the figure of SAS separately as its policy updates at a much slower frequency than RL methods, which update every a few environment steps.

5   Experiments

To evaluate our algorithm, we compare it with the baseline reinforcement learning algorithms in the Grid2Op environment, including DQN [19], APE-X [11], and Semi-Markov Afterstate Actor-Critic (SMAAC) [31] recent proposed for power grid management. SMAAC tackles the challenge of large action and state space by introducing the afterstate representation [26], which models the state after the agent has made the decision but before the environment has responded.

5.1   Experiment Setup

The experiment includes two tasks in NeurIPS 2020 L2RPN: robustness and adaptability tracks. The power grid in the robustness task has 36 substations, 59 power lines, 22 generators and 37 loads, providing the grid state of 1266 dimensions and 66918 possible topological actions. The most distinguishing feature in this task is the unexpected attack on power lines. Some of the power lines will be disconnected suddenly every day at different times, which will collapse the grid if the agent cannot overcome the attack in a short time.

The adaptability task has a larger power grid, approximately three times that of the robustness track, with 118 substations, 186 lines, 62 generators and 99 loads. The task reflects the emerging deployment of renewable energy generators, and it evaluates the agent with the environments containing different amounts of renewable energy generators. The agent has to adapt to the increasing proportion of renewable energy. Note that the control flexibility decreases as the number of less controllable renewable energy generators increases.

5.2   Implementation

The policy network contains four fully connected layers with RELU as the activation function and outputs the probabilities of each action given the current grid state. We use the same network structure for all the baseline algorithms for a fair comparison. The policy produces 100 action candidates for action selection at the planning stage (i.e., K = 100).

Following the parallel training implementation in ES [21], our implementation employs a number of CPUs to finish the exploration and search processes in parallel. At each iteration, we generate a large number of exploration policies and distribute them into different machines. The machines evaluate the performance of the noisy policies, compute the accumulative rewards and return them to the learner. Then we collect the feedback from the machines and compute the combined gradient for policy optimization. We use 500 CPUs and 1 GPU for the distributed version. Figure 3 shows the performance of various algorithms throughout training in robustness and adaptability tasks. The shaded area shows one standard deviation of scores. Each experiment was run four times with different random seeds.

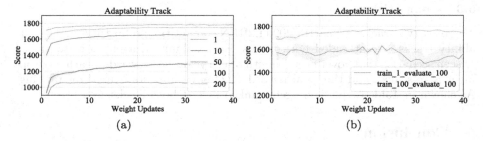

Fig. 4. (a) Training curves with different action set size (b) Re-evaluation of the learned policy with larger action set size.

SMAAC learns faster than DQN and APE-x in both tasks, as the after-state representation provides a sufficient representation of the grid than the traditional state-action pair representation [31]. However, its performance is worse than other methods at the adaptability task. The possible explanation is that the distribution shift of renewable energy production makes it more difficult to predict the future return. Though SMAAC provides better representation for the state and action pair, it cannot help model the distribution change that cannot be observed through the state. The SAS algorithm significantly outperforms the prior RL methods. Note that though SAS can achieve excellent performance in about one hundred iterations, a large amount of data (10 000 episodes) is required at each iteration. Since ES supports parallel training, we address the problem by using 500 CPUs for training, and it takes only about 1 h for each iteration.

To better understand SAS, we measure how SAS performs with respect to various action set sizes and present the result in Fig. 4 (a). While the set size is equal to 1, the performance of SAS is similar to the previous RL method. As SAS searches with a larger action set, its performance rises steadily. Does that mean our performance gain comes from more search times? We further re-evaluate the learned policy with a larger action set size. As shown in Fig. 4 (b), though the evaluation uses the same action set size (K = 100), the policy performs better while it is trained with a larger action set. The empirical result shows that the policy learns to produce high-quality action candidates, which can improve the search efficiency (i.e., higher return in the same search time).

Table 1. Top 5 teams in NeurIPS2020 L2RPN competition.

Team	1(Ours)	2	3	4	5
Robustness track	**59.26**	46.89	44.62	43.16	41.38
Adaptability track	**25.53**	24.66	24.63	21.35	14.01

5.3   Competition

There are two tracks in NeurIPS2020 L2RPN competition: robustness and adaptability tracks. We attended both tracks. Each submitted agent is tested in 24 unseen scenarios that cover every month of the year. The reward in each environment is re-scaled, and the total reward of 24 environments is used for ranking. As shown in Table 1, the SAS agent ranked first in both tracks.

6   Conclusion

In this paper, we propose a novel algorithm for grid management that searches within the action set decided by a policy network. By exploiting the simulation function to guarantee that selected actions strictly meet the constraints, the policy learn to adapts to the constraints while maximizing the reward in long run. To optimize the policy, we employed evolution strategies. With the proposed SAS algorithm, our agent outperformed prior RL approaches and won both tracks in the NeurIPS2020 L2RPN competition. Our work provides a novel approach to combine the complex environment constraints with policy optimization, which can potentially be applied to other real-world scenarios such as industry control and traffic control.

A   Grid2Op Environment

Grid2Op [18] is an open-source environment developed for testing the performance of controllers in power grid management. It simulates the physical power grid and follows the real-world power system operational constraints and distributions.

The environment provides interactive interfaces based on the gym library [3]. At each episode, it simulates a period of time (e.g., a week or month) at the time interval of 5 min. At each time step, the controller receives the state of the power grid and takes actions to operate the grid if necessary. The simulation terminates at the end of the period or terminates prematurely if the controller fails to operate the grid properly, which can occur under two conditions: (1) Some actions split the grid into several isolated sub-grids. (2) The electricity power transmitted from the stations cannot meet the consumption requirement of some loads. Too many disconnected lines will significantly increase the risk of causing these two conditions. A power line gets disconnected automatically if the current flow exceeds the maximum limit for 3 time steps (i.e., 15 min). In this case, the power line cannot be reconnected until the end of the recovery time of 12 time steps.

Grid2Op has a large state space and action space. In addition to the exponentially increasing possible grid topology we mentioned, the grid state contains other topology features such as the current, voltage magnitude of each power line, generation power of each power station, the required power of each load. Though only one substation can be reconfigured at each time step (to simulate that a human or an expert can perform a limited number of actions in a time period), the number of available actions for topology reconfiguration is $\sum_{i=1}^{i=M} 2^{|S_i|}$. In the NeurIPS2020 L2RPN competition, there are 118 substations with 186 power lines, which introduces over 70,000 discrete actions related to unique topology.

The reward setting in Grid2Op mainly relies on the reliability of the power grid. At each time step, the environment gives a bonus for safe management, and the controller will no longer gain positive rewards if it fails to manage the power network properly, which can lead to early termination of the episode. There are also costs (penalty) of operations. To encourage the controller to explore the operation flexibility on topology reconfiguration, the cost of topology change is much smaller than re-dispatching the power generation of the power plant.

In the physical world, the operators often use a simulation system to compute the possible outcomes of actions to control risk [1,18,28], and Grid2Op also provides a similar function named *simulate*, which can mimic the one-step operational process. It allows the user to check if the action violates the power network constraint (e.g., if the target power generation exceeds the maximum output of the power plant). Note that this function can only be called once at each time step (i.e., one-step simulation), and its prediction on future states may bias from the actual state.

References

1. Bernard, S., Trudel, G., Scott, G.: A 735 kV shunt reactors automatic switching system for hydro-Quebec network. IEEE Trans. Power Syst. **11**(CONF-960111-), 2024–2030 (1996)
2. Bishop, C.M.: Pattern Recognition and Machine Learning. Information Science and Statistics. Springer, New York (2006)
3. Brockman, G., et al.: Openai gym. arXiv preprint arXiv:1606.01540 (2016)

4. Coulom, R.: Efficient selectivity and backup operators in Monte-Carlo tree search. In: van den Herik, H.J., Ciancarini, P., Donkers, H.H.L.M.J. (eds.) CG 2006. LNCS, vol. 4630, pp. 72–83. Springer, Heidelberg (2007). https://doi.org/10.1007/978-3-540-75538-8_7

5. Dalal, G., Gilboa, E., Mannor, S.: Hierarchical decision making in electricity grid management. In: International Conference on Machine Learning, pp. 2197–2206 (2016)

6. Diao, R., Wang, Z., Shi, D., Chang, Q., Duan, J., Zhang, X.: Autonomous voltage control for grid operation using deep reinforcement learning. In: 2019 IEEE Power & Energy Society General Meeting (PESGM), pp. 1–5. IEEE (2019)

7. Eigen, M.: Ingo rechenberg evolutionsstrategie optimierung technischer systeme nach prinzipien der biologishen evolution. In: mit einem Nachwort von Manfred Eigen, vol. 45, pp. 46–47 (1973)

8. Ernst, D., Glavic, M., Wehenkel, L.: Power systems stability control: reinforcement learning framework. IEEE Trans. Power Syst. 19(1), 427–435 (2004)

9. Fisher, E.B., O'Neill, R.P., Ferris, M.C.: Optimal transmission switching. IEEE Trans. Power Syst. 23(3), 1346–1355 (2008)

10. Garcia, C.E., Prett, D.M., Morari, M.: Model predictive control: theory and practice-a survey. Automatica 25(3), 335–348 (1989)

11. Horgan, D., et al.: Distributed prioritized experience replay. In: International Conference on Learning Representations (2018). https://openreview.net/forum?id=H1Dy--0Z

12. Huang, Q., Huang, R., Hao, W., Tan, J., Fan, R., Huang, Z.: Adaptive power system emergency control using deep reinforcement learning. IEEE Trans. Smart Grid 11(2), 1171–1182 (2019)

13. Jin, L., Kumar, R., Elia, N.: Model predictive control-based real-time power system protection schemes. IEEE Trans. Power Syst. 25(2), 988–998 (2009)

14. Khodaei, A., Shahidehpour, M.: Transmission switching in security-constrained unit commitment. IEEE Trans. Power Syst. 25(4), 1937–1945 (2010)

15. Larsson, M., Hill, D.J., Olsson, G.: Emergency voltage control using search and predictive control. Int. J. Electr. Power Energy Syst. 24(2), 121–130 (2002)

16. Lee, J., Hwangbo, J., Wellhausen, L., Koltun, V., Hutter, M.: Learning quadrupedal locomotion over challenging terrain. Sci. Robot. 5(47), eabc5986 (2020)

17. Marot, A., et al.: Learning to run a power network challenge for training topology controllers. Electr. Power Syst. Res. 189, 106635 (2020)

18. Marot, A., et al.: L2RPN: learning to run a power network in a sustainable world NeurIPS2020 challenge design (2020)

19. Mnih, V., et al.: Human-level control through deep reinforcement learning. Nature 518(7540), 529–533 (2015)

20. Otomega, B., Glavic, M., Van Cutsem, T.: Distributed undervoltage load shedding. IEEE Trans. Power Syst. 22(4), 2283–2284 (2007)

21. Salimans, T., Ho, J., Chen, X., Sidor, S., Sutskever, I.: Evolution strategies as a scalable alternative to reinforcement learning. arXiv preprint arXiv:1703.03864 (2017)

22. Schrittwieser, J., et al.: Mastering atari, go, chess and shogi by planning with a learned model. Nature 588(7839), 604–609 (2020)

23. Schwefel, H.P.: Numerische optimierung von computer-modellen mittels der evolutionsstrategie. (Teil 1, Kap. 1–5). Birkhäuser (1977)

24. Shah, S., Arunesh, S., Pradeep, V., Andrew, P., Milind, T.: Solving online threat screening games using constrained action space reinforcement learning. In: Proceedings of the AAAI Conference on Artificial Intelligence, vol. 34, pp. 2226–2235 (2020)
25. Silver, D., et al.: Mastering the game of go with deep neural networks and tree search. Nature **529**(7587), 484–489 (2016)
26. Sutton, R.S., Barto, A.G.: Reinforcement Learning: An Introduction. MIT Press, Cambridge (2018)
27. Tomsovic, K., Bakken, D.E., Venkatasubramanian, V., Bose, A.: Designing the next generation of real-time control, communication, and computations for large power systems. Proc. IEEE **93**(5), 965–979 (2005)
28. Trudel, G., Bernard, S., Scott, G.: Hydro-Quebec's defence plan against extreme contingencies. IEEE Trans. Power Syst. **14**(3), 958–965 (1999)
29. Watkins, C.J., Dayan, P.: Q-learning. Mach. Learn. **8**(3–4), 279–292 (1992)
30. Yang, Q., Wang, G., Sadeghi, A., Giannakis, G.B., Sun, J.: Two-timescale voltage control in distribution grids using deep reinforcement learning. IEEE Trans. Smart Grid **11**(3), 2313–2323 (2019)
31. Yoon, D., Hong, S., Lee, B.J., Kim, K.E.: Winning the L2RPN challenge: power grid management via semi-Markov afterstate actor-critic. In: International Conference on Learning Representations (2021). https://openreview.net/forum?id=LmUJqB1Cz8

Conditional Neural Relational Inference
for Interacting Systems

Joao A. Candido Ramos[1,2(✉)], Lionel Blondé[1,2], Stéphane Armand[1],
and Alexandros Kalousis[2]

[1] University of Geneva, Geneva, Switzerland
joao.candido@etu.unige.ch
[2] Geneva School of Business Administration, HES-SO, Geneva, Switzerland

Abstract. In this work, we want to learn to model the dynamics of
similar yet distinct groups of interacting objects. These groups follow
some common physical laws that exhibit specificities that are captured
through some vectorial description. We develop a model that allows us
to do conditional generation from any such group given its vectorial
description. Unlike previous work on learning dynamical systems that
can only do trajectory completion and require a part of the trajectory
dynamics to be provided as input in generation time, we do generation
using only the conditioning vector with no access to generation time's
trajectories. We evaluate our model in the setting of modeling human
gait and, in particular pathological human gait.

1 Introduction

While modeling the evolution of an object in a physical dynamical system already
constitutes a tedious endeavor, modeling the evolution of a system of objects
interacting with each other is considerably more challenging. The complex physical laws describing the system are, in most cases, unknown to the learning
agent, who then only has access to observations depicting traces of interaction
of the whole physical system, called trajectories. Previous works have attempted
to learn the dynamics of systems involving interacting objects by injecting a
structural inductive bias in the model, allowing them to learn the inter-object
relationships [2,4,10,14,23,25,31,32]. When the relationships between the interacting objects are unknown *a priori*, there exist two approaches to leverage the
lack of structural information: modeling the interactions implicitly or explicitly. The first approach describes the physical system by a fully connected graph
where the message passing operations implicitly describe the interactions, hoping
that useful connections will carry more information [8,23,26,31]. Other works
add an attention mechanism to give more importance to some interactions in
the fully connected graph [10,25]. In the second approach, we have unsupervised models, such as NRI [14] and fNRI [32], which can explicitly predict the
interactions and dynamics of a physical system of interacting objects only from
their observed trajectories. When it comes to predicting the future states of the
system, previous works adopt different strategies.

© Springer Nature Switzerland AG 2021
Y. Dong et al. (Eds.): ECML PKDD 2021, LNAI 12979, pp. 182–197, 2021.
https://doi.org/10.1007/978-3-030-86517-7_12

In the prediction of the future states of the physical system, we find different strategies. Some works predict the next state from the previous ones [2,4]. Others, such as NRI, predict the continuation of the trajectories given a first observed part of the trajectories, essentially doing trajectory completion. All of them require access to a part of the trajectories to make the prediction of the next states [14,32]. To the best of our knowledge, there is no work that considers how the specificities of a given physical system impact the dynamics learned by such models, as well as how expliciting them through a conditioning feature vector can result in generated trajectories displaying the specific fingerprint behavior of the considered examples. In this work, we want to solve the problem of learning several slightly different dynamical systems, where the information differentiating them is contained in a description vector. To illustrate our setting, let us consider the modelling of human gait which has driven this work. Human gait follows a certain number of biomechanical rules that can be described in terms of kinetics and kinematics but also depends to a considerable extent on the individual. The neurological system and the person's past may influence the manner the individual walks significantly.

To generate trajectories for a given group of interacting objects, we introduce a conditional extension of NRI (cNRI) that can generate trajectories from an interaction graph given a conditioning vector describing that group. By providing the conditioning vector to the decoder, we allow the encoder to be any model that can output interactions. The decoder learns to generate the dynamics of the physical system from the conditioning vector. The encoder can be a fixed, known, graph, i.e. it is not learned, similar to the true graph in the original paper [14]. Our work differs considerably from NRI; we do not seek to learn the interactions explicitly. Instead, we want to use these interactions, whether they are given or learned, together with the conditioning vector to conditionally generate trajectories given only the conditioning vector.

We demonstrate our approach in the problem of learning to conditionally generate the gait of individuals with impairments. The conditioning vector describes the properties of an individual. Our ultimate goal is to provide decision support for selecting the appropriate treatment (surgery) for any given patient; this work is a stepping stone towards that direction. Selecting the most appropriate surgery for patients with motor and neurological problems such as cerebral palsy is a challenging task [21]. A tool that can model pathological gait and conditionally generate trajectories can allow physicians to simulate the outcomes of different operations on the patient's gait simply by modifying the conditioning vector. This will reduce in a considerable manner unnecessary operations and operations with adverse effects.

We will learn the dynamics of gait from the set of trajectories of the different body parties, described either in the form of euclidean coordinates or as joint angles. The conditioning vector will contain clinical information describing the patient's pathology, their anthropometric parameters, and measurements acquired during a physical screening. We experimentally show that our model achieves the best results in this setting, outperforming in a significant and consistent manner relevant baselines, providing thus a promising avenue for eventual decision support for treatment selection in the clinical setting.

2   Related Work

There are many works that tackle the problem of motion forecasting, using traditional methods such as hidden Markov models [15], Gaussian process latent variable models [27,30] or linear dynamical systems [20]. More recently, recurrent networks have been used to predict the future positions in a sequential manner [1, 6,7,11,16,17,19,28]. Imitation learning algorithms have also been used to model human motion [29]. However, all previous attempts use a part of the trajectories to predict their future. To the best of our knowledge, no work tackles the problem of full trajectory generation conditioned only on a description of the system for which we wish to generate trajectories.

3   The Conditional Neural Inference Model

We want to learn to model the dynamics of multi-body systems consisting of M interdependent and interacting bodies. Such a system when it evolves in time it generates a multi-dimensional trajectory $\mathbf{X} = [\mathbf{x}^1, ..., \mathbf{x}^T]$ (we assume trajectories of fixed length T), where the $\mathbf{x}^t$ element of that trajectory is given by $\mathbf{x}^t = [\mathbf{x}_1^t, ..., \mathbf{x}_M^t]^{\mathrm{T}}$ and $\mathbf{x}_i^t$ is the set of features describing the properties of the i body at time t. We will denote the complete trajectory of the body-part i by $\mathbf{x}_i^{1:T}$. In the following we will use boldface to indicate samples of a random variable and caligraphic for the random variable itself. One example of such an $\mathbf{x}_i^t$ can be the euclidean coordinates of the ith body if the trajectories track position of the body parts of a multi-body system. In addition each such system is also described by a set of properties $\mathbf{c} \in \mathbb{R}^d$ providing high level properties of the system that determine how its dynamics will evolve. Our goal is to learn the conditional generative model $p(\mathcal{X}|c)$ which will allow us to generate trajectories given only their conditioning property vector $\mathbf{c}$. Our training data consist of pairs $(\mathbf{X}_i, \mathbf{c}_i), i := 1 \ldots N$, produced by N different dynamical systems. Since we base our model on the NRI we will first provide a brief description of it.

In NRI the goal is to learn the dynamics of a *single* multi-body dynamical system and use the learned dynamics to forecast the future behavior of trajectories sampled from that system. To solve the forecasting problem it learns a latent-variable generative model of $P(\mathcal{X})$ where the latent variable captures the interactions. The training data $\mathbf{X}_i, i := 1, \ldots, N$, are thus samples from a fixed dynamical system whose dynamics NRI will learn. The basic NRI model is a Variational Auto-Encoder (VAE), [13]. The latent representation is a matrix-structured latent variable $\mathcal{Z} : N \times N$, where $z_{i,j}$ is a K-category categorical random variable describing the type of interaction, if one exists, between the i, j, bodies of the system. The approximate posterior distribution is given by $q_\phi(\mathcal{Z}|\mathcal{X}) = \prod_{i,j} q_\phi(z_{i,j}|\mathcal{X})$, where $z_{i,j} \sim q_\phi(z_{i,j}|\mathcal{X}) = \mathrm{Cat}(\mathbf{p} = [p_1, ..., p_K] = \boldsymbol{\pi}_{\phi_{i,j}}(\mathcal{X}))$. The encoder $\boldsymbol{\pi}_\phi(\mathcal{X})$ is a graph network that feeds on the trajectory data and outputs the probability vector for each i, j, interaction based on the learned representation of the respective i, j, edge of the graph network; more details on the encoder in Sect. 3.1.

The generative model has an autoregressive structure given by: $p_\theta(\mathcal{X}|\mathcal{Z}) = \prod_{t=1}^{T} p_\theta(x^{t+1}|x^t, ..., x^1, \mathcal{Z})$, where $p_\theta(x^{t+1}|x^t, ..., x^1, \mathcal{Z}) = \mathcal{N}(\boldsymbol{\mu}_\theta(x^t, ..., x^1, \mathcal{Z}), \sigma^2\mathbf{I})$. The $\boldsymbol{\mu}_\theta(x^t, ..., x^1, \mathcal{Z})$ is a graph network that feeds on the learned interaction matrix and the so far generated trajectory[1]. The autoregressive model in the generative distribution is trained using teacher forcing up to some step l in the trajectory after which the predictions are used to generate the remaining trajectory points from $l+1$ to T. This is a rather important detail because it also reflects how the decoder is used at test time to do trajectory forecasting. At test time in order for NRI to forecast the future of a given trajectory it will feed on the trajectory and then map it to its latent representation. Its decoder will feed on the real input trajectory and thanks to its autoregressive nature will generate its future states. By its conception NRI does not learn over different dynamical systems, nor can it generate trajectories from scratch, it has to feed on trajectory parts and then forecast. To address this setting we develop a conditional version of NRI.

The conditional-NRI (cNRI) has the same model architecture as NRI, i.e. it is a VAE with an encoder that outputs a latent space, structured as above, that describes the interactions and a decoder generates the complete trajectories. Unlike NRI which learns the distribution $p(\mathcal{X})$ of a fixed dynamical system here we want to learn over different dynamical systems and be able to generate from trajectories at will from each one of them. Thus in cNRI we model the conditional distribution $p(\mathcal{X}|c)$ where c provides the description of the conditional generation system from which we wish to sample. The posterior distribution is the same as that of NRI, while the generative distribution is now $p_\theta(\mathcal{X}|\mathcal{Z}, c) = \prod_{t=1}^{T} p_\theta(x^{t+1}|x^t, ..., x^1, \mathcal{Z}, c)$, where $p_\theta(x^{t+1}|x^t, ..., x^1, \mathbf{Z}, c) = \mathcal{N}(\boldsymbol{\mu}_\theta(x^t, ..., x^1, \mathcal{Z}, c), \sigma^2\mathbf{I})$. Unlike NRI we train the decoder without teacher forcing; at test time when we should conditionally generate a trajectory $\mathbf{X}$ from the description $\mathbf{c}$ of a dynamical system we do not require access to any trajectory from that system.

Our loss is the standard ELBO loss adjusted for the conditional setting and the optimization problem is:

$$\max_{\phi,\theta} \mathbb{E}_{\mathbf{X},\mathbf{c} \sim P(\mathcal{X},c)} \mathbb{E}_{\mathbf{Z} \sim q_\phi(\mathcal{Z}|\mathbf{X})} [\log p_\theta(\mathbf{X}|\mathbf{Z}, \mathbf{c})] - D_{\mathrm{KL}}[q_\phi(\mathbf{Z}|\mathbf{X})||p(\mathbf{Z})] \quad (1)$$

In the following sections we will review different options for the encoder architecture and we will described the decoder's architecture.

3.1 Encoding, Establishing the Body-Part Interactions

In NRI the role of the encoder is to learn the interaction network which is then used in the decoder to support the learning of the dynamics. However, in cNRI the primary goal is not to learn the interaction graph but to be able to conditionally generate trajectories from different dynamical systems. We will thus explore

---

[1] The first part of that trajectory will always be real data, even at test time, directly coming from the input trajectory as we will soon explain.

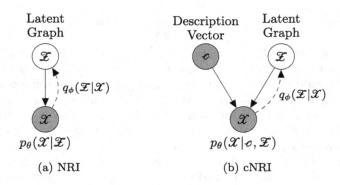

Fig. 1. The NRI and cNRI graphical models.

and evaluate different scenarios with respect to the prior knowledge we have about the interaction graph. In particular we will consider scenarios in which the real interaction graph is known and scenarios in which it is unknown and we learn it. Strictly speaking in the former case we do not have an encoder anymore and we are not learning a variational autoencoder but rather a conditional generative model that explicitly maximizes the data likelihood (Fig. 1).

Perfect Interaction Graph. In this scenario we assume that the interaction graph $\mathbf{Z}$ is known and it is the same for all our different dynamical systems. So in that setting there is no encoder involved, or alternatively we can think of the encoder as a constant function that maps all instances to the same latent vector. As an example in the gait modelling problem the $\mathbf{Z}$ matrix will be the adjacency matrix that describes the body-parts connectivities as these are given by the human skeleton. So in this setting the optimization problem reduces to:

$$\max_{\theta} \mathbb{E}_{\mathbf{X},\mathbf{c} \sim P(\mathcal{X},c)}[\log p_{\theta}(\mathbf{X}|\mathbf{Z},\mathbf{c})] \qquad (2)$$

Imperfect Interaction Graph. There are cases in which we have a good understanding of the interaction between the different body-parts but we do not have the complete picture. If we turn back to the example of the human gait modelling, the interactions between the body parts are not only sort range, through the immediate connections as above, but also longer range; while walking our arms in the opposite directions as the feet of the opposite body side. When we model the dynamics on the decoder side it might be beneficial for the generations to have explicitly in the interaction graph such longer dependencies. Remember that the decoder is a graph network whose adjacency matrix is given by $\mathbf{Z}$, having the longer dependencies explicitly modelled will not require the decoder graph network to transfer information over longer paths. To account for such a setting we now make $\mathbf{Z}$ a learnable parameter starting from the original interaction graph. As before there is no encoder and learning $\mathbf{Z}$ consists in making the generative model a function of the $\mathbf{Z}$ which is not sampled from the posterior distribution but treated as a deterministic variable that we learn with standard gradient descent. So our generative distribution is now $p_\theta(\mathcal{X}|\mathbf{Z})$. Such an approach has been also used in [24]. The optimization problem now is:

$$\max_{\mathbf{Z},\theta} \mathbb{E}_{\mathbf{X},\mathbf{c} \sim P(\mathcal{X},c)}[\log p_\theta(\mathbf{X}|\mathbf{Z},\mathbf{c})] \qquad (3)$$

Unknown Interaction Graph, the NRI Encoder. Often the interaction graph is not known. NRI was proposed for exactly such settings. Its encoder, $\pi_\phi(\mathbf{X})$, a fully connected graph network, learns the parameters of the categorical posterior distribution from which the latent interaction graph is sampled from the complete trajectories of the different body parts. In particular $\pi_\phi(\mathbf{X})$ consists of the following message passing operations:

$$\mathbf{h}_j^0 = f_{emb}(\mathbf{x}_j^{1:T}), \qquad \mathbf{h}_{(i,j)}^1 = f_e^1([\mathbf{h}_i^0, \mathbf{h}_j^0]),$$

$$\mathbf{h}_j^1 = f_v^1([\sum_{i \neq j} \mathbf{h}_{(i,j)}^1]), \qquad \mathbf{h}_{(i,j)}^2 = f_e^2([\mathbf{h}_i^1, \mathbf{h}_j^1]), \qquad \pi_{\phi_{i,j}}(\mathbf{X}) = \text{Softmax}(\mathbf{h}_{(i,j)}^2)$$

$f_{emb}(\mathbf{x}_j^{1:T})$ is a neural network that learns a hidden representation of the body-part (node) j from its full trajectory; $f_e^1([\mathbf{h}_i^0, \mathbf{h}_j^0])$ is a network that learns a hidden representation of the edge connecting nodes i and j; $f_v^1([\sum_{i \neq j} \mathbf{h}_{(i,j)}^1])$ updates the representation of the j node using information from all the edges in which it participates and finally $f_e^2([\mathbf{h}_i^1, \mathbf{h}_j^1])$ is a network that computes the final K-dimensional edge representation. This final representation of the i, j edge is passed from a softmax function to give the proportions $\mathbf{p}$ of the categorical distribution $q_\phi(z_{i,j}|\mathcal{X})$ from which we sample the type of the i, j edge.

With this formulation, the encoder has to assign an edge-type per pair of nodes, preventing the model from generalizing well on problems where the interaction graph should be sparse. As a solution [14] proposes defining an edge type as a non-edge, so no messages are passing through it.

Unknown Interaction Graph, the fNRI Encoder. In certain cases one might want more than a single edge type connecting at the same time a given pair of nodes. In the standard NRI approach this is not possible since the edge type is sampled from a categorical distribution. Instead we can model the $z_{i,j}$ variable as a K-dimensional random variable whose posterior $q_\phi(z_{i,j}|\mathcal{X})$ is given by a product of K Bernoulli distributions and have the graph network learn the parameters of these K distributions. More formally:

$$q_\phi(z_{i,j,k}|\mathcal{X}) = \text{Ber}(p_{k_{i,j}} = \pi_{\phi_{i,j,k}}(\mathcal{X}))$$

This is the approach taken in factorised NRI (fNRI) proposed in [32]. Instead of passing the result of the first message passing operation $\mathbf{h}_j^1$ through the second edge update function as NRI does, fNRI uses K edge update functions to get K different two-dimensional edge embeddings $h_{(i,j)}^2$ which are passed from a Softmax function to get the parameters of the K Bernoulli distributions:

$$\mathbf{h}_{(i,j)}^{2,l} = f_e^{2,l}([\mathbf{h}_i^1, \mathbf{h}_j^1]), \quad \mathbf{h}_{(i,j)}^2 = [\mathbf{h}_{(i,j)}^{2,1}, ..., \mathbf{h}_{(i,j)}^{2,K}], \quad \pi_{\phi_{i,j,k}}(\mathbf{X}) = \text{Softmax}(\mathbf{h}_{(i,j)}^{2,k})$$

When we learn the interaction graph using the NRI or the fNRI encoders we are sampling from a categorical distribution. In order to be able to backpropagate through the discrete latent variable $\mathbf{Z}$ we use their continuous relaxations given by the concrete distribution [18]:

$$z_{i,j} = \text{Softmax}(\frac{h_{(i,j)}^2 + \mathbf{g}}{\rho}) \quad z_{i,j,k} = \text{Softmax}(\frac{h_{(i,j)}^{2,k} + \mathbf{g}}{\rho}) \qquad (4)$$

where $\mathbf{g}$ is a vector of i.i.d samples from the Gumbel(0,1) distribution and ρ is the temperature term.

3.2 Decoding, Establishing the Dynamics

The role of the decoder is to learn the dynamics so that it can successfully generate trajectories for any given dynamical system. As already discussed the NRI architecture is designed for forecasting and does not address this task. This is because at test time in order to establish the interaction matrix its encoder needs to feed on a trajectory of the given system and the decoder needs this trajectory in order to achieve the forecasting task. In our setting at test time we do not have access to the trajectories but only to the condition vectors $\mathbf{c}$ of some dynamical system. The generative model of cNRI will only feed on the conditioning vector, the interaction matrix, and the initial state x^1 that provides a placement for the trajectory, and it will unroll its autoregressive structure only over generated data, more formally: $p_\theta(\mathcal{X}|\mathcal{Z}, c, x^1) = \prod_{t=1}^{T} p_\theta(x^{t+1}|\hat{x}^t, ..., \hat{x}^2, x^1, c, \mathcal{Z})$, where $\hat{x}^t$ is the t state of the trajectory sampled from the generative model.

To condition the generative model on the conditioning vector $\mathbf{c}$ we bring the information of the conditioning vector in two places within the generative model. First when to learn the initial hidden states of the different nodes (body-parts) we use an MLP that feeds on $\mathbf{c}$ and outputs an embedding $\mathbf{h}^0 = f_c^{\text{hid}}(\mathbf{c})$ of size $N \times H$ where H is the number of hidden dimensions we use to represent each one of the N nodes; as a result each i node has its own representation $\mathbf{h}_i^0$ which does not require the use of trajectory information. In NRI the node embeddings are initialized with zero vectors and the input trajectory is used as burn-in steps to update the state embeddings before forecasting the future trajectory.

One problem with the above conditioning is that it is used to compute only the initial hidden state of each node, whose effect due to the autoregressive nature of the decoder can be eventually forgotten. To avoid that we also use the conditioning vector $\mathbf{c}$ directly inside the message passing mechanism of the decoder. To do so we create a virtual edge that is a function of the conditioning vector and links to every node; essentially the conditioning vector becomes a global attribute of the graph that is then used by all update functions [3]. The virtual edge embedding is computed through an MLP as $\mathbf{h}^{\text{msgs}} = f_c^{\text{msgs}}$ and used in updating the stats of all nodes.

When we use the fNRI encoder the decoder $\mu_\theta(\hat{\mathbf{x}}^t, ..., \hat{\mathbf{x}}^2, \mathbf{x}^1, \mathbf{c}, \mathbf{Z})$ performs the following messages-passing and autoregressive operations to get the mean of the normal distribution from which the next trajectory state is sampled:

$$\mathbf{h}^t_{(i,j)} = \sum_k z_{ij,k} f_e^k([\mathbf{h}_i^t, \mathbf{h}_j^t]) \qquad\qquad \bar{\mathbf{h}}_j^t = \mathbf{h}^{\mathrm{msgs}} + \sum_{i \neq j} \mathbf{h}^t_{(i,j)}$$

$$\mathbf{h}_j^{t+1} = \mathrm{GRU}([\bar{\mathbf{h}}_j^t, \hat{\mathbf{x}}_j^t], \mathbf{h}_j^t) \qquad\qquad \mu_j^{t+1} = \mathbf{x}_j^t + f_{out}(\mathbf{h}_j^{t+1})$$

where $\mathbf{h}^t_{(i,j)}$ is the hidden representation of the i, j, edge at time t computed from the hidden representations of the i, j, nodes it connects. Note that this takes into account all different edge types that connect i and j through the use of one edge update function f_e^k per edge type. The $z_{ij,k}$ variable acts as a mask. If we use the NRI encoder then only one edge update is selected, since in that case there can be only one edge type connecting two nodes. $\bar{\mathbf{h}}_j^t$ is the aggregated edge information that arrives at node j at time t computed from all edges that link to it as well as the virtual edge. The new hidden state of the node j, $\mathbf{h}_j^{t+1}$, is given by a GRU which acts on the sampled $\hat{\mathbf{x}}_j^t$, the respective hidden representation $\mathbf{h}_j^t$, and the aggregated edge information, $\bar{\mathbf{h}}_j^t$. From this $\mathbf{h}_j^{t+1}$ we finally get the mean of the normal distribution from which we sample the next state of the trajectory as shown above; essentially we use the hidden representation to compute an offset from the previous state through the f_{out} MLP.

3.3   Conditional Generation

Once the model is trained we want to use the generative model $p_\theta(\mathcal{X}|\mathcal{Z}, c, x^1)$ to conditional generate trajectories from a dynamical system for which we only have access to $\mathbf{c}$ but not its trajectory, in such a case the interaction graph is not known. We sample the $\mathbf{Z}$ from the aggregate posterior : $q_\phi(z) = q_\phi^{\mathrm{avg}}(z) \triangleq \frac{1}{N} \sum_{n=1}^N q_\phi(z|x_n)$. Since we have a discrete distribution, the aggregated posterior is the probability to have a given edge-type in training samples. The sampling of the interaction graph only occurs in the unsupervised encoders. Finally to simplify our evaluation, we are not learning the probability of $p(x^1|c)$. We are giving this frame as the starting point of the generations. Nevertheless this probability can be learned by a neural network or by the decoder directly.

4   Experiments

As we have discussed in the introduction the main motivation for this work is the provision of decision support for the treatment of patients with motor impairments where the conditioning vector describes how an operation affects body structure and the generative model will show how such changes affect gait.

The data have been collected from a kinesiology laboratory and come from patients with hemiplegia. They contain the kinematics and clinical data of 72 patients for a total of 132 visits at the laboratory. The kinematics data are recorded by placing markers on the body patient who then walks on a corridor where infrared cameras record the motion. The clinical data, our conditioning vector **c**, are obtained by a physiotherapist and include parameters such as body measurements and evaluation of muscles' strength; we have a total of 84 such parameters. From the available data we obtain 714 gait cycles, where each cycle is a multidimensional trajectory giving the position of all body parts through time.

From these data we produce four different datasets which rely on a different interaction graph structure. Three of these dataset are based on the marker trajectories and one is based on the joint angle trajectories. We used three different graph structures which we will respectively call *complete skeleton, armless, lower body*. These graph structures are motivate by the fact that our skeleton provides a nature interaction graph. In the complete skeleton version we track 19 body parts by computing the center of mass of the sensors that are placed on each body part. In the armless version we track 15 body parts; we removed the elbow and hand markers because these are hard to predict and do not seem to influence the gait dynamics. In the lower body version we use all the available markers for the lower body part, i.e. we do not do body part aggregation as in the previous two. In all three datasets we normalise the trajectories by removing the pelvis position and dividing by the patient height. The result of this normalization is a patient that seems to walk on a treadmill with position values being in the range $[0, 1]$. Finally in the angle dataset instead of Euclidean trajectories we use the joint angle trajectories of the lower body resulting in the trajectories of the angles of eight joints. We normalise the angle dataset to the $\mathcal{N}(0, 1)$. Note that angles exhibit larger variability than marker position. We visualise the different structures in Fig. 2.

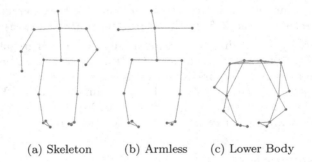

(a) Skeleton   (b) Armless   (c) Lower Body

Fig. 2. The interaction graphs we used to produce the trajectory datasets.

4.1 Experimental Setup

Depending on the dataset, we train cNRI with 128 (or 256) units per layer in the unsupervised encoders. The decoder performs the best with a hidden size of 256 or 384 units. This model overfits quickly on the angle dataset due to the small number of samples; we thus reduced the number of hidden units to 64 and 128 for the encoder and the decoder respectively. To avoid overfitting we use a 10% dropout. We use the Adam optimizer [12] with a time-based learning rate decay of $\frac{1}{2}$ every 50 epochs starting at 10^{-3}. For the unsupervised encoders, we found that our model generalizes better with two edge-types: a "non-edge" with a hard-coded prior of 0.91 (the non-edge) and 0.09 for the second edge-type (same as the original NRI). We evaluate the models using 3-fold cross validation were we divide the dataset to training, validation and test; we take care to keep all trajectories of a given patient/visit within one of these sets so that there is no information leakage. We tune the hyperparameters on the validation set. We report Mean squared error between the real and generated trajectories and its standard deviation that we compute over the denormalized generations; the markers' unit is millimeters and the angles' unit is degrees.

We will refer to the various combinations of encoder-decoder of our model as follows: PG-cNRI is the combination of the perfect interaction graph (PG) encoder with our conditional decoder; IG-cNRI uses the imperfect interaction graph (IG) encoder; NRI-cNRI combines the unsupervised encoder of the NRI with our decoder; and finally, fNRI-cNRI is the combination of fNRI encoder with cNRI.

We compare against several baselines. The two first baselines are based on the mean. Even though simple, they have excellent performance, and on the angles dataset, they are hard to beat. The first mean-based baseline predicts for each object its mean on the training set. The second uses more knowledge and predicts, for each object an object-based average over the side in which the patient is affected. To avoid errors coming from translation we slide the mean to start at the same position as the trajectory evaluated. In addition, we use three variants of recurrent neural networks (RNN): standard [22], GRU [5] and LSTM [9]. These are autoregressive models that tackle conditional generations heads-on. We condition their hidden states on the clinical features and train them to minimize the error between the generated trajectories and the true ones; we use no teacher forcing. We also add a reformulation of NRI that can generate the entire trajectory in which we sample the latent graph from the aggregated posterior. In our reformulation of the NRI there is no warm-up of the decoder state, and the decoder generates directly new states. In addition it uses no conditioning vector, we included in the experiments in order to verify that the conditional information does improve generation performance. The code associated with this work is available at https://github.com/jacr13/cNRI.

4.2   Results

The models that we propose here are the only ones that consistently beat the improved mean baseline (Table 1). From the other baselines only the RNN one is able to outperform the improved mean in three of the four datasets. On the skeleton dataset, the model that uses the real graph (PG-cNRI) achieves the lowest error. In PG-cNRI the decoder can only use the skeleton's links to propagate the information; this considerably reduces the model's power for reasoning on long relations. Since the arms are almost unpredictable, PG-cNRI has here the right inductive bias since it propagates less information through these nodes making overfitting less likely. When we remove the arms, (armless, lower body), as expected the performance improves. Our unsupervised models (NRI-cNRI and fNRI-cNRI) learn better the dynamics and their generations are very close to the real trajectories, and they outperform significantly all baselines. The angles dataset is the hardest to predict. Here the improved mean is an excellent approximation of the real trajectories. Here all our models are better than the improved mean (IG-cNRI being the best), though the performance gap is not as large as in the other three dataset.

Table 1. MSE and std of conditional generations.

Model	Skeleton	Armless	Lower body	Angles
Mean	477.71 ± 31.63	383.09 ± 36.00	988.00 ± 258.91	85.24 ± 7.79
Improved mean	461.04 ± 25.20	356.45 ± 16.68	815.58 ± 173.75	41.28 ± 3.11
RNN	437.82 ± 39.01	274.22 ± 15.01	776.55 ± 118.68	41.60 ± 2.87
GRU	527.43 ± 104.98	390.50 ± 70.08	868.63 ± 119.24	41.27 ± 4.71
LSTM	556.58 ± 22.28	384.49 ± 56.59	824.38 ± 94.46	41.85 ± 4.02
NRI	538.71 ± 5.95	354.67 ± 21.34	827.81 ± 74.12	41.42 ± 3.62
PG-cNRI	**380.42 ± 43.71**	302.68 ± 64.68	772.75 ± 113.13	38.19 ± 2.50
IG-cNRI	474.27 ± 122.62	351.33 ± 116.93	856.64 ± 150.88	**37.89 ± 2.30**
NRI-cNRI	399.83 ± 67.07	**212.97 ± 20.21**	696.09 ± 113.42	39.60 ± 2.35
fNRI-cNRI	433.87 ± 133.00	241.93 ± 19.23	**696.47 ± 58.93**	40.83 ± 3.11

We give examples of generations in Fig. 3 and Fig. 4, where we see that these are very close to the real ones. In Fig. 3 we report the angle trajectories, mean and standard deviation over the test set, for the real and generated data for the angles located in the right side of the body. Our model follows nicely the dynamics, but in some cases its trajectories have less variance than the real one. In Fig. 4 we provide snapshots of body positions for the real and generated data.

We notice that some of the baselines and cNRI models have high error variance. This is the result of the variable number of gait cycles we have per patient and the fact that we have patients that are affected on different body sides, left or right. When we split the data for evaluation we take care that a patient's are only present in one of the training, validation, test sets. As a result splits can be unbalanced with respect to the affected body side, which increases the risk of overfitting, with the underrepresented side in the training set leading to poor generations in the testing phase. This is something that we indeed verified by looking at the errors and distributions of the affected sides over the folds.

Finally in the top row of Fig. 5 we give the interaction maps that are used in PG-cNRI (given), IG-NRI (learned) and NRI-cNRI (aggregate posterior); we do not include fNRI-cNRI because it has one adjacency matrix per edge type and lack of space. As we see IG-NRI establishes a non-sparse interaction matrix where every part interacts with every other part. In NRI-cNRI the picture that

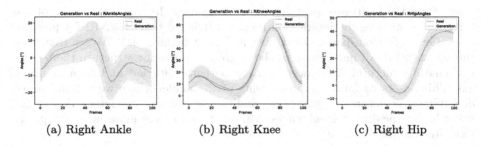

Fig. 3. Mean and variance of right side angle generations with IG-cNRI model.

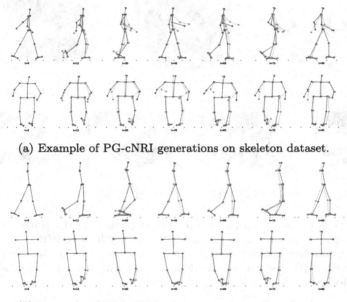

(a) Example of PG-cNRI generations on skeleton dataset.

(b) Example of NRI-cNRI generations on armless dataset.

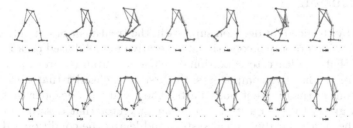

(c) Example of NRI-cNRI generations on lower body dataset.

Fig. 4. Generations (in blue) against real trajectories (in gray), the edges are from the real graph, not z. (Color figure online)

arises from the aggregate posterior is much sparser. In the bottom row of the same figure we give the interaction maps established by NRI-cNRI for particular patients (random patients with left and right hemiplegia). We see that even though they are all quite close to the aggregate posterior structure there exist systematic structural differences between patients with left and right hemiplegia. This points to future improvements of the model where we can introduce dependency structures between the condition vector and the interaction maps, using hierarchical models which we will allow as to have more informed priors that are conditioned on **c**.

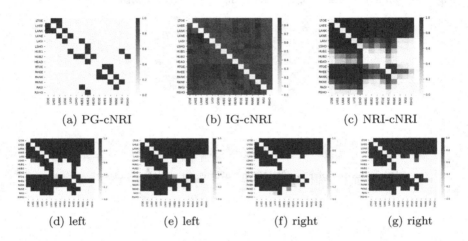

     (a) PG-cNRI   (b) IG-cNRI   (c) NRI-cNRI

    (d) left   (e) left   (f) right   (g) right

Fig. 5. Interaction maps. Top row interaction maps used in the three methods; for NRI-cNRI we give the aggregate posterior. We do not include fNRI-cNRI for space reasons due to the large number of edge types. Bottom row: patient specific interaction graphs for NRI-cNRI, first row two random patients with left hemiplegia, second row right hemiplegia.

5   Conclusion

Motivated by the need for decision support in the treatment of patients with gait impairments we propose a conditional generative model, based on an extension NRI [14], that can learn to conditionally generate from different physical systems. Our model has two components: the first is an encoder that learns a graph of interactions of the different body parts. The second is a decoder that uses the interaction graph together with a conditioning vector that describes the specificities of the particular dynamical system and learns the conditional dynamics. The experiments show that the proposed model outperforms the baselines in all dataset we experimented. Moreover the method achieves very good performance even though it has been trained on relatively small training datasets, in fact very small when it comes to the typical training size used in deep learning generative models. This is an important feature of the method since many applications, such as the one we explored here, the available training data will be very limited. As a future work we want to explore different structures in the inference and generative models and different dependence assumptions in order to increase further the generation quality, e.g. different dependency structures between the interaction matrix and the conditioning vector and/or learning to predict the latter from the former.

Acknowledgments. This work was supported by the Swiss National Science Foundation grant number CSSII5_177179 "Modeling pathological gait resulting from motor impairment".

References

1. Aliakbarian, M.S., Saleh, F., Salzmann, M., Petersson, L., Gould, S.: A stochastic conditioning scheme for diverse human motion prediction. In: 2020 IEEE/CVF Conference on Computer Vision and Pattern Recognition (CVPR), pp. 5222–5231 (2020)
2. Battaglia, P., Pascanu, R., Lai, M., Rezende, D., Kavukcuoglu, K.: Interaction networks for learning about objects, relations and physics. In: Advances in Neural Information Processing Systems, pp. 4509–4517. Neural information processing systems foundation (2016)
3. Battaglia, P.W., et al.: Relational inductive biases, deep learning, and graph networks. arXiv (2018)
4. Chang, M.B., Ullman, T., Torralba, A., Tenenbaum, J.B.: A compositional object-based approach to learning physical dynamics. In: International Conference on Learning Representations (ICLR) (2017)
5. Cho, K., et al.: Learning phrase representations using RNN encoder-decoder for statistical machine translation. In: EMNLP 2014–2014 Conference on Empirical Methods in Natural Language Processing, Proceedings of the Conference, pp. 1724–1734. Association for Computational Linguistics (ACL) (2014). https://doi.org/10.3115/v1/d14-1179
6. Fragkiadaki, K., Levine, S., Felsen, P., Malik, J.: Recurrent network models for human dynamics. In: IEEE International Conference on Computer Vision (ICCV) (2015)
7. Gopalakrishnan, A., Mali, A., Kifer, D., Giles, C.L., Ororbia, A.G.: A neural temporal model for human motion prediction. In: Proceedings of the IEEE Computer Society Conference on Computer Vision and Pattern Recognition, June 2019, pp. 12108–12117 (2018)
8. Guttenberg, N., Virgo, N., Witkowski, O., Aoki, H., Kanai, R.: Permutation-equivariant neural networks applied to dynamics prediction. arXiv (2016)
9. Hochreiter, S., Schmidhuber, J.: Long short-term memory. Neural Comput. **9**(8), 1735–1780 (1997). https://doi.org/10.1162/neco.1997.9.8.1735
10. Hoshen, Y.: VAIN: attentional multi-agent predictive modeling. In: Advances in Neural Information Processing Systems, December 2017, pp. 2702–2712. Neural information processing systems foundation (2017)

11. Jain, A., Zamir, A.R., Savarese, S., Saxena, A.: Structural-RNN: deep learning on spatio-temporal graphs. In: Proceedings of the IEEE Computer Society Conference on Computer Vision and Pattern Recognition, December 2016, pp. 5308–5317. IEEE Computer Society (2016). https://doi.org/10.1109/CVPR.2016.573

12. Kingma, D.P., Ba, J.L.: Adam: a method for stochastic optimization. In: 3rd International Conference on Learning Representations, ICLR 2015 - Conference Track Proceedings. International Conference on Learning Representations, ICLR (2015)

13. Kingma, D.P., Welling, M.: Auto-encoding variational Bayes. In: International Conference on Learning Representations (ICLR) (2014)

14. Kipf, T., Fetaya, E., Wang, K.C., Welling, M., Zemel, R.: Neural relational inference for interacting systems. In: International Conference on Machine Learning (ICML), pp. 2688–2697 (2018)

15. Lehrmann, A.M., Gehler, P.V., Nowozin, S.: Efficient nonlinear Markov models for human motion. In: Proceedings of the IEEE Computer Society Conference on Computer Vision and Pattern Recognition, pp. 1314–1321. IEEE Computer Society (2014). https://doi.org/10.1109/CVPR.2014.171

16. Li, M., Chen, S., Chen, X., Zhang, Y., Wang, Y., Tian, Q.: Symbiotic graph neural networks for 3D skeleton-based human action recognition and motion prediction. arXiv (2019)

17. Liu, Z., et al.: Towards natural and accurate future motion prediction of humans and animals. In: IEEE/CVF Conference on Computer Vision and Pattern Recognition (CVPR), pp. 10004–10012 (2019)

18. Maddison, C.J., Mnih, A., Teh, Y.W.: The concrete distribution: a continuous relaxation of discrete random variables. In: International Conference on Learning Representations (ICLR) (2016)

19. Martinez, J., Black, M.J., Romero, J.: On human motion prediction using recurrent neural networks. In: Proceedings - 30th IEEE Conference on Computer Vision and Pattern Recognition, CVPR 2017, January 2017, pp. 4674–4683 (2017)

20. Pavlovic, V., Rehg, J.M., Maccormick, J.: Learning switching linear models of human motion. In: Neural Information Processing Systems (NeurIPS), pp. 981–987 (2001)

21. Pitto, L., et al.: SimCP: a simulation platform to predict gait performance following orthopedic intervention in children with cerebral palsy. Front. Neurorobot. **13** (2019). https://doi.org/10.3389/fnbot.2019.00054

22. Rumelhart, D.E., Hinton, G.E., Williams, R.J.: Learning representations by back-propagating errors. Nature **323**(6088), 533–536 (1986). https://doi.org/10.1038/323533a0

23. Santoro, A., et al.: A simple neural network module for relational reasoning. In: Advances in Neural Information Processing Systems, December 2017, pp. 4968–4977 (2017)

24. Shi, L., Zhang, Y., Cheng, J., Lu, H.: Skeleton-based action recognition with directed graph neural networks. In: The IEEE Conference on Computer Vision and Pattern Recognition (CVPR) (2019)

25. van Steenkiste, S., Chang, M., Greff, K., Schmidhuber, J.: Relational neural expectation maximization: unsupervised discovery of objects and their interactions. In: 6th International Conference on Learning Representations, ICLR 2018 - Conference Track Proceedings (2018)

26. Sukhbaatar, S., Szlam, A., Fergus, R.: Learning multiagent communication with backpropagation. In: Neural Information Processing Systems (NeurIPS), pp. 2244–2252 (2016)

27. Urtasun, R., Fleet, D.J., Geiger, A., Popović, J., Darrell, T.J., Lawrence, N.D.: Topologically-constrained latent variable models. In: Proceedings of the 25th International Conference on Machine Learning, pp. 1080–1087. ACM Press, New York (2008). https://doi.org/10.1145/1390156.1390292

28. Walker, J., Marino, K., Gupta, A., Hebert, M.: The pose knows: video forecasting by generating pose futures. In: Proceedings of the IEEE International Conference on Computer Vision, October 2017, pp. 3352–3361 (2017)

29. Wang, B., Adeli, E., Chiu, H.K., Huang, D.A., Niebles, J.C.: Imitation learning for human pose prediction. In: Proceedings of the IEEE International Conference on Computer Vision, October 2019, pp. 7123–7132. Institute of Electrical and Electronics Engineers Inc. (2019). https://doi.org/10.1109/ICCV.2019.00722

30. Wang, J.M., Fleet, D.J., Hertzmann, A.: Gaussian process dynamical models for human motion. IEEE Trans. Pattern Anal. Mach. Intell. **30**, 283–298 (2008)

31. Watters, N., Tacchetti, A., Weber, T., Pascanu, R., Battaglia, P., Zoran, D.: Visual interaction networks: learning a physics simulator from video. In: Neural Information Processing Systems (NeurIPS), pp. 4539–4547 (2017)

32. Webb, E., Day, B., Andres-Terre, H., Lió, P.: Factorised neural relational inference for multi-interaction systems. In: ICML Workshop on Learning and Reasoning with Graph-Structured Representations (2019)

Recommender Systems and Behavior Modeling

MMNet: Multi-granularity Multi-mode Network for Item-Level Share Rate Prediction

Haomin Yu$^{(\boxtimes)}$, Mingfei Liang$^{(\boxtimes)}$, Ruobing Xie$^{(\boxtimes)}$, Zhenlong Sun, Bo Zhang, and Leyu Lin

WeChat Search Application Department, Tencent, China
haominyu@bjtu.edu.cn, {aesopliang,richardsun,
nevinzhang,goshawklin}@tencent.com

Abstract. Item-level share rate prediction (ISRP) aims to predict the future share rates for each item according to the meta information and historical share rate sequences. It can help us to quickly select high-quality items that users are willing to share from millions of item candidates, which is widely used in real-world large-scale recommendation systems for efficiency. However, there are several technical challenges to be addressed for improving ISRP's performance: (1) There is data uncertainty in items' share rate sequences caused by insufficient item clicks, especially in the early stages of item release. These noisy or even incomplete share rate sequences strongly restrict the historical information modeling. (2) There are multiple modes in the share rate data, including normal mode, cold-start mode and noisy mode. It is challenging for models to jointly deal with all three modes especially with the cold-start and noisy scenarios. In this work, we propose a multi-granularity multi-mode network (MMNet) for item-level share rate prediction, which mainly consists of a fine-granularity module, a coarse-granularity module and a meta-info modeling module. Specifically, in the fine-granularity module, a multi-mode modeling strategy with dual disturbance blocks is designed to balance multi-mode data. In the coarse-granularity module, we generalize the historical information via item taxonomies to alleviate noises and uncertainty at the item level. In the meta-info modeling module, we utilize multiple attributes such as meta info, contexts and images to learn effective item representations as supplements. In experiments, we conduct both offline and online evaluations on a real-world recommendation system in WeChat Top Stories. The significant improvements confirm the effectiveness and robustness of MMNet. Currently, MMNet has been deployed on WeChat Top Stories.

Keywords: Share rate prediction · Multi-granularity · Multi-mode

---

H. Yu, M. Liang and R. Xie—Contribute equally to this work.

© Springer Nature Switzerland AG 2021
Y. Dong et al. (Eds.): ECML PKDD 2021, LNAI 12979, pp. 201–217, 2021.
https://doi.org/10.1007/978-3-030-86517-7_13

1   Introduction

With the development of social media, people are becoming more enthusiastic about publishing their created contents and sharing their opinions on the internet, generating massive amounts of items. Users want to quickly obtain valuable information from massive items on social media platforms. Therefore, personalized recommendations are adopted to provide appropriate items effectively and efficiently for users to read and share.

Real-world large-scale recommendation systems should deal with millions of new items per day. It is essential to pre-select high-quality items for the following personalized matching and ranking modules in recommendation systems [6,16] for efficiency. In this work, we propose the **Item-level Share Rate Prediction (ISRP)** task, which aims to predict the future share rates for each item according to their meta information and historical share rate sequences. It can help us to quickly find appropriate items that users are interested in from millions of item candidates, which could be viewed as an item quality inspector that is essential in real-world large-scale recommendation systems.

In recent years, to better grasp the development trend of items, many scholars have studied popularity prediction by inferring the total counts of interactions between users and items (e.g., view, click and share). The popularity prediction approaches can be roughly divided into two categories, including social-based prediction methods [3] and item-based prediction methods [4]. Item-based prediction methods generally utilize item-related meta information such as images and contexts to predict popularity [15]. Inspired by this, ISRP can be regarded as a special item-based popularity prediction task in recommendation systems, which focuses on predicting item share rates only with the item-related information. Moreover, we creatively bring in the historical share rate sequence for each item containing the average item share rate at each time period. However, there are some challenges in combining different meta information and historical share rate information for ISRP in practice:

- **Item-related data uncertainty.** In ISRP, there are mainly two types of item-related data uncertainty, including the share rate uncertainty and the attribute uncertainty. Share rate uncertainty mainly occurs in the early stage of item release, which is caused by the insufficient item clicks. In addition, the share rates of an item may fluctuate greatly during the whole period, which makes it difficult for the model to obtain high-confidence information from the historical share rate trends. This uncertainty locates in every item's lifetime, since every item has a cold-start period and most items are long-tail. In contrast, attribute uncertainty derives from the noises or missing in item-related meta information. Therefore, it is essential to introduce an uncertainty eliminator to enable a robust ISRP framework.
- **Multi-mode share rate data.** In practice, there are multiple modes of the share rate data, including the normal mode, cold-start mode and noisy mode. *Normal mode* indicates that the share rate sequences are reliable with sufficient clicks, so the model can fully rely on historical share rates. In contrast,

cold-start mode refers to the mode influenced by the share rate uncertainty in the early stages of item release. *Noisy mode* represents that the share rate sequences are disturbed by noises and even data missing. Due to the unbalance in three modes, ISRP models can be easily dominated by the normal mode and is prone to over-rely on historical share rate data, losing the ability to model cold-start and noisy modes. Therefore, an intelligent multi-mode learner is needed to jointly handle all scenarios.

To address the above challenges, we propose a **Multi-granularity Multi-mode Network (MMNet)** for item-level share rate prediction. MMNet is composed of a coarse-granularity module, a fine-granularity module and a meta-info modeling module, where the first two modules aim to model the historical share rate sequences. Specifically, in the fine-granularity module, we design two disturbance blocks with different masking strategies to highlight all modes during training process. The coarse-granularity module is presented to alleviate share rate uncertainty by considering global preference features anchored by item taxonomies. The meta-info modeling module aims to introduce sufficient meta features to represent item information as a supplement to the historical share rate sequences. All three features are then combined for the ISRP task.

In experiments, we conduct extensive evaluations on three datasets with normal, cold-start and noisy modes. We also deploy MMNet on a widely-used recommendation system to evaluate its online effectiveness. In summary, the contributions of this work can be summarized as follows:

- We systematically highlight the challenges in real-world item-level share rate prediction, and propose a novel MMNet framework to address them.
- We design the multi-granularity share rate modeling to alleviate the uncertainty issues in cold-start and noisy scenarios, which helps to capture user preferences from both the global and local perspectives.
- We present a multi-mode modeling strategy in the fine-granularity module with dual disturbance blocks, which can jointly learn informative messages from all three modes to build a robust model in practice.
- MMNet achieves significant improvements in both offline and online evaluations. Currently, MMNet has been deployed on WeChat Top Stories, affecting millions of users.

2 Related Works

Time Series Modeling Techniques. Time series modeling techniques have been widely used in forecasting tasks. Autoregressive integrated moving average (ARIMA) [13] model, which is a classic statistical model in the time series field. However, this model requires the time series data stationary, or stationary after differencing steps. In recent years, various sequence modeling methods based on deep learning have emerged, such as recurrent neural network (RNN). Nevertheless, RNN suffers from gradient disappearance and explosion problems. To alleviate the problem, long short-term memory (LSTM) [9] and gated recurrent unit(GRU) [5] methods appeared. However, the inherently sequential nature of recurrent models limits the ability of parallelization ability. Temporal convolutional network (TCN) [1], which utilizes convolution algorithm to solving prediction problem. It can achieve parallelization computation. Since historical information can provide a certain degree of guidance for ISRP tasks, we introduce the historical share rate sequences to improve ISRP's performance.

Popularity Prediction. The popularity prediction task is generally to estimate how many attentions a given content will receive after it is published on social media. The task is mainly divided into two types: social-based prediction methods and item-based prediction methods.

The social-based methods aim to predict the popularity of item spread through social relationships in social networks. DeepCas [10] and DeepHawkers [2] are popularity prediction methods by modeling information cascade. DeepCas constructs a cascade graph as a collection of cascade paths that are sampled by multiple random walk processes, which can effectively predict the size of cascades. DeepHawkers learns the interpretable factors of Hawkers process to model information cascade. Cao et al. [3] proposed CoupledGNN, which uses two coupled graph neural networks to capture the interplay between nodes and the spread of influence. However, the social-based methods concentrate on the propagation on social networks, which have a great dependence on social relationships. This limits the application scenarios of these models.

In contrast, the item-based methods extract a large number of features related to contents for popularity prediction. UHAN [18] and NPP [4] design hierarchical attention mechanisms to extract representations of multi-modalities. Different from them, Wu et al. [14] and Liao et al. [12] paid more attention to the influence of temporal information on popularity prediction. The former utilizes neighboring temporal and periodic temporal information to learn sequential popularity in short-term and long-term popularity fluctuations. The latter leverages RNN and CNN to capture item-related long-term growth and short-term fluctuation. Xie [15] proposed a multimodal variational encoder-decoder (MMVED) framework, which is the most related model of our task. It introduces the uncertain factors as the randomness for the mapping from the multimodal features to the popularity. However, in ISRP, the item-related data uncertainty and multimode share rate data will strongly affect the performance of existing popularity prediction models. Consequently, we propose disturbance blocks based multimode modeling strategy with multi-granularity share rate modeling for ISRP.

3   Preliminary

In this section, we first introduce some important notions used in this work.

Share Rate. Given an item I, the share rate y_t at time period t is defined as its overall shared number r_t divided by its overall click (i.e., items being clicked by users) number p_t, as shown in the following formula:

$$y_t = \frac{r_t}{p_t} \times 100\%, \quad p_t > 0. \tag{1}$$

A smaller click number p_t will result in data uncertainty of the share rate.

Multi-mode Data. For the historical share rate sequences, we define three modes according to different scenarios as follows:

- **Normal mode.** It represents that historical share rate sequences are reliable with sufficient clicks. The sequences can provide strong guidance.
- **Cold-start mode.** It indicates that whole historical share rate sequences are unreliable or even missing. This mode is usually caused by insufficient clicks, especially in the early stages of item release.
- **Noisy mode.** It means that there is partial uncertainty in the historical share rate sequences, which is usually caused by partial missing or noises.

Input Features. We deploy MMNet on a video recommendation system. The input features we use in MMNet can be mainly grouped into three categories, namely the fine-grained sequential features, the coarse-grained sequential features and the multi-modal meta-information features.

- **Fine-grained sequential features.** We calculate the share rates at all time period for each item, and arrange them into historical share rate sequences.
- **Coarse-grained sequential features.** To improve the generalization ability and reduce potential uncertainty in item-level share rates, we further bring in the coarse-grained sequential features. Precisely, we build the share rate sequences for each taxonomy (e.g., tag, category) in this item, modeling the share rate trends at the taxonomy level.
- **Multi-modal meta-information features**. These features consist of three heterogeneous parts: context features, visual features and meta features. The first part contains textual features such as item title. The second part regards the cover images as the visual features. The last part is composed of many meta information including video taxonomies and duration.

Item-Level Share Rate Prediction. Formally, given the multi-modal feature set $\mathbf{C}$, and the historical share rate sequence $\{y\}_{t-\omega_1}^t$ with a time window of length ω_1, our goal at time t is to predict the share rate $\hat{y}_{t+h}$ at the next h time as:

$$F(\mathbf{C}, \{y\}_{t-\omega_1}^t) \rightarrow \hat{y}_{t+h}, \tag{2}$$

where $\hat{y}_{t+h}$ is the predicted share rate at $t + h$, and h is the desirable prediction horizon time stamp. In most situations, the horizon h of share rate prediction task is chosen according to the practical demands of the real-world scenario. $F(\cdot)$ is the mapping function we aim to learn via MMNet.

4　Methodology

4.1　Overall Framework

Figure 1 shows the overall framework of MMNet. It mainly consists of three parallel modules, including a fine-granularity module, a coarse-granularity module, and a meta-info modeling module. The fine-granularity module conducts a multimode strategy with two disturbance blocks to enable a robust share rate sequence modeling. The coarse-granularity module models the coarse-grained share rate sequential information brought by the corresponding item's taxonomies, which can alleviate potential noises and missing in item-level share rate sequences. The meta-info modeling module further combines heterogeneous item meta features together. All three modules are then combined and fed into a gated fusion layer and a MLP (multi-layer perceptron) layer for the following prediction.

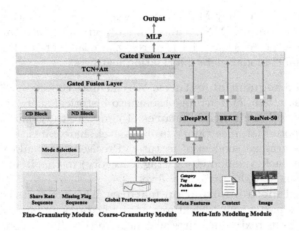

Fig. 1. Overall framework of the proposed MMNet.

4.2　Fine-Granularity Module

The fine-granularity module is responsible for encoding historical share rate sequences. However, there are two differences between the numerical share rate sequence in ISRP and other sequences (e.g., item sequences in session-based recommendation), which leads to the following challenges: (1) the share rate sequences are numerical sequences that suffer from data uncertainty and high variance caused by insufficient item clicks. (2) The fine-granularity module should jointly deal with three modes including the normal, cold-start and noisy modes. We conduct the multi-mode modeling to address these issues.

Multi-mode Modeling. In real-world scenarios, the multi-mode data are often unbalanced, and the normal mode data are far more than other two modes. Thus, if we directly use the original share rate sequence instances to train our MMNet, the model will be overfitting on the historical share rate information, regardless of other meta-information. Although the model can well predict the share rates of normal mode data with sufficient clicks and historical information, it cannot deal with items in cold-start and noisy scenarios, which heavily rely on MMNet for item pre-selection in real-world recommendation systems. Therefore, for all historical sequences during training process, we randomly feed them into three modes followed by different disturbance blocks with equal probability. To simulate different mode sequences, the proposed multi-mode modeling strategy introduces two disturbance blocks, including a cold-start disturbance block (CD block) and a noisy disturbance block (ND block), as shown in Fig. 2. Note that this strategy can be regarded as a form of data augmentation. To better represent the state of the historical sequence, we introduce a missing flag sequence $\mathbf{m} = \{m\}_{t-\omega_1}^t$, where $m \in \{0, 1\}$. If the missing flag is 1, it means that the data is missing or uncertain, and otherwise, it means that the data is normal.

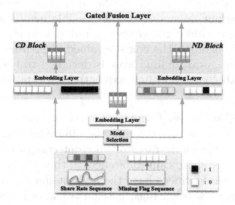

Fig. 2. The illustration of CD and ND disturbance blocks based multi-mode modelling.

Cold-Start Disturbance Block. CD block is in charge of simulating cold-start mode data. This block can make the model learn more comprehensive features and alleviate the model's excessive dependence on historical sequences. Formally, we exploit an all-one vector mask $\mathbf{m}^p = \{m^p\}_{t-\omega_1}^t$ ($m^p \in \{1\}$) as a missing flag vector, which means whole share rate sequence is missing or uncertain. Thus, the corresponding share rate sequence is erased by zero vector $\{y\}_{t-\omega_1}^t = \vec{\mathbf{0}}$. It aims to lead the model to focus more on other information rather than the share rates to improve the performance of all modes.

Noisy Disturbance Block. Similar to the CD block, we also design a ND block, which is responsible for simulating partial data uncertainty. For each missing flag in the sequence, we randomly sample a value T from the uniform distribution $U[0,1]$, and then set a threshold τ. When T is greater than τ, the missing flag is set as 0, otherwise, it is set to 1. Note that τ can be regarded as a missing rate. When τ is large, there are more missing. Considering the input sequence also contains missing data, we should keep the missing data of origin input sequence unchanged. Thus, the final missing flag sequence is $\mathbf{m}^c = \mathbf{m}^c \vee \mathbf{m}$, where $\vee$ represents logical or. Consequently, the corresponding input share rate sequence $\mathbf{y}$ is reset through $\mathbf{y} = \mathbf{m}^c \odot \mathbf{y}$, where $\odot$ denotes Hadamard product. After the input sequence is processed by a mode, it is sent into the embedding layer $Emb(\cdot)$ [12] to obtain the item-level share rate representation sequence $\{\mathbf{h}^m\}_{t-\omega_1}^t$ as:

$$\{\mathbf{h}^m\}_{t-\omega_1}^t = Emb(\text{CD}(\{y\}_{t-\omega_1}^t) \text{ or } \text{ND}(\{y\}_{t-\omega_1}^t) \text{ or } \{y\}_{t-\omega_1}^t). \tag{3}$$

4.3   Coarse-Granularity Module

The fine-granularity module focuses on the historical share rates at the item level, which is precise but noisy due to the possible insufficient clicks and even data missing. Hence, we build the coarse-granularity module as a supplement, which is in charge of encoding the coarse-grained sequential information at the taxonomy level. A temporal mining layer is designed to encapsulate the trend information from both coarse and fine sequential information in two modules.

Global Preference Features. Users have different priori preferences on different taxonomies. For example, considering the difference attractions of the item categories, we analyze the share rates of different categories in our system. As shown in Fig. 3, there are significant differences in the share rates of different categories (e.g., health-related videos have the highest share rate). Moreover, items with the same taxonomies (e.g., tags, categories) may have similar share rate trends. For instance, during the World Cup, the share rates of football-related videos generally grow higher than others. Therefore, it is essential to consider the share rate trends at the taxonomy level as a supplement to the item level.

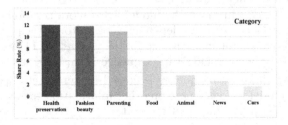

Fig. 3. The illustration of the global average share rates of different item categories.

Specifically, we introduce global preference features as a priori and generalized anchor to give coarse-granularity guidance for ISRP. Taking category for example, given an item I belonging to Category $Cat(I)$, the category global preference $g_t^{Cat(I)}$ at time t is obtained by calculating the global average share rate of whole item collection belonging to the $Cat(I)$ category:

$$g_t^{Cat(I)} = \underset{\hat{I} \in C(Cat(I))}{Mean} \{y_t(\hat{I})\}, \tag{4}$$

where $C(Cat(I))$ indicates the collection of all items which belongs to Category $Cat(I)$, $\hat{I}$ represents an item in collection $C(Cat(I))$, $y_t(\hat{I})$ stands for the share rate of item $\hat{I}$, and $Mean(\cdot)$ represents the average operation. Note that we calculate the global preferences for all time periods to build the global category-level share rate sequence. Other taxonomies' modeling is the same as the category's. Next, at each time step t, we acquire global preference features $\mathbf{g}_t$, where $\mathbf{g}_t = Concat\{g_t^{Cat(I)}, g_t^{Tag(I)}, ...\}$. Similarly, we use the same embedding layer on the global feature sequence $\{\mathbf{g}\}_{t-\omega_1}^t$ to obtain the taxonomy-level share rate representation sequence as follows:

$$\{\mathbf{h}^g\}_{t-\omega_1}^t = Emb(\{\mathbf{g}\}_{t-\omega_1}^t). \tag{5}$$

Temporal Mining Layer. Temporal mining layer is responsible for encoding both fine and coarse sequential information from item-level share rate representation $\{\mathbf{h}^m\}_{t-\omega_1}^t$ and taxonomy-level share rate representation $\{\mathbf{h}^g\}_{t-\omega_1}^t$. To better balance fine and coarse representations, we utilize an attention mechanism [17] to obtain the aggregated representation as:

$$\{\mathbf{h}^{mg}\}_{t-\omega_1}^t = Att(\{\mathbf{h}^m\}_{t-\omega_1}^t, \{\mathbf{h}^g\}_{t-\omega_1}^t). \tag{6}$$

To reveal the inherent regularity and encapsulate the share rate trend information of historical information, we adopt a temporal convolutional network (TCN) [1] to learn the final sequential representation $\{\mathbf{h}^q\}_{t-\omega_1}^t$ as:

$$\{\mathbf{h}^q\}_{t-\omega_1}^t = \text{TCN}(\{\mathbf{h}^{mg}\}_{t-\omega_1}^t). \tag{7}$$

Then, we utilize a temporal attention mechanism on the sequence $\{\mathbf{h}^q\}_{t-\omega_1}^t$ to automatically learn the impacts of share rate representations at different times:

$$\mathbf{q} = Temporal\_Att(\{\mathbf{h}^q\}_{t-\omega_1}^t), \tag{8}$$

where $\mathbf{q}$ is the final historical share rate representation we use for prediction.

4.4   Meta-info Modeling Module

Meta-info modeling module is exploited to capture multi-modal features from heterogeneous item profiles. Multi-modal information can introduce complementary information for ISRP, thereby alleviating attribute uncertainty.

Specifically, this module is responsible for extracting interactive meta feature representations, context representations and visual representations:

210 H. Yu et al.

- **Meta features**: there are potential relationships between different meta features. To capture effective feature interactions, we feed them into xDeepFM [11] for extracting the interactive representation $\mathbf{u}$.
- **Context features**: we directly use a pre-trained BERT [7], and acquire a context representation $\mathbf{c}$ from the input contexts (e.g., video titles).
- **Visual features**: we use a pre-trained ResNet-50 [8], which utilizes skip connections, or shortcuts to jump over some layers. Precisely, we feed the cover image into a ResNet-50 model, and obtain the visual representation $\mathbf{v}$.

These features are essential especially in the cold-start and noisy scenarios.

4.5 Optimization Objectives

We jointly consider the share rate sequential representation $\mathbf{p}$, meta representation $\mathbf{u}$, context representation $\mathbf{c}$, and visual representation $\mathbf{v}$ in ISRP. To automatically determine the influence of these representations, We concatenate these features, and send them into a gated fusion layer as follows:

$$\mathbf{h}^{\text{fuse}} = Gating([\mathbf{p}; \mathbf{u}; \mathbf{c}; \mathbf{v}]), \tag{9}$$

where $Gating(\cdot)$ is similar as the attention mechanism in [17]. $\mathbf{h}^{\text{fuse}}$ represents the aggregated feature which encloses multi-modal information. Finally, we feed the aggregated feature $\mathbf{h}^{\text{fuse}}$ into a MLP layer to generate the predicted share rate $\hat{y}_{t+h}$ at the $t + h$ time as follows:

$$\hat{y}_{t+h} = MLP(\mathbf{h}^{\text{fuse}}). \tag{10}$$

In this work, we optimize the proposed MMNet by minimizing a mean square error (MSE) between the predicted and real share rates $\hat{y}_{t+h}$ and y_{t+h} as follows:

$$MSE = \frac{1}{N} \sum_{t=1}^{N} (\hat{y}_{t+h} - y_{t+h})^2, \tag{11}$$

where N is the number of training samples. Note that it is also convenient to transfer MMNet to other rate prediction tasks (e.g., CTR, complete rate).

5 Online Deployment

We have deployed our MMNet model on a well-known real-world recommendation system, which is widely used by millions of users per day. This online system should deal with massive numbers of new items generated everyday. Therefore, based on the classical two-stage recommendation framework containing matching (i.e., candidate generation) and ranking modules introduced in [6], we further deploy MMNet on the pre-matching module to judge item quality according to item's meta-information and historical behaviours for efficiency. The predicted share rates of each item candidate are used in two manners: (1) we directly filter low-quality items according to a low-standard rule-based threshold, and (2) the predicted share rates are fed into ranking modules as features. For efficiency, we use 200 workers equipped with 1 core and 8 GB memory for online inference. The source code is in https://github.com/MingFL/MMNET.

6   Experiments

6.1   Datasets

To thoroughly evaluate the performance of our methods, we build an online video dataset from a widely-used recommendation system named WeChat Top Stories. The dataset contains nearly 40 million share instances on 35 thousand items. All data are pre-processed via data masking for user privacy. We divide the dataset by a ratio of 8:1:1 for training, validation and testing. Due to the uncertainty in share rates, we discard all items that have low clicks in the test set, for we want all instances in the test set to have high confidence. This test setting is named as the *Normal Dataset*, since it mainly contains items with sufficient clicks and reliable historical share rates. To further investigate the model abilities for multi-mode data, the normal dataset is further processed into two other datasets, namely the *Cold-start dataset* and the *Noisy dataset*. To simulate the cold-start mode where all historical share rate data is unreliable or empty, we mask out all historical share rates on the normal dataset for generating the cold-start dataset. Similarly, in order to verify that the model deals with the data of noisy mode, we mask out the historical data with a certain probability.

6.2   Baselines and Experimental Settings

Baselines. The main contributions of MMNet locate in the share rate sequence modeling. Therefore, we compare MMNet with five competitive baselines in the share rate modeling. For fair comparisons, all baselines also contain the same meta-info modeling module, where the encoding of multi-modal features is consistent with MMNet (i.e., BERT processing context features, ResNet-50 processing visual features, and xDeepFM processing meta features). All models including MMNet and baselines share the same input features. We have:

- **HA.** Historical average (HA) is a straightforward method. Here, we use the average share rate value of the most recent 6 time periods (the same as MMNet) to predict the share rates in the next horizon time.
- **GRU.** Gated recurrent unit [5] is a classical model that can alleviate the problem of vanishing gradient in RNN. It performs well in solving time series forecasting problems.
- **Encoder-Decoder.** The encoder-decoder model [5] is a classical sequence modeling method, which is widely utilized in real-world tasks.
- **MMVED.** Multimodal variational encoder-decoder framework [15] is designed for sequential popularity prediction task, which considers the uncertain factors as randomness for the mapping from the multimodal features to the popularity.
- **DFTC.** The approach of deep fusion of temporal process and content features [12] is utilized in online article popularity prediction. It utilizes RNN and CNN to capture long-term and short-term fluctuations, respectively.

Ablation Settings. Furthermore, to verify the advantages of each component of MMNet, we conduct four ablation versions of MMNet implemented as follows:

- **MMNet-M.** It is an incomplete MMNet, in which the multi-mode modeling strategy is removed, in order to verify the multi-mode modeling influence.
- **MMNet-C.** It is an incomplete MMNet, in which the coarse-granularity module is removed on the basis of MMNet-M, in order to verify the role of global preference features on three modes.
- **MMNet$_{norm/noisy}$.** It is a variant of MMNet, which lets the historical sequence select the normal mode and the noisy mode with equal probability without considering the cold-start mode.
- **MMNet$_{norm/cold}$.** It is a variant of MMNet, which lets the historical sequence select the normal mode and the cold-start mode with equal probability without considering the noisy mode.

Experimental Settings. The proposed method is implemented with Tensorflow. The learning rate is set as 0.001, the batch size is set as 64, and the model is trained by minimizing the mean squared error function. The historical sequence window lengths (i.e. ω_1) of the share rates and global preference features are set to 6. In MMNet, the representations after embedding layer are all set as 64, including sequence representation, visual representation, context representation and meta representation. For TCN, we set three channels, and the hidden layers of these channels are 256, 128 and 64 respectively. Meanwhile, the size of the convolution kernel in TCN is 2. The missing rate τ is set to be 0.5, and the parameter sensitivity experiment of τ can be seen in Sect. 6.6.

6.3   Offline Item-Level Share Rate Prediction

Evaluation Protocol. We adopt two representative evaluation metrics for ISRP, including mean squared error (MSE) and precision@N% (P@N%). (1) MSE is a classical metric that is calculated by the average squared error between predicted and real share rates. It aims to measure the ability of MMNet in predicting share rates. (2) As for P@N%, we first rank all items in the test set via their predicted share rates, and then calculate the precision of top N% items as P@N%. It reflects the real-world performance of ISRP in recommendation systems. To simulate the practical settings, we report P@5% and P@10% in evaluation.

Experimental Results. Table 1 presents the offline ISRP results of all models. We analyze the experimental results in details:

(1) MMNet achieves the best overall performance on all three datasets. The improvements of three metrics on the cold-start/noisy datasets, and the improvement of MSE on the normal dataset are significant with the significance level $\alpha = 0.01$. Since the proposed multi-granularity multi-mode

strategy mainly aims to solve the cold-start and noisy issues, it is natural that the improvements on the cold-start and noisy datasets are much more significant. It indicates that MMNet can well deal with all three scenarios in ISRP, especially in the cold-start and noisy scenarios.

(2) Comparing with baselines, we find that the results of baselines are not ideal in cold-start and noisy datasets. It is because that the multi-mode data is not balanced, where the normal mode is the dominating mode. Therefore, most baselines are strongly influenced by the normal mode data during training. In contrast, our MMNet is armed with the multi-granularity sequence modeling that can alleviate the cold-start and low click issues. Moreover, the multi-mode modeling also brings in robustness for these two scenarios. It can be regarded as a certain data argumentation, which can improve both the generalization ability of the share rate sequence modeling as well as the feature interactions between sequential and meta information in different scenarios.

Table 1. Calibration results for three datasets.

Method	Normal dataset			Cold-start dataset			Noisy dataset		
	MSE	P@5%	P@10%	MSE	P@5%	P@10%	MSE	P@5%	P@10%
HA	1.650	0.919	0.920	42.143	0.054	0.101	7.174	0.747	0.791
GRU	0.260	0.975	0.973	16.410	0.219	0.293	3.299	0.688	0.761
Encoder-Decoder	0.256	0.975	0.973	15.663	0.233	0.291	3.990	0.656	0.725
MMVED	1.431	0.882	0.880	22.072	0.052	0.108	2.109	0.840	0.847
DFTC	0.257	0.976	0.974	14.980	0.284	0.348	3.677	0.760	0.786
MMNet	**0.149**	**0.977**	**0.976**	**3.442**	**0.755**	**0.786**	**0.175**	**0.968**	**0.969**

6.4   Online A/B Tests

Evaluation Protocol. To further evaluate MMNet in practice, we deploy our model on a real-world recommendation system as introduced in Sect. 5. Specifically, MMNet is deployed in the pre-matching module and predicts item-level share rates for all items, which is used as (1) a coarse filter, and (2) features for the next matching and ranking modules. We conduct an online A/B test with other modules unchanged. The online base model is an ensemble of some rule-based filterers. In this online A/B test, we focus on two metrics: (1) average item-level share rate (AISR), (2) average dwell time per user (ADT/u).

Similarly, we further transfer the idea of MMNet on ISRP to the item-level complete rate prediction task. The complete rate is calculated by *user-finished duration divided by video's full duration*, which reflects the qualities of items from another aspect. Precisely, we build a similar MMNet model with different parameters, and train it under the supervision of item-level complete rates of videos. We deploy this MMNet as in Sect. 5, and focus on (1) average dwell time per user (ADT/u), and (2) average dwell time per item (ADT/i). We conduct this online A/B test for 14 days, affecting nearly 6 million users.

Table 2. Online A/B tests on a real-world recommendation system.

Settings	Supervised by share rates		Supervised by complete rates	
	AISR	ADT/u	ADT/u	ADT/i
MMNet	+0.91%	+0.93%	+1.02%	+1.31%

Experimental Results. Table 2 shows the relative improvements of MMNet over the online base model, from which we can observe that:

(1) MMNet achieves significant improvements in both item-level share rate and average dwell time. It indicates that our MMNet can well capture multi-granularity features, distinguish multi-mode share rate sequences, and combine multi-modal features for all normal, cold-start and noisy scenarios in ISRP.

(2) The successes in MMNet supervised by complete rates verify that our proposed framework is robust and easy to transfer to other scenarios.

6.5   Ablation Studies

Table 3 lists the results of the above-mentioned ablation settings with MSE, P@5% and P@10%. Note that since the multi-granularity and multi-mode modeling are mainly designed for the cold-start and noisy scenarios, we focus on these two datasets in ablation studies. We can observe that:

(1) MMNet achieves the best performance on all metrics in the noisy dataset and normal dataset, and the second best performance in the cold-start scenario. It verifies that all components in MMNet are essential in ISRP.

(2) Comparing with MMNet-C and MMNet-M, we can find that the global preference features are more suitable for cold-start and noisy scenarios. Meanwhile, the results also show that multi-granularity and multi-mode modeling are effective in capturing informative messages for all three modes in ISRP.

(3) Comparing with $\text{MMNet}_{norm/noisy}$ and $\text{MMNet}_{norm/cold}$, we find that both CD and ND disturbance blocks are effective for the cold-start and noisy scenarios respectively. It is worth noting that $\text{MMNet}_{norm/cold}$ focuses on the cold-start mode, so it is natural that it has better cold-start performance. In practice, we can flexibly set the weights of different disturbance blocks for specific motivations.

Table 3. Ablation study results for three datasets.

Method	Normal dataset			Cold-start dataset			Noisy dataset		
	MSE	P@5%	P@10%	MSE	P@5%	P@10%	MSE	P@5%	P@10%
MMNet-M	0.248	0.974	0.974	14.594	0.365	0.406	0.633	0.931	0.938
MMNet-C	0.671	0.975	0.975	15.369	0.316	0.373	1.537	0.907	0.922
MMNet$_{norm/noisy}$	0.210	0.974	0.973	11.944	0.466	0.503	0.315	0.964	0.966
MMNet$_{norm/cold}$	0.165	0.973	0.976	2.440	0.802	0.831	0.836	0.888	0.925
MMNet	0.149	0.977	0.976	3.442	0.755	0.786	0.175	0.968	0.969

6.6   Parameter Analyses

We further study the parameter sensitivity of MMNet. We vary the missing rate τ from 0.01 to 0.9, which is essential in model training. The results are reported in Table 4, from which we can find that: (1) The results of the parameter changes are relatively stable on the normal dataset. (2) In the cold-start dataset, the performance gradually improves as the missing rate increases. The main reason is that the missing rate is higher, and the data in the noisy dataset and the cold-start dataset will be more similar. (3) In the noisy dataset, as the missing rate increases, the performance has a gradual improvement followed by a slight decrease. The size of the missing rate can reflect the model's dependence on historical data to a certain extent, so it can be concluded that the appropriate dependence on historical data is helpful to the model performance improvement. We select $\tau = 0.5$ according to the overall performance on three modes.

Table 4. Parameter analysis with different missing rates τ.

Method	Normal dataset			Cold-start dataset			Noisy dataset		
	MSE	P@5%	P@10%	MSE	P@5%	P@10%	MSE	P@5%	P@10%
0.01	0.594	0.974	0.975	4.073	0.742	0.783	0.883	0.945	0.956
0.05	0.156	0.977	0.976	3.453	0.757	0.789	0.287	0.959	0.961
0.1	0.161	0.975	0.975	3.585	0.749	0.785	0.252	0.963	0.963
0.2	0.160	0.977	0.975	3.535	0.751	0.784	0.238	0.966	0.966
0.3	0.154	0.977	0.976	3.441	0.759	0.789	0.226	0.968	0.969
0.4	0.159	0.975	0.975	3.607	0.747	0.783	0.176	0.966	0.968
0.5	0.149	0.977	0.976	3.442	0.755	0.786	0.175	0.968	0.969
0.6	0.154	0.975	0.975	3.184	0.769	0.796	0.223	0.966	0.969
0.7	0.148	0.975	0.975	2.919	0.779	0.813	0.217	0.967	0.968
0.8	0.152	0.974	0.974	2.744	0.790	0.815	0.223	0.964	0.967
0.9	0.150	0.975	0.975	2.346	0.813	0.837	0.224	0.964	0.965

7   Conclusion and Future Work

In this paper, we present MMNet for ISRP. We propose a multi-granularity sequence modeling to improve the generalization ability from item taxonomies. Moreover, we design two multi-mode disturbance blocks to enhance the robustness of MMNet against potential data noises and uncertainty. Both offline and online evaluations confirm the effectiveness and robustness of MMNet in WeChat Top Stories. In the future, we will design an adaptive mode selection strategy based on the characteristics of the instance itself, so as to fully learn feature representations from existing instances. We will also explore more sophisticated feature interaction modeling between all types of features.

References

1. Bai, S., Kolter, J.Z., Koltun, V.: An empirical evaluation of generic convolutional and recurrent networks for sequence modeling. arXiv preprint arXiv:1803.01271 (2018)
2. Cao, Q., Shen, H., Cen, K., Ouyang, W., Cheng, X.: DeepHawkes: bridging the gap between prediction and understanding of information cascades. In: Proceedings of the 2017 ACM on Conference on Information and Knowledge Management, pp. 1149–1158 (2017)
3. Cao, Q., Shen, H., Gao, J., Wei, B., Cheng, X.: Popularity prediction on social platforms with coupled graph neural networks. In: Proceedings of the 13th International Conference on Web Search and Data Mining, pp. 70–78 (2020)
4. Chen, G., Kong, Q., Xu, N., Mao, W.: NPP: a neural popularity prediction model for social media content. Neurocomputing **333**, 221–230 (2019)
5. Cho, K., et al.: Learning phrase representations using RNN encoder-decoder for statistical machine translation. arXiv preprint arXiv:1406.1078 (2014)
6. Covington, P., Adams, J., Sargin, E.: Deep neural networks for Youtube recommendations. In: Proceedings of RecSys (2016)
7. Devlin, J., Chang, M.W., Lee, K., Toutanova, K.: BERT: pre-training of deep bidirectional transformers for language understanding. arXiv preprint arXiv:1810.04805 (2018)
8. He, K., Zhang, X., Ren, S., Sun, J.: Deep residual learning for image recognition. In: Proceedings of the IEEE Conference on Computer Vision and Pattern Recognition, pp. 770–778 (2016)
9. Hochreiter, S., Schmidhuber, J.: Long short-term memory. Neural Comput. **9**(8), 1735–1780 (1997)
10. Li, C., Ma, J., Guo, X., Mei, Q.: DeepCas: an end-to-end predictor of information cascades. In: Proceedings of the 26th International Conference on World Wide Web, pp. 577–586 (2017)
11. Lian, J., Zhou, X., Zhang, F., Chen, Z., Xie, X., Sun, G.: xDeepFM: combining explicit and implicit feature interactions for recommender systems. In: Proceedings of the 24th ACM SIGKDD International Conference on Knowledge Discovery & Data Mining, pp. 1754–1763 (2018)
12. Liao, D., Xu, J., Li, G., Huang, W., Liu, W., Li, J.: Popularity prediction on online articles with deep fusion of temporal process and content features. In: Proceedings of the AAAI Conference on Artificial Intelligence, vol. 33, pp. 200–207 (2019)

13. Makridakis, S., Hibon, M.: ARMA models and the box-Jenkins methodology. J. Forecast. **16**(3), 147–163 (1997)

14. Wu, B., Cheng, W.H., Zhang, Y., Huang, Q., Li, J., Mei, T.: Sequential prediction of social media popularity with deep temporal context networks. arXiv preprint arXiv:1712.04443 (2017)

15. Xie, J., et al.: A multimodal variational encoder-decoder framework for micro-video popularity prediction. In: Proceedings of the Web Conference 2020, pp. 2542–2548 (2020)

16. Xie, R., Qiu, Z., Rao, J., Liu, Y., Zhang, B., Lin, L.: Internal and contextual attention network for cold-start multi-channel matching in recommendation. In: Proceedings of IJCAI (2020)

17. Yang, Z., Yang, D., Dyer, C., He, X., Smola, A., Hovy, E.: Hierarchical attention networks for document classification. In: Proceedings of the 2016 Conference of the North American chapter of the Association for Computational Linguistics: Human Language Technologies, pp. 1480–1489 (2016)

18. Zhang, W., Wang, W., Wang, J., Zha, H.: User-guided hierarchical attention network for multi-modal social image popularity prediction. In: Proceedings of the 2018 World Wide Web Conference, pp. 1277–1286 (2018)

The Joy of Dressing Is an Art: Outfit Generation Using Self-attention Bi-LSTM

Manchit Madan$^{(\boxtimes)}$, Ankur Chouragade , and Sreekanth Vempati$^{(\boxtimes)}$

Myntra Designs, Bangalore, India
{manchit.madan,sreekanth.vempati}@myntra.com

Abstract. Fashion represents one's personality, what you wear is how you present yourself to the world. While in traditional brick & mortar stores, there is staff available to assist customers which results in increased sales, online stores rely on recommender systems. Proposing an outfit with-respect-to the desired product is one such type of recommendation. This paper describes an outfit generation framework that utilizes a deep-learning sequence classification based model. While most of the literature related to outfit generation is regarding model development, the segment describing training data generation is still not mature. We have proposed a novel approach to generate an accurate training dataset that uses the latent distance between positive and random outfits to classify negative outfits. Outfits are defined as a sequence of fashion items where each fashion item is represented by its respective embedding vector obtained from the Bayesian Personalised Ranking- Matrix Factorisation (BPR-MF) algorithm which takes user clickstream activity as an input. An outfit is classified as positive or negative depending on its Goodness Score predicted by a Bi-LSTM model. Further, we show that applying Self-Attention based Bi-LSTM model improved the performance (AUC), relevance (NDCG) by an average 13%, 16% respectively for all gender-categories. The proposed outfit generation framework is deployed on Myntra, a large-scale fashion e-commerce platform in India.

Keywords: Outfit recommendation · Self-attention · Bidirectional LSTM · Deep learning · Bayesian personalized ranking · Matrix factorization

1 Introduction

The worldwide revenue from fashion products is expected to rise from \$485.6 billion in 2020 to \$672.7 billion by 2023[1]. With the shutdown of retail stores due to coronavirus, online stores are attracting customers due to increased online

---

[1] https://www.shopify.com/enterprise/ecommerce-fashion-industry.

"The Joy of Dressing is an Art" is a famous quote by John Galliano
A. Chouragade—Work done while at Myntra.

access & smartphone penetration, enhanced user experience by personalizing recommendations. While similar product recommendations is one way where users can be suggested products according to their preference, recommending the whole outfit with-respect-to the desired product could be a game-changer. For our formulation, we define an outfit as a sequence of four products (top-wear, bottom-wear, footwear, accessory; where this order is necessarily preserved). Although there have been numerous studies [1–3] on clothing retrieval and recommendation, they fail to consider the problem of fashion outfit composition and creation. Extensive studies have been conducted on learning fashion compatibility with the objective to recommend products such that they are complementing each other to form an outfit [4–8]. While these studies do a good job at finding the compatibility of items pairwise and outfit as a whole, they do not explain the generation of accurate training data. For example, a recent paper by Bettaney *et al*. described a system for Generating Outfit Recommendations from Deep Networks (GORDN) using multi-modal data, but they assumed randomly generated outfits as the negative samples [9]. Our paper bridges that gap with a unique method of generating a positive & negative training dataset. The generated dataset is then used to train a sequence classification model which predicts the goodness of a given outfit. To build a generalized model, both positive and negative outfits are required (example is shown in Fig. 1). In the context of e-commerce, product catalogue images significantly describe the product to the customers, hence images are cataloged keeping the latest trend in mind. These catalog images can be considered to generate positive outfits for training. A full-shot image (See Fig. 3) is one of the images present in the catalog showcasing an outfit created using the primary product and other compatible products. We use this full-shot image of a product to generate positive outfits, but this process is computationally heavy & time-consuming, hence a classifier was built as a robust solution to outfit generation instead of recommending just the outfits generated using full-shot image. Negative outfits for training were generated such that they are farther from the positive outfits in a product embedding space, where any given outfit is represented as a sequence of product embeddings. Now since both positive and negative outfits are available, labeling can be done automatically, making our solution scalable. Other existing studies [3] suffer from this tedious task of labeling outfit as positive or negative. Also, set of outfits recommended for a product should be diverse in the sense- a women top could be paired with either jeans, skirt, shorts, etc. To address this, we have used product clusters which enables us to recommend a diverse set of outfits.

Each outfit (positive & negative) is represented by a sequence of product embeddings generated using Bayesian Personalized Ranking based Matrix Factorization approach (BPR-MF) [10]. These embeddings help us to take into account a user's preference over quality assortments, promotions, etc., and use implicit signals (views, clicks, orders) from their interaction on the platform. Since we are defining the outfit to be a sequence of fashion products where the position of categories are fixed- top-wear comes first, then bottom-wear, followed by footwear and accessories (e.g., t-shirt, jeans, shoes and watch), where each product is a time step, Bidirectional-LSTMs can be used here. At each time step,

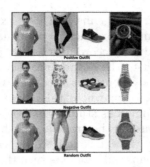

Fig. 1. This figure depicts a positive, negative & random outfit for women-tops

given the previous product, we train the Bi-LSTM model to learn the sequence of products in an outfit. This helps the model to identify the compatibility relationship of fashion products in an outfit. Self-Attention is used on top of Bi-LSTM to emphasize on important information present in the product embeddings by assigning attention weights. A Goodness Score (GS) for each outfit is generated using this model which quantifies the compatibility of products in that outfit. Self-Attention Bi-LSTM model is compared with its various variants in an offline experimentation space using NDCG [11] as the primary metric. We also compare the AUC score and ROC curves for these models on the test set. Our contributions are three-fold and are summarized as follows:

- A novel approach to generate an accurate and large scale training dataset (positive & negative) which helps the model to generalize their compatibility
- Self-attention based Bi-LSTM outfit classifier
- New outfits generation framework for all products on the platform

The rest of the paper is organized as follows, In Sect. 2, we briefly discuss the related work. We introduce the Methodology in Sect. 3, that comprises of creation of BPR-MF product embeddings, training dataset generation, and the explanation of the model architectures used, along with the generation of new outfits. In Sect. 4, we compare different variants of Bi-LSTM model, showcase the reproducibility of our work, and conclude the paper in the last section.

2   Related Work

There is a growing interest in using AI to identify fashion items in images due to the huge potential for commercial applications[2], some of which are identifying fashion fakes and counterfeit products. AI-enabled shopping apps allow customers to take screenshots of clothes they see online, identify shoppable apparels and accessories in that photo, and then find the same outfit and shop for similar styles. Several works in the fashion domain are closely related to ours. We

---

[2] https://www.forbes.com/sites/cognitiveworld/2019/07/16/the-fashion-industry-is-getting-more-intelligent-with-ai/.

first discuss fashion image retrieval, recognition, and image attribute learning. Recently, Z Kuang *et al*. introduced a novel Graph Reasoning Network, trained on a graph convolutional network which compares a query and a gallery image to search the same customer garment image as in the online store [12]. Liao *et al*. developed an EI (Exclusive & Independent) tree that organizes fashion concepts into multiple semantic levels and helps to interpret the semantics of fashion query [13]. Liu *et al*. showcased a deep model- FashionNet that learns fashion features by jointly predicting clothing attributes and landmarks [14]. Hadi *et al*. deployed deep learning techniques to learn a similarity measure between the street and shop fashion images [15]. Comparing with the previous works on fashion image retrieval, the goal of this work is to compose fashion outfit automatically, which has its own challenges in modeling many aspects of the fashion outfits, such as compatibility.

Literature on Attention-based LSTMs: Wang-Cheng *et al*. built a Self-Attention based Sequential Recommendation model (SASRec), which adaptively assigns weights to previous items at each time step [16]. In [17], Wang *et al*. proposed an Attention-based LSTM Network for aspect-level sentiment classification. In this work, we employ a Self-Attention based Bi-LSTM model to classify a sequence of fashion products into positive & negative outfits.

Thirdly, we discuss literature available in fashion recommendations. Hu *et al*. implemented a functional tensor factorization approach to give personalized outfit recommendations [5]. Li *et al*. adapted an RNN as a pooling model to encode the variable-length fashion items and predicted the popularity of a fashion set by fusing multiple modalities (text and image) [8]. Liu *et al*. proposed a latent Support Vector Machine (SVM) model that gives occasion-based fashion recommendation which relies on a manually annotated dataset [3]. In [18], item representations were generated using an attention-based fusion of product images & description and its effectiveness on polyvore dataset was showcased. In [4], researchers employed a Bi-LSTM to capture the compatibility relationships of fashion items by considering an outfit as a sequence. While there are some similarities, none of these works talk about how to generate an accurate negative sample for model creation. Also, the Self-Attention Bi-LSTM model for outfit generation has not been used in prior studies. Our paper proposes an algorithm to create pure negative outfits, which leads to improved performance as compared to randomly created negative outfits; as shown in Sect. 4. These negative outfits along with positive outfits are used to build an outfit classifier using a Self-Attention layer on top of a Bi-LSTM. This classifier is then used to generate new outfits.

3 Methodology

In this section, we present the key components of our Outfit Generation Framework (OGF). A fashion outfit is composed of multiple fashion items. These items are expected to share a similar style (design, color, texture, etc.). For the scope of this work, we have considered four items in an outfit, $O_i = \{s_1, s_2, s_3, s_4\}$. Figure 2

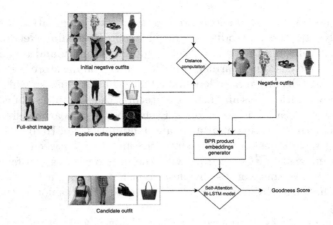

Fig. 2. Outfit generation framework

explains the OGF. Given some seed fashion items, positive outfits for them were generated using the full-shot image (Algorithm 1). An initial set of negative outfits were generated randomly. These were then compared with the positive ones to ensure they are distant enough (Algorithm 3). Fashion items in both positive & negative outfits were represented by their respective embedding vectors constructed using BPR-MF. By treating an outfit as an ordered sequence of items (top-wear always comes first, followed by bottom-wear, footwear, and accessory), we have built a Self-Attention based Bi-LSTM sequence classification model. New outfits were generated by passing candidate outfits (generated using Algorithm 2) into the trained model which predicts their Goodness Score (GS). Outfits having GS above a certain threshold GS^* were finally displayed on the platform.

All of the above exercises are done at a Gender-Category level to design different outfits for them. Eg., men might pair a t-shirt with jeans, casual or sports shoes, watches; whereas women might pair a t-shirt with jeans, boots, handbag. While one could always generate outfits using only a gender level model, we decided to go with the gender-category level approach to ensure that the outfits are present for all fashion items live on the platform. The category in the gender-category model refers to either a top-wear (shirts, t-shirts, tops, jackets, etc.) or a bottom-wear (jeans, trousers, track pants, shorts, etc.); but the order of outfit remains constant (top-wear, bottom-wear, footwear, accessory). For instance, while outfits for all Men-Shirts were generated using a Men-Shirts model wherein bottom-wear could either be jeans, trousers, etc.; it won't be able to capture all jeans products or all trousers. Hence, a different model for Men-Jeans helped to create outfits for all jeans products.

3.1 Bayesian Personalized Ranking (MF) Embedding

Being one of the largest e-commerce platform, there are significant amount of long-tail products in our catalogue since a user cannot possibly interact with all the products on our platform. To solve this problem, embeddings were created using Matrix Factorization (MF) approach [19].

Due to the absence of significant amount of product ratings, implicit signals (such as the number of views, clicks, add to carts, orders) were utilized. A user-product interaction matrix $(UPIM)$ was constructed using implicit signals from the user clickstream data. Each element of this $UPIM$ refers to the implicit rating of a product with respect to a user. It was calculated by the weighted sum of implicit signals. To generate embeddings, popular Bayesian Personalized Ranking (BPR) [10] based MF approach was used as the $UPIM$ was 99% sparse. The algorithm works by transforming the $UPIM$ into lower-dimensional latent vectors where BPR helps in pairwise ranking. Loss function of BPR:

$$- \sum_{(u,i,j)} \ln \sigma(x_{uij}) + \lambda_\Theta ||\Theta||^2 \tag{1}$$

where u, i, j are the triplets of product pairs (i, j) and user u available in the interactions dataset. $x_{uij} = p_{ui} - p_{uj}$; denotes the difference of preference scores for the user u, representing that the user u likes product i over product j. Θ are the model parameters and λ_Θ is model specific regularization parameter. Similarity $g(P_i, P_j)$ between product P_i and P_j was computed using BPR. Product embeddings for all the products were generated using this approach.

3.2 Training Dataset Generation

In this section, we describe our algorithms of generating positive, candidate and negative outfits.

Positive Outfits Generation. Let S denote the set of fashion items in any outfit O_i. Then, $O_i = \{s_1, s_2, s_3, s_4\}$ for all s_i belonging to S. Each item s_i belongs to different product category (top-wear, bottom-wear, footwear, and accessory). The steps for generating positive outfits are presented in Algorithm 1.

Given a seed fashion item s_1, we need to create positive outfits for training the model. Here s_1 is the primary category product. From the set of catalog images I of s_1, we identify the full-shot image $I*$ such that it has a maximum number of different categories (top-wear, bottom-wear, footwear, accessory) present, as shown in Fig. 3. Then $I* = [s_1, s_2, s_3, s_4]$. Note that the detector might not be able to detect all four categories in $I*$, but it certainly ensures that the selected image has at least three categories detected. Also, it's okay if no such image $I*$ is identified for any s_1 since we only need a sample of primary category products to generate positive outfits training dataset.

Apart from s_1, visually similar products were fetched for s_2, s_3 from set P_i* to get their visually similar products s_{v_2} and s_{v_3} respectively. N number of sequences of fashion items, G_n*, were created keeping s_1 (top-wear) always in the first position, random item from s_{v_2} (bottom-wear) and from s_{v_3} (footwear) in second and third position respectively. There were challenges in detecting the fourth item (accessory) and hence its visually similar products were not found.

Algorithm 1: Positive Outfits Generation

Input:
- Seed fashion item s_1
- Catalog images of $s_1 : I = (i_1, i_2, i_3, i_4)$
- Visually similar products set P_i*, i = (2, 3)
- Accessory products set A*
- Combination feature function $C(s_i)$

Output: Positive outfits set G_n*

1. Full-shot image, $I* \longleftarrow I$
2. Compute visually similar products, $s_{v_i} \ \forall \ i \in [P_2, P_3]$
3. Select accessory product s_{c4} from A* based on combination of features
4. $G_n* = \{s_1, s_{v_2}, s_{v_3}, s_{c_4}\}$

To complete the outfit, a combination of features (price, color, brand, etc.) was used to randomly sample an accessory item s_{c_4} from set $A*$. This way a single positive outfit $G*$ was created.

Keeping the primary product (either top-wear or bottom-wear depending on the Gender-Category for which model is to be generated) fixed, the rest of the products in an outfit were determined using the above approach to get a positive outfit set G_n*. The non-primary products can be sampled in any order since each of them come from independent sets. Since we hypothesized that $I* = [s_1, s_2, s_3, s_4]$ is a positive outfit, replacing s_2 with any s_{v_2} and keeping the rest of products will give another positive outfit. We pick only those visually similar products that have high visual similarity score. Figure 1 depicts a positive outfit $G*$ created using the above algorithm. There are in-house components built for the purpose of detecting different categories present in an image and creation of visually similar products set.

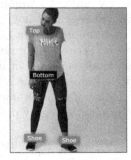

Fig. 3. Detection of categories in a full-shot image

Fig. 4. Each row depicts products present in that cluster

Algorithm 2: Candidate Outfits Generation

Input:
- Candidate fashion item s_i,
- Set of all Fashion items $F*$,
- BPR-MF product embeddings

Output: Candidate Outfits set R_n*

1. $s_i \longleftarrow F*, \forall \; i \in [1,n]$
2. Clusters $K_c \; \forall \; c \in \{\text{non-primary categories}\}$
3. Select s_c from $K_c \; \forall \; c$
4. $R_n* = \{s_i, s_{bottomwear}, s_{footwear}, s_{accessory}\}$ or $R_n* = \{s_{topwear}, s_i, s_{footwear}, s_{accessory}\}$

Candidate Outfits Generation. It is equally important to generate non-compatible or negative outfits in order to build an accurate classifier. Candidate outfits were used as input to generate negative outfits. They were also used to generate new outfits. The candidate outfits creation problem is formulated and presented in Algorithm 2.

Given a set of all fashion items $F*$, m items from the primary category were randomly selected for which candidate outfits were generated. For a candidate fashion item s_i, an outfit could comprise of total four items out of which s_i could be a top-wear or a bottom-wear depending on the Gender-Category level. Keeping s_i fixed, for the remaining three items- fashion items from multiple categories are eligible. For example: A women-tshirt could be paired with either a jeans, casual shoes, a watch or a shorts, sports shoes, handbag; and a women-jeans could be paired with either a tshirt, casual shoes, watch or a tshirt, sports shoes, handbag. Here we follow the same ideology of an outfit as described above- a sequence of four products (top-wear, bottom-wear, footwear, accessory). Within each non-primary category ({bottom-wear, footwear, accessory} for primary category {top-wear} and {top-wear, footwear, accessory} for primary category {bottom-wear}), fashion items were clustered using their embedding vectors as input. The optimal number of clusters, K was decided using the elbow method and the K-nearest neighbour approach was employed for clustering. Fashion items were randomly sampled from different clusters, K_c within a category to form a candidate outfit R_i*. For each s_i, n number of candidate outfits were generated to form set R_n*. Clustering is done to bring diversity in outfits. Figure 4 shows products in different clusters for category *bottomwear*. Since, out of all fashion items only m were selected to generate candidate outfits, we might have missed out on some genuinely good outfits.

Negative Outfits Generation. The goal of Algorithm 3 is to construct a set of negative outfits such that they are far away from the positive outfits in the embeddings space. To get the initial set of negative outfits $B_{In}*$, a set

---

Algorithm 3: Negative Outfits Generation

Input:
- Candidate Outfits set R_n*,
- Positive Outfits set G_n*,
- BPR-MF product embeddings

Output: Negative Outfits set B_n*

1. $B_{In}* = R_n* - G_n*$
2. $B_n* \subset B_{In}*$ if d(B_n*, G_n*) ¡= $M_d(100^{th}$ p), $M_d(95^{th}p)$

---

difference is taken between candidate outfits, R_n* (Algorithm 2) and positive outfits G_n* (Algorithm 1). Then, we transform each outfit into a sequence of their respective item embedding vectors so that a cosine distance metric can be computed between different outfits.

A distribution of deciles of cosine distance between G_n* and $B_{In}*$ was constructed. It was used to sample high confidence negative outfits B_n* from $B_{In}*$ using the following rules-

$$d(B_n*, G_n*), 100^{th}p <= M_d(100^{th}p) \tag{2}$$

$$d(B_n*, G_n*), 95^{th}p <= M_d(95^{th}p), \tag{3}$$

where d(B_n*, G_n*), $100^{th}p$ & d(B_n*, G_n*), $95^{th}p$ are the 100th & 95th percentiles of distance d(B, G_n*). And, $M_d(100^{th}p)$ is the Median of 100th percentile deciles and $M_d(95^{th}p)$ is the Median of 95th percentile deciles. Using the above approach, around 30% outfits from $B_{In}*$ got selected to form Negative Outfits Set B_n*. Figure 1 shows one such outfit which a human eye would also perceive as bad.

3.3 Bi-LSTM

Bidirectional-LSTM is a variant of Recurrent Neural Networks (RNNs) that was created as the solution to short-term memory. It has internal mechanisms called gates that can regulate the flow of information. These gates can learn which data-point in a sequence is important to retain or discard. This way, it can pass relevant information down the long chain of sequences to make predictions. Unidirectional LSTM can only preserve information of the past because the only input it has seen is from the past, whereas a Bidirectional LSTMs will run your inputs in two ways, one from past to future and another from future to past. They use the same hidden states while running front and back which helps to retain information from the future and past. Hence, we have used it to learn the sequence of fashion items in outfits. Bi-LSTMs have been successfully applied to temporal modeling tasks such as sequence tagging [20], speech recognition [21], sentiment classification [22], and image and video captioning [23].

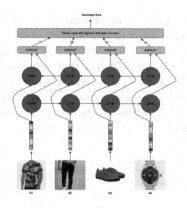

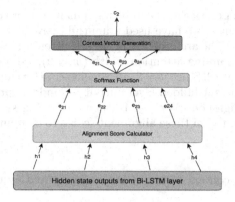

Fig. 5. Bi-LSTM model architecture   **Fig. 6.** Self-attention block

Figure 5 shows the architecture of the Bi-LSTM model used in this paper. Each fashion item in an outfit is represented by its respective embedding vector prepared using the BPR-MF technique. These embeddings were passed as a sequence into one Bi-directional LSTM layer. Hidden state outputs from the front LSTM and back LSTM were concatenated to give final hidden state outputs as (h_1, h_2, h_3, h_4). These outputs were then passed to a dense layer with *sigmoid* activation function which gives the Goodness Score of a sequence (outfit).

3.4 Self-attention Bi-LSTM

Attention is an algorithm used industry-wide to map important and relevant information from the input and assign higher weights to them, enhancing the accuracy of the output [24]. It uses hidden state output from encoder and state input from a decoder to form a context vector that gives the relative importance of each state. Self-attention [25], also called intra-attention is an attention mechanism relating to different positions of a single sequence in order to compute a representation of that sequence. As compared to vanilla attention, which uses inputs from hidden states of sequence data (encoder) and sequence data (decoder), the self-attention uses only the hidden states of the encoder. Depending on the attention width, w (which controls the width of the local context), hidden states are chosen. If $w = 3$, then to calculate the context vector for state 2; hidden state outputs (h_1, h_2, h_3) are considered.

Given current hidden state h_i and the previous hidden state s_j, let $f(h_i, s_j)$ be the attention function that calculates an unnormalized alignment score between h_i and s_j, and $a_i = \text{softmax}(f(h_i, s_j))$ be the attention scores. Then, Additive Self-Attention states that $f(h_i, s_j) = v_a^T \cdot \tanh(W_1 h_i + W_2 s_j)$, whereas Multiplicative Self-Attention states that $f(h_i, s_j) = h_i^T W_a s_j$; where v_a and W are learned attention parameters. Additive and multiplicative attention are similar in complexity, although multiplicative attention is faster and more space-efficient in practice as it can be implemented more efficiently using matrix multiplication. Hence, we have used a multiplicative self-attention layer in our model.

The architecture of the self-attention block (See Fig. 6) is discussed as follows- Assuming attention width ($w = 2$), we need to calculate the context vector for state 2. Since $w = 2$, alignment score for each state will be computed using only the current hidden layer output h_i, and the previous hidden state output s_j. Hence, alignment score e_{21} is calculated using s_1 and h_2. A softmax function is applied on top of these alignment scores to compute attention score (weight) a_{21}:

$$a_{21} = e_{21}/(e_{21} + e_{22} + e_{23} + e_{24}) \tag{4}$$

A context vector c_2 is then computed by the formula :

$$c_2 = (a_{21}.s_1) + (a_{22}.s_2) + (a_{23}.s_3) + (a_{24}.s_4) \tag{5}$$

The context vectors c_i enable us to focus on certain parts of the input to learn the outfit compatibility. These were passed into a dense layer with *sigmoid* activation to give the probability of goodness of an outfit.

3.5   Generation of New Outfits

In this sub-section, we describe the generation of new outfits using the architecture defined in Fig. 2. This whole module is divided into 2 parts - the creation of a training module, and new outfits generation. The creation of the training module is explained as follows- positive & negative outfits, created using Algorithm 1 and 3 respectively, were used to train a Self-Attention based Bi-LSTM model. Positive outfits were labeled as 1 while the negative ones as 0. These outfits were then randomly shuffled together to get a list of outfits (sequences). 75% of the outfits in the list were used as a training set, while 25% of them constituted the test set. The sequence of outfits was converted into a sequence of product embeddings so that they can be used by the model. Using grid-search, the set of optimal hyper-parameters {*epochs, batch size, learning rate*} was found automatically for each Gender-Category model. The best set of weights was chosen basis the *minimum validation loss* metric and saved for future prediction.

Candidate outfits (created using Algorithm 2) were passed into the training module to get the *Goodness Score* (GS) of each outfit. The outfits having score greater than a chosen threshold (GS^*) were selected to be displayed on the platform. Threshold GS^* varied for different gender-category since modeling is done at that level. Choice of GS^* was made based on models' precision & recall values.

Both of these components (training & prediction) are offline, taking away all the concerns related to online traffic. Training is done once a month, while prediction happens daily to account for new products added to the platform. This system has been live for more than 6 months on Myntra (http://www.myntra.com).

4   Results

This section demonstrates the goodness of the proposed Self Attention Bi-LSTM model (with varying attention-width parameter) as compared to a Bi-LSTM model without the self-attention layer. We have evaluated different approaches on the data taken from Myntra, one of the leading fashion e-commerce platform in India. To evaluate these models, we have used nearly 150K positive and negative outfits across four gender-categories (Men-Shirts, Women-Tops, Men-Jeans & Women-Jeans) as the test set. These outfits are a set of four fashion items, necessarily in the order- {top-wear, bottom-wear, footwear, accessory}, wherein each item is represented by its BPR embedding. Goodness Scores for these outfits were generated using the Training Module (defined in Sect. 3.5). Outfits having a score greater than a chosen threshold, GS^* were given the *Prediction Label*, L_P equal to 1, otherwise 0. L_P was compared with actual label L_A to evaluate different models.

Quantitative Evaluation: As this is a sequence classification problem, standard evaluation metrics are used, namely Area under the curve (AUC), Receiver operating characteristic (ROC) curve. AUC on the test set for some categories is presented in Table 1. For instance in the Men-Shirts gender-category, the Bi-LSTM model gave an AUC of 79% on the test set. On applying a Self-Attention layer (with attention-width, $w = 1$) on top of the Bi-LSTM layer improved the AUC by 3%. Further improvements in AUC were achieved by gradually increasing the value of w from 1 to 4, with $w = 4$ giving the best result out of all the models. Since with $w = 4$, the self-attention layer focused on all four product embeddings in a sequence; it delivered the best result in every Gender-Category. ROC curves for different classifiers for Men-Shirts can be seen in Fig. 7.

A comparison of the winning model, Self Attention Bi-LSTM ($w = 4$), considering the same set of positive outfits and different negative outfits is shown in Fig. 8. The model trained using negative samples generated by Algorithm 3 ($M1$) resulted in an increase in AUC value by 9% as compared to the model using randomly generated negative samples ($M2$) since $M1$ better differentiated between the positive and negative outfits. $M2$ wrongly classified some negative

Table 1. Evaluation metrics on different models for different gender-categories, where the Self-Attention layer based Bi-LSTM model performs the best

Models	Men-shirts		Women-tops		Men-jeans		Women-jeans	
	NDCG	AUC	NDCG	AUC	NDCG	AUC	NDCG	AUC
Bi-LSTM (BL)	0.50	0.79	0.54	0.83	0.52	0.81	0.53	0.80
Self-Attn. BL (w = 1)	0.52	0.82	0.55	0.85	0.54	0.84	0.54	0.82
Self-Attn. BL (w = 2)	0.54	0.86	0.58	0.88	0.57	0.87	0.56	0.85
Self-Attn. BL (w = 3)	0.57	0.89	0.62	0.90	0.60	0.90	0.57	0.87
Self-Attn. BL (w = 4)	**0.58**	**0.92**	**0.64**	**0.93**	**0.62**	**0.91**	**0.60**	**0.90**

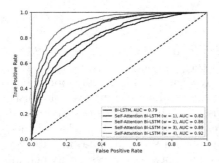

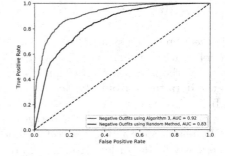

Fig. 7. ROC curves for men-shirts: self-attention Bi-LSTM ($w = 4$) dominate others

Fig. 8. Model trained using negative samples generated by Algorithm 3 performs better than the one using random samples

outfits as positive outfits leading to a drop in AUC. This result was generic across all gender-category level models- $M1$ performed better than $M2$. Hence, the approach proposed in this paper to generate accurate training outperforms the random dataset generation technique used by other works in the similar domain. Note that this result holds to our dataset (it cannot be made public due to legal constraints).

Offline Evaluation: We have compared our models offline using NDCG (Normalized Discounted Cumulative Gain) [11] as a metric. NDCG is a standard information retrieval measure used for evaluating the goodness of ranking a set. Here, the hypothesis was that using a Self-Attention layer on top of a Bi-LSTM layer increases Click Through Rate (CTR). For computing NDCG, we have used the CTR as the true score. Outfit's goodness score from the model is used as the predicted score. Since there were outfits present for different primary products, we computed NDCG for each primary product and took their average for comparison purposes across models. The results showed the NDCG score improved (in all Gender-Categories) as we moved from Bi-LSTM to higher Self-Attention width Bi-LSTM models, as shown in Table 1.

Qualitative Evaluation: When the model (trained on sequences of product embeddings) was used to predict the goodness score of the newly generated outfits, the results were quite similar to what a human eye would observe. Figure 9 shows some positive outfits & negative outfits as predicted by the winning model for four different gender-categories: men-shirts, women-tops, men-jeans and women-jeans. Fashion items in positive outfits complement each other, which is not observed in case of negative outfits. Here the fashion items sport dissimilar style/design, for example: the first negative outfit in Fig. 9a shows a combination of red & white checked shirt with black regular shorts and brown boat shoes, this was predicted as negative by the model. A human eye would also classify this combination as an incompatible outfit.

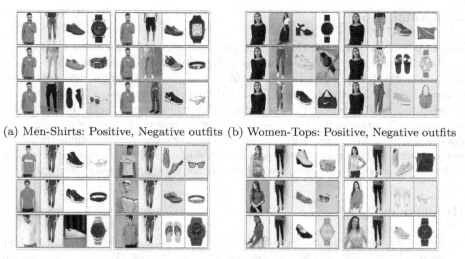

(a) Men-Shirts: Positive, Negative outfits (b) Women-Tops: Positive, Negative outfits

(c) Men-Jeans: Positive, Negative outfits (d) Women-Jeans: Positive,Negative outfits

Fig. 9. Positive and Negative outfits generated by the model (Color figure online)

Reproducibility: Though we cannot share the dataset due to legal constraints, the work presented in this paper can be replicated by following these steps. Generate positive outfits either at Gender or Gender-Category level using Algorithm 1, where each outfit follows the schema- top-wear, bottom-wear, footwear, accessory. Candidate outfits can be generated using Algorithm 2. From these, the negative outfits can be generated by Algorithm 3. Positive & negative outfits must be labeled as 1 & 0 respectively. These outfits are then randomly shuffled together to get a list of outfits (sequences). 75% of the outfits form the training set, while 25% of them lie in the test set. Each outfit can be represented as a sequence of product embeddings (we have used BPR-MF embeddings of size 64, generation discussed in Sect. 3.1; *implicit* python library used) since this helps a sequence classification model to learn their compatibility. A Self-Attention Bi-LSTM model can be built by passing these outfits into one Bi-directional LSTM layer with 150 hidden units, using *glorot normal* as the *kernal initializer* to set the initial random weights of Bi-LSTM layer. A dropout and recurrent dropout probability equal to 0.2 is applied to reduce overfitting in the model. Then, a Self-Attention layer is applied to focus on certain parts of the input sequence. We used multiplicative self-attention & experimented with varying values of *attention width* (w) parameter. Finally, a dense layer with *sigmoid* activation function is applied to return the Goodness Score of an outfit. Since there are only two labels (1 & 0), *binary cross entropy* loss can be used. The set of optimal hyper-parameters {*epochs, batch size, learning rate*} can be found using grid-search. Python libraries- *keras* & *keras-self-attention* were used for this implementation.

5   Conclusion

In this paper, we solve the challenging problem of creating an accurate training dataset for modeling fashion outfits and show its effectiveness compared to the random method. By considering an outfit as a sequence of fashion items, we have deployed a Self-Attention based Bi-LSTM model wherein each item has been represented by its respective embedding vector generated using the BPR-MF technique. This model has been used to generate new outfits by predicting their goodness score. As future work, we plan to improve outfit recommendations by personalizing outfits using user's preferences and also diversify the outfits using different aspects like occasion, theme, and other category-specific attributes.

References

1. Iwata, T., Watanabe, S., Sawada, H.: Fashion coordinates recommender system using photographs from fashion magazines. In: IJCAI (2011)
2. Veit, A., Kovacs, B., Bell, S., McAuley, J., Bala, K., Belongie, S.: Learning visual clothing style with heterogeneous dyadic co-occurrences. In: International Conference on Computer Vision (ICCV), Santiago, Chile (2015). *Equal Contribution
3. Liu, S., et al.: Hi, magic closet, tell me what to wear! In: Proceedings of the 20th ACM International Conference on Multimedia, MM 2012, pp. 619–628. Association for Computing Machinery, New York (2012)
4. Han, X., Wu, Z., Jiang, Y.-G., Davis, L.S.: Learning fashion compatibility with bidirectional LSTMs. In: Proceedings of the 25th ACM International Conference on Multimedia, MM 2017, pp. 1078–1086. Association for Computing Machinery, New York (2017)
5. Hu, Y., Yi, X., Davis, L.S.: Collaborative fashion recommendation: a functional tensor factorization approach. In: Proceedings of the 23rd ACM International Conference on Multimedia, MM 2015, pp. 129–138. Association for Computing Machinery, New York (2015)
6. Wang, X., Wu, B., Zhong, Y.: Outfit compatibility prediction and diagnosis with multi-layered comparison network. In: Proceedings of the 27th ACM International Conference on Multimedia, MM 2019, pp. 329–337. Association for Computing Machinery, New York (2019)
7. Lin, Y., Ren, P., Chen, Z., Ren, Z., Ma, J., de Rijke, M.: Explainable outfit recommendation with joint outfit matching and comment generation. IEEE Trans. Knowl. Data Eng. 32(8), 1502–1516 (2020)
8. Li, Y., Cao, L., Zhu, J., Luo, J.: Mining fashion outfit composition using an end-to-end deep learning approach on set data. IEEE Trans. Multimedia 19, 1946–1955 (2017)
9. Bettaney, E.M., Hardwick, S.R., Zisimopoulos, O., Chamberlain, B.P.: Fashion outfit generation for e-commerce. arXiv, abs/1904.00741 (2019)
10. Rendle, S., Freudenthaler, C., Gantner, Z., Schmidt-Thieme, L.: BPR: Bayesian personalized ranking from implicit feedback. In: Proceedings of the Twenty-Fifth Conference on Uncertainty in Artificial Intelligence, UAI 2009, pp. 452–461. AUAI Press, Arlington (2009)
11. Järvelin, K., Kekäläinen, J.: Cumulated gain-based evaluation of IR techniques. ACM Trans. Inf. Syst. (TOIS) 20(4), 422–446 (2002)

12. Kuang, Z., et al.: Fashion retrieval via graph reasoning networks on a similarity pyramid. In: 2019 IEEE/CVF International Conference on Computer Vision (ICCV), pp. 3066–3075 (2019)
13. Liao, L., He, X., Zhao, B., Ngo, C.-W., Chua, T.-S.: Interpretable multimodal retrieval for fashion products. In: Proceedings of the 26th ACM International Conference on Multimedia, MM 2018, pp. 1571–1579. Association for Computing Machinery, New York (2018)
14. Liu, Z., Luo, P., Qiu, S., Wang, X., Tang, X.: DeepFashion: powering robust clothes recognition and retrieval with rich annotations. In: CVPR, pp. 1096–1104. IEEE Computer Society (2016)
15. Hadi Kiapour, M., Han, X., Lazebnik, S., Berg, A.C., Berg, T.L.: Where to buy it: matching street clothing photos in online shops. In: International Conference on Computer Vision (2015)
16. Kang, W.-C., McAuley, J.J.: Self-attentive sequential recommendation. In: 2018 IEEE International Conference on Data Mining (ICDM), pp. 197–206 (2018)
17. Wang, Y., Huang, M., Zhu, X., Zhao, L.: Attention-based LSTM for aspect-level sentiment classification. In: Proceedings of the 2016 Conference on Empirical Methods in Natural Language Processing, Austin, Texas, November 2016, pp. 606–615. Association for Computational Linguistics (2016)
18. Laenen, K., Moens, M.-F.: Attention-based fusion for outfit recommendation. arXiv, abs/1908.10585 (2019)
19. Hoyer, P.O.: Non-negative matrix factorization with sparseness constraints. J. Mach. Learn. Res. 5, 1457–1469 (2004)
20. Huang, Z., Xu, W., Yu, K.: Bidirectional LSTM-CRF models for sequence tagging. arXiv, abs/1508.01991 (2015)
21. Graves, A., Mohamed, A., Hinton, G.: Speech recognition with deep recurrent neural networks. In: ICASSP, IEEE International Conference on Acoustics, Speech and Signal Processing - Proceedings, vol. 38, March 2013
22. Sharfuddin, A.A., Tihami, M.N., Islam, M.S.: A deep recurrent neural network with BilSTM model for sentiment classification. In: 2018 International Conference on Bangla Speech and Language Processing (ICBSLP), pp. 1–4 (2018)
23. Pan, Y., Mei, T., Yao, T., Li, H., Rui, Y.: Jointly modeling embedding and translation to bridge video and language. In: CVPR (2016)
24. Vaswani, A., et al.: Attention is all you need. In: Guyon, I., et al. (eds.) Advances in Neural Information Processing Systems 30, pp. 5998–6008. Curran Associates Inc (2017)
25. Lin, Z., et al.: A structured self-attentive sentence embedding. In: International Conference on Learning Representations 2017 (Conference Track) (2017)

On Inferring a Meaningful Similarity Metric for Customer Behaviour

Sophie van den Berg$^{(\boxtimes)}$ and Marwan Hassani$^{(\boxtimes)}$ [iD]

Eindhoven University of Technology, Eindhoven, The Netherlands
sophievandenberg4@gmail.com, m.hassani@tue.nl

Abstract. In omnichannel customer service environments, where no real process is enforced, a wide variety of customer journey variants exists. This variety makes it complex to find process improvement opportunities. Modeling the journeys as traces is an essential step before discovering an explainable model of various behaviours. Trace clustering helps improvement efforts by separating the journeys into homogeneous subsets in terms of behaviour and purpose. For this, a one-size-fits-all distance metric has been used so far in the literature. This paper shows that a domain-informed similarity metric will improve customer journey clustering compared to a generic one. We propose SIMPRIM framework, which uses clustering quality metrics to develop a similarity metric that maximizes the separability of the journeys in a low dimensional space while agreeing with existing process knowledge. Experimental evaluation on real life use cases of a large telecom company and a benchmark dataset show that, compared to a generic metric, respectively a 46% and 39% improvement can be obtained in terms of the internal clustering quality while keeping the external clustering quality equal. We also show that the inferred metric can be useful for prediction applications.

Keywords: Similarity metric · Customer journey clustering

1 Introduction

In today's business environment, delivering a superior customer experience is becoming a priority to compete. Customer experience is the result of every interaction a customer has with a business, from navigating the website to using the product or talking to a customer service agent. The sequential steps and interactions, i.e. touchpoints, that a customer goes through for accessing or using a product, is referred to as a customer journey. Analyzing customer journeys is an extremely useful exercise for companies that aim to understand and improve the customer experience for their users as interactions do not occur in isolation. Nowadays, many companies are adopting a data-driven way of working in which a lot of data about customer contact is collected. However, data-driven methods for customer *journey* analysis are still very limited. Recently, [1] highlighted that process mining techniques are suitable for exploring customer behaviour.

© Springer Nature Switzerland AG 2021
Y. Dong et al. (Eds.): ECML PKDD 2021, LNAI 12979, pp. 234–250, 2021.
https://doi.org/10.1007/978-3-030-86517-7_15

Such techniques aim to extract useful information about process execution by analyzing event logs. In the scope of customer journey analysis, the process can be considered the sequence of touchpoints in a customer journey for which at least its timestamp, activity and a journey identifier are stored. Traditional process mining approaches expect processes to be well-structured and limited in scope [8]. Customer journeys, on the other hand, are often derived from processes that have opposite characteristics. Customers can usually operate in very *flexible* environments where no process is enforced. To offer the best possible customer experience, companies provide omnichannel customer service which yields high customer journeys variability: for reaching each goal, a customer has countless ways. There are inherent problems of applying process mining techniques to flexible environments like these. The corresponding event log is very diverse as it captures a wide spectrum of behaviour, both in terms of topic and journey length, meaning journeys vary significantly from each other. This typically yields so-called spaghetti-like process models that are large, highly unstructured and essentially useless for further analysis [12]. To overcome this issue, [12] propose to use Trace Clustering in which the event log is split into homogeneous subsets. Separately, these processes are significantly more structured than the complete process and thus yield more interpretable process models. Using this approach, journeys could be clustered based on *customer intent*, i.e. the purpose of and the behaviour behind customer contact. Due to the high journey variety, both aspects are required to find clusters that reflect similar journeys. To cluster journeys in a meaningful way, an appropriate notion of journey similarity is of a critical importance. However, current approaches to trace clustering solely make use of standard distance metrics (e.g. Euclidean, Cosine or Jaccard). They are assumed to create meaningful clusters for journeys belonging to a wide spectrum of processes ranging from hospital patients to customer webclicks. This assumption seems to be violated in practise where for instance different pieces of information are relevant in varying percentages (cf. Fig. 1). Therefore, the suitability of a certain similarity function for a given business problem depends on the characteristics of the data and nature of the problem. Hence, we challenge

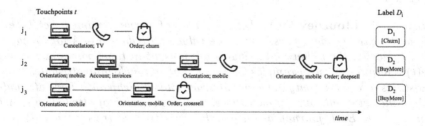

Fig. 1. Schematic overview of touchpoints t from 3 example journeys j_i in telecom journey log L (cf. Definition 1) and the corresponding journey labels D_i (cf. Definition 9). Standard distance metrics will fail in deciding which of j_i are closest to each other. The perspective V_i (cf. Definition 2) from which one looks at j_i matters too.

the current usage of 'naive' metrics and hypothesise that a meaningful similarity metric is much more suitable. Such a metric can be further useful in any prediction and/or recommendation task beyond clustering. As such, inferring a meaningful similarity metric is essential for advanced journey analysis in general.

In this paper, we propose a methodology for the development of a domain-informed similarity metric for customer journeys from *flexible* environments. Predominantly focusing on, but not limited to, *customer service* environments from which the trivial examples in Fig. 1 are illustrated. The main use case of this work comes from a leading telecommunication company that we will refer to as Anonycomm. The core of the framework is based on unsupervised learning (clustering), while domain knowledge in the form of journey annotation is used to support the development of the metric along the way as much as needed. This approach is hypothesized to yield journey analysis results that could not be found by domain experts but are in line with their existing knowledge. We can assume that a rough journey categorization is available since information of this kind is available in almost every data science problem or can be created using automatic labelling techniques. However, this categorization is often imperfect or totally wrong. Often, only a high-level journey categorization (e.g. topic-based) is available in which high variability of journeys is expected. Therefore, the similarity metric should not overfit on that domain knowledge.

The paper is organized as follows: Sect. 2 formalizes the addressed problem. Section 3 introduces the proposed framework to develop a meaningful customer journey similarity metric. In Sect. 4 we demonstrate the usefulness of the framework using two real-life event logs. In Sect. 5 we extensively discuss related work before concluding the paper in Sect. 6.

2   Problem Definition

Below, we introduce some definitions and subsequently we formally define the problem of inferring a meaningful similarity metric out of customer behaviour. Definition 1 & 2 are adopted from [8] but adjusted to the customer journey application.

Definition 1. (Journey log). *Let A be a set of attribute names. Let T be the set of all touchpoint identifiers. $a(t)$ is the value of attribute $a \in A$ for touchpoint $t \in T$. Typically, the following attributes are present in touchpoints: $activity(t)$, $time(t)$, $channel(t)$, see examples in Fig. 1. Other touchpoint attributes can be the cost, resource or activity outcome. Journeys, like touchpoints, have attributes. Let J be the set of all journey identifiers. $a(j)$ is the value of attribute $a \in A$ for journey $j \in J$. Each journey has a mandatory attribute 'trace': $trace(j) \in T^*$. A trace is a finite sequence of touchpoints $\sigma \in T^*$. If t or j do not have a, then $a(t) = \perp$ or $a(j) = \perp$ (null value). A journey log is a set of journeys $L \subseteq J$. A log is* complex *if it contains journeys from a less structured process, i.e. no enforced process, ('structured' defined in [18]), where the activity (and channel) set is large, due to which L contains many journey variants.*

Definition 2. (Perspective, journey profiles). *Let $V = \{V_1, V_2, ..., V_H\}$ be a set of perspectives, views on a journey. $map_V : J \to \mathbb{R}^n$ denotes the function that maps a journey to an n-dimensional vector according to perspectives V. $\mathbf{p_V}(\mathbf{j})$ denotes the projection of journey $j \in L$ to perspectives V. Furthermore, we let $\mathbf{p_V}(\mathbf{j}) = p_{\{V_1,V_2,...,V_h\}}(j) = map_{V_1}(j) \parallel map_{V_2}(j) \parallel ... \parallel map_{V_h}(j)$, i.e. $\mathbf{p_V}(\mathbf{j})$ is a journey profile vector in which all perspectives are concatenated.*

Definition 3. (Feature set). *Let $F(n) = \{f_1, f_2, ..., f_n\}$ be the full set of features in journey profile $\mathbf{p_V}(\mathbf{j})$. $F(m) \subseteq F(n)$ denotes a feature subset that contains m most relevant features according to a specific feature selection technique where $m \leq n$. Journey profiles reduced to this subset are denoted by $\mathbf{p_{F(m)}}(\mathbf{j})$.*

Definition 4. (Similarity metric). *We define a similarity function $S(j, j', w)$, S in short, is a linear function parameterized by a set of weights $\mathbf{w}$ that defines pairwise similarities of customer journeys j in a journey log $L \subseteq J$. Here, $\mathbf{w} = (w_1, w_2, ..., w_m)$ is a set of weights where $w_i \in [0, 1]$. S operates over journey profile vectors $\mathbf{p_{F(m)}}(\mathbf{j})$ and has the following properties:*

- *For the sake of interpretability, S is bounded between 0 and 1.*
- *When two journeys j and j' have exactly similar profiles, then $S(j, j') = 1$;*
- *If journey j is more similar to j' than to j'', then $S(j, j') > S(j, j'')$;*
- *S is symmetric, i.e. $S(j, j') = S(j', j)$.*

Definition 5. (Universal similarity metric). *A universal similarity metric S_U is a standard similarity metric, e.g. Cosine similarity, that uses a set of uniform weights $w_i = 1$ and operates over $F(n)$.*

Definition 6. (Trace clustering). *A trace clustering $TC = \{TC_1, TC_2, ..., TC_k\}$ is a set of k trace clusters over journey log L. We assume a hard clustering, i.e. every journey is part of exactly one trace cluster. TC is the result of a convex-based clustering algorithm with a pre-defined number of clusters k. In theory, other clustering techniques such as hierarchical clustering could also be used but these would not make use of the initial guess from domain experts about k that we have in our problem (Definition 9) and would yield other parameter problems.*

Definition 7. (Internal trace clustering quality). *Internal quality of a trace clustering is measured using an internal index validity statistic γ that expresses the quality of a clustering TC in terms of the cohesion of traces within the same cluster and the separation between traces in different clusters. For a clustering TC' that has a higher internal quality than TC, we write $\gamma(TC') > \gamma(TC)$.*

Definition 8. (External trace clustering quality). *External quality of a trace clustering is measured using an external validity statistic δ that expresses the quality of a clustering TC based on the correspondence between TC and a labeling D (cf. Definition 9). For a clustering TC' that has more over overlap with labeling D than TC, we write $\delta(TC') > \delta(TC)$.*

Definition 9. (Domain knowledge). *Domain knowledge provides a journey labelling $D = \{D_1, D_2, ..., D_b\}$ that assigns a journey $j \in L$ to exactly one label. This labeling represents domain experts' intuition about a (high-level) clustering structure of L. The domain knowledge also informs about the possibility of journeys being separated over much more labels than b, namely g, but $g << |J|$. Due to the existence of many journey variants, D_i could still include very different behaviour. Finally, it also specifies δ_{min}, the lowest acceptable $\delta(TC)$ with respect to the similarity metric learning problem.*

Definition 10. (Meaningful similarity metric). *An optimal similarity metric S^* maximizes $\gamma(TC)$ for $k = b$ clusters (here b is derived from D). This metric is considered meaningful if a TC' with a number of clusters k up to g is of comparable (or better) quality, i.e. $\gamma(TC')$ is comparable to $\gamma(TC)$.*

Definition 11. (Optimal feature set). *A feature set $F(m)$ is considered optimal if journey profiles $\mathbf{p_{F(m)}}(\mathbf{J})$ maximizes $\gamma(TC)$ while keeping $\delta(TC) > \delta_{min}$. This set is denoted by $F(*)$.*

This work aims at inferring a meaningful similarity function S^* that defines similarities between customer journeys in a complex journey log $L \subseteq J$ based on *customer intent*. Using S^*, a trace clustering TC can be created. For a specific clustering algorithm, S^* maximizes $\gamma(TC)$ for $k = b$ clusters while $\delta(TC) > \delta_{min}$.

Initially, domain knowledge D is helpful for the similarity metric development but at some point it becomes questionable. The journey labeling D is not blindly trusted, it is not considered the ground truth. This is a typical setting for a semi-supervised approach where the domain knowledge is considered a good starting point. The solution approach aims to collaborate with the existing knowledge but also extends this knowledge (in particular for $k > b$).

3 SIMPRIM Framework

We propose SIMPRIM, a framework for Similarity Metric learning for Process Improvement, that simultaneously learns a similarity metric S and a journey clustering TC. By integrating metric optimization techniques with trace clustering and domain knowledge in a joint framework, S maximizes the separability of the journeys in a low dimensional space while agreeing with existing domain knowledge D. Since clustering results are heavily influenced by the metric that is used, the quality of S is approximated with the quality of the corresponding TC. The methodology is formalized in Algorithm 1. For each step, a set of suitable techniques is selected to experiment with, such that no assumptions have to be made about the effectiveness of a technique for a specific dataset. This makes the framework applicable to manifold applications contexts.

3.1 Journey Log to Journey Profiles

As a first step, journeys $j \in L$ are translated into a format on which similarity can be calculated. SIMPRIM adopts the abstract representation approach

Algorithm 1. SIMPRIM: Finding a domain-informed similarity metric for customer journeys and a meaningful journey clustering

---
1: Determine journey perspectives V and find profiles $\mathbf{p_V}(\mathbf{J})$;   ▷ Cf. Sect. 3.1
2: Find $S_U{}^*$ that maximizes $\gamma(TC)$ for $\mathbf{p_V}(\mathbf{J})$;   ▷ Cf. Sect. 3.2
3: $F(*) \leftarrow FeatureSelection(\mathbf{p_V}(J),\ D,\ S_U{}^*)$;   ▷ Cf. Sect. 3.3
4: $\mathbf{w}^*, TC^* \leftarrow WeightOptimization(\mathbf{p}_{F(*)}(J),\ D,\ S_U{}^*)$;   ▷ Cf. Sect. 3.4 &
   Algorithm 2
5: $S^* = S(\mathbf{p}_{\mathbf{F}(*)}(\mathbf{j}),\ \mathbf{p}_{\mathbf{F}(*)}(\mathbf{j}'),\ \mathbf{w}^*)$;
6: Qualitatively evaluate S^* and TC^* and adjust previous steps if
   required.   ▷ Cf. Sect. 3.5
---

from [12] in which journey profiles $p_V(J)$ are used. While traditional approaches only describe a journey from the control-flow perspective [7], this approach also allows for the inclusion of other trace perspectives V. Figure 1 shows the necessity of including different perspectives to create a meaningful journey separation. Perspectives can be based on both journey and touchpoint attributes. Common perspectives are the Activity, Originator, Transition, Event-Attributes, Case-Attributes and Performance perspective [12,15]. Custom perspectives can be added if preferred. Carefully designing the journey profiles can improve the quality of TC and thus S.

SIMPRIM uses function $map_V : J \rightarrow \mathbb{R}^n$ to map the journeys into journey profiles (Definition 2). Most perspectives V mark the presence of a certain attribute (hot-encoding). For the mapping function, two representation techniques are compared within the framework: Bag-Of-Activities (BOA) [2], and Set-Of-Activities (SOA), indicating a real and a binary attribute count respectively. For complex journey logs, this mapping typically yields numerical trace vectors that are very high dimensional and sparse. This is caused by the large set of attributes available of which journeys often only contain a very small subset.

3.2 Measuring Similarity

To express the similarity between any two journey profiles that are represented as n-dimensional vectors, a number of universal similarity metrics S_U can be used. The *weighted* Jaccard similarity [4] and Cosine similarity are considered most suitable for our problem. They operate efficiently on sparse and high-dimensional vectors, are bounded between $[0,1]$ and proven effective in existing work [8,12]. We define $Cosine\ similarity(j,j') = \frac{\mathbf{p}_{\mathbf{F(m)}}(j)\cdot\mathbf{p}_{\mathbf{F(m)}}(j')}{||\mathbf{p}_{\mathbf{F(m)}}(j)||_2||\mathbf{p}_{\mathbf{F(m)}}(j')||_2}$ and $Jaccard\ similarity(j,j') = \frac{\sum_{i=1}^{m} min(\mathbf{p}_{\mathbf{F(i)}}(j),\ \mathbf{p}_{\mathbf{F(i)}}(j'))}{\sum_{i=1}^{m} max(\mathbf{p}_{\mathbf{F(i)}}(j),\ \mathbf{p}_{\mathbf{F(i)}}(j'))}$ which is set to 1 if the denominator is 0. Their different notion of similarity makes them interesting to compare. While Jaccard similarity is originally designed for binary sets, here we use the *weighted* version. No actual weights are assigned to features but it means that the metric can also be used on non-negative real vectors $\mathbb{R}$. This ensures that there is a difference in measuring the similarity using the BOA versus SOA trace representation.

3.3   Dimensionality Reduction

Many issues arise when applying clustering and weight optimization techniques on high dimensional vectors. Therefore, *feature selection* is applied to reduce a journey profile to the most effective feature subset. Such techniques are found more suitable here than *feature extraction* techniques (e.g. PCA) since the interpretability of the metric is considered highly important for the business use case. SIMPRIM allows for experimentation with a variety of techniques to find $F(*) \subseteq F(n)$ on which S will operate. In theory, any *filter* or *wrapper* feature selection method can be incorporated. However, techniques that are *embedded* in a clustering algorithm are not suitable for this framework as no dependence on specific techniques is desired. Besides supervised techniques, we suggest experimenting with unsupervised techniques since the quality of labeling D might be poor. Besides, an optimal clustering from a label perspective, i.e. $TC = D$, could have a very low $\gamma(TC)$ in which case γ and δ are competitors. In that case, supervised feature selection might not lead to the desired optimization.

Feature Set Size Restrictions. To be able to develop a *meaningful* similarity metric, the size of feature set m is restricted by a lower bound. A very small m yields a non-meaningful metric because (1) journey profiles $\mathbf{p}_{\mathbf{F(m)}}(\mathbf{J})$ might contain too little information to separate journeys correctly for $k > b$, and (2) it can yield a large set of exactly similar journey profiles which causes all these journeys to be put in the same cluster $TC_i \in TC$ that cannot be separated by increasing k since for these profiles $S = 1$. Therefore, SIMPRIM puts an indirect lower bound on the feature set size using Definition 12. On the other hand, when dimensionality increases the difference between the min and max pairwise similarity, i.e. *similarity contrast*, becomes really small which makes the similarity values indistinctive. If similarity is only expressed in a small part of the $[0, 1]$ range, its interpretation is not intuitive and S would not be a useful stand-alone product. Therefore, Definition 13 puts a minimum on the similarity contrast which indirectly upper bounds m. All remaining values for m are considered for finding $F(*)$.

Definition 12. (Optimal Feature Set Lower Bound). *A feature set $F(m)$ can only be optimal, i.e. $F(*)$, if the corresponding number of exactly similar journey profiles $\mathbf{p}_{\mathbf{F(m)}}(\mathbf{j})$ for journey $j \in L$ does not exceed $max(\alpha|D_i|)$ for $D_i \in D$ where $\alpha \in [0, 1]$ is a framework parameter.*

Definition 13. (Optimal Feature Set Upper Bound). *Let $H \in J$ be the set of journey with pairwise distances in J that lay within two standard deviations from the mean distance between any two points in J, i.e. $\mu_J \pm 2 * \sigma_J$. We define the similarity contrast c as $max(H) - min(H)$. Now, a feature set $F(m)$ can only be optimal, i.e. $F(*)$, if $c > \beta$ where $\beta \in [0, 1]$ is a framework parameter.*

3.4   Co-learning of Metric Weights and Journey Clustering

Unweighted metrics assume all features are of equal importance to distinguish traces, while in reality their importance for finding a good clustering structure

Algorithm 2. Weight optimization using an SMBO method [10]

Input: uniform weights w, surrogate model f, objective function O, acquisition function Γ, stopping criteria
Output: optimal weights $\mathbf{w}^*$

1: $\mathcal{R} \leftarrow \{\}$
2: **while** stopping criteria are not met **do**
3:     $\mathbf{w} \leftarrow SMBO(\mathcal{R})$ {Fit f and maximize Γ}
4:     $\lambda_{\mathbf{w}} \leftarrow O(TC_{\mathbf{w}})$ {Evaluate weight set}
5:     $\mathcal{R} \leftarrow \mathcal{R} \cup \{(\mathbf{w}, \lambda_{\mathbf{w}})\}$ {Add to results}
6: $\mathbf{w}^* \leftarrow argmax_{(\mathbf{w}, \lambda_{\mathbf{w}}) \in \mathcal{R}} \ \lambda_{\mathbf{w}}$
7: **return** $\mathbf{w}^*$

---

differs. Handpicking weights in a meaningful way is not a trivial task. To tune distance metrics automatically, feature weighting techniques can be used that learn a set of weights $\mathbf{w}^*$ in domain W that optimizes a given objective function. We use the clustering quality as objective such that the weights and the clustering are optimized simultaneously. We refer to this as a *co-learning* approach.

Weight Optimization. For weight optimization we use Sequential-Model-Based-Optimization (SMBO), i.e. Bayesian optimization. SMBO is commonly used for hyper parameter tuning and is very efficient for expensive objective functions O [10]. Besides, it can take any objective function and the resulting weights w^* can be added to an arbitrary metric to adapt the scaling of dimensions. The SBMO algorithm is described in Algorithm 2. An acquisition function Γ is used to maximize over a cheap surrogate model and find a new set of candidate weights. Because of our expensive objective (clustering) function, informed weight sampling could yield great efficiency by finding the optimum with as few evaluations as possible. The algorithm stops when (1) x sequential function calls did not improve the clustering quality or (2) after a specified maximum number of calls.

We aim to develop a similarity metric S^* that maximizes $\gamma(TC)$ while keeping $\delta(TC)$ at an acceptable level. Therefore, both quality criteria are included in the optimization objective (Eq. 1). The contribution of δ is parameterized by $\theta \in [0, 1]$, and can be determined based on the quality of journey labeling D.

$$O(TC_{\mathbf{w}}) = \theta \cdot \delta(TC_{\mathbf{w}}) + (1 - \theta) \cdot \gamma(TC_{\mathbf{w}}) \tag{1}$$

The surrogate model should reflect the actual objective as much as possible. Within SIMPRIM, we experiment with two surrogate models: (1) a Gaussian Process (GP) (most common) [10] and a Gradient Boosting Regression Trees (GBT) model [3]. Both are available in the `scikit-optimize` Python library.

Clustering. While any convex-based clustering algorithm can be used, k-Medoids is selected based on its simplicity and speed. It is preferred over k-Means as its clusters are represented by actual journeys, which allows for more intuitive interpretation. Throughout the optimization, both the clustering algorithm and k remain fixed. This allows to optimize TC by varying S merely.

3.5   Evaluation

The quality of S is approximated with the quality of its corresponding TC. Many clustering validation indices exist. Based on [11], we use the revised validity index (S_Dbw) for γ, a metric that resembles the well-known Silhouette index. S_Dbw is a summation of inner-cluster compactness and inter-cluster separation. A lower value indicates a better quality. For δ, the adjusted rand index ARI and adjusted mutual information AMI are commonly used (a higher value indicates better quality). In the assessment of a more fine-grained cluster output ($k > b$), for which no labelled data exists, we test the robustness of the approach by involving domain experts. This yields a *qualitative evaluation* instead of a quantitative one.

4   Experimental Evaluation

The usefulness of SIMPRIM is demonstrated by applying it to a real-life use case with a leading telecom provider in Sect. 4.1. Additionally, we evaluate our framework over another real world benchmark dataset in Sect. 4.2 which is hosted by its owners here: https://data.4tu.nl/articles/dataset/BPI_Challenge_2012/ 12689204. Our Python implementation of SIMPRIM is accessible via https:// github.com/sophievdberg/SIMPRIM.

4.1   Customer Service Process at Anonycomm

Application Scenario. Our work is performed in collaboration with a large telecom provider and is inspired by their data. Since they would like to stay anonymous, we will refer to them as Anonycomm. Hundreds of thousands of customers interact with Anonycomm each month using multiple channels. Offering a superior customer experience is a high priority for Anonycomm as competitors have increasingly similar service offers and devices. As such, they aim to improve their business process towards a more efficient and self-service user portal. However, the unstructured nature of their customer support process allows for a wide variety of journey types, which makes it very difficult to find behavioural improvement opportunities. A meaningful journey clustering could provide a more complete picture of the average journey per customer intent. This allows for better sizing and prioritization of improvement efforts.

Journey log L used for this research is a collection of customer contact moments (touchpoints t) in a 3-month period. A customer is getting into contact for many different reasons ($activity(t)$), such as having a question about a service, acquiring a new subscription or the installation of a new piece of hardware. They do so using a contact type such as call, mechanic visit or order placement ($channel(t)$). Finally, for all touchpoints t we have a more detailed description/reason of $activity(t)$ stored in $eventtype(t)$, e.g. 'disruption after installation'. The log is considered *complex* because of the large number of channels, activities and event-types (18, 80 and 767 respectively) and heavily varying journey lengths (2–183 touchpoints). Journeys with only one touchpoints are removed. In total, this leaves us with around half a million journeys in the dataset.

Journey labeling D distinguishes 53 journey categories. For example, 'WiFi disruption' or 'Extra TV subscription' or 'Move'. The labeling is done by domain experts based on a large set of business rules that assign a touchpoint to a journey (=class based on the presence of specific channels, activities, event types and time). Domain experts are not completely sure about their labeling. Besides, the labeling is performed from a reason/topic-based perspective on journeys which motivated us to find similarities of journeys amongst two axes: topic and behaviour. This indicates that the metric S should not overfit on the labeling D (i.e. relatively small δ_{min} and limited contribution of δ in the weight optimization process). The distribution of labeling D is skewed and since some of the categories are still relatively big, domain experts indicated that dividing the journeys up to 100 clusters could be meaningful ($g \approx 100$).

Implementation Details. Journeys are clustered using the Partitioning Around Medoids (PAM) algorithm. Since this clustering algorithm is sensitive to the set of initial medoids, we run each experiment 5 times and report the average clustering quality. Furthermore, a label-based medoid initialization is used (1 journey per class). This initialization technique outperformed the kmeans++ initializer for our data. For the external cluster validation, AMI index is used since it better suits the unbalanced labeling D (small clusters exist). We experiment with 2 supervised feature selection techniques, **Fast-Based-Correlation-Filter** and **L1-regularization**, and with 2 unsupervised techniques, **Variance** and **Laplacian score**. These are selected since they are common and relatively efficient. Note that SIMPRIM allows for the usage of other feature selection techniques too. The journey profiles are scaled in the range $[0, 1]$. Since the unsupervised techniques do not take correlation into account, features with a covariance ≥ 0.8 are removed. A regularization of 0.1 is used for $L1$.

We have recommendations about parameters: we set $\alpha = 0.5$ (Definition 12), $\beta = 0.2$ (Definition 13) and $\theta = 0.3$ (Eq. 1). A lower value for θ is used due to the limitations of labeling D. Stopping criteria are 60 iterations without an improvement or a maximum of 150 iterations. Note that neither the scope of the paper nor the space capacity allow us to do an extensive parameter sensitivity evaluation.

The similarity metric is optimized on 3 training sets to assess overfitting and tested on 3 test sets to evaluate its stability. Splits are stratified on class labels. Due to memory issues, clustering could only be done on 20.000 journeys simultaneously using our machine (2.3 GHz Intel Core, 16 GB RAM). Other than that, with this implementation we did not experience any run out of memory issues or freezing of the experimentation. However, if one decides to use another technique and SIMPRIM is selecting multiple iterations of the same task, then the

244   S. van den Berg and M. Hassani

Table 1. Clustering quality results for universal metrics S_U in the Anonycomm dataset. Error indicates the σ over the training and test sets (i.e. stability).

Metric	Trace type	$S\_Dbw\ (\gamma)$	$AMI\ (\delta)$
Cosine	**BOA**	**1.078** $\pm$ 0.015	**0.526** $\pm$ 0.010
	SOA	1.049 $\pm$ 0.030	0.531 $\pm$ 0.045
Jaccard	BOA	1.046 $\pm$ 0.054	0.403 $\pm$ 0.062
	SOA	1.051 $\pm$ 0.089	0.482 $\pm$ 0.005

user should select an efficient model: our framework does not make the approach more efficient. The total running time was around 20 h.

Experimental Results. For Step 1 in Algorithm 1, we translate journeys in L into numerical journey profiles using 4 perspectives V. V_1 is the Activities profile that hot-encodes event and sub-event types. V_2 is the Event Attribute profile that does the same thing for the contact types of a journey (i.e. *channel*). V_3 is the Transition profile that only includes 2-grams for channel types, not for activities and event types as they are extremely diverse and dimensions would explode. Finally, the Performance perspective V_4 includes the duration of the journey, the number of different channel types and the number of touchpoints in a journey. This mapping yields journey profiles with $n = 2756$ features.

To find the optimal universal similarity metric $S_U{}^*$ for the resulting profiles $\mathbf{p_V(J)}$, in Step 2 of Algorithm 1 we compare the quality of the Jaccard and Cosine metric on both BOA and SOA journey profiles (4 metrics). Table 1 shows that, in terms of γ, all metrics have similar performance (differences within error bandwidth) but the Cosine metrics have slightly better stability. Furthermore, the Cosine metrics yield better results in terms of δ. The BOA representation is preferred as it shows a slightly more stable quality than SOA and it is preferred by domain experts since it aligns with their current view on journeys. We therefore find $S_U{}^*$ to be based on the Cosine metric and BOA journey profiles. Now that we have found $S_U{}^*$, we can start optimizing it. First, we reduce the dimensions it is operating on by evaluating the feature selection techniques (Step 3 in Algorithm 1). Figure 2 visualizes the clustering results on different feature sets F. The largest evaluated set has $m = 400$ since the results stabilize from that point. We find that L_1-regularization and Variance are not yielding any candidate feature sets for $m \leq 400$ (dashed lines). Additional evaluation indicates they yield too many similar journey vectors and thus do not meet Definition 12. Based on Fig. 2, the Laplacian feature set with 60 features can be considered $F(*)$. Using this feature set, γ is improved with 41% compared to the baseline $S_U{}^*$ and the clustering remains stable over the different train and test sets used. The external clustering quality δ for $F(*)$ remains comparable to $S_U{}^*$: only a 2.6% reduction is observed, which is still considered to overlap sufficiently with D.

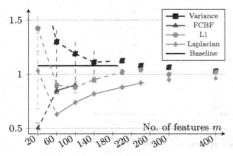

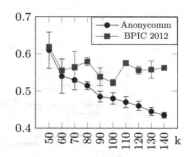

Fig. 2. Clustering quality γ for S operating over different $F(m)$ in Anonycomm dataset. Error bars indicate the σ over the training and test sets. Dashed lines indicate that the corresponding $F(m)$ does not fulfill Definition 12 & 13 and thus cannot be considered $F(*)$.

Fig. 3. Clustering quality γ for different k using S^*. For each k, 10 clusterings with different initial medoids sets are compared. Note: for the BPIC 2012 dataset, the x-axis should be divided by 10, i.e. read as $5 \geq k \leq 14$.

Table 2. Clustering quality γ and δ before (surrogate = None) and after weight optimization using two surrogate models in Anonycomm dataset.

Surrogate	Train γ	Test γ	%Δ	Train δ	Test δ	%Δ
None		0.634 ± 0.011			0.520 ± 0.010	
GP	0.594 ± 0.049	0.641 ± 0.032	-1.1	0.507 ± 0.028	0.493 ± 0.031	-5.2
GBT	0.597 ± 0.021	**0.583 ± 0.013**	**+8.0**	0.518 ± 0.025	**0.523 ± 0.026**	**+0.6**

Using only the features in $F(*)$, we can optimize the weights (w) of S (Algorithm 1, Step 4). Figure 4 shows how the weight optimization techniques converge to a final set of weights. The GP surrogate model is able to minimize the objective function (Eq. 1) best, indicating it is better able to approximate the expensive clustering objective than the GBT model. Also, it shows a very efficient optimization process as it reaches its optimum very quickly. However, Table 2 indicates that the GBT weight sets yield better clustering quality when evaluated on the test sets. GP weights seem to suffer from slight overfitting on the training set, while this is not the case for GBT. A possible explanation for this could be that GP shrinks more weights to zero than GBT, for all training sets, while for generalization it is more safe to not completely remove them. Although the feature space is reduced, the feature set requirements are still met (Definition 12 & 13). The optimal metric S^* is thus based on GBT weights and operates on only 36 out of the original 2756 features. Overall, optimizing the weights yields an additional improvement of γ of 8%, while keeping δ at a comparable quality. In total, compared to S_U^*, S^* is able to improve the $\gamma(TC)$ with 46%, while

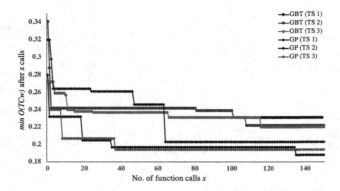

Fig. 4. Weight convergence per training set (TS) for weight optimization based on Gaussian Process (GP) or Gradient Boosting Trees (GBT) (Anonycomm).

maintaining a similar $\delta(TC)$. Figure 3 shows that S^* is indeed *meaningful* as it shows a better γ for k up to g clusters. Around $k = 90\text{--}100$, γ stabilizes, indicating this could be a more natural number of clusters (Elbow method). Finally, the results are qualitatively evaluated (last step of Algorithm 1). In collaboration with a domain expert it is assessed whether journey clustering TC obtained with S^* makes sense. In general, the results were appreciated. Most clusters in TC show very similar behaviour and deal with the same problem type, especially for a more fine-grained clustering ($k = 100$). To improve the incorrect clusters, further experimentation can be performed by manually adding features that include information for better separation.

Next Event Prediction. The added value of S^* can also be demonstrated using a *model-based* approach. For any model that uses some notion of similarity, including meaningful similarity information might improve the prediction quality. For Anonycomm, we experimented with a model that predicts the contact type of next touchpoint t for a running journey j, to be able to get ahead of mistakes and prevent them beforehand ('on the fly') [16]. A baseline prediction model trained on all journeys is compared to a cluster-based approach in which one model is build per cluster. A prediction is then made by the model that corresponds to the cluster closest to the journey at that point in time according to S^*. This approach is hypothesized to have higher predictive abilities since the selected model selected is trained on journeys with similar behaviour.

To predict the next contact type of t_{i+1}, we can include all available information up until t_i. We included the information of the previous 4 touchpoints (i.e. 4 'lagged' features). A Random Forest model is used (for the baseline and cluster-based approach) and its predictive capability is expressed using the weighted F_1 score. Table 3 shows that the cluster-based approach does not improve the prediction quality when S^* is used. This could indicate that the clusters in TC are not highly discriminative in terms of contact types. Since S^* operates over a very small feature set ($m = 36$), it might not contain sufficient information regarding contact types. Hence, the performance of a S with more features $m = 140$, i.e. S_{140}, is also tested. Table 3 shows that this increases the prediction quality with 1.4%, which is small but significant. Note that the performance could be further improved by properly tuning the model but this is out of the scope of this paper.

4.2 BPIC 2012 Real Dataset

Since the data of Anonycomm is confidential, the SIMPRIM methodology is replicated on the publicly available BPIC 2012 event log, using Annonycom implementation details. This real-life event log contains 13087 journeys in the loan application process of a Dutch Financial Institute. It should be mentioned that SIMPRIM adds the most value for journeys from very flexible environments, while this event log has a much more sequential (structured) nature than Anonycomm's event log. Since no labeling D for this log exists, classes are derived based on the loan application outcome. We distinguish 6 classes: applications that are (1) accepted directly (only 1 offer); (2) accepted after some optimization of the offer; (3) rejected straight away, (4) rejected after an offer was drafted, (5) cancelled before an offer was drafted and (6) cancelled after an offer was drafted. D is heavily imbalanced: most applications are rejected straight away. The separation between with or without offer is created as corresponding journeys show very different journeys based on the absence or presence of "O_" states. Journeys without an event related to being accepted, rejected or cancelled are considered 'running' journeys and thus incomplete. Hence, they are removed from the dataset (2%). In the first step of SIMRPIM (Algorithm 1), the journey profile mapping is done similarly as on the Anonycomm dataset. We only have *resources* instead of *contact types* in V_2 and since the number of activities is significantly smaller we also include 2-grams for these in V_3. This yields a total of $n = 670$ features. Only 197 of these features are considered relevant, i.e. not fully correlating and a variance score below 1. We make use of one training set (70%) and two test sets (15% each). Table 4 shows the sequential optimal

Table 3. Weighted F-scores of the next contact type prediction models, comparing a baseline with cluster-based approaches (CBA). An 80/20 train-test split is made and 5-fold CV is used for the train scores. (Anonycomm dataset).

Prediction model	Baseline	CBA with S^*	CBA with S_{140}
Training set	0.571	0.563	0.579
Test set	0.580	0.579	**0.594**

Table 4. Optimal clustering results of Step 2–3 in Algorithm 1 (BPIC 2012 dataset).

	S_Dbw (γ)	AMI (δ)	Techniques
S_U^* (Step 2)	0.777 ± 0.016	0.639 ± 0.005	Cosine, BOA
S on $F(*)$ (Step 3)	0.704 ± 0.002	0.645 ± 0.003	Variance, $m = 50$

Table 5. Clustering quality after weight optimization using two surrogate models (BPIC 2012 dataset). $\%\Delta$ relates to the *test* results of Step 3 in Table 4.

Surrogate	m	Train γ	Test γ	$\%\Delta$	Train δ	Test δ	$\%\Delta$
GP	42	0.461	**0.471**	**+33**	0.644	0.658	+2
GBT	50	0.593	0.555	+21	0.673	0.614	−5

results for Step 2–3 in Algorithm 1. As can be seen, $S_U{}^*$ is again based on Cosine similarity and a BOA feature representation. Variance is found to be the most suitable dimensionality reduction technique, with $F(*)$ consisting of 50 features. We also tested $m = 100$ and $m = 150$. Table 5 shows the weight optimization results (Step 4 in Algorithm 1). The metric that uses GP-based weights can be considered S^*. The metric operates on 42 features, the other 8 were shrank to zero. Again, the test score is slightly better than the training score, which could indicate overfitting but to a smaller degree than on the Anonycomm dataset. The weight optimization here was expected to suffer less from overfitting since a larger percentage of journeys is used to train the weights on. Again, for GBT, that does not shrink any feature to zero, no overfitting is seen. In total, γ is improved with 39% while keeping δ similar. Different to the first experiment, the largest improvement of γ here is obtained with the weight optimization. Figure 3 shows that S^* is *meaningful* for larger k and that $k = 10$ might be a more natural number of clusters than 6.

5   Related Work

Trace clustering in process mining is discussed in several works e.g. [12]. Defining an appropriate feature space and distance metric are still key challenges in trace clustering. The work of [6] and [2] contribute to this by developing syntactical techniques based on which appropriate feature spaces are derived using an edit-based distance. However, our work is focusing on vector-based approaches since syntactic techniques do not yield a standalone metric S that we are looking for. This paper contributes to trace clustering techniques that aim to differentiate business processes rather than reducing the complexity of the underlying process models. Specifically, a contribution is made to *distance-based* approaches to trace clustering. Model-based approaches, such as [5], are not considered suitable for our setting since no similarity between vectors is defined and thus no similarity metric can be tested. The methodology proposed is unique because it integrates metric optimisation techniques with clustering in one framework. A similar setting was discussed recently in [13] but for developing a hierarchical distance metric to measure the similarity between different market baskets where neither the behaviour nor the order of items matter. SIMPRIM adopts a semi-supervised approach to metric learning while existing frameworks, with a specific application to clustering, either are completely optimizing on some sort of ground truth [17] or do not include domain knowledge at all [9]. Most feature weighting methods employ some variation of gradient descent. This works for distance metrics from the Euclidean family. However, for other distance metrics, such as Cosine, this task is not so trivial. Especially when the dimensionality of trace vectors is high, the complexity of the gradient is large due to which this feature weight learning approach will be inefficient and ineffective. The dimensionality reduction techniques used in this paper are widely adopted in the field of data mining but are not typically used in trace clustering literature.

6   Conclusion

In this paper, we proposed SIMPRIM, a framework for inferring an appropriate domain-informed similarity metric that outperforms standard similarity metrics in the clustering task. The developed metric can also be used for further customer journey analysis or to improve the accuracy of other predication or recommendation tasks. A co-learning approach is adopted that simultaneously learns metric weights and optimizes the journey clustering. Several components used in our approach can easily be replaced with others equivalent. SIMPRIM has shown to be useful for two real-life event logs. A 46% and 39% improvement of the internal clustering quality is obtained, while agreeing with existing process knowledge in the form of journey labeling. Furthermore, an acceptable improvement of a next touchpoint prediction model was achieved. An interesting future direction is the added value of the metric to recommender systems that recommend a next best action for a running customer journey [14].

References

1. Bernard, G., Andritsos, P.: A process mining based model for customer journey mapping. AISE, vol. 1848, pp. 49—56 (2017)
2. Bose, R., Van der Aalst, W.: Context Aware Trace Clustering: Towards Improving Process Mining Results. SDM, pp. 401–412 (2009)
3. Breiman, L.: Classification and Regression Trees. Routledge, Milton Park (1984)
4. Chierichetti, F., Kumar, R., Pandey, S., Vassilvitskii, S.: Finding the Jaccard median. In: 21st ACM-SIAM Symposium on Discrete Algorithms, pp 293—311 (2010)
5. De Weerdt, J., Van den Broucke, S., Van Thienen, J., Baesens, B.: Active trace clustering for improved process discovery. TKDE **25**(12), 2708—2720 (2013)
6. Evermann, J., Thaler, T., Fettke, P.: Clustering traces using sequence alignment. In: Reichert, M., Reijers, H.A. (eds.) BPM 2015. LNBIP, vol. 256, pp. 179–190. Springer, Cham (2016). https://doi.org/10.1007/978-3-319-42887-1_15
7. Greco, G., Guzzo, A., Pontieri, L., Sacca, D.: Discovering expressive process models by clustering log traces. TKDE **18**(8), 1010–1027 (2006)
8. Hompes, B., Buijs, J., Van der Aalst, W., Dixit, P., Buurman, J.: Discovering deviating cases and process variants using trace clustering. BNAIC (2015)
9. Huang, J., Ng, M., Rong, H., Li, Z.: Automated variable weighting in k-means type clustering. Patt. Anal. Mach. Intell. **27**(5), 657–668 (2005)
10. Lacoste, A., Larochelle, H., Laviolette, F., Marchand, M.: Sequential Model-Based Ensemble Optimization. UAI (2014)

11. Liu, Y., Li, Z., Xiong, H., Gao, X., Wu, J.: Understanding of internal clustering validation measures. IEEE International Conference on Data Mining (2010)
12. Song, M., Günther, C., Van der Aalst, W.: Trace Clustering in Process Mining. Lecture Notes in Business Information Processing, vol. 17, pp. 109–120 (2008)
13. Spenrath, Y., Hassani, M., Van Dongen, B., Tariq, H.: Why did my consumer shop? Learning an efficient distance metric for retailer transaction data. In: ECML PKDD 2020, vol. 12461, pp. 323—338 (2020)
14. Terragni, A., Hassani, M.: Optimizing Customer Journey Using Process Mining and Sequence-Aware Recommendation. Ass. for Computing Machinery, NY (2019)
15. Thaler, T., Ternis, S., Fettke, P., Loos, P.: A Comparative Analysis of Process Instance Cluster Techniques (2015)
16. van der Aalst, W.M.P., Pesic, M., Song, M.: Beyond process mining: from the past to present and future. In: Pernici, B. (ed.) CAiSE 2010. LNCS, vol. 6051, pp. 38–52. Springer, Heidelberg (2010). https://doi.org/10.1007/978-3-642-13094-6_5
17. Xing, E., Ng, A., Jordan, M., Russell, S.: Distance metric learning with application to clustering with side-information. NIPS, vol. 15, MIT Press (2003)
18. Van der Aalst, W.: Data Science in Action. In: Process Mining, pp. 3–23. Springer, Heidelberg (2016). https://doi.org/10.1007/978-3-662-49851-4_1

Quantifying Explanations of Neural Networks in E-Commerce Based on LRP

Anna Nguyen$^{(\boxtimes)}$ ⓘ, Franz Krause ⓘ, Daniel Hagenmayer,
and Michael Färber ⓘ

Karlsruhe Institute of Technology, Karlsruhe, Germany
{anna.nguyen,michael.faerber}@kit.edu,
{franz.krause,daniel.hagenmayer}@student.kit.edu

Abstract. Neural networks are a popular tool in e-commerce, in particular for product recommendations. To build reliable recommender systems, it is crucial to understand how exactly recommendations come about. Unfortunately, neural networks work as black boxes that do not provide explanations of how the recommendations are made.

In this paper, we present *TransPer*, an explanation framework for neural networks. It uses novel, explanation measures based on *Layer-Wise Relevance Propagation* and can handle heterogeneous data and complex neural network architectures, such as combinations of multiple neural networks into one larger architecture. We apply and evaluate our framework on two real-world online shops. We show that the explanations provided by *TransPer* help (i) understand prediction quality, (ii) find new ideas on how to improve the neural network, (iii) help the online shops understand their customers, and (iv) meet legal requirements such as the ones mandated by GDPR.

1 Introduction

The breakthrough with neural networks as a pattern recognition technique has lead its way into many industry sectors. Especially in e-commerce, it can be used as recommender system for advanced searches [12], personalization of shopping experiences and direct marketing [20], or advanced sales forcasting and predictions [14]. Improving the predictions and the usefulness of those recommenders can increase sales and customer satisfaction. Additionally, there is increasing legal pressure in favor of privacy and data protection. For example, the General Data Protection Regulation [10] (GDPR) states that data subjects should be enabled to check the collection, processing, or use of their data. Thus, businesses may be legally required to make their recommender systems transparent.

Multilayer Perceptrons (MLP) have been applied in recommender systems learning feature representations as an extension to collaborative filtering [11]. In combination with convolutional layers, they are applied to generate fashion

Supported by the German Research Ministry (BMBF), the Smart Data Innovation Lab (01IS19030A), and the company econda GmbH.

Y. Dong et al. (Eds.): ECML PKDD 2021, LNAI 12979, pp. 251–267, 2021.
https://doi.org/10.1007/978-3-030-86517-7_16

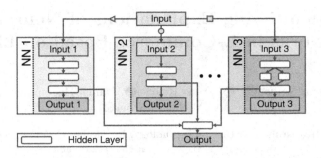

Fig. 1. Model of a neural network with different input data types

outfits for e-commerce or to personalize outfit recommendations based on learned embeddings in Convolutional Neural Networks (CNN) [3,7]. Recurrent Neural Networks (RNN) have shown success in modelling sequential data and have been used for personalized product recommendations based on the purchase patterns of customers [17], learning embeddings of fashion items [13] and modelling user behaviour to predict clicks [5].

However, neural networks are black box models, i.e., the predictions can not be explained. In order to tackle this, it is beneficial to make them more transparent and therefore, more human-understandable. Typically, the Gradient-based Sensitivity Analysis [21] is used to explain the predictions of neural networks. By optimizing the gradient ascent in the input space, it is possible to determine which inputs lead to an increase or decrease of the prediction score when changed [23,25]. Although applications based on this method enable a statement regarding positive or negative influence of an input on a prediction, they do not reveal a quantitative decision-relevant input score such as Guided Backpropagation [24], DeconvNet [19], or DeepLIFT [22]. These algorithms use the trained weights and activations within the forward pass to propagate the output back to the input. This way, it is possible to determine which features in an input vector contribute to the classification and to what extent. Exploiting this, ObAlEx [18] is an explanation quality metric which measures to what extent the classified object is aligned to the mentioned explanations. Nonetheless, all these methods are solely applied to CNNs with image data where single pixels are then highlighted. Another back-propagating algorithm is the Layer-Wise Relevance Propagation (LRP) that has already been successfully used in interaction with MLPs and CNNs [1,2,15]. LRP computes the relevance of each input neuron to an output by performing a value-preserving backpropagation of the output. Furthermore, this method is even applicable on RNNs with sequential data [4,16] which often occurs in processing customer profiles in e-commerce.

Contribution. Our contribution is threefold. First, we provide an explanation framework called TRANSPER[1] for e-commerce businesses in online shopping

---

[1] We provide the source code online at https://github.com/Krusinaldo9/TransPer.

(e.g., for product recommendation) to provide transparency to the neural networks used. Based on a custom implementation of *Layer-Wise Relevance Propagation*, our approach can not only handle individual neural networks types, but also more complex architectures that contain multiple neural subnetworks, such as shown in Fig. 1. This is required in the presence of highly heterogeneous input data (e.g., product images, chronological shopping interactions, personal information) where different neural network types are necessary (e.g., CNN, RNN, MLP). We not only take into account the relevance of the activations of the neurons, but also the bias. This has not been considered in depth in the literature. Second, we define quantity measures to evaluate the helpfulness of these explanations. The individuality measure can be used to determine those parts of the input that are particularly relevant for the decision. The certainty measure quantifies how certain the system is about its prediction. The diversity measure states whether there are clear top predictions. Third, we evaluate our approach on real-world scenarios. To this end, we used data from two real-world online shops provided by our partner *econda*, an e-commerce solution provider. We show that TRANSPER helps in (i) understanding the prediction quality, (ii) finding ideas to improve the neural network, and (iii) understanding the customer base. Thus, TRANSPER brings *trans*parency to *per*sonally individualised automated neural networks and provides new knowledge about customer behaviour. We believe that this helps to fulfill GDPR requirements.

The remainder of this paper is structured as follows. After introducing preliminary definitions and concepts in Sect. 2, we go on to describe the problem setting and formally define an online shop in Sect. 3, to introduce our quantity measures in Sect. 4. We evaluate our approach on the basis of a real-world scenario in Sect. 5 before ending with some concluding remarks.

2   Preliminaries

In this section, we present the fundamentals for the application of our approach. To begin with, we consider a trained neural network with $K \in \mathbb{N}$ layers as shown on the left-hand side of Fig. 2. We refer to Π_k as the set of all neurons in the k-th layer, σ as a nonlinear monotonously increasing activation function, z_i^k as the activation of the i-th neuron in the k-th layer, $w_{ij}^{k,k+1}$ as the weight between the neurons z_i^k and z_j^{k+1}, and b_j^k as the bias term w.r.t. z_j^{k+1}. Assuming that we know the activations in Π_k, the activations in Π_{k+1} can be determined via forward pass as follows:

$$z_j^{k+1} = \sigma \left(\left(\sum_{i \in \Pi_k} z_i^k w_{ij}^{k,k+1} \right) + b_j^k \right) \tag{1}$$

For non-connected neurons z_i^k and z_j^{k+1} we assume $w_{ij}^{k,k+1} = 0$. If a network has no bias, then $b_j^k = 0$.

Layer-Wise Relevance Propagation is a method that represents a backward analysis method [2]. Knowing the activations z_j^{k+1} in layer $k+1$, we can determine

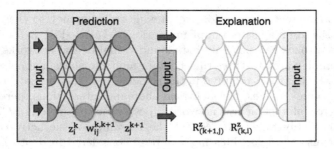

Fig. 2. Exemplary run of LRP. The left-hand side shows the calculation of neuron activations in the forward pass. These activations are then part of the calculation of its relevances in the backward analysis depicted on the right-hand side.

to what extent the neurons in Π_k and the biases b_j^k have contributed, or how relevant they were. The idea behind the standard implementation of the LRP algorithm can be found on the right-hand side of Fig. 2 and is defined as

$$R_{(k,i)}^z = \sum_{j \in \Pi_{k+1}} \frac{z_i^k w_{ij}^{k,k+1}}{\left(\sum_{i \in \Pi_k} z_i^k w_{ij}^{k,k+1}\right) + b_j^k} R_{(k+1,j)}^z, \tag{2}$$

$$R_{(k,j)}^b = \frac{b_j^k}{\left(\sum_{i \in \Pi_k} z_i^k w_{ij}^{k,k+1}\right) + b_j^k} R_{(k+1,j)}^z. \tag{3}$$

For a layer $k+1$, we assume for each neuron j that a relevance can be assigned in the form of a real-valued number $R_{(k+1,j)}^z$. Using Eq. 2, we obtain the relevance, i.e., quantitative contribution, of the i-th neuron in the k-th layer to the overall relevance of layer $k+1$. Furthermore, Eq. 3 provides the relevance of the bias b_j^k of the j-th neuron in layer $k+1$.

In certain applications, customized variations of the standard LRP algorithm presented above can be considered to increase the performance. In particular, with respect to the explainability of CNNs, it has been found that adapted LRP methods lead to better results than the standard LRP method [1,2,15]. These are characterized, e.g., by the use of tuning parameters or penalty terms for negative neuron activations. Regarding RNNs, however, hardly any results exist concerning the use of such variations. Therefore, in relation with the use cases in Sect. 5, we provide results of a test study comparing well-known customizations with the standard method.

3   Formal Model of an Online Shop

In this section, we define an online shop with regard to a suitable neural network which can handle specific characteristics. Especially, we include heterogeneous input data such as interest in products or interactions with products which additionally can have different input lengths. In order to generalize our definition, we consider a neural network consisting of several neural subnetworks to cover different cases as can be seen in Fig. 1. Considering all this, we define our online shop as follows.

Definition 1 (Online Shop Model). *We define an online shop T as a tuple*

$$T = (C, P, (P^*, \Phi), \Lambda, \Lambda^*, S, (\Omega_c)_{c \in C}, (\omega_c)_{c \in C}, (f_c)_{c \in C})$$

with the following entries:

a) *We denote C as the finite set of all customers of the shop.*
b) *Let P be the finite set of all products that the shop offers.*
c) *Then, let P^* be a subset of P or P itself, i.e., $P^* \subseteq P$, and Φ denotes the real-valued output space $[0,1]^{|P^*|}$.*
d) *We denote Λ as the set of information types that the shop T can have about one of its customers $c \in C$ and assume that this amount is finite.*
e) *We define Λ^* as a finite set of disjoint subsets $\Lambda_1, ..., \Lambda_n$ of Λ which corresponds to neural networks $S = \{s_1, \cdots, s_n\}$.*
f) *For a customer $c \in C$ we define an associated real-valued input space*

$$\Omega_c = \mathbb{R}^{m_1(c)} \times ... \times \mathbb{R}^{m_n(c)}$$

with the mappings $m_i : C \to \mathbb{N}$ for $i \in \{1, .., n\}$ with respect to s_i.
g) *Considering a particular customer $c \in C$, we define his input as $\omega_c \in \Omega_c$.*
h) *For a customer $c \in C$, we also define the mapping $f_c : \Omega_c \to \Phi$ where $f_c(x)$ is the recommender's output vector for an input $x \in \Omega_c$.*

Assume we have an online shop T with customers C. The online shop has a catalogue of offered products P. Though, not all products are predicted for example only seasonally available ones or most purchased ones in the last week denoted by P^*. These are used as output space Φ in the neural network, i.e., if $\Phi(p) > \Phi(p')$ then product p is recommended. Now, consider the types of information Λ the online shop can have about their customers such as already purchased products, interactions, or ratings. As mentioned in Sect. 1, certain network types are more suitable for specific data types. Therefore, this information is then classified into disjoint information types, such as sequential data Λ_1, graphical data Λ_2, etc., and summarized in Λ^*. So, if an online shop T has heterogeneous user data, Fig. 1 would consist of neural subnetworks $s_1, \cdots, s_n$. With homogeneous data, we would have a special case of the previous one. Hence, we have:

1. $\Lambda = \Lambda'$ and $\Lambda^* = \{\Lambda_1', ..., \Lambda_n'\}$.   (heterogeneous data)   (4)
2. $\Lambda = \Lambda_i'$ for some i and $\Lambda^* = \{\Lambda_i'\}$.   (homogeneous data)   (5)

The different Λ_i' can have different input lengths depending on the sequence length of the interactions or the size of the images. So, we use the mappings m_i

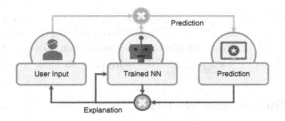

Fig. 3. TRANSPER Overview.

to deal with it and summarize them in Ω. Thus, for a customer $c \in C$, we obtain the neural network's output vector $y = f_c(\omega_c)$.

Considering the different data types, the online shop has three possibilities to define a suitable neural network: (i) The online shop uses n different data types, i.e., heterogeneous data, and needs n different neural subnetworks. An overall decision is obtained by concatenating the hidden layers at a suitable positions, see Fig. 1. (ii) Second, the online shop decides to just use one data class, i.e., homogeneous data, and therefore has just one neural subnetwork in Fig. 1. However, important information can be lost from the other data classes. (iii) Third, it is possible to define suitable neural subnetworks for $n > 1$ data classes, train them separately and then save their weights. These n trained neural subnetworks can be concatenated and trained again with the entire data, using the already trained weights and biases as initial values. This approach is therefore a combination of the two mentioned possibilities above. Thus, $n + 1$ neural subnetworks are obtained in total, with one resulting from the concatenation of the n individual neural subnetworks. The output vector then depends on whether one uses the concatenated network s_{n+1} or one of the neural subnetworks $s_1, ..., s_n$. This third possibility will be relevant for our use case.

4   Explanation Approach

The goal of our approach is to evaluate the explanation of product recommendations of a shop-adapted neural network in order to better understand the decision. Given an input from a user of an online shop and a trained neural network as recommender, TRANSPER performs a backward analysis based on an individual prediction. In this way, it can be explained to what extent components of the trained network or certain inputs were relevant. This process can be seen in Fig. 3. In the following, we will (i) describe how these explanations can be gained with LRP, (ii) specify how to analyze the input with Leave-One-Out method and (iii) define quantity measures to evaluate the explanations.

4.1   Explanation via Layer-Wise Relevance Propagation

Following the notation of Sect. 2 and Definition 1, we assume that $K \in \mathbb{N}$ denotes the number of layers in the neural network, i.e., the first layer is the input layer and the K-th layer is the output layer. Furthermore, for $k \in \{1, ..., K\}$ let $|\Pi_k| = I_k \in \mathbb{N}$ be the number of neurons in the k-th layer, i.e., I_1 describes the number of input neurons and I_K the number of output neurons. Indeed, in the context of classifiers, each neuron of the output layer represents one element of the target set. For example, for an input x, the neuron (K, i^*) with the highest prediction score $f(x)_{i^*}$ as output is the actual recommendation. In this context, it is then of interest to find out to what extent the neurons of the lower layers contributed to the decision $f(x)_{i^*}$. For our approach, we define the initial relevance vector $R^z_{(K,\cdot)} := (R^z_{(K,i)})_{i \in \{1,...,I_K\}}$ with

$$R^z_{(K,i)} = \begin{cases} f(x)_{i^*} & \text{if } i = i^* \\ 0 & \text{otherwise} \end{cases}$$

which can be used to iteratively compute the relevance for layers $K-1, ..., 1$ using Eq. 2 and Eq. 3. Finally, we obtain $R^z_{(1,\cdot)}$ as the input layer's relevance vector and can thus determine to what extent an input neuron is decision-relevant (see Fig. 2). Note that a negative relevance in an input neuron diminishes the prediction i^* whereas a positive relevance underpins it. In contrast to most LRP approaches, we also consider the relevance of the bias $R^b_{(k,j)}$ of the j-th neuron of the $(k+1)$-th layer. Our LRP method is characterized as follows:

$$\sum_{i \in \Pi_k} R^z_{(k,i)} + \sum_{j \in \Pi_{k+1}} R^b_{(k,j)} = \sum_{i \in \Pi_k} \sum_{j \in \Pi_{k+1}} \frac{z_i^k w_{ij}^{k,k+1}}{\left(\sum_{i \in \Pi_k} z_i^k w_{ij}^{k,k+1}\right) + b_j^k} R^z_{(k+1,j)}$$

$$+ \sum_{j \in \Pi_{k+1}} \frac{b_j^k}{\left(\sum_{i \in \Pi_k} z_i^k w_{ij}^{k,k+1}\right) + b_j^k} R^z_{(k+1,j)}$$

$$= \sum_{j \in \Pi_{k+1}} \frac{\sum_{i \in \Pi_k} z_i^k w_{ij}^{k,k+1}}{\left(\sum_{i \in \Pi_k} z_i^k w_{ij}^{k,k+1}\right) + b_j^k} R^z_{(k+1,j)}$$

$$+ \sum_{j \in \Pi_{k+1}} \frac{b_j^k}{\left(\sum_{i \in \Pi_k} z_i^k w_{ij}^{k,k+1}\right) + b_j^k} R^z_{(k+1,j)}$$

$$= \sum_{j \in \Pi_{k+1}} \frac{\left(\sum_{i \in \Pi_k} z_i^k w_{ij}^{k,k+1}\right) + b_j^k}{\left(\sum_{i \in \Pi_k} z_i^k w_{ij}^{k,k+1}\right) + b_j^k} R^z_{(k+1,j)}$$

$$= \sum_{j \in \Pi_{k+1}} R^z_{(k+1,j)}.$$

As $f(x)_{i^*} = \sum_{j \in \Pi_K} R^z_{(K,j)}$ is satisfied by assumption, we obtain

$$
\begin{aligned}
f(x)_{i^*} &= \sum_{i \in \Pi_{K-1}} R^z_{(K-1,i)} + \sum_{j \in \Pi_K} R^b_{(K-1,j)} \\
&= \sum_{i \in \Pi_{K-2}} R^z_{(K-2,i)} + \sum_{j \in \Pi_{K-1}} R^b_{(K-2,j)} + \sum_{j \in \Pi_K} R^b_{(K-1,j)} \\
&= \cdots \\
&= \underbrace{\sum_{i \in \Pi_1} R^z_{(1,i)}}_{=:R^z} + \underbrace{\sum_{k=1}^{K-1} \sum_{j \in \Pi_{k+1}} R^b_{(k,j)}}_{=:R^b},
\end{aligned} \tag{6}
$$

i.e., the sum of the final relevancies R^z and R^b equals the original output score. By comparing the two summands in Eq. 6, the LRP algorithm also provides a method to find out how much relevance R^z, R^b can be assigned to the input neurons and the trained bias, respectively.

4.2   Input Analysis with Leave-One-Out Method

In this section, we want to find out why well-functioning recommenders actually work and provide new insights into the customers' shopping behavior. Additionally, we want to know why an insufficiently functioning recommender delivers meaningless predictions. Therefore, we need to further analyze the explanations gained from LRP regarding their helpfulness, i.e., the impact of an input on the prediction. Using the Leave-One-Out method [26], we evaluate the input relating to the explanations. By consistently leaving one product out by setting its input value to zero, we can observe its effect on the predictions and explanations, see Fig. 4. Assuming a trained neural network, we perform the following steps:

(i) We start with a particular customer and the associated input x which is mapped to an output vector y via the trained network.

(ii) According to Eq. 6, for a given output neuron y_{i^*} with $i^* \in \{1,..,I_K\}$ (e.g., the one with the highest prediction score), we compute the associated input relevancies $(R^z_{(1,j)})_{j \in \{1,...,I_1\}}$ and the overall relevance of the bias R^b.
Thus, we consider the set of relevancies $R := \{R^b\} \cup \{R^z_{(1,j)} : 1 \le j \le I_1\}$.

(iii) For a salient subset of the relevancies $R^* \subset R$ (e.g., the inputs with the highest/lowest relevancies), we set the associated input neurons (marked red in Fig. 4) in x to 0 and obtain the adapted input vector x^*.

(iv) As in Step (i), we map the input x^* to the corresponding output y^* via the same trained network and obtain the test output y^*.

Thus, with steps (i)–(iv), we obtain the input vectors x and x^*, the output vectors y and y^*, and the set of relevancies R. They are used in Sect. 4.3 to enable the explainability of neural network predictions according to the online shop in Definition 1.

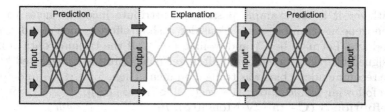

Fig. 4. Selection and analysis of the most relevant inputs via LRP

4.3   Explanation Quantity Measures

Methods such as A/B testing exist to test the performance of a recommender system [6,9]. They aim at evaluating the predictions trained on a fixed group of customers with new test customers. Ideally, positive feedback on the training process is obtained. However, the results can be unsatisfactory as well. In both cases, it is of interest to know how the predictions come about and how certain inputs influence them specifically. Using Eq. 6 and the definitions

$$R_+^z := \sum_{i \in \Pi_1} \max\{0, R_{(1,i)}^z\}, \quad R_-^z := \sum_{i \in \Pi_1} \min\{0, R_{(1,i)}^z\},$$

we obtain the network's top prediction within the setting of Definition 1

$$y_{i^*} := f_c(\pi_c)_{i^*} = R^z + R^b = R_+^z + R_-^z + R^b. \tag{7}$$

In the following we consider two disjoint subsets $C_1, C_2 \subset C$. C_1 represents a set of customers where the inconsistencies to be analysed occur. In contrast, this is not the case for customers from C_2. With Eq. 7, it is then possible to define measures that can be used to analyse such irregularities in specific test cases. W.l.o.g we always assume for the output value $y_{i^*} > 0$. Based on these considerations, we define three measures to quantify the relevance of the input.

(i) **Definition 2 (Individuality Measure).** $\sigma_T : C \to \mathbb{R}$ *with*

$$\sigma_T(c) := \frac{R^z}{R^z + R^b} = \frac{R^z}{y_{i^*}}.$$

The individuality measure can be used to determine to what extent the input was relevant for the decision. Via Eq. 7, we obtain $1 = R^z/y_{i^*} + R^b/y_{i^*}$ and define that a prediction y_{i^*} is *maximally individual*, if $\sigma_T(c) = 1$ holds. In contrast, y_{i^*} is considered to be *minimally individual*, if $\sigma_T(c) = 0$ holds. In this case only the bias was relevant. For $\sigma_T(c) \in (0,1)$ we generally have $R^z, R^b > 0$, so both of these components contribute positively to y_{i^*}. If R^z or R_b are negative, this component argues against prediction y_{i^*} and we either have $\sigma_T(c) \in (-\infty, 0)$ or $\sigma_T(c) \in (1, \infty)$. Note that due to $y_{i^*} > 0$ it can not occur that R^z and R_b are negative.

With σ_T it is for example possible to attribute inconsistencies to overly homogeneous training data. Consider a shop offering men's and women's products. Let men be C_1 and women be C_2. If the training data is largely assigned to men, women could often get men's products suggested because the recommender's bias was trained on men. Then, for $c_1 \in C_1$ and $c_2 \in C_2$, the following would apply: $|1 - \sigma_T(c_1)| < |1 - \sigma_T(c_2)|$.

(ii) **Definition 3 (Certainty Measure).** $\nu_T : C \to (0,1]$ *with*

$$\nu_T(c) := \begin{cases} R^z/R^z_+, & if \ R^z > 0 \\ R^z/R^z_-, & if \ R^z < 0. \end{cases}$$

The certainty measure can be used to make a quantitative statement about the deviation of the individual relevancies from the overall relevance. Considering definitions of R^z_+, R^z_-, and Eq. 7, we have $R^z_+ \in [R^z, \infty)$ and $R^z_- \in (-\infty, R^z]$. Depending on the sign of R^z, one can determine whether the input neurons as a whole had a positive or negative relevance for the decision made. We restrict ourselves to the case of $R^z > 0$. However, the results apply to $R^z < 0$, respectively. Thus, we can deduce that a value of $\nu_t(c) = 1$ means that no negative relevancies were assigned to the input neurons. A value close to zero, on the other hand, indicates a strong dispersion of the relevancies.

(iii) **Definition 4 (Diversity Measure).** $\zeta_T, \zeta_T^+, \zeta_T^- : C \to [0, \infty)$ *with*

$$\zeta_T(c) := \max_{r \in \mathcal{R}} \left| \frac{r - \mu_{\mathcal{R}}^r}{\mu_{\mathcal{R}}^r} \right| \quad and \quad \mu_{\mathcal{R}}^r := \frac{1}{|\mathcal{R}| - 1} \sum_{r' \in \mathcal{R} \setminus \{r\}} r',$$

$$\zeta_T^+(c) := \max_{r \in \mathcal{R}_+} \left| \frac{r - \mu_{\mathcal{R}_+}^r}{\mu_{\mathcal{R}_+}^r} \right| \quad and \quad \mu_{\mathcal{R}_+}^r := \frac{1}{|\mathcal{R}_+| - 1} \sum_{r' \in \mathcal{R}_+ \setminus \{r\}} r',$$

$$\zeta_T^-(c) := \max_{r \in \mathcal{R}_-} \left| \frac{r - \mu_{\mathcal{R}_-}^r}{\mu_{\mathcal{R}_-}^r} \right| \quad and \quad \mu_{\mathcal{R}_-}^r := \frac{1}{|\mathcal{R}_-| - 1} \sum_{r' \in \mathcal{R}_- \setminus \{r\}} r'$$

for a customer $c \in C$ and top prediction y_{i^*}. We additionally introduce the set of input relevancies $\mathcal{R} := R^z_{(1,\cdot)}$, which we divide as follows:

$$\mathcal{R}_0 := \{r \in \mathcal{R} : r = 0\}, \ \mathcal{R}_+ := \{r \in \mathcal{R} : r > 0\}, and \ \mathcal{R}_- := \{r \in \mathcal{R} : r < 0\}.$$

The diversity measure finds outliers within certain input relevancies. For example, considering $r \in \mathcal{R}$, then $(r - \mu_{\mathcal{R}}^r)/\mu_{\mathcal{R}}^r$ is the proportional deviation between the values in $\mathcal{R}$ except for r. For $r \in \mathcal{R}_+$ or $r \in \mathcal{R}_-$ one proceeds analogously. Note that the calculation of diversity measures does not apply to empty sets $\mathcal{R}, \mathcal{R}_+$, and $\mathcal{R}_-$, respectively. Furthermore, the zero is always obtained for one-element sets. For two customers c_1, c_2 with $\zeta_T^+(c_1) \ll \zeta_T^+(c_2)$, we can thus state that the prediction for c_2 depends more on a single input neuron than the prediction for c_1.

5 Evaluation

In this section, we demonstrate the benefits and application of our approach in three use cases. First, our explanation approach can help in understanding fluctuations in the recommender's quality. Second, TRANSPER can help in finding ideas on how to improve the recommender. Third, our contribution can help to improve the understanding of the customer base. In the course of this research, we kindly received permission from the e-commerce service provider econda [8] and two of its partner companies to use their customer data. These partner companies are a jewellery shop and an interior design shop.

5.1 Evaluation Setting

At this point, we show that both online shops fit the formal model from Definition 1 and are thus applicable to the TRANSPER framework. We assume that T^1 is the jewellery shop and T^2 the interior design shop. As shortly mentioned in Sect. 3, the neural network econda uses for T^1 and T^2 comply with the third neural network type with three neural subnetworks s_1, s_2, s_3 in Fig. 1.

Online Shop Models. We now illustrate how the shops satisfy Definition 1:

a) Both shops provide anonymized information about a variety of their customers $\widetilde{C}^1 \subseteq C^1, \widetilde{C}^2 \subseteq C^2$, for example shopping history,

b) and their offered products P^1, P^2.

c) The targets P^*, in our use case a subset of selected products of the offered products, define the real output space Φ^1 and Φ^2, respectively.

d) The available customer information types are based on the information sets Λ^1 and Λ^2, respectively.

e) The information from Λ^1 (Λ^2) is classified according to its characteristic properties. In our case, the disjoint subsets are the same for both shops, i.e., $\Lambda^* = \Lambda^{1*} = \Lambda^{2*}$. Especially, T^1 and T^2 have three disjunctive information types, i.e., $|\Lambda^*| = 3$, which result in three neural subnetworks s_1, s_2, s_3.

f) According to Λ^*, any customer c has therefore the associated input space denoted by $\Omega_c = \mathbb{R}^{m_1} \times \mathbb{R}^{m_2} \times \mathbb{R}^{m_3(c)}$. The first two neural subnetworks s_1, s_2 have a fixed number of input neurons independent of the customer, so in a slight abuse of notation we write m_1 and m_2 instead of $m_1(c)$ and $m_2(c)$, respectively. The third subnetwork has a number of neurons dependent on the number of interactions of c.

g) Via preprocessing, the information about a user $c \in C$ is converted into an input $\omega_c \in \Omega_c$.

h) The function f_c represents the recommender's implicit process of decision making. Given an input ω_c, the vector $f_c(\omega_c)$ contains an entry for each product in P^* and the product with the corresponding highest prediction score is recommended.

The neural networks are trained in two steps, respectively. First, the neural subnetworks s_1, s_2, s_3 are trained independently. Based on the trained weights and biases, the subnetworks are concatenated according to Fig. 1 in their hidden layers and trained again to obtain the combined decision function f_c. This also means, each of the subnetworks s_1, s_2, s_3 individually fits Definition 1 and processes the following information types which we will further analyze in Sect. 5.3. (i) s_1 processes information regarding general interactions, whereby the input vector is an embedding of a user profile. For example, an input neuron can represent the purchase of a certain product or interest in a product category. This neural subnetwork is designed as a multi-layer perceptron. (ii) s_2 processes personal information not related to former product interactions. A multi-layer perceptron is used as well. (iii) s_3 processes the most recent customer interactions as sequences, whose lengths may be different for each customer. An action performed by a user is embedded and considered as a part of the interaction sequence. An RNN approach with Gated Recurrent Unit layers is used here.

5.2 Evaluation Data Set

The data set used in this work consists of the online shops T^1 and T^2 as instantiations of the model from Definition 1. For each online shop, the corresponding recommender is provided in the form of a trained neural network. Furthermore, we receive the profile stream, which contains the user information about the customers which were previously considered as training and test data. econda updates the respective recommender at regular time intervals based on current purchasing behaviour. Therefore, the data set used includes several profile streams and recommenders per online shop. In total, we use 8 (10) profile streams for T^1 (T^2). A profile stream contains on average 524 (1004) customers and per customer we have on average 33 (64) customer interactions. All recommenders were realised in Python 3.7 with Tensorflow v2.1.0.

5.3 Evaluation Results

In Sect. 2, we have defined the standard LRP method. However, there are also variants of this methods which outperform the standard on some architectures. To the best of our knowledge, it is not known which of these methods works best for RNNs. As a preliminary step, we therefore fill in this gap by evaluating the performance of the standard LRP and some of its most popular variants using our algorithm from Sect. 4.2. As a reference, we switch off each input neuron once at a time to find the neuron that is actually most relevant to the decision. This is the case, when the change of the original prediction value is maximal by leaving out this specific input. Finally, per LRP variant, we determine the relative frequency with respect to detecting the most relevant input neuron. Regarding the mentioned LRP methods, we first consider all possible parameter combinations with respect to the values $0.01, 0.1, 1, 5, 10$, and then choose the best combination. We obtained the scores *standard* [2] **0.9800**, *epsilon* [1] 0.9560, *gamma* [15] 0.9080, *alpha-beta* [16] 0.7720, and *non-negative* [16] 0.5040. Based

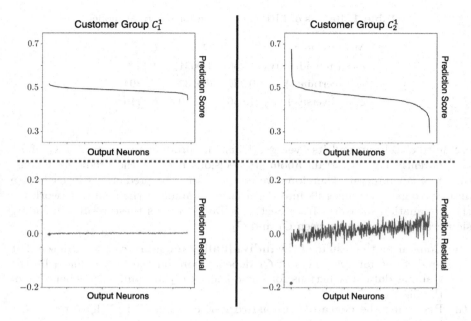

Fig. 5. Two exemplary output layers for shop T^1. Output vectors of the NN ranked in descending order for customer groups C_1 and C_2 in the upper part including corresponding residual plots after setting the most relevant input neuron to zero in the lower part. The residual of the original top prediction is marked red. (Color figure online)

on these and due to the fact that the standard method achieved a hit rate of 100% in the case of MLPs, we will limit ourselves to this method. In the following, we describe three use cases that can be achieved with our explanation quantity measures defined in Sect. 4.3.

Understanding the Recommendation Quality. To tackle this, we have to examine discrepancies between prediction and input. We found one within the predictions provided by econda for the jewellery shop T^1 that could not be explained intuitively. Therefore, we apply the measures from Sect. 4.3 to obtain explanations regarding the recommender's decisions. The upper part of Fig. 5 shows two exemplary output layers of the neural subnetwork s_1, where $C_1, C_2 \subset C^1$ are disjoint subsets of customers C^1 of T^1. The exemplary customers were each randomly selected from 25 customers in C_1 and 29 customers in C_2, respectively. The output neurons are ranked in descending order regarding their prediction score. It can be seen that the preferred outputs for customers from C_1 are almost indistinguishable. In contrast, the scores for customers from C_2 imply clear top predictions. Considering the lower part of Fig. 5, we plot the residuals after setting the most relevant input neuron to zero to show the discrepancy. For customers from C_1, the discrepancy between the top prediction and the average prediction score is much smaller than for customers from C_2 because the entire curve hov-

Table 1. Results of LRP-comparison for recurrent model

Measure\user	C_1	C_2	C^1
σ_{T^1} (individuality)	1.2668	1.0021	1.1409
ν_{T^1} (certainty)	0.7302	0.9733	0.8804
$\zeta_{T^1}^{+}$ (diversity)	1.5564	143.1009	65.7103

ers quite closely around its average. Thus, the product recommender s_1 of T^1 is apparently not as certain about its decisions because the predictions range over a small interval. Therefore, we consider the top predictions in each case and try to gain new insights into the decision-making of the neural network via the explanation measures from Sect. 4.3. Table 1 shows these results including significant differences between C_1 and C_2:

(i) Comparing the results of the **individuality measure** σ_{T^1}, we can see that predictions for customers of C_1 depend more on the bias induced by the training data. Predictions for customers of C_2 are almost independent of the bias.

(ii) Regarding the **certainty measure** ν_{T^1}, customers of C_1 have more contradictory input neurons with negative relevance.

(iii) Since we are interested in the positive influence of input neurons on the overall decision, we consider the **diversity measure** $\zeta_{T^1}^{+}$. We can see the greatest divergence between customers of the two classes C_1 and C_2. Regarding the inputs with positive relevance, customers of C_2 have an input with a relevance that is significantly greater than the other relevancies. This means that there are inputs that speak in favour of the decision made which is not the case for customers from C_1.

All three measures reveal differences between the two customer groups. The diversity measure stands out particularly prominently. The key figures listed here reflect a well explainable prediction of the recommender for customers from C_2. This means that few input neurons had the strongest influence on the prediction made which is not the case for customers from C_1. This discrepancy can also be seen very well if we switch off the input with the highest relevance and plot the residuals of the output vectors, see the lower part of Fig. 5. The input with highest relevance is marked red. It has a significantly stronger influence on the prediction for customers from C_2 than C_1. For the latter, switching off this input causes almost no deviation in the predictions. Using the LRP approach and the explanatory measures, it has thus been possible to establish that the clear predictions for customers from C_2 are quite simple to explain. Namely, these customers have activated input neurons that contribute massively to the prediction made. For the customers from C_1 on the other hand, the decision-making is rather based on the entire interaction of the input neurons.

Ideas to Improve the Recommender. A closer look at the most relevant inputs reveals a certain pattern. We have two different types of input neurons: (a) input neurons representing the interaction with a product from P^* and (b) input neurons representing an interaction with a certain product category. In the latter case, an interaction with a category can only take place via an interaction with a product from the associated category. The activation of the categories occurs for each product interaction, regardless of whether or not it is contained in P^*. Now, when looking at the input relevancies for customers from C_1 or C_2, the following is noticeable: Firstly, for customers from C_1 there are no activations of products. The most relevant inputs are therefore categories and the relevancies hardly differ. Secondly, customers from C_2 always have product activations. In these cases, the most relevant input is always a neuron belonging to a product interaction and these relevancies are significantly higher than those of the likewise activated categories. We were thus able to determine that the activation of products as input neurons leads to more unambiguous decision-making. In particular, these represent a better explanatory power as the neural network predicter can identify certain information that significantly influenced the decision made. It would therefore make sense to separate the user information even further and define the products or categories as separate subnetworks. In this way, the decision-making process for user profiles that only contain categories as input neurons could be given a stronger explanatory power.

Understanding the Customer Base. We also performed an evaluation on the interior design shop T^2. Our diversity measures σ_{T^2} and $\zeta^+_{T^2}$ revealed that the trained bias and outliers within the positive input relevancies of the neural subnetwork s_2 were particularly relevant for the decisions made. Thus, it was found that buying interest is based on daily trends rather than past interactions. Unfortunately, we cannot explain this in more detail here due to space constraints.

6   Conclusion

In this paper, we have presented TRANSPER, an explanation framework for neural networks used in online shopping.

We used the LRP method to define three explanation measures, namely the individuality measure, used to determine those parts of the input that are particularly relevant for the decision; the certainty measure, which measures how certain the system is about its prediction; and the diversity measure, which measures whether there are clear top predictions. These measures can be defined on complex neural networks which process heterogeneous input data.

We have demonstrated the usefulness of our metrics in three explanation use cases. First, we explained fluctuations in the prediction qualities. Second, TRANSPER explanations can help find ideas on how to improve the neural network. Third, our explanations can help online shops better understand their customer base. These explanations also play an important role in fulfilling legal requirements such as the ones mandated by GDPR.

References

1. Ancona, M., Ceolini, E., et al.: Towards better understanding of gradient-based attribution methods for Deep Neural Networks. In: ICLR (2018)
2. Bach, S., Binder, A., et al.: On pixel-wise explanations for non-linear classifier decisions by layer-wise relevance propagation. PLoS ONE **10**(7) (2015)
3. Bettaney, E.M., Hardwick, S.R., et al.: Fashion outfit generation for e-commerce. In: eCom@SIGIR. CEUR, vol. 2410 (2019)
4. Bharadhwaj, H.: Layer-wise relevance propagation for explainable deep learning based speech recognition. In: IEEE ISSPIT, pp. 168–174 (2018)
5. Borisov, A., Markov, I., et al.: A neural click model for web search. In: WWW Conference, pp. 531–541 (2016)
6. Chen, M., Liu, P.: Performance evaluation of recommender systems. Int. J. Performability Eng. **13**, 1246 (2017)
7. Chen, W., Huang, P., et al.: POG: personalized outfit generation for fashion recommendation at Alibaba iFashion. In: ACM SIGKDD, pp. 2662–2670 (2019)
8. econda: Personalization & Analytics. https://www.econda.de. Accessed 29 Mar 2021
9. Gebremeskel, G.G., de Vries, A.P.: Recommender systems evaluations: offline, online, time and A/A test. In: CLEF. CEUR, vol. 1609, pp. 642–656 (2016)
10. General Data Protection Regulation: Art. 12 GDPR. https://gdpr-info.eu/art-12-gdpr/. Accessed 29 Mar 2021
11. Khoali, M., Tali, A., et al.: Advanced recommendation systems through deep learning. In: NISS, pp. 51:1–51:8 (2020)
12. Laenen, K., Moens, M.: A comparative study of outfit recommendation methods with a focus on attention-based fusion. Inf. Process. Manag. **57**(6), 102316 (2020)
13. Li, Y., Cao, L., et al.: Mining fashion outfit composition using an end-to-end deep learning approach on set data. IEEE Trans. Multim. **19**(8), 1946–1955 (2017)
14. Loureiro, A.L.D., Miguéis, V.L., et al.: Exploring the use of deep neural networks for sales forecasting in fashion retail. Decis. Support Syst. **114**, 81–93 (2018)
15. Montavon, G., Binder, A., Lapuschkin, S., Samek, W., Müller, K.-R.: Layer-wise relevance propagation: an overview. In: Samek, W., Montavon, G., Vedaldi, A., Hansen, L.K., Müller, K.-R. (eds.) Explainable AI: Interpreting, Explaining and Visualizing Deep Learning. LNCS (LNAI), vol. 11700, pp. 193–209. Springer, Cham (2019). https://doi.org/10.1007/978-3-030-28954-6_10
16. Montavon, G., Samek, W., et al.: Methods for interpreting and understanding deep neural networks. Digital Signal Process. **73**, 1–15 (2018)
17. Nelaturi, N., Devi, G.: A product recommendation model based on recurrent neural network. Journal Européen des Systèmes Automatisés **52**, 501–507 (2019)
18. Nguyen, A., Oberföll, A., Färber, M.: Right for the right reasons: making image classification intuitively explainable. In: Hiemstra, D., Moens, M.-F., Mothe, J., Perego, R., Potthast, M., Sebastiani, F. (eds.) ECIR 2021. LNCS, vol. 12657, pp. 327–333. Springer, Cham (2021). https://doi.org/10.1007/978-3-030-72240-1_32
19. Noh, H., Hong, S., et al.: Learning deconvolution network for semantic segmentation. In: IEEE ICCV, pp. 1520–1528 (2015)
20. Park, S.: Neural networks and customer grouping in e-commerce: a framework using fuzzy ART. In: AIWoRC, pp. 331–336 (2000)
21. Rumelhart, D.E., Hinton, G.E., et al.: Learning representations by back-propagating errors. Nature **323**(6088), 533–536 (1986)

22. Shrikumar, A., Greenside, P., et al.: Learning important features through propagating activation differences. In: ICML, vol. 70, pp. 3145–3153 (2017)
23. Simonyan, K., Vedaldi, A., et al.: Deep inside convolutional networks: visualising image classification models and saliency maps. In: ICLR (2014)
24. Springenberg, J.T., Dosovitskiy, A., et al.: Striving for simplicity: the all convolutional net. In: ICLR (2015)
25. Sundararajan, M., Taly, A., et al.: Axiomatic attribution for deep networks. In: ICML, vol. 70, pp. 3319–3328 (2017)
26. Yuan, J., Li, Y.-M., Liu, C.-L., Zha, X.F.: Leave-one-out cross-validation based model selection for manifold regularization. In: Zhang, L., Lu, B.-L., Kwok, J. (eds.) ISNN 2010. LNCS, vol. 6063, pp. 457–464. Springer, Heidelberg (2010). https://doi.org/10.1007/978-3-642-13278-0_59

Natural Language Processing

Balancing Speed and Accuracy in Neural-Enhanced Phonetic Name Matching

Philip Blair$^{(\boxtimes)}$ⒾⒹ, Carmel EliavⒾⒹ, Fiona HasanajⒾⒹ, and Kfir BarⒾⒹ

Basis Technology, 1060 Broadway, Somerville, MA 02144, USA
{pblair,carmel,fiona,kfir}@basistech.com
https://www.basistech.com

Abstract. Automatic co-text free name matching has a variety of important real-world applications, ranging from fiscal compliance to border control. Name matching systems use a variety of engines to compare two names for similarity, with one of the most critical being phonetic name similarity. In this work, we re-frame existing work on neural sequence-to-sequence transliteration such that it can be applied to name matching. Subsequently, for performance reasons, we then build upon this work to utilize an alternative, non-recurrent neural encoder module. This ultimately yields a model which is 63% faster while still maintaining a 16% improvement in averaged precision over our baseline model.

Keywords: Name matching · Transliteration · Natural language processing · Sequence-to-sequence · Multilingual · Performance

1 Introduction

Names are an integral part of human life. From people to organizations and beyond, understanding what things are called is a critical aspect of natural language processing. A significant challenge in many applications is the fact that systems of appellation vary widely across cultures and languages, meaning that it can be challenging for a human, let alone a machine, to determine that two names are equivalent. Automatic evaluation of pairs of names has many important real-world applications, ranging from border control to financial know-your-customer ("KYC") compliance, which both involve searching for a name inside of large databases.

Computerized name matching systems attempt to accomplish this through a variety of statistical measurements that compare different properties of the given names. These systems are built with the goal of assigning a score to a pair of names that reflects the likelihood that those two names are "equivalent." For example, a name matching system should assign a high score to the input pair "Nick" and "Nicholas". Similarly, it ideally is able to competently handle a variety of other name-specific phenomena, such as missing components (e.g. "Franklin D. Roosevelt" and "Franklin Roosevelt"), initialisms (e.g. "J. J. Smith" and "John Joseph

© Springer Nature Switzerland AG 2021
Y. Dong et al. (Eds.): ECML PKDD 2021, LNAI 12979, pp. 271–284, 2021.
https://doi.org/10.1007/978-3-030-86517-7_17

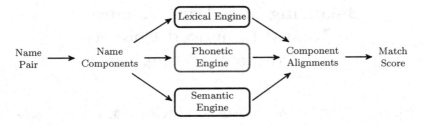

Fig. 1. Architecture of an enterprise name matching system.

Smith"), transliteration variations (e.g. "Abdul Rasheed" and "Abd al-Rashid"), different scripts (e.g. "Caesar" and "シーザー" (*Shīzā*)), and so on. In principle, this is accomplished via partitioning a full name into components (i.e. turning "John Smith" into ["John", "Smith"]) and comparing the components of each name via a variety of engines. Using those comparisons, the components are aligned appropriately and combined into a single name match score (Fig. 1).

One engine that is particularly important is phonetic similarity, which measures how close the pronunciation of the names are to one another. For example, a phonetic name engine would assign a low similarity to the input pair ("John", "J."), but it would assign a high similarity to the input pair ("Chris", "クリス" (*kurisu*)). This could be powered by a number of different technologies, ranging from Soundex-based [35] indexing to statistical modeling techniques. In this paper, we focus on this phonetic-based engine and how it can be improved with neural techniques.

1.1 Challenges

When dealing with name matching across different languages, there is often a difference in writing scripts, which presents unique challenges. For some writing systems this is not particularly difficult, whereas with others (such as with the Latin alphabet and Japanese syllabaries), it is more challenging. This is because there is not a one-to-one correspondence in the characters used for each alphabet, meaning that it can be difficult to transliterate Latin alphabet-based languages into Japanese. For example, the word "photo" would be transliterated as "フォト" (*foto*). Not only is this transliteration two characters shorter than the English word, but the interactions between different parts of the English word inform the transliteration in a non-trivial fashion. Specifically, the small "オ " (*o*) character can only be used when forming a digraph with another character (in this case, the "フ " (*fu*) character).

Thus, a statistical name matcher must take into account a nontrivial scope of contextual information when doing name matching. In our work, we explore the relationship between name matching and the related task of name *transliteration*, which seeks to produce the corresponding transliteration of a name in one language, given that name in another language.

Moreover, as our objective is to deploy a system in a production environment, speed becomes a concern. While recurrent neural networks are extremely powerful tools for sequence modeling, they incur a significant amount of over-head in comparison to non-neural based techniques. To combat this, we explore an alternative, non-recurrent architecture in search of a model which balances the improved modeling capacity of neural networks with the superior speed of non-neural graphical models.

To summarize, our contributions in this work are the following: (1) we utilize prior work on neural transliteration in order to perform name matching, (2) we address performance issues surrounding the deployment of neural network-based name matching systems in an industrial context, and (3) we do so by suggesting an alternative neural architecture for name transliteration.

2   Related Work

Approaches to matching names across languages using non-neural machine learn-ing techniques have a long history in literature. A variety of existing work opts for a cost-based approach to the problem [1,2,25,31], which computes various similarity metrics between a given pair of names. For example, when doing mono-lingual name matching, thresholding a simple Levenshtein distance [21] may be a sufficient approach. While much of this cost-based work has focused around languages with large amounts of written variation, such as Arabic [1,2], only a smaller amount of it has focused on cross-orthographic name matching [25]. The challenge with applying cost-based algorithms to cross-script name matching is that it often relies, at some level, on normalization techniques such as romaniza-tion or Soundex-based [35] indexing, which can introduce noise into the name matching process.

A closely related problem to name matching in NLP research is that of entity linking. The primary differentiating aspect in this scenario is the presence of co-text; that is, the problem is that of linking an entity mentioned in a larger piece of text to an entity present in some larger knowledge base. While the scope of our research could certainly enhance real-world applications of these systems in order to grapple with representations of entities in unseen scripts, research in this area [19,24,41] typically (implicitly or otherwise) assumes that the entity in the document and the knowledge base have been written in the same script. This is why the task is sometimes referred to as "named-entity disambiguation," as the focus tends to be more centered around leveraging co-text in order to disambiguate amongst a list of candidate entities.

Name matching is a subset of the broader problems of entity matching and record linkage. Neural networks have been applied to this area [11,23,30], but this prior work does not specifically focus on the name matching problem. Research in this area often emphasizes the structured nature of records (e.g. attempting to correlate many distinct schemas together) so a direct comparison to this work is made difficult. Further complicating the issue is the fact that record linkage research is, to the authors' knowledge, always an end-to-end process which operates on *full* names, not specific name components. This work focuses on the integration of neural-based techniques into an *existing*, larger enterprise name matching system, so a meaningful direct comparison to the performance of our specific sub-engines is challenging if not impossible.

Direct statistical modeling of name matching has been published as well. Most similar to what we discuss in our baseline system outlined Sect. 3 is Nabende et al.'s work [27] on using Pair-HMMs [10] to model cross-lingual name matching and transliteration. Moreover, there is a body of work that has applied deep neural networks to the task of name matching [20,43]; however, these do so by directly training a discriminative model that classifies names as matching or not. In contrast, our work is based on solving this problem by modeling the corresponding generative distribution, which can yield benefits ranging from lower data requirements [29] to lower amounts of supervision (we need only collect pairs of known transliterations, without needing to collect known non-transliterations).

As discussed in Sect. 3, modeling name transliteration is an important step in our system's process, so it is apt to review prior art in this domain as well. Neural name transliteration based on sequence-to-sequence models has been described in a limited amount of prior work [14]. Furthermore, there has been further research on neural techniques for transliteration in general [26,34,38]. Beyond this, there have been a variety of non-neural approaches to transliteration, ranging from systems based on conditional random fields [9] to local classification of grapheme clusters [33,36], statistical machine translation techniques [6,32,40], and modeling transliteration as a mixture distribution [22].

In this work, we take this existing work on transliteration and re-frame it such that it can be applied to name matching. We then, for performance reasons, build upon this work to utilize an alternative, non-recurrent neural encoder module.

3   Phonetic Name Matching Systems

In this work, we explore the capabilities of an enterprise statistical phonetic name matching engine, with a focus on matching names across English and Japanese. More specifically, we will be taking names written in the Standard Latin script [16] and comparing equivalent names written in the Japanese Katakana script, which is typically used for foreign names. This is an interesting problem, as we must deal with different scripts that are not one-to-one with each other (for example, the two Latin letters "na" correspond to the single Katakana character "ナ"). Our goal is to have a system that can assign a high score to a pair of names that are pronounced similarly, and a low score to name pairs that are not. One idea for English-Japanese name matching is to develop a probabilistic model describing the likelihood of one name being a transliteration of another; this probability would then directly reflect the same semantics that we are trying to capture with the aforementioned score. More concretely, for a given English name n_{en} and Japanese name n_{ja}, this can be modeled probabilistically as $P_{ja-en}(n_{en}|n_{ja})$ (or vice versa, via an appropriate P_{en-ja} model). Per the chain rule, one approach to learning this conditional distribution is by modeling the following fully-generative distribution:

$$P(n_{en}, n_{ja}) = P_{ja-en}(n_{en}|n_{ja})P(n_{ja}) = P_{en-ja}(n_{ja}|n_{en})P(n_{en}). \qquad (1)$$

Semantically, this has an interesting interpretation: If we can fit a generative model that allows us to transliterate (phonetically) a name from one Japanese to English, we are then able to use this to determine the quality of a given potential transliteration. In other words, by probabilistically modeling the task of name transliteration, we are able to directly model name matching by reading these probabilities.

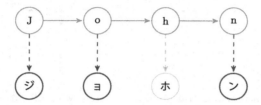

Fig. 2. A Hidden Markov Model-based phonetic name matching system.

For our baseline system, we make use of a Hidden Markov Model [4], as this directly models this distribution. Our model operates on sequences of characters, with the character sequence in one language being modeled as the hidden state sequence and the character sequence in the other language being modeled as emissions. A simplified version of this process is shown in Fig. 2. In order to compensate for scripts that are not one-to-one (such as Latin and Katakana), we extend the character alphabet to contain a closed set of digraphs (such as "na", shown above), which are discovered automatically via Expectation Maximization (EM) [8]. For our neural model, we would like to model name matching in a largely similar fashion to our HMM-based technique. To this end, we approach the problem by first developing a neural network that models the process of English-Japanese name transliteration and then use the probability distributions computed by this model to facilitate name matching. Our name transliteration model was inspired by work done on neural machine translation [42]. In this setup, we utilize a sequence-to-sequence architecture [37], translating from a "source" domain of English name character sequences to a "target" domain of the corresponding character sequences for the Japanese transliteration. We explore two variants of this architecture: In the first, seq2seq-LSTM, the encoder module of our model is implemented using a Long Short-Term Memory (LSTM)-based [15], while the second, seq2seq-CNN, uses an encoder based on a CNN [7,12]. For both of these architectures, we use an LSTM-based decoder module.

3.1 Neural Name Transliteration

Our first approach to neural name transliteration was directly based on the architecture used in Sutskever et al.'s work [37]. We used a bidirectional LSTM encoder module, with the output being fed into an LSTM-based decoder module, augmented with a basic attention mechanism [3], to produce the Japanese transliteration. This architecture, illustrated in Fig. 3, is referred to as seq2seq-LSTM. This architecture is effectively the same as used for general transliteration by Rosca and Breuel [34].

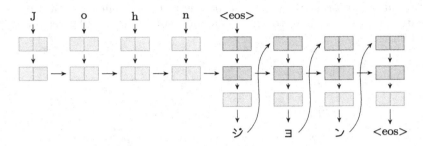

Fig. 3. A seq2seq-based name transliteration system with LSTM-based encoder and decoder (unidirectional encoder shown).

As discussed in Sect. 4, we observed this architecture's throughput to be too low for our use case. We hypothesized that this slowdown was largely due to two aspects of the seq2seq-LSTM network: (1) the expense of computing the attention step at each point of the decoding process and (2) the lack of parallelizability of the recurrent architecture used in the encoder and decoder modules. When developing our second model, known as seq2seq-CNN, we addressed the former by simply eschewing the attention mechanism. Additionally, drawing from character encoding successes utilizing convolutional neural networks (CNNs) [7,12] in other text applications [5,17], this second architecture uses a CNN for the encoder module, based on work done by Gehring et al. [13].

As in Gehring et al.'s work, we utilize a series of CNN kernels that span the full character embedding dimension and process different-sized windows of characters. In contrast to their work, we run the channels through a max-pooling layer, providing us with an efficiently-computable, fixed-size representation of the full sequence (with one dimension for every kernel used). This representation is then passed to an LSTM decoder in the same way that the encoder representation was in the seq2seq-LSTM network.

3.2   Neural Name Matching

Running the network for name *matching* is very similar to running it for transliteration, but, instead of using the probability distribution produced by the decoder module at each time step to determine the next Japanese character to produce, we simply use the *known* Japanese character sequence to select it. As is traditionally done when working with language models, we then take the observed probabilities assigned to this known character sequence in order to compute the perplexity

$$PP(n_{ja}|n_{en}) = 2^{H(n_{en}, n_{ja})},$$ (2)

where

$$H(n_{en}, n_{ja}) = -\frac{1}{N} \sum_{i=1}^{N} \log P(n_{ja}|n_{en})[i],$$

$$P(n_{ja}|n_{en})[i] = P_{en-ja}(n_{ja,i}|n_{en}; n_{ja,1...i-1}),$$

$n_{ja,k}$ is the kth character of n_{ja}, and N is the length of the target Japanese character sequence. We note that, in this mode of operation, we feed in the next character from the target sequence at each time step (i.e., we will keep going, even if the decoder would have predicted <eos> at a given time step). Now, conceptually, when comparing a name with a list of potential names, we would like to assign the highest score to the one that is *closest* to what the model predicts. Therefore, we define the reciprocal of the perplexity as the scoring function:

$$S(n_{en}, n_{ja}) = (PP(n_{ja}|n_{en}))^{-1}.$$ (3)

4   Experimental Results

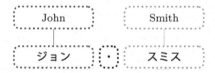

Fig. 4. Illustration of the token alignment procedure used for data generation between "John Smith" and "ジョン・スミス" (*Jon Sumisu*). Note that the "·" character is treated as a special token, and ignored during the matching procedure.

4.1   Training and Hyperparameters

We train our system with a list of 33,215 component pairs; these were produced by collecting aligned English-Japanese full name pairs from Wikipedia, tokenizing them, and recording the aligned tokens in both scripts (throwing away items with non-Katakana names and pairs with differing numbers of tokens), as shown in Fig. 4. This was possible due to the regularity of the dataset we collected and the nature of English-Japanese transliterations; however, to be safe, a bilingual Japanese-English speaker reviewed 6% of the data and confirmed that the process had produced properly-aligned word pairs.

This training set was divided into a 90% train and 10% validation split. All neural networks were trained with the Adam optimizer [18]. We trained the model for 100 epochs. The LSTM was trained with a single layer (for both the encoder and decoder), using a dropout rate of 0.5 on the decoder outputs, an embedding size of 60 dimensions, and a hidden layer size of 100. Our CNN-based system utilizes six two-dimensional convolutional blocks with 100 output channels and kernel sizes of 2, 3, 4, 5, 6, and 7. The input embeddings (of dimension 60) are summed with trained position embeddings (up to a maximum length of 23, as seen in the training data) and passed through a dropout layer with a dropout rate of 0.25. These inputs are then passed in parallel to each of the convolutional blocks with ReLU activations [28], whose outputs are then max-pooled into six 100-dimensional vectors and concatenated. This is then passed to a linear layer to reduce it to a 200-dimensional vector, which is concatenated to each decoder input. The LSTM decoder is otherwise the same as the model using the LSTM-based encoder, except for the fact that it uses a hidden size of 100.

The primary purpose of this component-level engine is to exist as the keystone of our *full name* matching system. As such, we felt it would be most appropriate to measure its utility in that context for our evaluation. Rather than evaluating on another split-off piece of our word-level training data, we evaluate our models on a separate Wikipedia-sourced test set of 60,706 full name pairs. We measure Averaged Precision (AP) by calculating a match score between two full names using different underlying phonetic engines. Additionally, to simulate real-world usage in an information retrieval system, we use them to query a database of 14,941 names with a list of 500 names, and re-score the resulting name matches, measuring the Mean Averaged Precision (MAP), as is common practice.

4.2 Results

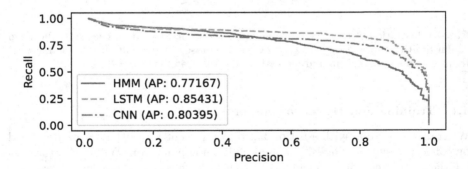

Fig. 5. Precision-Recall trade-off of different algorithms at various thresholds.

The models described in this work are now used to serve clients in real-life scenarios. Specifically, one client provided us with a list of 98 names that were incorrectly matched using the baseline HMM algorithm, so we began to explore neural techniques for name matching in order to reduce these false positives.

When looking at this list of names, we determined that the mistakes were likely caused by the HMM overly biasing towards frequencies in the training data, so an improvement would reflect an algorithm which better incorporates context. The first algorithm we tried was the seq2seq-LSTM architecture.

Table 1. Selected examples of name pair scores using different phonetic engines.

Name 1	Name 2	HMM-based	LSTM-based	CNN-based
Ada Lovelace	エイダ・ラヴレス (*Eida Raburesu*)	0.48	**0.74**	0.63
Albert Schweitzer	アルベルト・シュバイツエル (*Aruberuto Shubaitseru*)	0.69	**0.86**	0.81
Christopher Marlowe	クリストファー・マーロー (*Kurisutofā Mārō*)	0.76	**0.95**	0.83
Easy Goer[a]	イージーゴーアー (*Ījīgōā*)	0.39	0.66	**0.82**
James Whale	ジェームズ・ホエール (*Jēmuzu Hoēru*)	0.62	0.80	**0.88**
Alexandre Trauner	トラウネル・シャーンドル (*Torauneru Shāndoru*)	**0.84**	0.49	0.49
Allez France[a]	アレフランセ (*Arefuranse*)	**0.80**	0.46	0.46
U Nu	ウ・ヌー (*U Nū*)	**0.96**	0.93	0.68

[a]These are the names of racehorses.

Table 2. Accuracy and speed results.

Matching engine	MAP	AP	Speed (sec/1000 tokens)	Slowdown
HMM	63.61	77.17	**0.76**	**1x**
seq2seq-LSTM	**69.47**	**85.43**	16.3	21.26x
seq2seq-CNN	66.69	80.40	5.9	7.7x

Qualitatively, we found that the neural models are able to reduce false positives in our matching software on inputs that require more nuanced interpretation and composition of contextual information. This is empirically supported by the data in Fig. 5, which shows a larger area under the precision-recall curve for our neural models than our baseline HMM model. A manual inspection of test cases, seen in Table 1, with large differences in scores from the baseline gives some insight into the strengths and weaknesses of the neural-based matcher. The primary strength appears to be improved performance on particularly difficult transliterations. There are relatively few examples of the neural models doing markedly worse than the HMM-based model, but they broadly appear to fall under two categories. First, the neural engines seem to handle transliterated nicknames poorly (e.g. "Alexandre" and "シャーンドル" (*Shāndoru*); the latter being a transliteration of "Sandra"). Second, for extremely infrequent characters, such as ヌ (*nu*) (which appears in only around 1% of the training data),

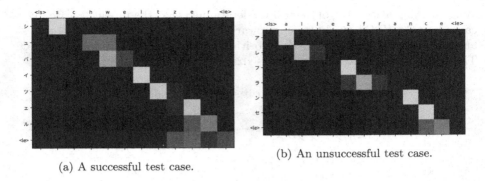

(a) A successful test case.

(b) An unsuccessful test case.

Fig. 6. Selected attention heatmaps from our LSTM-based name matcher architecture when measuring perplexity. In subfigure (b), the original input (as shown in Table 1) has two English tokens, but the name matching framework attempts to concatenate them when evaluating whether they match the given Japanese token.

the neural models are less confident than the HMM. We hypothesize that this could also be compounded by the high amount of variance in how ヌ is transliterated (for example, the "ne" suffix on words such as "neptune" and "arsene" is often transliterated as ヌ), which could lead our model unable to learn a high-probability transliteration for this type of character. Finally, as a sanity-check, we inspect the attention heatmaps produced by the LSTM-based sequence-to-sequence models, shown in Fig. 6. We can see that these indicate which sequences are not in line with what the model would expect to see: in subfigure (a), which shows an input pair that the model correctly assigns a high score to, we see that the primary activations align neatly with the corresponding Katakana characters in a manner similar to what one would intuitively expect. Conversely, in subfigure (b), which shows an input pair which was incorrectly assigned a low score, we see that the French phonology is the source of the confusion: at the position of the "フ" (*fu*) in the Katakana sequence, the attention is focused on the "z" in the Latin sequence. This indicates that the model is not expecting to treat the "z" as silent when performing the transliteration, meaning that it is surprised by the fact that the Katakana sequence is already on the "france" portion of the Latin name.

Quantitatively, as demonstrated in Table 2 and Fig. 5, our initial seq2seq-LSTM architecture led to a dramatic improvement in accuracy. One will also notice the downside of such an approach: a roughly 21x speed degradation. While it is expected that a neural algorithm will incur some amount of additional performance overhead, the scale of data processed by our customers made this level of slowdown too great for us to accept. This is what led us to take a look at alternative seq2seq approaches, which are more efficiently computable thanks to improved parallelization. After developing the seq2seq-CNN network, we found that it lied directly in this sought-after "sweet spot" of having improved evaluation set performance over the HMM without the dramatic slowdown that the LSTM suffers from or sacrificing the qualitative improvements we observed with the seq2seq-LSTM model.

5 Conclusion and Future Work

Our work demonstrates that neural machine translation techniques can be applied to the problem of name matching, with notable success over a traditional graphical-based technique. Moreover, we show that modifying the encoder module of this neural network to use a convolutional neural network yields a significant speed improvement with only a moderate sacrifice in accuracy (while still outperforming the baseline algorithm).

There are a number of future steps we wish to explore with this work. The primary bottleneck remaining in our system's speed is the decoder. Due to the recurrent nature of our LSTM decoder, the perplexity of an input name component must be computed in $\mathcal{O}(n)$ time, character-by-character. We wish to remove this limitation by exploring alternative encoder architectures which predict the output sequence simultaneously, such as the work in [13]. Another natural extension of our experiments would be to additionally try a Transformer-based [39] encoder module, as one would expect this to yield speed improvements over the LSTM without the same degree of sacrifice in accuracy. Finally, this work explores how we can apply neural networks to a phonetic name matching engine, but we would like to explore what opportunities exist in other types of name matching engines.

References

1. Al-Hagree, S., Al-Sanabani, M., Alalayah, K.M., Hadwan, M.: Designing an accurate and efficient algorithm for matching Arabic names. In: 2019 First International Conference of Intelligent Computing and Engineering (ICOICE), pp. 1–12 (2019).https://doi.org/10.1109/ICOICE48418.2019.9035184
2. Al-Hagree, S., Al-Sanabani, M., Hadwan, M., Al-Hagery, M.A.: An improved n-gram distance for names matching. In: 2019 First International Conference of Intelligent Computing and Engineering (ICOICE), pp. 1–7 (2019). https://doi.org/10.1109/ICOICE48418.2019.9035154
3. Bahdanau, D., Cho, K., Bengio, Y.: Neural machine translation by jointly learning to align and translate. In: Bengio, Y., LeCun, Y. (eds.) 3rd International Conference on Learning Representations, ICLR 2015, San Diego, CA, USA, 7–9 May, 2015, Conference Track Proceedings (2015), http://arxiv.org/abs/1409.0473
4. Baum, L.E., Petrie, T.: Statistical inference for probabilistic functions of finite state Markov chains. Ann. Math. Statist. **37**(6), 1554–1563 (1966). https://doi.org/10.1214/aoms/1177699147
5. Belinkov, Y., Durrani, N., Dalvi, F., Sajjad, H., Glass, J.: What do neural machine translation models learn about morphology? In: Proceedings of the 55th Annual Meeting of the Association for Computational Linguistics (Volume 1: Long Papers), pp. 861–872. Association for Computational Linguistics, Vancouver, Canada, July 2017. https://doi.org/10.18653/v1/P17-1080. https://www.aclweb.org/anthology/P17-1080

6. Chen, Y., Skiena, S.: False-friend detection and entity matching via unsupervised transliteration. CoRR abs/1611.06722 (2016). http://arxiv.org/abs/1611.06722
7. Cun, Y.L., et al.: Handwritten digit recognition with a back-propagation network. In: Advances in Neural Information Processing Systems 2, pp. 396–404. Morgan Kaufmann Publishers Inc., San Francisco (1990)
8. Dempster, A.P., Laird, N.M., Rubin, D.B.: Maximum likelihood from incomplete data via the EM algorithm. J. Roy. Stat. Soc. Series B (Methodological) **39**(1), 1–38 (1977), http://www.jstor.org/stable/2984875
9. Dhore, M., Shantanu, K., Sonwalkar, T.: Hindi to English machine transliteration of named entities using conditional random fields. Int. J. Comput. Appl. **48**, July 2012. https://doi.org/10.5120/7522-0624
10. Durbin, R., Eddy, S.R., Krogh, A., Mitchison, G.J.: Biological Sequence Analysis: Probabilistic Models of Proteins and Nucleic Acids. Cambridge University Press, Cambridge (1998)
11. Ebraheem, M., Thirumuruganathan, S., Joty, S., Ouzzani, M., Tang, N.: Distributed representations of tuples for entity resolution. Proc. VLDB Endow. **11**(11), 1454–1467 (2018). https://doi.org/10.14778/3236187.3236198
12. Fukushima, K.: Neocognitron: a self-organizing neural network model for a mechanism of pattern recognition unaffected by shift in position. Biol Cybern **36**(4), 193–202 (1980). https://doi.org/10.1007/bf00344251, https://doi.org/10.14778/3236187.3236198
13. Gehring, J., Auli, M., Grangier, D., Yarats, D., Dauphin, Y.N.: Convolutional sequence to sequence learning. In: Proceedings of the 34th International Conference on Machine Learning - Volume 70, ICML 2017, pp. 1243–1252. JMLR.org (2017)
14. Gong, J., Newman, B.: English-Chinese name machine transliteration using search and neural network models (2018)
15. Hochreiter, S., Schmidhuber, J.: Long short-term memory. Neural Comput. **9**(8), 1735–1780 (1997)
16. ISO: ISO Standard 646, 7-Bit Coded Character Set for Information Processing Interchange. International Organization for Standardization, second edn. (1983). http://www.iso.ch/cate/d4777.html, also available as ECMA-6
17. Kim, Y., Jernite, Y., Sontag, D., Rush, A.M.: Character-aware neural language models. In: Proceedings of the Thirtieth AAAI Conference on Artificial Intelligence, AAAI 2016, pp. 2741–2749. AAAI Press (2016)
18. Kingma, D.P., Ba, J.: Adam: a method for stochastic optimization. In: Bengio, Y., LeCun, Y. (eds.) 3rd International Conference on Learning Representations, ICLR 2015, San Diego, CA, USA, 7–9 May, 2015, Conference Track Proceedings (2015). http://arxiv.org/abs/1412.6980
19. Kolitsas, N., Ganea, O.E., Hofmann, T.: End-to-end neural entity linking. In: Proceedings of the 22nd Conference on Computational Natural Language Learning. pp. 519–529. Association for Computational Linguistics, Brussels, Belgium, October 2018. https://doi.org/10.18653/v1/K18-1050, https://www.aclweb.org/anthology/K18-1050
20. Lee, C., Cheon, J., Kim, J., Kim, T., Kang, I.: Verification of transliteration pairs using distance LSTM-CNN with layer normalization. In: Annual Conference on Human and Language Technology, pp. 76–81. Human and Language Technology (2017)
21. Levenshtein, V.I.: Binary codes capable of correcting deletions, insertions and reversals. Soviet Physics Doklady **10**, 707 (1966)

22. Li, T., Zhao, T., Finch, A., Zhang, C.: A tightly-coupled unsupervised clustering and bilingual alignment model for transliteration. In: Proceedings of the 51st Annual Meeting of the Association for Computational Linguistics (Volume 2: Short Papers), pp. 393–398. Association for Computational Linguistics, Sofia, Bulgaria, August 2013. https://www.aclweb.org/anthology/P13-2070
23. Li, Y., Li, J., Suhara, Y., Doan, A., Tan, W.C.: Deep entity matching with pretrained language models. Proc. VLDB Endow. **14**(1), 50–60 (2020). https://doi.org/10.14778/3421424.3421431. https://doi.org/10.14778/3421424.3421431
24. Martins, P.H., Marinho, Z., Martins, A.F.T.: Joint learning of named entity recognition and entity linking. In: Proceedings of the 57th Annual Meeting of the Association for Computational Linguistics: Student Research Workshop, pp. 190–196. Association for Computational Linguistics, Florence, Italy, July 2019. https://doi.org/10.18653/v1/P19-2026, https://www.aclweb.org/anthology/P19-2026
25. Medhat, D., Hassan, A., Salama, C.: A hybrid cross-language name matching technique using novel modified Levenshtein distance. In: 2015 Tenth International Conference on Computer Engineering Systems (ICCES), pp. 204–209 (2015). https://doi.org/10.1109/ICCES.2015.7393046
26. Merhav, Y., Ash, S.: Design challenges in named entity transliteration. In: Proceedings of the 27th International Conference on Computational Linguistics, pp. 630–640 (2018)
27. Nabende, P., Tiedemann, J., Nerbonne, J.: Pair hidden Markov model for named entity matching. In: Sobh, T. (ed.) Innovations and Advances in Computer Sciences and Engineering, pp. 497–502. Springer, Netherlands (2010)
28. Nair, V., Hinton, G.E.: Rectified linear units improve restricted Boltzmann machines. In: ICML (2010)
29. Ng, A.Y., Jordan, M.I.: On discriminative vs. generative classifiers: a comparison of logistic regression and naive Bayes. In: Proceedings of the 14th International Conference on Neural Information Processing Systems: Natural and Synthetic, NIPS 2001, pp. 841–848. MIT Press, Cambridge (2001)
30. Nie, H., et al.: Deep sequence-to-sequence entity matching for heterogeneous entity resolution. In: Proceedings of the 28th ACM International Conference on Information and Knowledge Management, CIKM 2019, pp. 629–638. Association for Computing Machinery, New York (2019). https://doi.org/10.1145/3357384.3358018
31. Peled, O., Fire, M., Rokach, L., Elovici, Y.: Matching entities across online social networks. Neurocomputing **210**, 91–106 (2016)
32. Priyadarshani, H., Rajapaksha, M., Ranasinghe, M., Sarveswaran, K., Dias, G.: Statistical machine learning for transliteration: Transliterating names between Sinhala, Tamil and English. In: 2019 International Conference on Asian Language Processing (IALP), pp. 244–249 (2019). https://doi.org/10.1109/IALP48816.2019.9037651
33. Qu, W.: English-Chinese name transliteration by latent analogy. In: Proceedings of the 2013 International Conference on Computational and Information Sciences, ICCIS 2013, pp. 575–578. IEEE Computer Society, USA (2013). https://doi.org/10.1109/ICCIS.2013.159
34. Rosca, M., Breuel, T.: Sequence-to-sequence neural network models for transliteration. arXiv preprint arXiv:1610.09565 (2016)
35. Russell, R.C.: Index (April 1918), US Patent 1,261,167
36. Sarkar, K., Chatterjee, S.: Bengali-to-english forward and backward machine transliteration using support vector machines. In: Mandal, J.K., Dutta, P., Mukhopadhyay, S. (eds.) CICBA 2017. CCIS, vol. 776, pp. 552–566. Springer, Singapore (2017). https://doi.org/10.1007/978-981-10-6430-2_43

37. Sutskever, I., Vinyals, O., Le, Q.V.: Sequence to sequence learning with neural networks. In: Proceedings of the 27th International Conference on Neural Information Processing Systems - Volume 2, NIPS 2014, pp. 3104–3112. MIT Press, Cambridge (2014)

38. Upadhyay, S., Kodner, J., Roth, D.: Bootstrapping transliteration with constrained discovery for low-resource languages. In: Proceedings of the 2018 Conference on Empirical Methods in Natural Language Processing, pp. 501–511. Association for Computational Linguistics, Brussels, Belgium, October–November 2018. https://doi.org/10.18653/v1/D18-1046, https://www.aclweb.org/anthology/D18-1046

39. Vaswani, A., et al.: Attention is all you need. In: Proceedings of the 31st International Conference on Neural Information Processing Systems, NIPS 2017, pp. 6000–6010. Curran Associates Inc., Red Hook (2017)

40. Wang, D., Xu, J., Chen, Y., Zhang, Y.: Monolingual corpora based Japanese-Chinese translation extraction for kana names. J. Chinese Inf. Process. **29**(5), 11 (2015)

41. Wu, L., Petroni, F., Josifoski, M., Riedel, S., Zettlemoyer, L.: Scalable zero-shot entity linking with dense entity retrieval. In: Proceedings of the 2020 Conference on Empirical Methods in Natural Language Processing (EMNLP), pp. 6397–6407. Association for Computational Linguistics, Online, November 2020. https://doi.org/10.18653/v1/2020.emnlp-main.519

42. Wu, Y., et al.: Google's neural machine translation system: Bridging the gap between human and machine translation. CoRR abs/1609.08144 (2016). http://arxiv.org/abs/1609.08144

43. Yamani, Z., Nurmaini, S., Firdaus, R, M.N., Sari, W.K.: Author matching using string similarities and deep neural networks. In: Proceedings of the Sriwijaya International Conference on Information Technology and Its Applications (SICONIAN 2019), pp. 474–479. Atlantis Press (2020). https://doi.org/10.2991/aisr.k.200424.073

Robust Learning for Text Classification with Multi-source Noise Simulation and Hard Example Mining

Guowei Xu, Wenbiao Ding[✉], Weiping Fu, Zhongqin Wu, and Zitao Liu

TAL Education Group, Beijing, China
{xuguowei,dingwenbiao,fuweiping1,wuzhongqin,liuzitao}@tal.com

Abstract. Many real-world applications involve the use of Optical Character Recognition (OCR) engines to transform handwritten images into transcripts on which downstream Natural Language Processing (NLP) models are applied. In this process, OCR engines may introduce errors and inputs to downstream NLP models become noisy. Despite that pre-trained models achieve state-of-the-art performance in many NLP benchmarks, we prove that they are not robust to noisy texts generated by real OCR engines. This greatly limits the application of NLP models in real-world scenarios. In order to improve model performance on noisy OCR transcripts, it is natural to train the NLP model on labelled noisy texts. However, in most cases there are only labelled clean texts. Since there is no handwritten pictures corresponding to the text, it is impossible to directly use the recognition model to obtain noisy labelled data. Human resources can be employed to copy texts and take pictures, but it is extremely expensive considering the size of data for model training. Consequently, we are interested in making NLP models intrinsically robust to OCR errors in a low resource manner. We propose a novel robust training framework which 1) employs simple but effective methods to directly simulate natural OCR noises from clean texts and 2) iteratively mines the hard examples from a large number of simulated samples for optimal performance. 3) To make our model learn noise-invariant representations, a stability loss is employed. Experiments on three real-world datasets show that the proposed framework boosts the robustness of pre-trained models by a large margin. We believe that this work can greatly promote the application of NLP models in actual scenarios, although the algorithm we use is simple and straightforward. We make our codes and three datasets publicly available (https://github.com/tal-ai/Robust-learning-MSSHEM).

Keywords: Robust representation · Text mining

1 Introduction

With the help of deep learning models, significant advances have been made in different NLP tasks. In recent years, pre-trained models such as BERT [4] and its

© Springer Nature Switzerland AG 2021
Y. Dong et al. (Eds.): ECML PKDD 2021, LNAI 12979, pp. 285–301, 2021.
https://doi.org/10.1007/978-3-030-86517-7_18

variants achieved state-of-the-art performance in many NLP benchmarks. While human being can easily process noisy texts that contain typos, misspellings, and the complete omission of letters when reading [13], most NLP systems fail when processing corrupted or noisy texts [2]. It is not intuitive, however, if pre-trained NLP models are robust under noisy text setting.

There are several scenarios in which noise could be generated. The first type is user-generated noise. Typos and misspellings are the major ones and they are commonly introduced when users input texts through keyboards. Some other user-generated noise includes incorrect use of tense, singular and plural, etc. The second type of noise is machine-generated. A typical example is in the essay grading system [18]. Students upload images of handwritten essays to the grader system in which OCR engines transform images to structured texts. In this process, noise is introduced in texts and it can make downstream NLP models fail. We argue that the distribution of user-generated errors is different from that of OCR errors. For example, people often mistype characters that are close to each other on the keyboards, or make grammatical mistakes such as incorrect tense, singular and plural. However, OCR is likely to misrecognize similar handwritten words such as "dog" and "dag", but it it unlikely to make mistakes that are common for humans.

There are many existing works [15,16] on how to improve model performance when there are user-generated noises in inputs. [15] studied the character distribution on the keyboard to simulate real user-generated texts for BERT. [16] employed masked language models to denoise the input so that model performance on downstream task improves. Another existing line of work focuses on adversarial training, which refers to applying a small perturbation on the model input to craft an adversarial example, ideally imperceptible by humans, and causes the model to make an incorrect prediction [6]. It is believed that model trained on adversarial data is more robust than model trained on clean texts. However, adversarial attack focuses on the weakness in NLP models but does not consider the distribution of OCR errors, so the generated sample is not close to natural OCR transcripts, making adversarial training less effective in our problem.

Despite that NLP models are downstream of OCR engines in many real-world applications, there are few works on how to make NLP models intrinsically robust to natural OCR errors. In this paper, we discuss how the performance of pre-trained models degrades on natural OCR transcripts in text classification and how can we improve its robustness on the downstream task. We propose a novel robust learning framework that largely boosts the performance of pre-trained models when evaluated on both noise-free data and natural OCR transcripts in text classification task. We believe that this work can greatly promote the application of NLP models in actual noise scenarios, although the algorithm we use is simple and straightforward. Our contributions are:

- We propose three simple but effective methods, rule-based, model-based and attack-based simulation, to generate natural OCR noises.

– In order to combine the noise simulation methods, we propose a hard example mining algorithm so that the model focuses more on hard samples in each epoch of training. We define hard examples as those whose representations are quite different between noise-free inputs and noisy inputs. This ensures that the model learns more robust representations compared to naively treating all simulated samples equally.
– We evaluate the framework on three real-world datasets and prove that the proposed framework outperforms existing robust training approaches by a large margin.
– We make our code and data publicly available. To the best of our knowledge, we are the first to evaluate model robustness on OCR transcripts generated by real-world OCR engines.

2   Related Work

2.1   Noise Reduction

An existing approach to deal with noisy inputs is to introduce some denoising modules into the system. Grammatical Error Correction (GEC) systems have been widely used to address this problem. Simple rule-based and frequency-based spell-checker [12] are limited to complex language systems. More recently, modern neural GEC systems are developed with the help of deep learning [3,23]. Despite that neural GEC achieves SOTA performance, there are at least two problems with using GEC as a denoising module to alleviate the impact of OCR errors. Firstly, it requires a massive amount of parallel data, e.g., [17] to train a neural GEC model, which is expensive to acquire in many scenarios. Secondly, GEC systems can only correct user-generated typos, misspellings and grammatical errors, but the distribution of these errors is quite different from that of OCR errors, making GEC limited as a denoiser. For example, people often mistype characters that are close to each other on the keyboards, or make grammatical mistakes such as tense, singular and plural. However, OCR is likely to misrecognize similar handwritten words such as "dog" and "dag", but it is unlikely to make mistakes that are common for humans. Another line of research focuses on how to use language models [22] as the denoising module. [16] proposed to use masked language models in an off-the-shelf manner. Although this approach does not rely on massive amount of parallel data, it still oversimplifies the problem by not considering OCR error distributions. More importantly, we are interested in boosting intrinsic model robustness. In other words, if we directly feed noisy data into the classification model, it should be able to handle it without relying on extra denoising modules. However, both GEC and language model approaches are actually pre-processing modules, and they do not improve the intrinsic robustness of downstream NLP models. Therefore, we do not experiment on denoising modules in this paper.

2.2 Adversarial Training

Adversarial attack aims to break down neural models by adding imperceptible perturbations on the input. Adversarial training [10,21] improves the robustness of neural networks by training models on adversarial samples. There are two types of adversarial attacks, the white-box attack [5] and the black-box attack [1,24]. The former assumes access to the model parameters when generating adversarial samples while the latter can only observe model outputs given attacked samples. Recently, there are plenty of works on attacking NLP models. [14] found that NLP models often make different predictions for texts that are semantically similar, they summarized simple replacement rules from these semantically similar texts and re-trained NLP models by augmenting training data to address this problem. [15] proved that BERT is not robust to typos and misspellings and re-trained it with nature adversarial samples. Although it has been proved that adversarial training is effective to improve the robustness of neural networks, it searches for weak spots of neural networks but does not consider common OCR errors in data augmentation. Therefore, traditional adversarial training is limited in our problem.

2.3 Training with Noisy Data

Recent work has proved that training with noisy data can boost NLP model performance to some extent. [2] pointed out that a character-based CNN trained on noisy data can learn robust representations to handle multiple kinds of noise. [9] created noisy data using random character swaps, substitutions, insertions and deletions and improved model performance in machine translation under permuted inputs. [11] simulated noisy texts using a confusion matrix and employed a stability loss when training models on both clean and noisy samples.

In this paper, our robust training framework follows the same idea to train models with both clean and noisy data. The differences are that our multi-source noise simulation can generate more natural OCR noises and using hard example mining algorithm together with stability loss can produce optimal performance.

3 Problem

3.1 Notation

In order to distinguish noise-free texts, natural handwritten OCR transcripts and simulated OCR transcripts, we denote them by $\mathcal{X}$, $\mathcal{X}'$ and $\widetilde{\mathcal{X}}$ respectively. Let $\mathcal{Y}$ denote the shared labels.

3.2 Text Classification

Text classification is one of the most common NLP tasks and can be used to evaluate the performance of NLP models. Text classification is the assignment of documents to a fixed number of semantic categories. Each document can be in multiple or exactly one category or no category at all. More formally, let $\mathbf{x} = (\mathbf{w_0}, \mathbf{w_1}, \mathbf{w_2}, \cdots, \mathbf{w_n})$ denote a sequence of tokens and $\mathbf{y} = (\mathbf{y_0}, \mathbf{y_1}, \cdots, \mathbf{y}_m)$ denote the fixed number of semantic categories. The goal is to learn a probabilistic function that takes $\mathbf{x}$ as input and outputs the probability distribution over $\mathbf{y}$. Without loss of generality, we only study the binary text classification problem under noisy setting in this work.

3.3   A Practical Scenario

In the context of supervised machine learning, we assume that in most scenarios, we only have access to labelled noise-free texts. There are two reasons. Firstly, most open-sourced labelled data do not consider OCR noises. Secondly, manual labelling usually also labels clean texts, and does not consider OCR noise. One reason is that annotating noisy texts is difficult or ambiguous. Another reason is that labelling becomes subject to changes in OCR recognition. For different OCR, we need to repeat the labelling multiple times.

In order to boost the performance of model when applied on OCR transcripts, we can train or finetune the model on labelled noisy data. Then the question becomes how to transform labelled noise-free texts into labelled noisy texts. Due to the fact that labelled texts do not come with corresponding images, it is impossible to call OCR engines and obtain natural OCR transcripts. Human resources can be employed to copy texts and take pictures, but it is extremely expensive considering the size of data for model training. Then the core question is how to inject natural OCR noise into labelled texts efficiently.

3.4   OCR Noise Simulation

When OCR engine transforms images into texts, we can think of it as a noise induction process. Let $\mathbf{I}$ denote a handwritten image, $\mathbf{x}$ denotes the text content on image $\mathbf{I}$, OCR would transform the noise-free text $\mathbf{x}$ into its noisy copy $\mathbf{x}'$.

The problem is then defined as modeling a noise induction function $\widetilde{\mathcal{X}} = \mathcal{F}(\mathcal{X}, \theta)$ where θ is the function parameters and $\mathcal{X}$ is a collection of noise-free texts. A good simulation function makes sure that the simulated $\widetilde{\mathcal{X}}$ is close to the natural OCR transcripts $\mathcal{X}'$. It should be noted that noise induction should not change the semantic meaning of content so that $\mathcal{X}$, $\mathcal{X}'$ and $\widetilde{\mathcal{X}}$ share the same semantic label in text classification task.

3.5   Robust Training

In this work, we deal with off-line handwritten text recognition. We do not study how to improve the accuracy of recognition, but only use the recognition model as a black box tool. Instead, we are interested in how to make downstream NLP models intrinsically robust to noisy inputs.

Let $\mathcal{M}$ denote a pre-trained model that is finetuned on a noise-free dataset $(\mathcal{X}, \mathcal{Y})$, firstly we investigate how much performance degrades when $\mathcal{M}$ is applied on natural OCR transcripts $\mathcal{X}'$. Secondly, we study on how to finetune $\mathcal{M}$ on simulated noisy datasets $(\widetilde{\mathcal{X}}, \mathcal{Y})$ efficiently to improve its performance on input $\mathcal{X}'$ that contains natural OCR errors.

4 Approach

4.1 OCR Noise Simulation

In this section, we introduce the multi-source noise simulation method.

Rule-Based Simulation. One type of frequent noise introduced by OCR engines is the token level edit. For example, a word that is not clearly written could be mistakenly recognized as other synonymous word, or in even worse case, not recognized at all. In order to synthesize token level natural OCR noise from noise-free texts, we compare and align parallel data of clean and natural OCR transcript pairs $(\mathcal{X}, \mathcal{X}')$ using the Levenshtein distance metric (Levenshtein, 1966). Let $\mathcal{V}$ be the vocabulary of tokens, we then construct a token level confusion matrix $\mathcal{C}_{conf}$ by aligning parallel data and estimating the probability $P(\mathbf{w}'|\mathbf{w})$ with the frequency of replacing token $\mathbf{w}$ to $\mathbf{w}'$, where $\mathbf{w}$ and $\mathbf{w}'$ are both tokens in $\mathcal{V}$. We introduce an additional token ϵ into the vocabulary to model the insertion and deletion operations, the probability of insertion and deletion can then be formulated as $P_{ins}(\mathbf{w}|\epsilon)$ and $P_{del}(\epsilon|\mathbf{w})$ respectively. For every clean sentence $\mathbf{x} = (\mathbf{w_0}, \mathbf{w_1}, \mathbf{w_2}, \cdots, \mathbf{w_n})$, we independently perturb each token in $\mathbf{x}$ with the following procedure, which is proposed by [11]:

- Insert the ϵ token before the first and after every token in sentence $\mathbf{x}$ and acquire an extended version $\mathbf{x}_{ext} = (\epsilon, \mathbf{w_0}, \epsilon, \mathbf{w_1}, \epsilon, \mathbf{w_2}, \epsilon, \cdots, \epsilon, \mathbf{w_n}, \epsilon)$.
- For every token $\mathbf{w}$ in sentence $\mathbf{x}_{ext}$, sample another token from the probability distribution $P(\mathbf{w}'|\mathbf{w})$ to replace $\mathbf{w}$.
- Remove all ϵ tokens from the sentence to obtain the rule-based simulated noisy sentence $\widetilde{\mathbf{x}}$.

Attack-Based Simulation. The attack-based method greedily searches for the weak spots of the input sentence [20] by replacing each word, one at a time, with a "padding" (a zero-valued vector) and examining the changes of output probability. After finding the weak spots, attack-based method replaces the original token with another token. One drawback of greedy attack is that adversarial examples are usually unnatural [7]. In even worse case, the semantic meaning of the original text might change, this makes the simulated text a bad adversarial example. To avoid such problem, we only replace the original token with its synonym. The synonym comes from the confusion matrix $\mathcal{C}_{conf}$ by aligning clean texts and OCR transcripts. This effectively constrains the semantic drifts and makes the simulated texts close to natural OCR transcripts.

Model-Based Simulation. We observe that there are both token level and span level noises in natural OCR transcripts. In span level noises, there are dependencies between the recognition of multiple tokens. For example, a noise-free sentence "乌龟默默想着" (translated as "The tortoise meditated" by Google Translate[1]) is recognised as "乌乌黑黑的箱子" (translated as "Jet black box" by Google Translate). A possible reason is that the mis-recognition of "龟" leads to recognizing "默" into "黑" because "乌黑" is a whole word in Chinese. The rule-based and attack-based simulation mainly focuses on token-level noise where a character or token might be edited. It makes edits independently and does not consider dependency between multiple tokens. As a consequence, both ruled-based and attack-based simulation are not able to synthesize the span level noise.

We proposed to model both token level and span level noise using the encoder-decoder architecture, which is successful in many NLP tasks such as machine translation, grammatical error corrections (GEC) and etc. While a GEC model takes noisy texts as input and generates noise-free sentences, our model-based noise injection model is quite the opposite. During training, we feed parallel data of clean and OCR transcripts $(\mathcal{X}, \mathcal{X}')$ into the injection model so that it can learn the frequent errors that OCR engines will make. During inference, the encoder first encode noise-free text into a fix length representation and the decoder generates token one step a time with possible noise in an auto-regressive manner. This makes sure that both token level and span level noise distribution can be captured by the model. We can use the injection model to synthesize a large number of noisy texts that approximate the natural OCR errors. It should be noted that the injection model is not limited to a certain type of encoder-decoder architecture. In our experiment, we employ a 6-layer vanilla Transformer (base model) as in [19].

4.2 Noise Invariance Representation

[25] pointed out the output instability issues of deep neural networks. They presented a general stability training method to stabilize deep networks against small input distortions that result from various types of common image processing. Inspired by [11,25] adapted the stability training method to the sequence labeling scenario. Here we adapt it to the text classification task. Given the standard task objective $\mathcal{L}_{stand}$, the clean text $\mathbf{x}$, its simulated noisy copy $\widetilde{\mathbf{x}}$ and the shared label $\mathbf{y}$, the stability loss is defined as

$$\mathcal{L} = \alpha * \mathcal{L}_{stand} + (1 - \alpha) * \mathcal{L}_{sim} \tag{1}$$

$$\mathcal{L}_{sim} = Distance(\mathbf{y}(\mathbf{x}), \mathbf{y}(\widetilde{\mathbf{x}})) \tag{2}$$

where $\mathcal{L}_{sim}$ is the distance between model outputs for clean input $\mathbf{x}$ and noisy input $\widetilde{\mathbf{x}}$, α is a hyper-parameter to trade off $\mathcal{L}_{stand}$ and $\mathcal{L}_{sim}$. $\mathcal{L}_{sim}$ is

---

[1] https://translate.google.cn.

expected to be small so that the model is not sensitive to the noise disturbance. This enables the model to obtain robust representation for both clean and noisy input. Specifically, we use Euclidean distance as our distance measure.

4.3 Hard Example Mining

The proposed noise simulation methods could generate quadratic or cubic number of parallel samples compared to the size of original dataset. It is good that we now have sufficient number of training data with noises and labels. Nevertheless, the training process becomes inefficient if we naively treat each simulated sample equally and feed all the samples into the classifier. This makes the training process extremely time-consuming and does not lead to an optimal performance. Consequently, we need a strategy to sample examples from large volumes of data for optimal performance. Ideally, a robust model should learn similar representations for all possible noise-free text $\mathbf{x}$ and its corresponding noisy copy $\widetilde{\mathbf{x}}$. In reality, however, the model can only capture noise-invariance representations for some of the simulated samples, for some other samples, the representations of the clean text and its noisy copy are still quite different. For any given model $\mathcal{M}$, we define a sample $\mathbf{x}$ as a hard example for $\mathcal{M}$ if the representations of $\mathbf{x}$ and $\widetilde{\mathbf{x}}$ are not similar. We believe that at different training iterations, the hard examples are different, and the model should focus more on the hard ones. We propose a hard example mining algorithm that dynamically distinguishes hard and easy samples for each training epoch as follows:

- Step 1. Initialize the classifier by finetuning a pre-trained model on the noise-free training data $\mathcal{D}_{clean} = \{\mathbf{x}_i\}_{i=1,2,...N}$
- Step 2. Generate a large number of simulated noisy texts $\mathcal{D}_{noisy} = \{\widetilde{\mathbf{x}}_i\}_{i=1,2,...M}$ and construct a collection of all training samples $\mathcal{D} = \{\mathcal{D}_{clean}, \mathcal{D}_{noisy}\}$
- Step 3. For each iteration t, we feed training samples $\mathcal{D}$ to the classifier and obtain their representations $\mathcal{E}_t = \{\mathbf{e}_i, \widetilde{\mathbf{e}}_i\}_{i=1,2,...M}$ from classifier.
- Step 4. Calculate the cosine distance of $\mathbf{e}_i$ and $\widetilde{\mathbf{e}}_i$. Rank all the distances, i.e., $Distance = \{cosine(\mathbf{e}_i, \widetilde{\mathbf{e}}_i)\}_{i=1,2,...M}$, and only keep samples with the top largest distance. These are the hard examples and we use $\mathcal{D}_{hard}$ to denote it. We use a hyper-parameter $\beta = |D_{hard}|/M$ to control the number of hard examples.
- Step 5. Train classifier on $\mathcal{D}_t = \{\mathcal{D}_{hard}, \mathcal{D}_{clean}\}$ and update model by minimizing $\mathcal{L} = \alpha * \mathcal{L}_{stand} + (1 - \alpha) * \mathcal{L}_{sim}$

4.4 The Overall Framework

The overall framework is shown in Fig. 1. Let $\mathbf{x}_i$, $i = 0, 1, 2, ...N$ denote the noise-free text, where $\mathbf{x}_i$ is a sequence of tokens, and $\widetilde{\mathbf{x}}_i$ is the simulated noisy copy. $\mathbf{e}_i$ and $\widetilde{\mathbf{e}}_i$ are the model representations for $\mathbf{x}_i$ and $\widetilde{\mathbf{x}}_i$, we calculate the cosine distance between $\mathbf{e}_i$ and $\widetilde{\mathbf{e}}_i$ and select those pairs with largest distance as the hard examples. Then hard examples together with original noise-free data are used to train the model. For each iteration, we select hard examples dynamically.

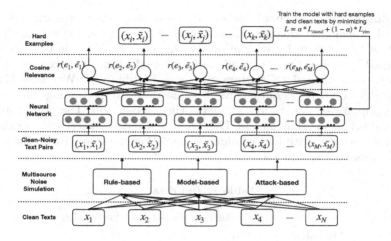

Fig. 1. The overview of the robust training framework.

5 Experiment

5.1 Dataset

We describe three evaluation datasets and the parallel data for training the model-based noise simulation model below.

Test Data. To comprehensively evaluate pre-trained models and the proposed framework, we perform experiments on three real-world text classification datasets, e.g., Metaphor, Personification and Parallelism detection. In each dataset, the task is to assign a positive or negative label to a sentence.

- Metaphor, is a figure of speech that describes an object or action in a way that is not literally true, but helps explain an idea or make a comparison.
- Personification, is a figure of speech when you give an animal or object qualities or abilities that only a human can have.
- Parallelism, is a figure of speech when phrases in a sentence have similar or the same grammatical structure.

In order to obtain the above three datasets, we hired five professional teachers to annotate essays of primary school students. We broke down essays into sentences and each sentence was annotated as one of the three rhetoric or did not belong to any rhetoric. We aggregated crowd-sourced labels into ground-truth label using majority voting. Each task contains over 2000 sentences and the number of positive examples is between 48 to 156. It should be noted that this imbalance is caused by the fact that rhetoric is not so common in students' essays. We simply keep the natural positive and negative sample ratio in the test set for objectiveness. Details about the test data are in Table 1.

OCR Engine and Natural Noise. Different from existing work [11] which evaluated model performance on simulated OCR transcripts, we constructed six real OCR test data for evaluation. We hired over 20 people to write down the original noise-free texts, take pictures and feed images to commercial OCR engines so that natural OCR transcripts can be obtained. We chose Hanvon OCR[2] and TAL OCR[3] as our engines because they are the leading solutions for Chinese primary school student's handwriting recognition. The noise rates are 3.42% and 6.11% for Hanvon and TAL OCR test data respectively. Because we can only experiment with limited number of OCR engines, we discuss the impact of different noise levels in Sect. 6.1.

Table 1. Test data.

Dataset	#sentences	#positives	AvgSentLen
Metaphor	2064	156	37.5
Personification	2059	64	37.6
Parallelism	2063	48	37.5

Parallel Data for Noise Simulation. In order to train the model-based noise simulation model, we collect about 40,000 parallel data[4] of human transcripts and OCR transcripts as our training data. We believe that 40,000 is a reasonable amount to train a high quality model-based noise generator. More importantly, once trained, the model can serve as a general noise generator regardless of specific tasks. In other words, we can use it to quickly convert annotated clean text into annotated noisy text in all sorts of tasks.

5.2   Implementation

For each classification task, we first finetune pre-trained models on noise-free training data $\mathcal{D}_{clean}$, save models with the best validation loss as $\mathcal{M}^*_{clean}$. To perform robust training, we synthesize noisy copies of the original training data and then finetune $\mathcal{M}^*_{clean}$ on both clean and noisy data as denoted by $\mathcal{M}^*_{noisy}$. Both $\mathcal{M}^*_{clean}$ and $\mathcal{M}^*_{noisy}$ are tested on original noise-free test data and noisy copies of the test data.

We implement the framework using PyTorch and train models on Tesla V100 GPUs. We use an opensource release[5] of Chinese BERT and RoBERTa as the pre-trained models. We tune learning rate $\in \{5e^{-8}, 5e^{-7}\}$, batch size $\in \{5, 10\}$, $\alpha \in \{1.0, 0.75, 0.50\}$ where $\alpha = 1.0$ indicates no stability loss is employed. We keep all other hyper-parameters as they are in the release. We report precision, recall and F1 score as performance metrics.

---

[2] https://www.hw99.com/index.php.

[3] https://ai.100tal.com/product/ocr-hr.

[4] Parallel data do not have task specific labels, so they are not used as training data.

[5] https://github.com/ymcui/Chinese-BERT-wwm.

5.3   Results

Robust Training on Simulated Texts. Instead of naively combining multi-source simulation data and finetuning model $\mathcal{M}^*_{clean}$ on it, we employ the hard example mining algorithm in Sect. 4.3 and the stability loss in Sect. 4.2 for robust training. We compare the proposed robust training framework with several strong baselines.

- Random. We randomly select several tokens and make insertion, deletion or substitution edits to generate permuted data. We then combine the permuted and clean data and finetune models on it.
- Noise-aware Training, i.e., NAT [11], noise-aware training for robust neural sequence labeling, which proposes two objectives, namely data augmentation and stability loss, to improve the model robustness in perturbed input.
- TextFooler, [8], a strong baseline to generate adversarial text for robust adversarial training.
- Naively Merge. We finetune $\mathcal{M}^*_{clean}$ on clean and noisy samples generated by all three simulation methods, but without hard example mining and stability loss.

The results are in Table 2. Ours is the proposed robust training framework that finetunes $\mathcal{M}^*_{clean}$ on clean and noisy samples generated by all three simulation methods, together with hard example mining and stability loss. We have the following observations:

- Compared with $\mathcal{M}^*_{clean}$, all robust training approaches, Random, NAT, TextFooler, Naively Merge and our robust training framework (Ours) improve the F1 score on both noise-free test data and OCR test data on all three tasks.
- Compared with Naively Merge, Ours demonstrates improvements in both precision and recall in all test data, which proves that hard example mining and stability loss are vital to the robust training framework.
- When compared with existing baselines, Ours ranks the first place eight times and the second place once out of all nine F1 scores (three tasks, three test data for each task). This proves the advantages of using the proposed robust training framework over existing approaches.

We think of two reasons. Firstly, the proposed noise simulation method generates more natural noisy samples than baselines do. Baselines might introduce plenty of unnatural noisy samples, making precision even lower that of $\mathcal{M}^*_{clean}$. Secondly, hard example mining algorithm enables the model to focus on hard examples whose robust representation has not been learned. NAT and TextFooler finetunes models by naively combing clean and noisy samples.

Table 2. Evaluation results of BERT on metaphor, personification and parallelism.

	Task	Noise-free data			Hanvon OCR			TAL OCR		
		P	R	F1	P	R	F1	P	R	F1
$\mathcal{M}^*_{clean}$	Metaphor	**0.897**	0.833	0.864	0.888	0.814	0.849	**0.886**	0.795	0.838
Random	Metaphor	0.873	**0.885**	0.879	0.864	0.853	0.858	0.868	0.840	0.854
NAT	Metaphor	0.871	0.866	0.868	0.868	0.846	0.857	0.877	0.821	0.848
TextFooler	Metaphor	0.883	0.872	0.877	0.874	0.846	0.860	0.872	0.833	0.852
Naively Merge	Metaphor	0.877	0.872	0.875	0.880	0.846	0.863	0.873	0.833	0.852
Ours	Metaphor	0.890	**0.885**	**0.887**	**0.889**	**0.872**	**0.880**	0.877	**0.865**	**0.871**
$\mathcal{M}^*_{clean}$	Personification	0.855	0.828	0.841	0.868	0.719	0.787	0.825	0.734	0.777
Random	Personification	0.831	**0.844**	0.837	0.814	0.750	0.781	0.842	0.750	0.793
NAT	Personification	0.925	0.766	0.838	0.904	0.734	0.810	0.917	0.688	0.786
TextFooler	Personification	0.831	**0.844**	0.837	0.803	**0.766**	0.784	0.831	**0.766**	0.797
Naively merge	Personification	0.895	0.797	0.843	0.875	0.766	0.817	0.885	0.719	0.793
Ours	Personification	**0.927**	0.797	**0.857**	0.923	0.750	**0.828**	**0.926**	0.734	**0.817**
$\mathcal{M}^*_{clean}$	Parallelism	0.720	0.750	0.735	0.756	0.646	0.697	0.725	0.604	0.659
Random	Parallelism	0.717	**0.792**	0.753	0.714	**0.729**	0.721	0.717	0.688	0.702
NAT	Parallelism	**0.814**	0.729	**0.769**	**0.821**	0.667	0.736	**0.795**	0.646	0.713
TextFooler	Parallelism	0.731	**0.792**	0.760	0.733	0.688	0.710	0.767	0.688	0.725
Naively merge	Parallelism	0.777	0.729	0.753	0.781	0.667	0.719	0.781	0.667	0.719
Ours	Parallelism	0.783	0.750	0.766	0.773	0.708	**0.739**	0.778	**0.729**	**0.753**

6 Analysis

6.1 Naive Training with a Single Noise Simulation Method

We introduce our multi-source noise simulation methods in Sect. 4.1. Using these methods, we can generate a large number of noisy texts from noise-free data. In this section, we evaluate the effectiveness for each method independently. We reload $\mathcal{M}^*_{clean}$ and finetune it combining clean texts and noisy texts generated by a single noise simulation method. At this stage, neither hard example mining nor stability loss is employed. The results of using a single noise simulation method are listed in Tables 3, 4, 5. $\mathcal{M}^*_{clean}$ is finetuned on noise-free data. Rule-based, Model-based and Attack-based are finetuned with a single noise simulation method without hard example mining and stability loss

Firstly, we observe that both recall and F1 score decrease significantly on two noisy test sets compared to performance on noise-free test set. For example, on TAL OCR test set, F1 score of BERT decreases 6.4% and 7.6% for Personification and Parallelism detection and F1 score of RoBERTa decreases 8.5% and 4.0% respectively. This proves that pre-trained models trained on noise-free data are not robust to OCR noises.

Secondly, all three noise simulation methods can improve the F1 scores of BERT and RoBERTa for all three tasks. However, when we naively combine multi-source simulations and finetune models on it ("Naively Merge" in Table 2), the performance does not exceed the effect of using a single noise simulation method. This motivates us to introduce hard example mining and stability loss into the proposed robust training framework.

Table 3. Performance on metaphor detection with a single noise simulation.

Simulation	Model	Noise-free data			Hanvon OCR			TAL OCR		
		P	R	F1	P	R	F1	P	R	F1
$\mathcal{M}^*_{clean}$	BERT	**0.897**	0.833	0.864	**0.888**	0.814	0.849	0.886	0.795	0.838
Rule-based	BERT	0.877	**0.872**	**0.874**	0.874	**0.846**	0.860	0.872	**0.833**	**0.852**
Model-based	BERT	0.882	0.865	0.873	0.885	0.840	**0.862**	0.878	0.827	**0.852**
Attack-based	BERT	0.887	0.859	0.873	0.879	0.840	0.859	**0.894**	0.808	0.849
$\mathcal{M}^*_{clean}$	RoBERTa	0.872	**0.917**	**0.894**	0.862	0.878	0.870	**0.873**	0.878	0.875
Rule-based	RoBERTa	0.836	**0.917**	0.875	0.821	0.910	0.863	0.844	**0.904**	0.873
Model-based	RoBERTa	0.872	**0.917**	**0.894**	0.856	**0.917**	**0.885**	0.859	0.897	**0.878**
Attack-based	RoBERTa	**0.889**	0.872	0.880	**0.879**	0.840	0.859	0.872	0.827	0.849

6.2   The Impact of Different Noise Level

We prove that the proposed robust training framework can largely boost model performance when applied on noisy inputs generated by real OCR engines. Since the noise rate in both Hanvon and TAL OCR test data is relatively low, we have not evaluated the effectiveness of the proposed robust training framework under different noise rates, especially when there are significant number of noises in the inputs. In this section, we investigate this problem and show the results in Fig. 4. We introduce different levels of noises by randomly inserting, deleting or replacing tokens in noise-free texts with equal probability.

As shown in Fig. 4, we can observe that F1 score decreases as the noise rate increases. When noise rate is less than 25%, F1 score decreases slowly for Parallelism and Metaphor detection, and drops significantly when noise rate exceeds 30%. Another observation is that performance of Personification detection degrades faster than the other two tasks, as reflected in a sharper slope in Fig. 4.

Table 4. Performance on personification detection with a single noise simulation.

Simulation	Model	Noise-free data			Hanvon OCR			TAL OCR		
		P	R	F1	P	R	F1	P	R	F1
$\mathcal{M}^*_{clean}$	BERT	0.855	0.828	0.841	**0.868**	0.719	0.787	0.825	0.734	0.777
Rule-based	BERT	0.818	**0.844**	0.831	0.817	**0.766**	0.791	0.833	**0.781**	**0.806**
Model-based	BERT	**0.862**	0.781	0.820	0.855	0.734	0.790	**0.855**	0.734	0.790
Attack-based	BERT	0.844	**0.844**	**0.844**	0.831	**0.766**	**0.797**	0.831	0.765	0.797
$\mathcal{M}^*_{clean}$	RoBERTa	0.764	**0.859**	0.809	0.754	0.812	0.782	0.730	0.719	0.724
Rule-based	RoBERTa	0.775	**0.859**	0.815	0.783	**0.844**	**0.812**	0.739	**0.797**	0.767
Model-based	RoBERTa	0.776	0.812	0.794	0.785	0.797	0.791	**0.817**	0.766	**0.791**
Attack-based	RoBERTa	**0.850**	0.797	**0.823**	**0.828**	0.750	0.787	0.808	0.656	0.724

Table 5. Performance on parallelism detection with a single noise simulation.

Simulation	Model	Noise-free data			Hanvon OCR			TAL OCR		
		P	R	F1	P	R	F1	P	R	F1
$\mathcal{M}^*_{clean}$	BERT	0.720	0.750	0.735	0.756	0.646	0.697	0.725	0.604	0.659
Rule-based	BERT	0.679	**0.792**	0.731	0.700	**0.714**	**0.758**	0.739	**0.708**	0.723
Model-based	BERT	**0.771**	0.771	**0.771**	0.733	0.688	0.710	0.786	0.688	**0.734**
Attack-based	BERT	0.766	0.750	0.758	**0.789**	0.625	0.698	**0.800**	0.667	0.727
$\mathcal{M}^*_{clean}$	RoBERTa	**0.795**	0.729	0.761	**0.838**	0.646	0.730	**0.816**	0.646	0.721
Rule-based	RoBERTa	0.780	**0.812**	**0.796**	0.800	**0.750**	**0.774**	0.795	**0.729**	**0.761**
Model-based	RoBERTa	0.792	0.792	0.792	0.814	0.729	0.769	0.810	0.708	0.756
Attack-based	RoBERTa	0.787	0.771	0.779	0.829	0.708	0.764	0.805	0.688	0.742

6.3   The Impact of Hard Example Mining

Hard example mining algorithm allows the model to dynamically pay more attention to hard examples $(\mathbf{x}_i, \widetilde{\mathbf{x}}_i)$ whose representations $(\mathbf{e}_i, \widetilde{\mathbf{e}}_i)$ are still quite different. We believe that it is vital for the model to learn robust representations. In this section, we investigate the performance difference with and without hard example mining. As shown in Fig. 2, F1 score consistently increases for both noise-clean and noisy OCR test data when hard example mining is employed. For example, hard example mining improves F1 by 2% on Metaphor and 3.4% on Parallelism using TAL OCR. This indicates the importance of hard example mining in the proposed framework.

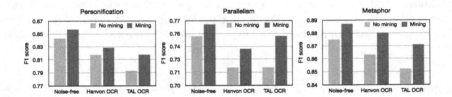

Fig. 2. The impact of hard example mining.

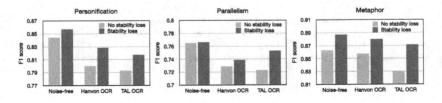

Fig. 3. The impact of stability loss.

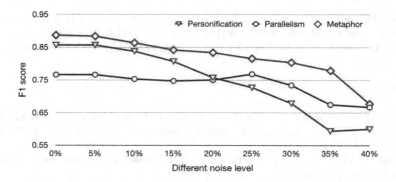

Fig. 4. The impact of different noise levels.

6.4 The Impact of Stability Loss

The use of stability loss guarantees that model can learn similar representations for clean text $\mathbf{x}$ and its noisy copy $\mathbf{x}'$. In this section, we investigate the performance difference with and without stability loss. As shown in Fig. 3, F1 score decreases when there are no stability loss for all three datasets. On Metaphor detection, using stability loss improves F1 by 4.1% and 2.3% for TAL OCR and Hanvon OCR. This indicates that stability loss is vital to the proposed framework.

7 Conclusion

In this paper, we study the robustness of multiple pre-trained models, e.g., BERT and RoBERTa, in text classification when inputs contain natural OCR noises. We propose a multi-source noise simulation method that can generate both token-level and span-level noises. We finetune models on both clean and simulated noisy data and propose a hard example mining algorithm so that during each training iteration, the model can focus on hard examples whose robust representations have not been learned. For evaluation, we construct three real-world text classification datasets and obtain natural OCR transcripts by calling OCR engines on real handwritten images. Experiments on three datasets proved that the proposed robust training framework largely boosts the model performance for both clean texts and natural OCR transcripts. It also outperforms all existing robust training approaches. In order to fully investigate the effectiveness of the framework, we evaluate it under different levels of noises and study the impact of hard example mining and stability loss independently. In the future, we will experiment the proposed framework on other NLP tasks and more languages. In the meanwhile, we will study the problem under automatic speech recognition (ASR) transcripts.

Acknowledgment. This work was supported in part by National Key R&D Program of China, under Grant No. 2020AAA0104500 and in part by Beijing Nova Program (Z201100006820068) from Beijing Municipal Science & Technology Commission.

References

1. Alzantot, M., Sharma, Y., Elgohary, A., Ho, B.J., Srivastava, M., Chang, K.W.: Generating natural language adversarial examples. In: Proceedings of EMNLP, pp. 2890–2896 (2018)
2. Belinkov, Y., Bisk, Y.: Synthetic and natural noise both break neural machine translation. In: Proceedings of ICLR (2018)
3. Chollampatt, S., Ng, H.T.: Neural quality estimation of grammatical error correction. In: Proceedings of EMNLP, pp. 2528–2539 (2018)
4. Devlin, J., Chang, M.W., Lee, K., Toutanova, K.: BERT: pre-training of deep bidirectional transformers for language understanding. In: Proceedings of NAACL-HLT, pp. 4171–4186 (2019)
5. Ebrahimi, J., Rao, A., Lowd, D., Dou, D.: HotFlip: white-box adversarial examples for text classification. In: Proceedings of ACL, pp. 31–36 (2018)
6. Goodfellow, I.J., Shlens, J., Szegedy, C.: Explaining and harnessing adversarial examples. In: Proceedings of ICLR (2015)
7. Hsieh, Y.L., Cheng, M., Juan, D.C., Wei, W., Hsu, W.L., Hsieh, C.J.: On the robustness of self-attentive models. In: Proceedings of ACL, pp. 1520–1529 (2019)
8. Jin, D., Jin, Z., Zhou, J.T., Szolovits, P.: Is BERT really robust? A strong baseline for natural language attack on text classification and entailment. In: Proceedings of AAAI, vol. 34, 7–12 February 2020, New York, NY, USA, pp. 8018–8025 (2020)
9. Karpukhin, V., Levy, O., Eisenstein, J., Ghazvininejad, M.: Training on synthetic noise improves robustness to natural noise in machine translation. In: Proceedings of W-NUT, pp. 42–47 (2019)
10. Miyato, T., Dai, A.M., Goodfellow, I.J.: Adversarial training methods for semi-supervised text classification. In: Proceedings of ICLR (2017)
11. Namysl, M., Behnke, S., Köhler, J.: NAT: Noise-aware training for robust neural sequence labeling. In: Proceedings of ACL, pp. 1501–1517 (2020)
12. Ndiaye, M., Faltin, A.V.: A spell checker tailored to language learners. Comput. Assist. Lang. Learn. **2–3**, 213–232 (2003)
13. Rawlinson, G.: The significance of letter position in word recognition. IEEE Aerosp. Electron. Syst. Mag. **1**, 26–27 (2007)
14. Ribeiro, M.T., Singh, S., Guestrin, C.: Semantically equivalent adversarial rules for debugging NLP models. In: Proceedings of ACL, pp. 856–865 (2018)
15. Sun, L., et al.: Adv-BERT: BERT is not robust on misspellings! generating nature adversarial samples on BERT. arXiv preprint arXiv:2003.04985 (2020)
16. Sun, Y., Jiang, H.: Contextual text denoising with masked language models. In: Proceedings of W-NUT, pp.286–290. arXiv preprint arXiv:1910.14080 (2019)
17. Tjong Kim Sang, E.F.: Introduction to the CoNLL-2002 shared task: language-independent named entity recognition. In: COLING (2002)
18. Valenti, S., Neri, F., Cucchiarelli, A.: An overview of current research on automated essay grading. J. Inf. Technol. Educ. Res. **1**, 319–330 (2003)
19. Vaswani, A., et al.: Attention is all you need. In: Proceedings of NIPS, 4–9 December 2017, Long Beach, CA, USA, pp. 5998–6008 (2017)

20. Yang, P., Chen, J., Hsieh, C.J., Wang, J.L., Jordan, M.I.: Greedy attack and Gumbel attack: generating adversarial examples for discrete data. J. Mach. Learn. Res. **43**, 1–36 (2020)
21. Yasunaga, M., Kasai, J., Radev, D.: Robust multilingual part-of-speech tagging via adversarial training. In: Proceedings of NAACL-HLT, pp. 976–986 (2018)
22. Zhai, C.: Statistical language models for information retrieval. In: Proceedings of NAACL-HLT, pp. 3–4 (2007)
23. Zhao, W., Wang, L., Shen, K., Jia, R., Liu, J.: Improving grammatical error correction via pre-training a copy-augmented architecture with unlabeled data. In: Proceedings of NAACL-HLT, pp. 156–165 (2019)
24. Zhao, Z., Dua, D., Singh, S.: Generating natural adversarial examples. In: Proceedings of ICLR (2018)
25. Zheng, S., Song, Y., Leung, T., Goodfellow, I.J.: Improving the robustness of deep neural networks via stability training. In: Proceedings of CVPR, CVPR 2016, 27–30 June 2016, Las Vegas, NV, USA, pp. 4480–4488 (2016)

Topic-to-Essay Generation with Comprehensive Knowledge Enhancement

Zhiyue Liu, Jiahai Wang$^{(\boxtimes)}$, and Zhenghong Li

School of Computer Science and Engineering, Sun Yat-sen University,
Guangzhou, China
{liuzhy93,lizhh98}@mail2.sysu.edu.cn, wangjiah@mail.sysu.edu.cn

Abstract. Generating high-quality and diverse essays with a set of topics is a challenging task in natural language generation. Since several given topics only provide limited source information, utilizing various topic-related knowledge is essential for improving essay generation performance. However, previous works cannot sufficiently use that knowledge to facilitate the generation procedure. This paper aims to improve essay generation by extracting information from both internal and external knowledge. Thus, a topic-to-essay generation model with comprehensive knowledge enhancement, named TEGKE, is proposed. For internal knowledge enhancement, both topics and related essays are fed to a teacher network as source information. Then, informative features would be obtained from the teacher network and transferred to a student network which only takes topics as input but provides comparable information compared with the teacher network. For external knowledge enhancement, a topic knowledge graph encoder is proposed. Unlike the previous works only using the nearest neighbors of topics in the commonsense base, our topic knowledge graph encoder could exploit more structural and semantic information of the commonsense knowledge graph to facilitate essay generation. Moreover, the adversarial training based on the Wasserstein distance is proposed to improve generation quality. Experimental results demonstrate that TEGKE could achieve state-of-the-art performance on both automatic and human evaluation.

Keywords: Topic-to-essay generation · Knowledge transfer · Graph neural network · Adversarial training

1 Introduction

Topic-to-essay generation (TEG) is a challenging task in natural language generation, which aims at generating high-quality and diverse paragraph-level text under the theme of several given topics. Automatic on-topic essay generation would bring benefits to many applications, such as news compilation [10], story generation [3], and intelligent education. Although some competitive results for TEG have been reported in the previous works using deep generative models [4,15,20], the information gap between the source topic words and the targeted essay blocks

© Springer Nature Switzerland AG 2021
Y. Dong et al. (Eds.): ECML PKDD 2021, LNAI 12979, pp. 302–318, 2021.
https://doi.org/10.1007/978-3-030-86517-7_19

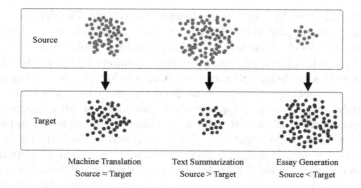

Fig. 1. Toy illustration of the information volume on different text generation tasks.

their models from performing well. The comparison of information flow between TEG and other text generation tasks is illustrated in Fig. 1 [20]. For machine translation and text summarization, the source provides enough information to generate the targeted text. However, for the TEG task, the information provided by only the topic words is much less than that contained in the targeted text during generation, making the generated essays low-quality.

The proper utilization of various topic-related knowledge is essential to enrich the source information, which has not been sufficiently explored. Incorporating the external knowledge from related common sense into the generation procedure is an efficient way to improve the TEG performance. However, in the common-sense knowledge graph, previous works [15,20] only consider the nearest neighbor nodes of topic words, and neglect the multi-hop neighbors which would bring more structural and semantic information. Moreover, without considering external knowledge, their models cannot fully exploit the relation between topics and essays to assist the generation procedure.

This paper proposes a topic-to-essay generation model with comprehensive knowledge enhancement, named TEGKE. By extracting both internal and external knowledge, TEGKE greatly enriches the source information. Besides, the adversarial training based on the Wasserstein distance is proposed to further enhance our model. Thus, there are three key parts, including internal knowledge enhancement, external knowledge enhancement, and adversarial training.

For internal knowledge enhancement, our model is based on the auto-encoder framework including a teacher network and a student network. Inspired by the conditional variational auto-encoder (CVAE) framework, the teacher network takes both topic words and related essays as source information to get informative latent features catching the high-level semantics of the relation between topics and essays. Then, a decoder could better reconstruct the targeted essay conditional on these features. Since only topic words could be used as the input source during inference, the informative features (i.e., internal knowledge) from the teacher network would be transferred to the student network. Different from CVAE that trains the recognition network and the prior network to be close to each other in the latent space, the teacher network in TEGKE maintains an independent training procedure. Then, the student network is forced to be close to the teacher network. That is, the student could take only topics as input but output comparable informative latent features compared with the teacher.

For external knowledge enhancement, ConceptNet [17] is employed as the commonsense knowledge base. Different from the previous works only using the nearest neighbors of topics, a topic-related knowledge graph is extracted from ConceptNet, which consists of multi-hop neighbors from the source topic words. Then, a topic knowledge graph encoder is proposed to perform on the multi-hop knowledge graph. It employs a compositional operation to obtain graph-aware node representations (i.e., external knowledge), which could conclude the structural information and the semantic information. The external knowledge is involved in the essay generation and helps select a proper decoding word.

Moreover, a discriminator is introduced for adversarial training. For alleviating the discrete output space problem of text, previous works adopt the adversarial training based on reinforcement learning (RL), which has the drawbacks of less-informative reward signals and high-variance gradients [1]. In contrast, this paper proposes to directly optimize the Wasserstein distance for the adversarial training, which avoids the problem of vanishing gradients and provides strong learning signals [5]. Based on the Wasserstein distance, the discriminator could operate on the continuous valued output instead of discrete text [18]. For aligning essays with the related topics, topics are combined with generated essays as generated samples and combined with targeted essays as real samples. By the minimax game, the discriminator would provide an informative learning signal guiding our model to generate high-quality essays.

In summary, our contributions are as follows:

- A topic-to-essay generation model is proposed based on the knowledge transfer between a teacher network and a student network. The teacher network could obtain informative features for the student network to learn, making the student network provide abundant information with only topics as the input source.
- A topic knowledge graph encoder is proposed to perform on the multi-hop knowledge graph extracted from the commonsense base. It helps our model exploit the structural and semantic information of the knowledge graph to facilitate essay generation. Moreover, a discriminator is introduced to improve generation quality by the adversarial training based on the Wasserstein distance.
- Experimental results on both automatic evaluation and human evaluation demonstrate that TEGKE could achieve better performance than the state-of-the-art methods.

2   Related Work

As a text generation task, TEG aims at generating high-quality and diverse paragraph-level text with given topics, which has drawn more attention. This task is first proposed by Feng et al. [4], and they utilize the coverage vector to integrate topic information. For enriching the input source information, external commonsense knowledge has been introduced for TEG [15,20]. Besides, Qiao et al. [15] inject the sentiment labels into a generator for controlling the sentiment of a generated essay. However, during essay generation, previous works [15,20] only consider the nearest neighbors of topic nodes in the commonsense knowledge graph. This limitation blocks their models from generating high-quality essays. For better essay generation, this paper makes the first attempt to utilize both structural and semantic information from the multi-hop knowledge graph.

Poetry generation is similar to TEG, which could be regarded as a generation task based on topics. A memory-augmented neural model is proposed to generate poetry by balancing the requirements of linguistic accordance and aesthetic innovation [24]. The CVAE framework is adopted with adversarial training to generate diverse poetry [12]. Yang et al. [21] use hybrid decoders to generate Chinese poetry. RL algorithms are employed to improve the poetry diversity criteria [22] directly. Different from poetry generation showing obvious structured rules, the TEG task needs to generate a paragraph-level unstructured plain text, and such unstructured targeted output brings severe challenges for generation.

The RL-based adversarial training [6,23] is used to improve essay quality in previous works [15,20]. However, the noisy reward derived from the discriminator makes their models suffer from high-variance gradients. In contrast, our model directly optimizes the Wasserstein distance for the adversarial training without RL, achieving better generation performance.

3 Methodology

3.1 Task Formulation

Given a dataset including pairs of the topic words $\mathbf{x} = (x_1, ..., x_m)$ and the related essay $\mathbf{y} = (y_1, ..., y_n)$, for solving the TEG task, we want a θ-parameterized model to learn each pair from the dataset and generate a coherent essay under given topic words, where the number of essay words n is much larger than that of topic words m. Then, the task could be formulated as obtaining the optimal model with $\hat{\theta}$ which maximizes the conditional probability as follows:

$$\hat{\theta} = \arg \max_{\theta} P_{\theta}(\mathbf{y}|\mathbf{x}). \tag{1}$$

3.2 Model Description

Our TEGKE is based on the auto-encoder framework, utilizing both internal and external knowledge to enhance the generation performance. As shown in Fig. 2, the model mainly contains three encoders (i.e., a topic encoder, an essay encoder, and a topic knowledge graph encoder) and an essay decoder. A discriminator is introduced at the end of the essay decoder for adversarial training.

For internal knowledge enhancement, the topic encoder and the essay encoder encode the topic words and the targeted essay sequence as x_{enc} and y_{enc}, respectively. The latent features z_1 and z_2 are obtained from a teacher network taking

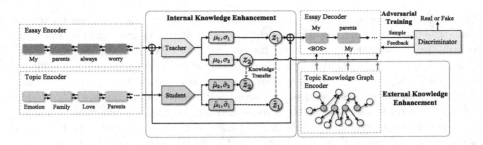

Fig. 2. Overview of the proposed model. Our model uses the teacher network for training (black solid arrows), and the student network for inference (black dotted arrows). The student network learns the latent features from the teacher network (red dotted arrows) for internal knowledge enhancement. The information from the topic knowledge graph encoder is integrated at each decoding step (red solid arrows) for external knowledge enhancement. During adversarial training, the generated essays are fed to the discriminator which provides learning signals as feedback. (Color figure online)

both x_{enc} and y_{enc} as input. Then, a student network, which takes x_{enc} solely as input, produces $\tilde{z}_1$ and $\tilde{z}_2$ to learn from z_1 and z_2 as internal knowledge, respectively. The essay decoder would generate a topic-related essay by receiving the latent features from the teacher network during training or those from the student network during inference.

For external knowledge enhancement, the multi-hop topic knowledge graph is constructed from the commonsense knowledge base, ConceptNet. Then, the topic knowledge graph encoder could represent the topic-related structural and semantic information as external knowledge to enrich the source information. The extracted external knowledge is attentively involved in each decoding step of the essay decoder to help select proper words and boost generation performance.

Through the adversarial training based on the Wasserstein distance, the discriminator could make the generated essay more similar to the targeted essay, which improves essay quality.

Topic Encoder and Essay Encoder. The topic encoder employs a bidirectional gated recurrent unit (GRU) [2], which integrates the information of the topic sequence from both forward and backward directions. The topic encoder reads the embeddings of topic words **x** from both directions and obtains the hidden states for each topic word as follows:

$$\overrightarrow{h_i^x} = \overrightarrow{\text{GRU}}(\overrightarrow{h_{i-1}^x}, e(x_i)), \quad \overleftarrow{h_i^x} = \overleftarrow{\text{GRU}}(\overleftarrow{h_{i+1}^x}, e(x_i)), \tag{2}$$

where $e(x_i)$ is the embedding of x_i. The representation of the i-th topic is obtained as $h_i^x = [\overrightarrow{h_i^x}; \overleftarrow{h_i^x}]$, and ";" denotes the vector concatenation. The mean-pooling operation is conducted on the representations of all topics to represent **x** as $x_{\text{enc}} = \text{mean}(h_1^x, ..., h_m^x)$. Similarly, another bidirectional GRU is adopted as the essay encoder. The representation of the essay **y** could be obtained in the same way as the topic encoder does, which is denoted as y_{enc}.

Internal Knowledge Enhancement. Although the auto-encoder framework has shown competitive performance in many text generation tasks, the limited source information of the TEG task cannot provide sufficient information for the decoder to reconstruct the targeted output essay. This paper notices that informative latent features produced by the encoder are essential for a better decoding procedure. Inspired by the CVAE framework taking both the source and the target to train a recognition network, a teacher network is proposed by taking both the topics and essay as source information to get informative latent features for the essay decoder. Since only topics could be accessed during inference, a student network taking topic words solely as input is designed to learn from the teacher network's latent features as internal knowledge. Different from CVAE that trains both the recognition network and the prior network to be close to each other in the latent space, the teacher network in our model maintains an independent training procedure following minimizing the reconstruction error. Because the teacher network is expected to provide strong supervision without being influenced by the student network. The student network would generate latent features which learn the information from the teacher network's latent features through knowledge transfer. That is, the student network is pushed to be close to the teacher network in the latent space.

The teacher network consists of two feed-forward networks, and each network takes x_{enc} and y_{enc} as input to produce the mean and the diagonal covariance by two matrix multiplications, respectively. The latent features z_1 and z_2 are sampled from two Gaussian distributions defined by the above two feed-forward networks, respectively. During training, z_1 is used as a part of the essay decoder's initial hidden state, and z_2 is used as a part of the essay decoder's input at each step to provide more source information. The decoder receives z_1 and z_2 to optimize the training objective. Similarly, there are two feed-forward networks in the student network, where each network takes x_{enc} solely as input to sample a latent feature. Then, the student network's latent features $\tilde{z}_1$ and $\tilde{z}_2$ could be obtained. During inference, the decoder decodes $\tilde{z}_1$ and $\tilde{z}_2$ into a essay. Hence, above latent features could be obtained as follows:

$$\begin{matrix} z_1 \sim \mathcal{N}(\mu_1, \sigma_1^2 \mathbf{I}) \\ z_2 \sim \mathcal{N}(\mu_2, \sigma_2^2 \mathbf{I}) \end{matrix}, \quad \left(\begin{bmatrix} \mu_1 \\ \log\left(\sigma_1^2\right) \end{bmatrix}, \begin{bmatrix} \mu_2 \\ \log\left(\sigma_2^2\right) \end{bmatrix} \right) = \text{Teacher}(x_{\text{enc}}, y_{\text{enc}}), \quad (3)$$

$$\begin{matrix} \tilde{z}_1 \sim \mathcal{N}(\tilde{\mu}_1, \tilde{\sigma}_1^2 \mathbf{I}) \\ \tilde{z}_2 \sim \mathcal{N}(\tilde{\mu}_2, \tilde{\sigma}_2^2 \mathbf{I}) \end{matrix}, \quad \left(\begin{bmatrix} \tilde{\mu}_1 \\ \log\left(\tilde{\sigma}_1^2\right) \end{bmatrix}, \begin{bmatrix} \tilde{\mu}_2 \\ \log\left(\tilde{\sigma}_2^2\right) \end{bmatrix} \right) = \text{Student}(x_{\text{enc}}), \quad (4)$$

where $\mathbf{I}$ is an identity matrix, and the reparametrization trick is used to sample the latent features. For enhancing the generation performance, the teacher network is trained to reconstruct the target, while the internal knowledge from the teacher network is transferred to the student network by minimizing the Kullback-Leibler (KL) divergence between the teacher's distributions and the student's distributions in the latent space as follows:

$$\mathcal{L}_{\text{trans}} = \text{KL}(\mathcal{N}(\tilde{\mu}_1, \tilde{\sigma}_1^2 \mathbf{I}) || \mathcal{N}(\mu_1, \sigma_1^2 \mathbf{I})) + \text{KL}(\mathcal{N}(\tilde{\mu}_2, \tilde{\sigma}_2^2 \mathbf{I}) || \mathcal{N}(\mu_2, \sigma_2^2 \mathbf{I})). \quad (5)$$

Topic Knowledge Graph Encoder. Incorporating external commonsense knowledge is important to bridge the information gap between the source and the target. Unlike previous works only considering the nearest neighbor nodes of topics, this paper constructs a topic knowledge graph queried by the topic words over a few hops from ConceptNet to assist the generation procedure. Then, a topic knowledge graph $\mathbf{G} = (\mathbf{V}, \mathbf{R}, \mathbf{E})$ could be obtained, where $\mathbf{V}$ denotes the set of vertices, $\mathbf{R}$ is the set of relations, and $\mathbf{E}$ represents the set of edges. The topic knowledge graph encoder is designed to integrate the topic-related information from $\mathbf{G}$. By considering the topic knowledge graph, the objective of the TEG task could be modified as follows:

$$\hat{\theta} = \arg\max_{\theta} P_{\theta}(\mathbf{y}|\mathbf{x}, \mathbf{G}). \tag{6}$$

External Knowledge Enhancement. Appropriate usage of the structural and semantic information in the external knowledge graph plays a vital role in the TEG task. Each edge (u, r, v) in $\mathbf{G}$ means that the relation $r \in \mathbf{R}$ exists from a node u to a node v. This paper extends (u, r, v) with its reversed link (v, r_{rev}, u) to allow the information in a directed edge to flow along both directions [13], where r_{rev} denotes the reversed relation. For instance, given the edge $(worry, isa, emotion)$, the reversed edge $(emotion, isa\_r, worry)$ is added in $\mathbf{G}$. Our topic knowledge graph encoder is based on the graph neural network (GNN) framework, which could aggregate the graph-structured information of a node from its neighbors. Specifically, a graph convolution network (GCN) with L layers is employed. For jointly embedding both nodes and relations in the topic knowledge graph, this paper follows Vashishth et al. [19] to perform a non-parametric compositional operation ϕ for combining the neighbor node and the relation of a central node. As shown in Fig. 3, for a node $v \in \mathbf{V}$, its embedding would be updated at the $l+1$-th layer by aggregating information from its neighbors $N(v)$. The topic knowledge graph encoder treats incoming edges and outgoing edges differently to sufficiently encode structural information. Specifically, the related edges of the node v could be divided into the set of incoming edges and that of outgoing edges, denoted as $\mathbf{E}_{\text{in}(v)}$ and $\mathbf{E}_{\text{out}(v)}$, respectively. Then, the node embedding of v could be updated as follows:

$$o_v^l = \frac{1}{|N(v)|} \sum_{(u,r) \in N(v)} W_{\text{dir}(r)}^l \phi(h_u^l, h_r^l), \tag{7}$$

$$h_v^{l+1} = \text{ReLU}(o_v^l + W_{\text{loop}}^l h_v^l), \tag{8}$$

where h_v^0 is initialized by the original word embedding, and h_r^0 is initialized by the relation embedding. The weight matrix $W_{\text{dir}(r)}^l$ is a relation-direction specific parameter at the l-th layer as follows:

$$W_{\text{dir}(r)}^l = \begin{cases} W_{\text{in}}^l, & (u, r, v) \in \mathbf{E}_{\text{in}(v)} \\ W_{\text{out}}^l, & (v, r, u) \in \mathbf{E}_{\text{out}(v)} \end{cases}. \tag{9}$$

The compositional operation employs $\phi(h_u^l, h_r^l) = h_u^l + h_r^l$ when incoming edges are considered, and $\phi(h_u^l, h_r^l) = h_u^l - h_r^l$ when outgoing edges are considered [19].

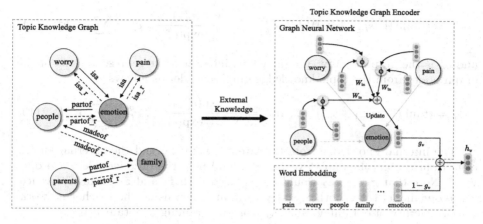

Fig. 3. Topic knowledge graph encoder. The graph neural network performs a compositional operation for a central node (e.g., emotion). Only incoming edges are considered in the diagram for clarity. The information from the topic knowledge graph is aggregated to update the embedding of the central node. Then, the final updated embedding is combined with the original word embedding to assist the essay decoding.

o_v^l is the aggregated information from the l-th layer, and the weight matrix W_{loop}^l is used to transform v's own information from the l-th layer. For the relation r, its embedding is updated as follows:

$$h_r^{l+1} = W_r^l h_r^l, \tag{10}$$

where W_r^l is a weight matrix. A gate mechanism is designed to combine h_v^L containing graph knowledge and h_v^0 containing original semantic knowledge by:

$$g_v = \text{Sigmoid}(W_{\text{gate}}[h_v^L; h_v^0]), \tag{11}$$

$$h_v = g_v \odot h_v^L + (1 - g_v) \odot h_v^0, \tag{12}$$

where W_{gate} is a weight matrix, and $\odot$ is the element-wise multiplication.

Finally, the node embedding h_v is obtained to encode both structural and semantic information of the knowledge graph as external knowledge, involved in each decoding step for better essay generation.

Essay Decoder. The essay decoder employs a single layer GRU. The initial hidden state s_0 is set with $s_0 = [x_{\text{enc}}; z_1]$ containing the topics' representation and the latent feature. Both internal and external knowledge should be involved in each decoding step. Specifically, the hidden state s_t of the decoder at time step t is obtained as follows:

$$s_t = \text{GRU}(s_{t-1}, [e(y_{t-1}); z_2; c_t^x; c_t^g]), \tag{13}$$

where $e(y_{t-1})$ is the embedding of the essay word y_{t-1} at the time step $t-1$, c_t^x is the topic context vector at the time step t, which integrates the output representations from the topic encoder by the attention mechanism as follows:

$$e_{t,i}^x = (\tanh(W_x s_{t-1} + b_x))^T h_i^x, \ \alpha_{t,i}^x = \frac{\exp(e_{t,i}^x)}{\sum_{j=1}^{m} \exp(e_{t,j}^x)}, \ c_t^x = \sum_{i=1}^{m} \alpha_{t,i}^x h_i^x, \quad (14)$$

and c_t^g is the graph context vector, which integrates the representations of the graph nodes from the topic knowledge graph encoder as follows:

$$e_{t,v}^g = (\tanh(W_g s_{t-1} + b_g))^T h_v, \ \alpha_{t,v}^g = \frac{\exp(e_{t,v}^g)}{\sum_{u \in \mathbf{V}} \exp(e_{t,u}^g)}, \ c_t^g = \sum_{v \in \mathbf{V}} \alpha_{t,v}^g h_v. \quad (15)$$

The internal knowledge from the latent feature z_2, and the external knowledge from the graph context vector c_t^g would help the decoder select a proper word. Note that z_1 and z_2 would be replaced with $\tilde{z}_1$ and $\tilde{z}_2$ during inference. Since our model takes both $\mathbf{x}$ and $\mathbf{y}$ as input when using the teacher network, the probability of obtaining an essay word for training is obtained by:

$$P_\theta(y_t|y_{<t}, \mathbf{x}, \mathbf{y}, \mathbf{G}) = \text{Softmax}(W_o s_t + b_o). \quad (16)$$

Discriminator. A ψ-parameterized CNN-based discriminator [8] D_ψ is introduced in our model for adversarial training which would improve essay quality.

Adversarial Training. Due to the discrete output space problem of text generation, previous works heavily rely on the RL-based adversarial training which has less-informative reward signals and high-variance gradients. In contrast, this paper proposes the adversarial training through the Wasserstein distance for TEG. Based on the Wasserstein distance, the discriminator could operate on continuous valued output and provide strong learning signals by distinguishing between a real text sequence of one-hot vectors and a generated text sequence of probabilities. Specifically, the hidden state s_t of the essay decoder is employed to generate a probability output $y_t^\theta = \text{Softmax}(W_o s_t + b_o)$. Then, a sequence of outputs $\mathbf{y}_\theta = (y_1^\theta, ..., y_n^\theta)$ could be regarded as the generated essay for adversarial training. For aligning the generated essay with the related topics, the pair of the topics $\mathbf{x}$ and the ground truth essay $\mathbf{y}$ is fed to D_ψ as the real sample, while the pair of $\mathbf{x}$ and $\mathbf{y}_\theta$ is treated as the generated sample. Then, the adversarial training objective based on the Wasserstein distance for D_ψ is formulated by:

$$\mathcal{L}_{D_\psi} = D_\psi(\mathbf{x}, \mathbf{y}_\theta) - D_\psi(\mathbf{x}, \mathbf{y}) + \lambda(||\nabla_{\hat{\mathbf{y}}} D_\psi(\mathbf{x}, \hat{\mathbf{y}})||_2 - 1)^2, \quad (17)$$

where the gradient penalty $(||\nabla_{\hat{\mathbf{y}}} D_\psi(\mathbf{x}, \hat{\mathbf{y}})||_2 - 1)^2$ weighted by λ is imposed on the discriminator to enforce the Lipschitz constraint, and $\hat{\mathbf{y}} = \alpha\mathbf{y} + (1 - \alpha)\mathbf{y}_\theta$ with $\alpha \sim \text{Uniform}(0, 1)$. The auto-encoder framework in our model could act as a generator to minimize the following adversarial training objective as:

$$\mathcal{L}_{adv} = -D_\psi(\mathbf{x}, \mathbf{y}_\theta) - \beta \log[P_\theta(\mathbf{y}|\mathbf{x}, \mathbf{y}, \mathbf{G})], \quad (18)$$

where the log-likelihood term $\log[P_\theta(\mathbf{y}|\mathbf{x}, \mathbf{y}, \mathbf{G})]$ weighted by β would help align the generated essay with the topics further and keep generation diversity. The generator and the discriminator D_ψ are alternately trained to play a minimax game, where D_ψ assists the generator to obtain high-quality essays.

3.3   Training and Inference

For the training procedure, the latent features for decoding an essay are computed by the teacher network. Two training stages are employed in TEGKE. At the first training stage, the negative log-likelihood is minimized to reconstruct the ground truth essay $\mathbf{y} = (y_1, ..., y_n)$ as follows:

$$\mathcal{L}_{\text{rec}} = \sum_{t=1}^{n} -\log[P_\theta(y_t|y_{<t}, \mathbf{x}, \mathbf{y}, \mathbf{G})], \tag{19}$$

where all parameters except the student network's parameters are optimized in an end-to-end manner. For transferring internal knowledge from the teacher network to the student network, the KL divergence between the student's distributions and the teacher's distributions is minimized by $\mathcal{L}_{\text{trans}}$ of Eq. (5) to optimize the student network's parameters.

At the second training stage, the auto-encoder framework in our model acts as a generator which is trained by $\mathcal{L}_{\text{adv}}$ of Eq. (18). The discriminator is trained by $\mathcal{L}_{D_\psi}$ of Eq. (17) to provide a learning signal for the generator. Note that the student network is still optimized by $\mathcal{L}_{\text{trans}}$ during the second stage. For the inference procedure, the latent features for decoding are computed by the student network. The input to our model is the topics $\mathbf{x}$ and the topic knowledge graph $\mathbf{G}$, and then the decoder would generate a related essay. The pseudo code of TEGKE is shown in the supplementary material: https://arxiv.org/abs/2106.15142.

4   Experiments

4.1   Datasets

Experiments are conducted on the ZHIHU corpus [4] consisting of real-world Chinese topic and essay pairs. The number of topic words is between 1 and 5. The length of an essay is between 50 and 100. For extracting external knowledge sufficiently, this paper constructs the topic knowledge graph from ConceptNet over 5 hops, and then 40 nodes are reserved per hop [7]. The constructed topic knowledge graph is a subgraph of ConceptNet. For this knowledge graph, the maximum number of nodes is 205, and the maximum number of edges is 912. The training set and the test set contain 27,000 samples and 2,500 samples, respectively. We set 10% of training samples as the validation set for hyperparameters tuning. Besides, the experimental results on the ESSAY corpus [4] are shown in the supplementary material.

4.2   Settings

The essay decoder is a GRU with a hidden size of 1024. Both the topic encoder and the essay encoder are implemented as a bidirectional GRU with a hidden size of 512. The size of latent features is 512 in the teacher network and the student network. For the discriminator, the weight λ of the gradient penalty is set to 10. The weight β is set to 10. The vocabulary size is 50,000, and the batch size is set to 32. The 200-dim pretrained word embeddings [16] are shared by topics, essays, and initial graph nodes. The 200-dim randomly initialized vectors are used as initial graph relation embeddings. Adam optimizer [9] is used to train the model with the learning rate 10^{-3} for the first training stage, and the learning rate 10^{-4} for the second training stage.

4.3 Baselines

TAV [4] encodes topic semantics as the average of the topic's embeddings and then uses an LSTM as a decoder to generate each word.
TAT [4] enhances the decoder of TAV with the attention mechanism to select the relevant topics at each step.
MTA [4] extends the attention mechanism of TAT with a topic coverage vector to guarantee that every single topic is expressed by the decoder.
CTEG [20] introduces commonsense knowledge into the generation procedure and employs adversarial training to improve generation performance.
SCTKG [15] extends CTEG with the topic graph attention and injects the sentiment labels to control the sentiment of the generated essay. The SCTKG model without sentiment information is considered as a baseline, since the original TEG task does not take the sentiment of ground truth essays as input.

4.4 Evaluation Metrics

In this paper, both automatic evaluation and human evaluation are adopted to evaluate the generated essays.

Automatic Evaluation. Following previous works [4, 15, 20], there are several automatic metrics considered to evaluate the model performance.

BLEU [14]: The BLEU score is widely used in text generation tasks (e.g., dialogue generation and machine translation). It could measure the generated essays' quality by computing the overlapping rate between the generated essays and the ground truth essays.

Dist-1, Dist-2 [11]: The Dist-1 and Dist-2 scores are the proportion of distinct unigrams and bigrams in the generated essays, respectively, which measure the diversity of the generated essays.

Novelty [20]: The novelty is calculated by the difference between the generated essay and the ground truth essays with similar topics in the training set. A higher score means more novel essays would be generated under similar topics.

Table 1. Automatic and human evaluation results. ↑ means higher is better. * indicates statistically significant improvements ($p < 0.001$) over the best baseline.

Method	Automatic evaluation				Human evaluation			
	BLEU(↑)	Novelty(↑)	Dist-1(↑)	Dist-2(↑)	T-Con.(↑)	Nov.(↑)	E-div.(↑)	Flu.(↑)
TAV	6.05	70.32	2.69	14.25	2.32	2.19	2.58	2.76
TAT	6.32	68.77	2.25	12.17	1.76	2.07	2.32	2.93
MTA	7.09	70.68	2.24	11.70	3.14	2.87	2.17	3.25
CTEG	9.72	75.71	5.19	20.49	3.74	3.34	3.08	3.59
SCTKG	9.97	78.32	**5.73**	23.16	3.89	3.35	3.90	3.71
TEGKE	**10.75***	**80.18***	5.58	**28.11***	**4.12***	**3.57***	**4.08***	**3.82***

Human Evaluation. Following previous works [15, 20], in order to evaluate the generated essays more comprehensively, 200 samples are collected from different models for human evaluation. Each sample contains the input topics and the generated essay. All 3 annotators are required to score the generated essays from 1 to 5 in terms of four criteria: **Topic-Consistency (T-Con.)**, **Novelty (Nov.)**, **Essay-Diversity (E-div.)**, and **Fluency (Flu.)**. For novelty, the TF-IDF features of topic words are used to retrieve the 10 most similar training samples to provide references for the annotators. Finally, each model's score on a criterion is calculated by averaging the scores of three annotators.

4.5 Experimental Results

Automatic Evaluation Results. The automatic evaluation results over generated essays are shown in the left block of Table 1. Compared with TAV, TAT, and MTA, TEGKE consistently achieves better results on all metrics. This illustrates that, without introducing sufficient knowledge, their models obtain unsatisfactory performance due to the limited source information. CTEG and SCTKG consider the nearest neighbor nodes of topics from ConceptNet as external information. In contrast, the multi-hop topic knowledge graph provides more structural and semantic information which is extracted by our topic knowledge graph encoder. Hence, our model outperforms the best baseline by 0.78 on the BLEU score, demonstrating that the potential of our model to generate high-quality essays. Moreover, TEGKE could obtain competitive results on the Dist-1 scores, while greatly improving the Dist-2 and novelty scores by 4.95 and 1.86 over SCTKG, respectively. That is, the essays generated from our model would be more diverse and different from the essays in the training corpus. In general, by integrating various internal and external knowledge into generation, TEGKE could achieve better quality and diversity simultaneously.

Human Evaluation Results. The human evaluation results are shown in the right block of Table 1, and TEGKE could obtain the best performance on all metrics. The external knowledge incorporated by the topic knowledge graph encoder would help the decoder select topic-related words, and the adversarial training

Table 2. Ablation study results.

Method	BLEU($\uparrow$)	Novelty($\uparrow$)	Dist-1($\uparrow$)	Dist-2($\uparrow$)
TEGKE	**10.75**	**80.18**	5.58	28.11
TEGKE w/o EX	10.18	78.67	5.38	21.16
TEGKE w/o AD	10.63	80.09	**5.65**	**28.33**
TEGKE w/o EX & AD	9.78	79.42	5.46	21.30

could further align generated essays with related topics. Thus, our model outperforms the best baseline by 0.23 on the topic-consistency score, showing that the generated essays are more closely related to the given topics. The improvement over the novelty, essay-diversity, and fluency scores demonstrates that TEGKE could obtain better samples in terms of quality and diversity. This conclusion is similar to that drawn from the automatic evaluation.

Ablation Study. To illustrate the effectiveness of our model's key parts, this paper performs an ablation study on three ablated variants: TEGKE without external knowledge enhancement (TEGKE w/o EX), TEGKE without adversarial training (TEGKE w/o AD), and TEGKE with only internal knowledge enhancement (TEGKE w/o EX & AD). The results are shown in Table 2.

Internal Knowledge Enhancement. Based on only the internal knowledge from the teacher network, TEGKE w/o EX & AD achieves the worst results among variants. However, its performance is still comparable to CTEG adopting both adversarial training and commonsense knowledge, showing that the latent features produced by the TEGKE w/o EX & AD's topic encoder benefit essay generation. Specially, TEGKE w/o EX & AD increases Dist-1 by 0.27 and Dist-2 by 0.81 over CTEG. This improvement comes from the teacher and student networks' various outputs, because our decoder generates essays depending on two latent features sampled from different Gaussian distributions. The above results illustrate that utilizing a student to learn from a teacher makes our model learn the relation between topics and essays better, which enhances the model performance.

External Knowledge Enhancement. Compared with TEGKE, TEGKE w/o EX shows much inferior performance on all metrics. Specifically, TEGKE w/o EX drops 0.57 on the BLEU score, since the external knowledge would help the model select a topic-related word by exploring the topic words and their neighbors in the multi-hop topic knowledge graph. Besides, the diversity of generated essays from TEGKE w/o EX degrades, which is shown by the decline on the novelty, Dist-1, and Dist-2 scores. Specially, TEGKE w/o EX greatly drops 6.95 on Dist-2, due to lacking the commonsense knowledge to provide background information and enrich the input source. By utilizing external knowledge, TEGKE w/o AD still outperforms SCTKG on most metrics. That is, our graph encoder could extract more informative knowledge from the multi-hop knowledge graph.

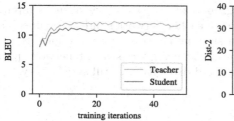
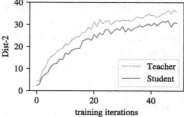

Fig. 4. Training curves. The BLEU score and the Dist-2 score are employed to measure quality and diversity, respectively. For both BLEU and Dist-2, the higher the better.

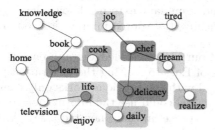

Input topics: Learn, Life, Delicacy
Output essay: In fact, my dream is to be a chef. But my major is not learning to cook a delicacy. I hope that someone can recommend some suitable daily part-time jobs. I want to realize my dream.

Fig. 5. Case study. The attention scores over the topic knowledge graph are shown on the left side. Deeper green indicates higher attention scores. The input topics and the generated essay are shown on the right side, where the selected words with higher attention scores are highlighted in blue. The original Chinese is translated into English. (Color figure online)

Adversarial Training. Based on the adversarial training, TEGKE w/o EX boosts the BLEU score by 0.4 over TEGKE w/o EX & AD, and only slightly sacrifices the novelty, Dist-1, and Dist-2 scores due to the inherent mode collapse problem in adversarial training. It demonstrates that the proposed adversarial training could effectively improve the essay quality. Compared with TEGKE w/o AD, TEGKE increases the BLEU score by 0.12, illustrating that our adversarial training could cooperate with the external knowledge enhancement. Since the external knowledge greatly enriches the source information and boosts the model performance, the improvement brought by the adversarial training is somewhat weakened when the topic knowledge graph is introduced.

4.6   Validity of Knowledge Transfer

To illustrate the validity of transferring knowledge from the teacher network to the student network, the performance of our model using the teacher network and that using the student network is shown in Fig. 4. The quality is measured by BLEU, and the diversity is measured by Dist-2. When our model uses the teacher network, the teacher network's latent features are fed to the decoder for generating essays. The model could maintain a stable training procedure and obtain excellent results since the ground truth essays are taken as input. The student network would learn from the teacher network's latent features. For the performance, the model using the student network closely follows that using the teacher network. Although the model using the student network performs slightly worse, the results on quality and diversity are still satisfactory.

4.7   Case Study

A case generated by our model is shown on the right side of Fig. 5. Under the given topics "learn", "life", and "delicacy", TEGKE obtains a high-quality essay that mainly covers the semantics of input topics. The reason is that our model could integrate internal knowledge and abundant external knowledge into the generation procedure. By greatly enriching the source information, our model would generate novel and coherent essays.

To further illustrate the validity of our topic knowledge graph encoder, this paper visualizes the attention weights of Eq. (15) during the generation procedure on the left side of Fig. 5. Compared with the previous works only considering the 1-hop neighbors of topics, our model could use the information from the multi-hop topic knowledge graph. For instance, in the generated essay, "dream" is a 2-hop neighbor of the topic "delicacy", and "realize" is a 3-hop neighbor of the topic "delicacy". It is observed that all nodes from the path ("delicacy", "chef", "dream", and "realize") get higher attention scores during the generation procedure, indicating that the structural information of the graph is helpful. The generated essay is consistent with the topics' semantics since the topics "learn" and "delicacy" both obtain higher attention scores. Although the topic "life" does not appear in the generated essay, its 1-hop neighbor "daily" injects the corresponding semantic information about "life" into the generated essay.

5   Conclusion

This paper proposes a topic-to-essay generation model with comprehensive knowledge enhancement, named TEGKE. For internal knowledge enhancement, the teacher network is built by taking both topics and related essays as input to obtain informative features. The internal knowledge in these features is transferred to the student network for better essay generation. For external knowledge enhancement, the topic knowledge graph encoder is proposed to extract both the structural and semantic information from commonsense knowledge, which significantly enriches the source information. Moreover, the adversarial training based on the Wasserstein distance is introduced to improve generation quality further. Experimental results on real-world corpora demonstrate that TEGKE outperforms the state-of-the-art methods.

Acknowledgements. This work is supported by the National Key R&D Program of China (2018AAA0101203), and the National Natural Science Foundation of China (62072483).

References

1. Caccia, M., Caccia, L., Fedus, W., Larochelle, H., Pineau, J., Charlin, L.: Language GANs falling short. In: ICLR (2020)
2. Cho, K., et al.: Learning phrase representations using RNN encoder-decoder for statistical machine translation. In: EMNLP, pp. 1724–1734 (2014)
3. Fan, A., Lewis, M., Dauphin, Y.: Hierarchical neural story generation. In: ACL, pp. 889–898 (2018)
4. Feng, X., Liu, M., Liu, J., Qin, B., Sun, Y., Liu, T.: Topic-to-essay generation with neural networks. In: IJCAI, pp. 4078–4084 (2018)
5. Gulrajani, I., Ahmed, F., Arjovsky, M., Dumoulin, V., Courville, A.: Improved training of Wasserstein GANs. In: NeurIPS, pp. 5769–5779 (2017)
6. Guo, J., Lu, S., Cai, H., Zhang, W., Yu, Y., Wang, J.: Long text generation via adversarial training with leaked information. In: AAAI, pp. 5141–5148 (2018)
7. Ji, H., Ke, P., Huang, S., Wei, F., Zhu, X., Huang, M.: Language generation with multi-hop reasoning on commonsense knowledge graph. In: EMNLP, pp. 725–736 (2020)
8. Kim, Y.: Convolutional neural networks for sentence classification. arXiv preprint arXiv:1408.5882 (2014)
9. Kingma, D.P., Ba, J.: Adam: A method for stochastic optimization. In: ICLR (2015)
10. Leppänen, L., Munezero, M., Granroth-Wilding, M., Toivonen, H.: Data-driven news generation for automated journalism. In: ICNLG, pp. 188–197 (2017)
11. Li, J., Galley, M., Brockett, C., Gao, J., Dolan, W.B.: A diversity-promoting objective function for neural conversation models. In: NAACL, pp. 110–119 (2016)
12. Li, J., et al.: Generating classical Chinese poems via conditional variational autoencoder and adversarial training. In: EMNLP, pp. 3890–3900 (2018)
13. Marcheggiani, D., Titov, I.: Encoding sentences with graph convolutional networks for semantic role labeling. In: EMNLP, pp. 1506–1515 (2017)
14. Papineni, K., Roukos, S., Ward, T., Zhu, W.J.: BLEU: a method for automatic evaluation of machine translation. In: ACL, pp. 311–318 (2002)
15. Qiao, L., Yan, J., Meng, F., Yang, Z., Zhou, J.: A sentiment-controllable topic-to-essay generator with topic knowledge graph. In: EMNLP: Findings, pp. 3336–3344 (2020)
16. Song, Y., Shi, S., Li, J., Zhang, H.: Directional skip-gram: Explicitly distinguishing left and right context for word embeddings. In: NAACL, pp. 175–180 (2018)
17. Speer, R., Havasi, C.: Representing general relational knowledge in ConceptNet 5. In: LREC, pp. 3679–3686 (2012)
18. Subramanian, S., Rajeswar, S., Dutil, F., Pal, C., Courville, A.: Adversarial generation of natural language. In: The 2nd Workshop on Representation Learning for NLP, pp. 241–251 (2017)
19. Vashishth, S., Sanyal, S., Nitin, V., Talukdar, P.: Composition-based multi-relational graph convolutional networks. In: ICLR (2020)
20. Yang, P., Li, L., Luo, F., Liu, T., Sun, X.: Enhancing topic-to-essay generation with external commonsense knowledge. In: ACL, pp. 2002–2012 (2019)
21. Yang, X., Lin, X., Suo, S., Li, M.: Generating thematic Chinese poetry using conditional variational autoencoders with hybrid decoders. In: IJCAI, pp. 4539–4545 (2018)
22. Yi, X., Sun, M., Li, R., Li, W.: Automatic poetry generation with mutual reinforcement learning. In: EMNLP, pp. 3143–3153 (2018)

23. Yu, L., Zhang, W., Wang, J., Yu, Y.: SeqGAN: sequence generative adversarial nets with policy gradient. In: AAAI, pp. 2852–2858 (2017)
24. Zhang, J., et al.: Flexible and creative Chinese poetry generation using neural memory. In: ACL, pp. 1364–1373 (2017)

Analyzing Research Trends in Inorganic Materials Literature Using NLP

Fusataka Kuniyoshi[1,3](✉) [iD], Jun Ozawa[1,3] [iD], and Makoto Miwa[2,3] [iD]

[1] Panasonic Corporation, 1006, Oaza Kadoma, Kadoma-shi, Osaka 571-8501, Japan
[2] Toyota Technological Institute, 2–12–1, Hisakata,
Tempaku-ku, Nagoya 468–8511, Japan
[3] National Institute of Advanced Industrial Science and Technology (AIST),
2–3–26, Aomi, Koto-ku, Tokyo 135–0064, Japan
{ozawa.jun,makoto.miwa}@aist.go.jp

Abstract. In the field of inorganic materials science, there is a growing demand to extract knowledge such as physical properties and synthesis processes of materials by machine-reading a large number of papers. This is because materials researchers refer to produce promising terms of experiments for material synthesis. However, there are only a few systems that can extract material names and their properties. This study proposes a large-scale natural language processing (NLP) pipeline for extracting material names and properties from materials science literature to enable the search and retrieval of results in materials science. Therefore, we propose a label definition for extracting material names and properties and accordingly build a corpus containing 836 annotated paragraphs extracted from 301 papers for training a named entity recognition (NER) model. Experimental results demonstrate the utility of this NER model; it achieves successful extraction with a micro-F1 score of 78.1%. To demonstrate the efficacy of our approach, we present a thorough evaluation on a real-world automatically annotated corpus by applying our trained NER model to 12,895 materials science papers. We analyze the trend in materials science by visualizing the outputs of the NLP pipeline. For example, the country-by-year analysis indicates that in recent years, the number of papers on "MoS_2," a material used in perovskite solar cells, has been increasing rapidly in China but decreasing in the United States. Further, according to the conditions-by-year analysis, the processing temperature of the catalyst material "PEDOT:PSS" is shifting below 200 °C, and the number of reports with a processing time exceeding 5 h is increasing slightly.

Keywords: Natural language processing · Text mining · Materials informatics

1 Introduction

Materials science literature includes considerable information such as material names and their properties described in natural language. Therefore, the

© Springer Nature Switzerland AG 2021
Y. Dong et al. (Eds.): ECML PKDD 2021, LNAI 12979, pp. 319–334, 2021.
https://doi.org/10.1007/978-3-030-86517-7_20

automatic extraction of the details necessary to reproduce and validate materials synthesis processes in a materials science laboratory remains difficult and requires extensive human intervention. The automatic compilation of such literature into a structured form could enable realizing a data-driven materials discovery system that does not require human intervention; such a system could become a key enabler in the design and discovery of novel materials. In this regard, named entity recognition (NER) is helpful, as it seeks to locate spans and classify named entities in unstructured text into predefined categories such as material names.

NER has already found many applications in materials science. For example, material names have been linked to their properties, such as characteristic values or their structures, through a combination of database lookup and the parsing of systematic nomenclature to create reader-friendly semantically enhanced literature [21,30]. Further, NER has been linked to material information retrieval techniques to search for materials similar to a query material from corpora [6,23–25] or to predict the characteristic values of a query material [15,22].

A technique that can extract natural language characteristic values and link a material name to a machine-readable representation will find importance in many practical applications. We believe that current research in this area is hampered by the lack of available annotated corpora.

In this study, we propose a natural language processing (NLP)-based approach to analyze trends in materials for developments in materials science (see Fig. 1). Toward this end, we propose a pipeline that integrates an NER model and a numeric normalization module. To evaluate this pipeline, we annotate 836 paragraphs extracted from 301 papers to extract material terminology and conduct initial analyses to extract material data from 12,895 unlabeled full-text literature. Through this evaluation, we demonstrate the reliability of our NLP framework by presenting the detailed NER model training process and by showing the detailed evaluation of the trained NER model. Our NER model can extract material names and several important properties such as temperature, time, conductivity, and activation energy. To demonstrate the utility of our annotated corpus and analyze the research trends, we explore the extracted outputs of our NLP framework from 12,895 unlabeled materials science papers. This study makes the following contributions:

- We propose a manually annotated corpus and an NLP framework for extracting material names with properties using an NER tagger and apply a numeric normalization module to NER outputs.
- We evaluate the reliability of the NLP system using by showing the detailed process of training the NER model with sufficient evaluation.
- We demonstrate the analysis for observing the research trends using our NLP outputs for material knowledge discovery from scientific literature.

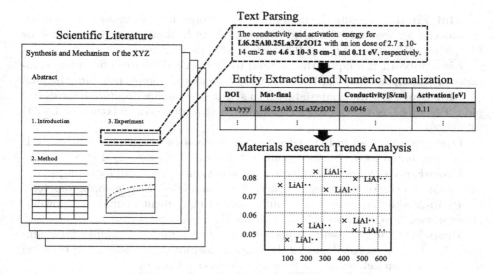

Fig. 1. Overview of pipeline for extraction.

2   Related Work

Many NLP systems and language resources are available for extracting different types of information from scientific literature, such as identifying drug names [13], discovering drugs [28], examining the side-effects of drugs [27], extracting biomedical terminology [19, 29, 32] or events [20, 34, 35], and extracting wet-lab protocols [7] are some of these examples.

In inorganic materials science, text mining is mainly used to search for a domain-specific material name or for classifying materials by their type, such as inorganic materials in general [10, 23, 25, 30, 33, 36], oxides [24], superconductors [15, 26], zeolites [22], and battery materials [2, 6, 31].

However, in inorganic materials science, few practical systems have been proposed to extract material names from a large number of papers by associating them with their property values. In this study, we propose an NLP system for extracting material names and properties.

3   Corpus Preparation

3.1   Definition of Types

Our proposed annotation scheme is based on Kuniyoshi's annotation scheme for materials synthesis processes [2]. We used 12 labels that were defined to annotate spans of text; these represent the materials, operations, and properties. In the list, we segmented the roles of materials (MAT), operations (OPE), properties (PROP), and characteristics (CHARA).

Mat-Final represents the final material (or product) of the material synthesis process; for example, "A solid solution of the lithium superionic conductor $Li_{10+\delta}Ge_{1+\delta}P_{2-\delta}S_{12}$ ($0 \leq \delta \leq 0.35$) was synthesized ..."

Mat-Solvent is a liquid that is used to dissolve substances and create solutions; for example, "Ga_2O_3 (99.999%) were ground and homogenized in <u>ethanol</u>."

Mat-Start is a raw material used to synthesize the final material; for example, "Precursor powders (10 g) containing a stoichiometric mixture of <u>La_2O_3</u> (99.998%)."

Ope represents an individual action performed by the experimenters. It is often represented by verbs; for example, "Carbon black was <u>dried</u> at 80 °C."

Prop-Equip represents equipment for analyzing a material; for example, "... spectrum analysis of the films was carried out on a <u>UV-Vis spectrophotometer</u>."

Prop-Maker represents a manufacturer of equipment or material powder; for example, "m-Cresol was obtained from <u>Sigma-Aldrich</u>."

Prop-Method represents a method to analyze a material sample; for example, "The surface morphologies of the relevant membranes were studied by using a high-resolution <u>field-emission scanning electron microscopy</u>."

Prop-Temp represents a temperature condition associated with an operation; for example, "... finally dried at <u>80 °C</u> in vacuum for 5 h."

Prop-Time represents a time condition associated with an operation; for example, "... finally dried at 80 °C in vacuum for <u>5 h</u>."

Chara-Name represents a characteristic name to classify characteristic values; for example, "... the glass ceramic has a room-temperature <u>ionic conductivity</u> as high as 3×10^{-5} S cm^{-1}."

Chara-Act represents a characteristic value of activation energy. For example, the unit of **activation energy** is eV; then, "The activation energy as a function of the vacancy concentration exhibits a minimum of <u>0.7 eV</u> ..."

Chara-Cond represents a characteristic value of conductivity. For example, the unit of **conductivity** is S/cm; then, "The ionic conductivity of the prepared pellets is <u>1.03×10^{-3} S/cm</u>."

3.2 Collecting Literature

Our corpus was constructed from papers published in the Journal of Material Chemistry A (JMCA; Royal Society of Chemistry (RSC))[1] from 2015 to 2019. JMCA focuses on energy and sustainability, and it publishes papers discussing materials such as solar cells, thermoelectric conversion materials, liquid lithium ion batteries (LIBs), and all-solid-state batteries. The RSC provides papers in XML format, in which contents have a hierarchical structure within different nested tags. For example, the <section> tag contains information such as the section title and paragraph. To extract plain text from such XML files, we created an extraction tool that exploits RSC's semantic markup features to extract information such as the title, abstract, and main contents. Then, we stripped

---

[1] https://www.rsc.org/journals-books-databases/about-journals/journal-of-materials-chemistry-a/.

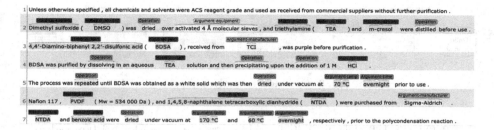

Fig. 2. Example of annotation of experimental section. Text labeling interface to annotate material names and their properties. Example of annotation of experimental sections. This text is referred from the study by Yuan [9].

out the embedded markups to produce the plain text and to create a linear stream of elements containing all data in the papers. These text data were then transferred into a document object comprising subobjects such as title, heading, and paragraph. Further, we automatically extracted paragraphs with their section and subsection titles by using regular expressions for target titles such as "Abstract," "Introduction," "Experimental," and "Conclusion."

3.3 Annotation

One Master's degree staff in the materials science department annotated labels on the 836 paragraphs extracted from 301 papers. Figure 2 illustrates annotations made to the text in the experimental section by using the brat annotation toolkit [8]. The annotated data are converted into the Inside, Outside, Beginning (IOB) scheme, where a token is labeled as I-$*$ if it is inside a named entity of type $*$, O if it is outside of named entities, and B-$*$ if it is at the beginning of an $*$ entity. Therefore, the model is trained to classify each word in a sequence into 25 different labels consisting of one O label and B-$*$ and I-$*$ labels for each of the 12 entity labels. Our corpus is shared at a github repository[2].

4 Approach

This section explains our framework for extracting material data, such as names and property values, from a large number of papers. Our framework (see Fig. 3) consists of an NER-based sequence labeling tool and a module that converts a natural language phrase to numeric values.

4.1 Sequence Labeling Architecture

First, we briefly describe bidirectional long short-term memory (BiLSTM), a type of recurrent neural network, and a subsequent conditional random field (CRF). Then, we explain the hybrid labeling architecture that is based on a previous study [17,18].

---

[2] https://github.com/BananaTonic/Material_Synthesis_Corpus.git.

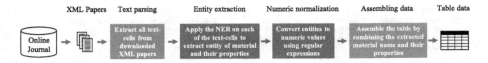

Fig. 3. Overview of analysis pipeline.

We extract the output hidden state after and before the word's token in a sentence from the corresponding forward and backward LSTMs to capture semantic-syntactic information from the beginning and ending of the sentence to the token, respectively. Both output hidden states are concatenated to form the final embedding and to capture the semantic-syntactic information of the word itself as well as its surrounding context. Figure 4 shows our proposed NER architecture.

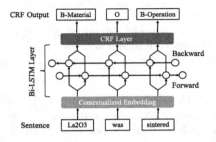

Fig. 4. Proposed NER architecture.

Let the individual tokens in a sentence be $t_0, t_1, ..., t_n$. We define the contextual string embeddings of these tokens as $h_0, h_1, ..., h_n$, where h_t represents the output hidden state of a token t. The final word embeddings are passed to a BiLSTM-CRF sequence labeling module to address downstream sequence labeling tasks.

Calling the inputs to the BiLSTM gives

$$\mathbf{r}_i \simeq [\mathbf{r}_i^f ; \mathbf{r}_i^b],$$

where $\mathbf{r}_i^f$ and $\mathbf{r}_i^b$ are forward and backward output states of the BiLSTM, respectively. The final sequence probability is then given by a CRF over the possible sequence labels $\mathbf{y}$:

$$\hat{P}(\mathbf{y}|\mathbf{r}) \propto \prod_{i=1}^{n} \phi(\mathbf{y}, \mathbf{r}),$$

where $\phi(\cdot)$ is a variation Markov Random Field of all clique potentials. Finally, the prediction of the label is given by

$$P(\mathbf{y}_i = j|\mathbf{r}_i) = softmax(\mathbf{r}_i)[j]$$

4.2 Numeric Normalization

We normalize the numeric values in a post-processing step. Although the phrases of the characteristic values extracted by the aforementioned entity extractor represent numerical values, they are annotated in various ways. Therefore, they

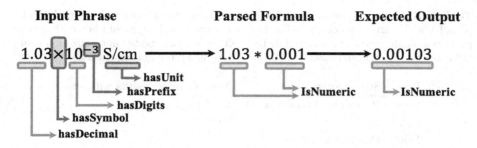

Fig. 5. Example of numeric normalization using pattern matching.

were normalized into a unified format that allows for comparisons and statistical processing. For example, when "1.03×10^{-3} S/cm" was extracted, it was normalized to a value of 0.00103 [S/cm]. In the text of the RSC paper, "10^{-3}" was described as "10⁻³." However, in this study, the XML tag was removed beforehand for simplicity, and the text was converted to the plain text "10 -3." In the case of this string, the string "0.1" was extracted as a value, "×" was extracted as a multiplication sign, and "10 -3" was extracted as the 3rd negative exponent of 10. Then, the extracted numbers were multiplied and normalized into the value "0.00103." Figure 5 shows an example of our numeric normalization, and Table 1 shows the string patterns that can be normalized by this system and their normalization results. The practical workflow of our numeric normalization is as follows: first, when we find measurement patterns in texts, we separate specific expressions into the numeric and unit parts. For example, when "14 °C – room temperature" was found, we replaced it with "14 °C – 22 °C." Next, we split extracted phrases into specific units such as "S/cm" and "°C." For example, when "irradiation times of 3 s to 8 min" was extracted, we split this as "["irradiation times of 3", "to 8"]" and "["s", "min"]". Finally, we extracted each value from split phrases when a unit had numeric patterns before it. For example, when we extracted "["0.53–0.58"]" and "["eV"]" through the previous operation, we extracted the values as the numeric values 0.53 and 0.58 and the unit "eV." In addition, there are variations in the expressions of characteristic values and units in each paper, such as the use of the ± symbol to indicate conductivity. Therefore, we used regular expressions to write down patterns of values, multiplication symbols, and powers and normalized phrases matching the written patterns into values.

5 Results

5.1 Inter-Annotator Agreement

The inter-annotator agreement (IAA) was evaluated to assess the reliability of the corpus. The IAA is calculated based on the matching of the spans of labels between two annotators who have Master's degrees in materials science. The

Table 1. Example of normalization results for temperature (Temp.), time, conductivity (Cond.), and activation energy (AE). The type indicates the type of phrase, the string pattern indicates the text extracted by the NER extractor, and the normalization result indicates the value normalized to numerical data using regular expressions.

Type	String pattern	Normalization results with unit
Temp.	Room temperature or RT	$22\,^{\circ}\mathrm{C}$
Temp.	500 K	$227\,^{\circ}\mathrm{C}$
Time	Overnight	8 h
Time	Half an hour or half a day	0.5 h or 12 h
Time	Two hours or 2 h	2 h
Cond.	1.66×10^{-4} S/cm	0.000166 S/cm
Cond.	4.2 mS/cm	0.0042 S/cm
Cond.	$4.28 \pm 0.41 \times 10^{-2}$	3.87 S/cm, 4.69 S/cm
AE	0.93–1.04	0.93 eV, 1.04 eV
AE	2.00(5) eV	1.95 eV, 2.05 eV
AE	$0.44 < Ea(eV) < 0.46$	0.44 eV, 0.46 eV

agreement score was calculated by considering the labels identified by one annotator as the gold label and those identified by the other annotator as the prediction. To evaluate the extraction performance, we performed a binary evaluation that classified all entities into either positive or negative. The precision was defined as the fraction of entities predicted as positive that are in fact positive, and recall is defined as the fraction of positive entities that are correctly predicted as positive. More precisely, for true-positive (TP), false-positive (FP), and false-negative (FN) entities, based on the entities extracted by the model, we define precision = TP/(TP + FP), recall = TP/(TP + FN), and F1-score = $2 \times$ precision $\times$ recall/(precision + recall). These validations were used to evaluate the machine extraction performance when worker A's labeling was considered the correct answer. To verify that the definitions of the types extracted are consistent among the annotators, we used the recall as an evaluation metric. Figure 6a and b show the calculated IAA scores and confusion matrix of each label, respectively, by using 60 paragraphs from 10 papers in our corpus. The result showed that the overall recall of IAA was 0.736, indicating good agreement between the two annotators. The confusion matrix showed that there were no type errors; however, there were many discrepancies owing to misses.

5.2 Comparing Language Models

Our NER Tagger trained on the corpus in this study was compared with four different language models: ELMo for materials synthesis (MatELMo) [10][3],

---

[3] https://github.com/olivettigroup/materials-synthesis-generative-models.

Label	Recall
Mat-Start	0.720
Mat-Solvent	0.529
Mat-Final	0.697
Ope	0.709
Prop-Equip	0.537
Prop-Maker	0.870
Prop-Method	0.569
Prop-Temp	0.722
Prop-Time	0.792
Chara-Name	0.867
Chara-Act	0.846
Chara-Cond	0.975
ALL	0.736

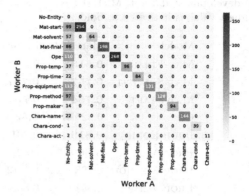

(a) Recall of each label. ALL is the overall macro-recall score.

(b) Confusion matrix of IAA.

Fig. 6. IAA results.

BERT [4][4], SciBERT [5][5], and PubmedBERT [11][6] for token embedding. For SciBERT and BERT, we used transformers [12] to obtain embeddings and to connect to the NER Tagger extractor created in Flair [1], a framework for using state-of-the-art NLP models. For evaluations, the dataset was divided in a ratio of 6:2:2 for training, development, and testing, respectively. Table 2 shows the obtained results. The Mat-ELMo language model had the highest overall micro-F1 score of 0.778. SciBERT and BERT had higher F1-scores for the extraction of conductivity and activation energy, respectively. Although the present study aims to extract material names and property values, Mat-ELMo was selected for further analysis considering raw materials, temperature, and time conditions as it had the highest overall micro-F1 score.

5.3 Tuning Hyperparameters

We performed hyperparameter tuning for the NER model employing Mat-ELMo, which showed the best extraction performance as described in Sect. 5.2. We used optuna [14], a sophisticated optimization tool, for exploring the parameter space. Figure 7b summarizes the evaluated hyperparameter space and the best parameters used for the final evaluation on the test dataset. After obtaining the optimal hyperparameter values, the model was trained again to evaluate its final performance. From the results shown in Fig. 7a, after tuning the hyperparameters, the micro-F1 score was improved from 77.8% to 78.1%.

---

[4] https://huggingface.co/bert-base-cased.

[5] https://huggingface.co/allenai/scibert_scivocab_cased.

[6] https://huggingface.co/microsoft/BiomedNLP-PubMedBERT-base-uncased-abstract-fulltext.

Table 2. F1 scores of sequence-labeling models with different base representations on development dataset. Micro-F1 scores were calculated using all labels (ALL). The highest value is indicated in bold.

Model	MatELMo	PubmedBERT	SciBERT	BERT
MAT-FINAL	**0.613**	0.572	0.595	0.543
MAT-SOLVENT	**0.757**	**0.757**	0.724	0.705
MAT-START	**0.754**	0.726	0.688	0.651
OPE	0.825	**0.835**	0.832	0.826
PROP-EQUIP	**0.819**	0.801	0.798	0.795
PROP-MAKER	**0.869**	0.827	0.794	0.816
PROP-METHOD	0.798	**0.802**	0.785	0.793
PROP-TEMP	0.851	0.855	**0.875**	0.867
PROP-TIME	0.867	**0.892**	0.883	0.887
CHARA-NAME	**0.918**	0.912	0.914	0.914
CHARA-ACT	0.593	0.571	**0.654**	0.509
CHARA-COND	0.605	0.630	0.649	**0.685**
ALL	**0.778**	0.767	0.761	0.749

5.4 Evaluation of Extracted NE Result

To verify the extraction performance of the tuned NER model, 100 paragraphs from 10 papers were labeled by NER. These paragraphs cover all sections that can be extracted from a paper. The labels were checked by the same annotator who labeled the corpus to correct any extraction errors or omissions. These validations were performed to evaluate the machine extraction performance when the machine's labeling was considered the correct answer, and the recall metric was used to evaluate the extraction performance of correctly extracted entities among the entities labeled by the NER model. Figure 8a shows the agreement between our NER model and the human annotators for each label, and Fig. 8b shows the confusion matrix.

6 Research Trends Analysis

In this section, we analyze the NLP outputs to understand the trend of materials by year. Figure 9a summarizes several key statistics of the NLP outputs, such as the number of papers, entities, and distribution of converted values. Further, Fig. 9b shows a country-by-country tabulation. Only abstracts extracted from papers were used for this tabulation, and the first author's country was counted.

Label	F1-score
MAT-FINAL	0.625
MAT-SOLVENT	0.771
MAT-START	0.771
OPE	0.827
PROP-EQUIP	0.827
PROP-MAKER	0.870
PROP-METHOD	0.813
PROP-TEMP	0.857
PROP-TIME	0.869
CHARA-NAME	0.917
CHARA-ACT	0.593
CHARA-COND	0.637
ALL	0.781

(a) Final result. ALL is the micro-F1 score.

Parameter	Range	Best
Learning rate	[0.05, 0.3]	0.15
Dropout	[0.3, 0.6]	0.3
Locked dropout	[0.3, 0.6]	0.4
Word dropout	[0.05, 0.15]	0.1
Hidden size	[32, 256]	256
RNN layers	[1, 3]	2
Weight decay	[0.0001, 0.0005]	0.0005

(b) Hyperparameters

Fig. 7. Micro-F1 score after hyperparameter tuning.

Label	Recall
MAT-FINAL	0.873
MAT-SOLVENT	0.956
MAT-START	0.751
OPE	0.997
PROP-EQUIP	0.990
PROP-MAKER	1.000
PROP-METHOD	0.987
PROP-TEMP	0.924
PROP-TIME	0.964
CHARA-NAME	0.994
CHARA-ACT	1.000
CHARA-COND	0.926
ALL	0.920

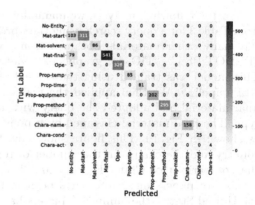

(a) Recall of each label. ALL is the macro recall score.

(b) Post-evaluation results

Fig. 8. Post-evaluation

Next, we aggregated the extracted final materials by year for a quick analysis of the trends by year, as shown in Table 3. Only the paragraph section name "Abstract" was used in the extracted papers to prevent double-counting of papers. This result shows that the trend of frequently used materials differs by year. For example, "TiO_2" is ranked fourth in 2016–2017; however, it is ranked first in 2018–2019.

Item	Count
Papers	12,895
Paragraphs	57,783
Mat-Final	919,645
Mat-Solvent	63,437
Mat-Start	406,387
Ope	277,418
Prop-Equip	175,299
Prop-Maker	55,018
Prop-Method	454,945
Prop-Temp	75,004
Prop-Time	77,889
Chara-Name	67,530
Chara-Act	14,596
Chara-Cond	31,005

Country	Count
China	7593
United States	947
Korea	884
Japan	384
India	365
UK	342
Germany	305
Australia	289
Spain	207
Singapore	204
Taiwan	150
France	133
Canada	113
Sweden	99

(a) Extracted data statistics.   (b) Top 14 countries.

Fig. 9. Base statistics

For the following analysis, we manually selected the following final materials from Table 3 for a query search using the word2vec model [37] to efficiently screen the many other materials obtained from extracted outputs: "CH$_3$NH$_3$PbI$_3$," "PEDOT:PSS," "TiO$_2$," "graphene," "ZnO," "MoS$_2$," "MOF," and "CNT." This model was trained using the same 12,895 papers that were input to the NLP pipeline. We then applied the trained word2vec model to the extracted final material and adopted the query with the highest cosine similarity as the type. The final material classified by type is used for trend analysis by country, as shown in Fig. 10. This figure shows the features of developed materials by country and the change in the number of reported materials by year. Consider the comparison of China Fig. 10a and United States Fig. 10b: in China, the number of papers on "MoS$_2$" has been increasing in recent years, whereas in the United States, the number of papers has been decreasing. Further, the most frequently reported material in China in 2015 was "TiO$_2$," whereas that in 2019 was "MoS$_2$," indicating the shifting trend in materials science research.

Table 3. Reported material final aggregation by year. Number in the brackets means number of extracted phrases.

2016 – 2017	2018 – 2019
reduced graphene oxide(21)	TiO_2(16)
$CH_3NH_3PbI_3$(20)	reduced graphene oxide(15)
graphene(18)	graphene(13)
TiO_2(17)	$CH_3NH_3PbI_3$(10)
carbon(17)	SnO_2(10)
graphene oxide(10)	carbon(9)
ZnO(9)	$MAPbI_3$(9)
PEDOT:PSS(8)	MoS_2(8)
activated carbon(7)	covalent organic frameworks(8)
$BiVO_4$(6)	MOFs(8)

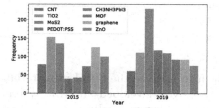 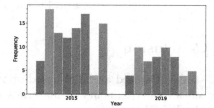

(a) Reported materials in China   (b) Reported materials in United States

Fig. 10. The year of transition of material by country and by year.

We also visualized the condition-by-year for temperature and time, as shown in Fig. 11. If multiple properties are extracted from a single paragraph, we select the property with the highest characteristic value. These results show that the trends of temperature and time when synthesizing "PEDOT:PSS" and "TiO_2" vary by year. In particular, the processing temperature of "PEDOT:PSS" shifted below $200\,^\circ C$. Further, the processing times of "PEDOT:PSS" and "TiO_2" differed in 2015, and the processing temperatures were similar in 2019. This indicates that there are similarities in the synthesis methods of "PEDOT:PSS" and "TiO_2." We have received comments from one material researcher that the results reported in this study are useful when investigating competitors and when designing material synthesis processes outside the laboratory.

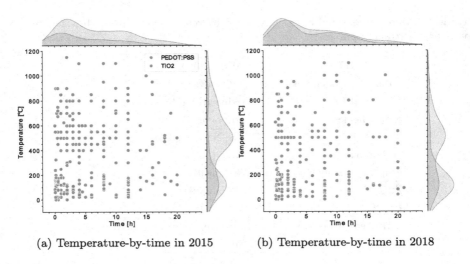

(a) Temperature-by-time in 2015   (b) Temperature-by-time in 2018

Fig. 11. Condition-by-year.

7 Conclusion

This study proposed an NLP-based approach to analyze the trend in materials research for developments in materials science. We developed an NLP system with the BiLSTM-CRF model that was trained using manually labeled literature for extracting material properties from scientific literature. We conducted experiments to verify the effectiveness of the proposed NER method in the field of materials science.

The present study has two limitations: (1) linking multiple materials and property values when they are written in a single document, and (2) extracting characteristic values from nontextual components such as charts, diagrams, and tables that provide key information in many scientific documents. We aim to overcome these limitations through our ongoing studies.

In future work, we will aim to predict characteristic values such as conductivity and materials research trends from previous scientific literature to achieve computational materials synthesis.

References

1. Akbik, A., Bergmann, T., Blythe, D., Rasul, K., Schweter, S., Vollgraf, R.: FLAIR: an easy-to-use framework for state-of-the-art NLP. In: Annual Conference of the North American Chapter of the Association for Computational Linguistics (NAACL-HLT) (2019)
2. Kuniyoshi, F., Makino, K., Ozawa, J., Miwa, M.: Annotating and Extracting Synthesis Process of All-Solid-State Batteries from Scientific Literature. In: International Conference on Language Resources and Evaluation (LREC) (2020)

3. Tshitoyan, V., et al.: Unsupervised word embeddings capture latent knowledge from materials science literature. Nature **571**, 95–98 (2019)
4. Devlin, J., Chang, M., Lee, K., Toutanova, K.: BERT: pre-training of deep bidirectional transformers for language understanding. In: Annual Conference of the North American Chapter of the Association for Computational Linguistics (NAACL-HLT) (2019)
5. Beltagy, I., Lo, K., Cohan, A.: SciBERT: a pretrained language model for scientific text. In: International Joint Conference on Natural Language Processing(IJCNLP), 2019
6. Huang, S., Cole, J.: A database of battery materials auto-generated using ChemDataExtractor. Sci. Data **7**, 260 (2020)
7. Chaitanya, K., Wei, X., Alan, R., Raghu, M.: An annotated corpus for machine reading of instructions in wet lab protocols. In: Annual Conference of the North American Chapter of the Association for Computational Linguistics (NAACL-HLT) (2018)
8. Stenetorp, P., Pyysalo, S., Topic, G., Ohta, T., Ananiadou, S., Tsujii, J.: Brat: a Web-based Tool for NLP-Assisted Text Annotation. In: European Chapter of the Association for Computational Linguistics (EACL) (2012)
9. Yuan, Q., Liu, P., Baker, G.: Sulfonated polyimide and PVDF based blend proton exchange membranes for fuel cell applications. J. Mater. Chem. **3**, 3847–3853 (2015)
10. Kim, E., et al.: Inorganic materials synthesis planning with literature-trained neural networks. J. Chem. Inf. Modeling **60**(3), 1194–1201 (2020)
11. Gu, Y., et al.: Domain-Specific Language Model Pretraining for Biomedical Natural Language Processing. In: ArXiv, abs/2007.15779 (2020)
12. Wolf, T., et al.: Transformers: state-of-the-art natural language processing. In: Empirical Methods in Natural Language Processing (EMNLP) (2020)
13. Krallinger, M., et al.: The CHEMDNER corpus of chemicals and drugs and its annotation principles. J. Cheminformatics **7**, S2–S2 (2015)
14. Takuya, A., Shotaro, S., Toshihiko, Y., Takeru, O., Masanori, K.: Optuna: a next-generation hyperparameter optimization framework. In: International Conference on Knowledge Discovery and Data Mining (SIGKDD) (2019)
15. Court, C.J., Cole, J.: Magnetic and superconducting phase diagrams and transition temperatures predicted using text mining and machine learning. npj Comput. Mater. 6, 1–9 (2020)
16. Kononova, O., et al.: Opportunities and challenges of text mining in aterials research. iScience **24**, 3 (2021)
17. Akbik, A., Blythe, D., Vollgraf, R.: Contextual string embeddings for sequence labeling. In: International Conference on Computational Linguistics (COLING) (2018)
18. Huang, Z., Xu, W., Yu, K.: Bidirectional LSTM-CRF Models for Sequence Tagging. In: ArXiv, abs/1508.01991 (2015)
19. Bada, M., et al.: Concept annotation in the CRAFT corpus. BMC Bioinform. **13**, 161 (2011)
20. Kim, J.D., Ohta, T., Tateisi, Y., Tsujii, J.: GENIA corpus - a semantically annotated corpus for bio-textmining. BMC Bioinformatics **19**(Suppl 1), i180-2 (2003)
21. Tshitoyan, V., et al.: Unsupervised word embeddings capture latent knowledge from materials science literature. Nature **571**, 95–98 (2019)
22. Jensen, Z., et al.: A Machine Learning Approach to Zeolite Synthesis Enabled by Automatic Literature Data Extraction. ACS Central Science **5**, 892–899 (2019)

23. Kim, E., Huang, K., Saunders, A., McCallum, A., Ceder, G., Olivetti, E.: Materials Synthesis Insights from Scientific Literature via Text Extraction and Machine Learning. Chemistry of Materials **29**, 9436–9444 (2017)
24. Kim, E., et al.: Machine-learned and codified synthesis parameters of oxide materials. Scientific Data **4**, 170127 (2017)
25. Young, S.R., et al.: Data Mining for better material synthesis: the case of pulsed laser deposition of complex oxides. J. Appl. Phys. **123**, 115303 (2018)
26. Yamaguchi, K., Asahi, R., Sasaki, Y.: SC-CoMIcs: a superconductivity corpus for materials informatics. In: International Conference on Language Resources and Evaluation (LREC) (2020)
27. Jeong, Y.K., Xie, Q., Yan, E., Song, M.: Examining drug and side effect relation using author-entity pair bipartite networks. J. Informetrics **14**, 100999 (2020)
28. Hansson, L., et al.: Semantic text mining in early drug discovery for type 2 diabetes. PLoS ONE 15(6) (2020)
29. Rebholz-Schuhmann, D., Oellrich, A., Hoehndorf, R.: Text-mining solutions for biomedical research: enabling integrative biology. Nature Reviews Genetics **13**, 829–839 (2012)
30. Kononova, O., et al.: Text-mined dataset of inorganic materials synthesis recipes. Scientific Data **6**, 203 (2019)
31. Mahbub, R., Huang, K., Jensen, Z., Hood, Z.D., Rupp, J., Olivetti, E.: Text mining for processing conditions of solid-state battery electrolyte. Electrochemistry Commun. **121**, 106860 (2020)
32. Weber, L., Sänger, M., Munchmeyer, J., Habibi, M., Leser, U.: HunFlair: an easy-to-use tool for state-of-the-art biomedical named entity recognition. BMC Bioinfomatics (2021)
33. Huo, H., et al.: Semi-supervised machine-learning classification of materials synthesis procedures. npj Computational Materials 5, 1–7 (2019)
34. Miwa, M., Thompson, P., Ananiadou, S.: Boosting automatic event extraction from the literature using domain adaptation and coreference resolution. BMC Bioinf. **28**, 1759–1765 (2012)
35. Björne, J., Salakoski, T.: Biomedical event extraction using convolutional neural networks and dependency parsing. In: Workshop on Biomedical Natural Language Processing (BioNLP) (2018)
36. Mysore, S., et al.: The materials science procedural text corpus: annotating materials synthesis procedures with shallow semantic structures. In: Linguistic Annotation Workshop (LAW) (2019)
37. Mikolov, T., Sutskever, I., Chen, K., Corrado, G.S., Dean, J.: Distributed representations of words and phrases and their compositionality. In: Neural Information Processing Systems (NeurIPS) (2013)

An Optimized NL2SQL System
for Enterprise Data Mart

Kaiwen Dong(✉) ⓘ, Kai Lu, Xin Xia, David Cieslak, and Nitesh V. Chawla ⓘ

Aunalytics, South Bend, IN 46545, USA
{kevin.dong,kai.lu,xin.xia,david.cieslak,nitesh.chawla}@aunalytics.com

Abstract. Natural language interfaces to databases is a growing field that enables end users to interact with relational databases without technical database skills. These interfaces solve the problem of synthesizing SQL queries based on natural language input from the user. There are considerable research interests around the topic but there are few systems to date that are deployed on top of an active enterprise data mart. We present our NL2SQL system designed for the banking sector, which can generate a SQL query from a user's natural language question. The system is comprised of the NL2SQL model we developed, as well as the data simulation and the adaptive feedback framework to continuously improve model performance. The architecture of this NL2SQL model is built on our research on WikiSQL data, which we extended to support multitable scenarios via our unique table expand process. The data simulation and the feedback loop help the model continuously adjust to linguistic variation introduced by the domain specific knowledge.

Keywords: Semantic parsing · Natural language interface · Database · Language model

1 Introduction

Natural language interfaces to databases (NLIDB) [1] provide a way of interacting with relational databases by simply typing a question or Statement in natural language. This problem has been studied extensively, with early work in this field focusing on rule-based [4,15,21,22] semantic parsing. The rule-based methods proved to be effective but lacked the ability to cover the linguistic variation and sophistication of end users. As deep neural networks have achieved state-of-the-art performance on numerous tasks around unstructured data [11,23], the research interests of NLIDB have shifted to incorporate deep learning based approaches. Recent advances [8,16,26,32] in the field have leveraged the release of large scale human-labeled datasets [33,36] for model training and evaluation. However, there are few [34] that can be deployed on an actual enterprise data mart in production.

From a system and algorithmic perspective, NLIDB are difficult to develop and maintain as they require a substantial amount of expertise in machine learning methods, database architecture, microservices frameworks, infrastructure

© Springer Nature Switzerland AG 2021
Y. Dong et al. (Eds.): ECML PKDD 2021, LNAI 12979, pp. 335–350, 2021.
https://doi.org/10.1007/978-3-030-86517-7_21

management, and DevOps practice. This type of application presents additional, non-obvious challenges for the machine learning practitioner. To begin with, an NLIDB faces an absolute dearth of available training data. Most ML methods require thousands – if not tens of thousands – of examples for reliable training. The most common bootstrap approach is to develop manually handcrafted example-and-label pairs; however, in this application building a suitable corpus upfront is extremely expensive and sometimes practically impossible. Should an NLIDB system complete initial training to satisfaction, it will face numerous ongoing operational issues. In some instances, new users will enter queries in unexpected ways and the system must accommodate feedback in order to continuously improve performance over time. Likewise, new versions of the underlying data model may incorporate new fields or tables, and it is critical that the system maintain expected performance on the original information while simultaneously integrating new database structures.

In this paper, we describe an NL2SQL system for the enterprise data mart [9] that can democratize access to relational databases for users without technical skills like SQL by allowing them to find meaningful insights and decisions with natural language. We will unfold how we developed this model in four sections. In the first section, we provide a functional description of the NL2SQL system including its web-based user interface, schema system and feedback logging mechanism. In the second section, based on the previous work of word contextualization [8] methods with a BERT-based encoder [3,25], we discuss our design methodology and model architecture including its distinct subtasks and the novel table expand methodology we developed to support queries for a multi-table data mart. In the third, we discuss capability of being continuously optimized by a template-based data simulation which grows with the history of users' interactions with the model, which can generate data that improves performance on domain-specific language patterns and adapt to changes in the underlying data mart it is intended for. Fourth and finally, we offer the results of our benchmarking experiments that showed our model performing comparably to the other state-of-the-art models when trained on generic datasets from WikiSQL and Spider train, but significant improvements over that model when trained with our own template-simulated datasets.

2 Related Work

The release of the WikiSQL dataset [36] has raised interest in applying deep learning models to solve the text-to-SQL problem. Zhong et al. [36] introduced Seq2SQL, a sequence-to-sequence neural network. Xu et al. [32] proposed SQL-Net structure using a sequence-to-set approach, which solves the order issue of the conditions in a SQL query. It also offers a column attention mechanism to identify the most relevant column for the natural language question. With transformer-based language models dominating most NLP tasks, Hwang et al. [8] leveraged the BERT [3] pretrained model and the stacked bidirectional LSTM [5] to construct a two-layer encoder and contextualize the natural language question with the headers. However, the research on WikiSQL is limited to one-table scenarios, due to the set structure of the dataset itself. Spider [33] proposes cross-domain text-to-SQL datasets across 200 databases, each database involving multiple tables with foreign keys. With the wide adoption of BERT in the encoding layer, Wang et al. [26] and Lin et al. [16] choose to concate-nate the question tokens along with the table and column name tokens, which are fed into the encoder to contextualize the word representation.

To provide a comprehensive solution for NLIDB, Li et al. [15] and Setlur et al. [21, 22] described systems with a rule-based parser and Dhamdhere et al. [4] discussed several implementation lessons and key design decisions for an industrial text-to-SQL tool. Zeng et al. [34] introduced a system consisting of a neural semantic parser, a question corrector, a SQL executor, and a response generator to tackle the task. For medical records information retrieval, Wang et al. [27] proposed a text-to-SQL system for relational databases at clinical centers. Data synthesis is essential when adapting the pretrained model to a specific domain; others have discussed several approaches [10, 27, 28, 35] to simulate the data consistent with the target domain where the system is being deployed.

3 NL2SQL System

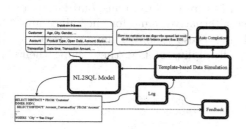

Fig. 1. Overview of the system workflow

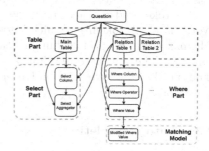

Fig. 2. The syntax-guided sketch and modules dependency

The design of our system is intended for practical industry application. Figure 1 shows the overview of the entire workflow. The system takes the user's question and the database schema as input and generates an executable query sent to a SQL engine to return the query result. While the user is typing the question, an auto-completion feature helps the user phrase their question. The NL2SQL model will then process the question and the schema. If the user is satisfied with the result, the query will be executed, otherwise the user can optionally submit feedback indicating the inaccuracy of the result.

The log is a place where we monitor the health of the model and seek opportunities to improve the model performance. A significant impact of the log is that it can guide the template writing for the data simulation process. The data simulation is the source for the training set of the NL2SQL model as well as the vocabulary for the auto-completion module. The template is updated continuously based on the logs generated by the users. This feedback loop helps us adjust the system to better adapt to users' language in a specific domain.

3.1 Question Textbox

The question text box is a standard text box, where the user inputs their question and submits it to the system. Hitting the submit button will launch a HTTP POST request to the server side and start the processing. There is also a collapsible schema viewer to remind the user of supported tables/columns in the database.

3.2 Schema

A relational database schema provides metadata like table names, column names, column types and foreign keys. Due to the abbreviated naming and blank space issues in the database world, the words of a column name are more likely to be concatenated together or linked by underscore and hyphen. This string format can cause problems when tokenizing. For instance, "AccountProductType" will be tokenized as "account", "#pro", "##du", "##ct", "##type" by the WordPiece [32] tokenizer, which leads to misinterpretations. Thus, our database schemas hold an optional human-readable alternative name for the tables and columns. The database developer can change these synonym names based on their need. This gives the end user more flexibility of how they shape their questions.

3.3 Auto Completion

Auto completion is another add-on feature embedded in the question textbox (Sect. 3.1) Apart from the general advantage of helping users formulate their questions and reduce user-introduced typos, our design also guides the user to compose a question more likely to be recognized by the text-to-SQL parser. The engine we use is Elasticsearch [6]. We index the suggestion with the phrases (discussed in Sect. 5.2) generated by the data simulation process, which will be further used as a source of training data.

3.4 Log

With the consent of the user, the logging system actively collects all incoming requests, the model execution log, and any feedback submitted by the user. The log can help quantify trends in the system usage and user satisfaction over time.

Log information is an essential source for continuous model improvement. Following a human-in-the-loop strategy, we pull the logs and analyze the system performance regularly. There are several aspects we look for in the logs: 1) the most asked-for tables and columns in the users' questions; 2) the questions that failed to be recognized by the system and are reported by users using the thumbs-down feedback mechanism; 3) the error messages; 4) the average latency of the request processing time.

The first two types of log data helps shape the template writing for data simulation, which provides a valuable channel to correct the bias of a language model like BERT [13,24]. Log of error messages can capture the unexpected runtime errors. The average latency is also an important measurement indicating whether there is a need to scale up the service cluster.

4   Method

4.1   Problem Statement

We wanted to build an end-to-end model which takes a natural language question and database schemas (and potentially the data of the database if applying matching process Sect. 4.5) as the input and generate SQL output. The query to be parsed is multi-table SQL without nested queries. To simplify the data structure of the SQL query, the SQL is converted into a logical form following the sketch style of SQLNET [32]. The sketch can ensure that the model always formulates the SQL query in a correct syntax. We have extended this sketch to support multi table samples. The query's component in this paper is always of the logical form.

Complexity always comes with flexibility in the SQL language. Even though the syntax-guided sketch [32] cannot completely cover the functionality of SQL, we still decided to employ it because of its standard structure. We are using this sketch to display the SQL query in an easier format for the user to understand. This can help the user make better decisions about whether the results from the system are desired.

The model is applied to the data model where foreign keys are predefined by a star schema [20]. Therefore, when generating the SQL query from the logical form, the database schema, instead of the NL2SQL model itself, will provide the necessary foreign keys to compose the SQL query for execution.

4.2   Model Overview

Following SQLOVA [8], the NL2SQL model is a sequence of sub-task classification models including SELECT column (sc), SELECT aggregation (agg), WHERE number (length of "conds", wn), WHERE column (wc), WHERE operator (wo) and WHERE value index (wvi). The tasks of WHERE value are tackled as a classification problem of locating the start and end tokens within the user's question as SQLOVA [8]. Besides the tasks above, we introduce 3 new tasks at the database level: main table (mt), relation table (rt), and relation number (length of "rt", rn), to parse the tables needed in the query.

There are 9 tasks in total with each model and these are formed in an encoder-decoder structure. All components share the same language model, BERT [3], as the first encoding layer, but the input can be different. Each module has its own bidirectional LSTM encoding layer. On top of the encoder, each task has its own classification layer. The tasks can be categorized into 3 parts based on their SQL clause: 1) Table part, including mt,rn,rt; 2) SELECT part, including sc,agg; 3) WHERE part, including wn,wc,wo,wvi. Categorizing modules into 3 parts gives us the advantage of only training a single part of modules when the dataset consists of both one-table and multi-table samples.

All the individual task module are similar to SQLOVA [8]. However, to accommodate multi-table requirements, we formulate a novel method called Table Expand to convert multiple tables per sample to one table in multiple samples.

4.3   Table Part

Similar to SQLOVA, Table part's input X_{Table} to the language model is composed of the tokenized natural language question and all the tokenized table names in the database. Question tokens and table name tokens are separated by a special token $[SEP]$. The tasks mt, rn and rt of the table part can be seen as the sc, wn and wc of the where part at the table level.

4.4   Table Expand

Once the relation tables $\hat{rn}s$ of the database are selected, we can expand one sample to multiple samples by selected relational tables, each of which comprises a single selected table and the same question.

For instance, if two tables are selected after the Table Part module, then two inputs are fed to the BERT and the WHERE Part module. One includes the question and the column names from the first selected relation table, and the other includes the same question and the column names from the second selected relation table.

After the WHERE Part module processes all the expanded inputs and generates the WHERE number, WHERE column, WHERE Operator, and WHERE Value Index for each table, they are assembled back to be one sample again.

One sample of multiple tables can be expanded to several samples as below:

$$X_{Column} = [CLS], Q_1, \ldots, Q_{L_Q}, [SEP], H_{1,1}, H_{1,2}, \ldots, H_{1,L_1}, [SEP], \ldots, [SEP], H_{N,1}, \ldots, H_{N,L_N}, [SEP]; (\hat{rt}_j); j = 1, 2, \ldots, \hat{rn} \quad (1)$$

where $\hat{rt}_j$ represents the index of the j-th selected relation table. $H_{i,L}$ is the L-th name token of the i-th column of table $\hat{rt}_j$. There are $\hat{rn}$ inputs in total.

The table expand can shrink the column search space for the model by splitting the selected relation tables to individual bins and limiting the number of columns to rank. Per each expanded input, the WHERE Number module will look for the number of WHERE columns $\hat{wn}_j$ from the current relation table

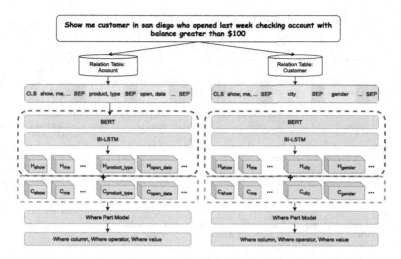

Fig. 3. Process of table expand and encoding

associated with the natural language question, instead of the total number of where conditions (Fig. 3).

Another advantage of table expand is that it allows the schema to have more columns and more tokens for each column, despite limitation of BERT [3] input size. In the regime of the enterprise data mart, the size of a schema is usually larger compared to the Spider dataset [33], which is collected from public resources online. The schema of the database can impact the feasibility of the model and its performance. As most of the recent approaches to the Spider dataset share the idea of applying language models like BERT to contextualize the token sequences, it can potentially limit the number of columns to be recognized by the model. However, our method, by splitting the whole schema into individual tables when feeding the word sequence, can significantly increase the size of the schema supported by the model.

4.5 Where Value Matching

WHERE value can be obtained by locating the tokens in the substring of the original question with the start and end of WHERE value index. However, the extracted WHERE value usually does not match the exact cell value in the database, which can cause the query to be non-executable. For different data types, we set up different solutions to map the substring to the cell value in the database in Table 1.

Categorical Column. FWe employ approximate string matching using Levenshtein distance [14,18] to find the closest cell value in the predicted column compared to the extracted substring, which can help correct the syntactically similar string.

Table 1. Matching process for different types of data

Date type	Problem	Question	Substring	Table cell
Categorical	Case or form doesn't match	Show me **mortgages**	Account_Type = "mortgages"	Account_Type = "Mortgage"
	No cell value present in question	Which customer **doesn't have** mobile bank?	HasMobileBank = "doesn't have"	HasMobileBank = "No"
		Customers with **dda**	ProductCategory = "dda"	ProductCategory = "Demand Deposit Account"
Datetime	The datetime format doesn't match	Accounts opened since **2018**	OpenDate> "2018"	OpenDate> "2018-01-01T00:00:00.000Z"
	Can't parse relative time expression	Accounts opened **this year**	OpenDate ="this year"	OpenDate≥ "2021-01-01T00:00:00.000Z" And OpenDate< "2022-01-01T00:00:00.000Z"
Numeric	The data type doesn't match	Accounts with balance more than **$100**	CurrentBalance > "$100"	CurrentBalance> 100.0

Because users usually don't type in the cell value explicitly in their question, it can raise the semantically close but syntactically different issue. We have an interim step when applying the approximate string matching. Instead of directly converting a substring to the cell value, we build a map dictionary between the cell value and the alternative strings. For instance, the binary value "Yes" or 1 will be mapped to a set of affirmation words like "with", "have", "has", "is", "are", while the binary value "No" or 0 will be mapped to a set of negation words like "without" , "don't", "doesn't", "isn't", "aren't". Cell value "dda" can be mapped to "Demand Deposit Account". When correcting the substring, we just find closest alternative value and then map it to the real cell value.

The building of the map dictionary is a rule-based iterative process. Even though we tried to apply other word representation technique like [19] to automate it, we found it is easier and more efficient to involve human-in-the-loop when solving these synonym wording requiring domain knowledge.

As the cell value needs to be preloaded, due to latency concerns, we will only string match for columns which have a relatively stable range of values. For example, the value set of column "Transaction_Type" is more consistent than "Transaction_Merchant_Name" over time.

In practice, we also set an empirical threshold to the distance depending on the sensitivity required for the matching process in case that the user indeed needs to query values not existing in the database.

Datetime Column. Parsing datetime text into structured data is a challenging problem. A question that includes a datetime value can be either absolute or relative. We utilize the Duckling [7] library to parse the extracted value to a formatted datetime interval following ISO8061 standard.

Numeric Column. A simple regular expression is applied to remove any non-number character in the substring, except the decimal point and the minus sign.

5 Template-Based Data Simulation

Previous work in the healthcare domain [27] has dealt with specific medical terminology and the lack of questions to a SQL dataset in that domain. As the banking sector is the target domain to introduce our NL2SQL interface, optimization for this specific domain is required. In both domains, the cost of acquiring new training samples which pair natural language and queries is very expensive or difficult to acquire. The template-based data simulation provides a way to directly intervene with the model's capability for a specific domain. In production, the process can serve as a powerful tool to quickly correct the model when the user gives feedback about queries that could not be recognized. We also collect more language templates from the feedback and add them to our existing data simulation template.

The end goal of the data simulation is to generate pairs of the natural language questions and their corresponding queries. There are two steps to simulate the training samples, creating the query and creating the corresponding natural language question.

5.1 SQL Query

The SQL query samples are generated based on the real databases where the NL2SQL interface is going to deploy. However, production data containing sensitive information must be substituted with dummy data.

The sample generation is a fill-in-slot process. After the number of conditions is randomly assigned, a permutation of wn will be picked of which the sum is the number of conditions. Then, all the other slots will be filled in randomly based on the database. mt is selected from the table names and the sc will be selected from table mt. agg can be no operation, AVG, COUNT, MAX, or MIN when sc is numerical column but can only be no operation or COUNT when sc is of categorical type. wc are selected from table rt based on the number wn. wo is selected from "=",">","<","! =" for the numerical column and "=", "! =" for string type. The wv will be sampled from the table cells of the database. A random state parameter of the generator is also required for purposes of reproducibility.

5.2 Natural Language Question

The template-based natural language generator can compose a question sentence corresponding to a SQL query. The template of the natural language needs to be customized for each data mart domain. Without applying any autonomous data augmentation on the natural language, we apply a rule-based simulation process for better control over what language expressions are generated. Thus, every word, except those coming from database values, is from the template we provide.

Each synthetic natural language question can be seen as a **Sentence** composed of **Phrases**. Each Phrase only holds the information of one condition in the **conds**, while a Sentence accounts for the whole query. The entire simulation process is to generate the Phrases and assemble them into a Sentence. The assumption of Phrase generation is that each column of the database can have its own expression style.

For each Phrase template, there are 2 major components: the column template and the value placeholder. The column template is composed of constant strings and placeholders for one column. The value placeholder will be substituted based on the elements of the condition in the query.

The column template needs to be set up for every field of all the tables in the database. If no template is specified for a column, a default one is used. The general workflow of composing a natural language question "Show me customers in San Diego who opened a checking account with balance greater than $100 last week" can be illustrated as follows:

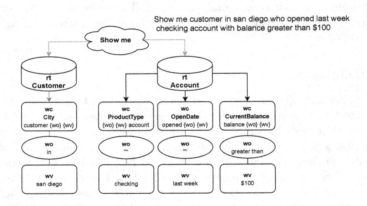

Fig. 4. Natural language generation from template-based data simulation

First, we need to generate each column phrase. The example contains four columns "City", "ProductType", "OpenDate", and "CurrentBalance" from "Account" and "Customer" tables. For "ProductType", it selects the "wo wv account" column template and then fills an empty string to the wo and the "checking" product type to the wv. For "CurrentBalance", the "balance wo $wv" template is chosen where wo and wv are filled with "greater than" and "100" accordingly. Similar approaches are applied for the remaining columns. After all conditions' Phrases have been realized, we assemble them into a whole Sentence (see Fig. 4).

5.3 Iterative Template Writing

In practice, the data simulation is an iterative process. Before exposing the model to any end user, we set up an initial version of the data simulation templates and trained the alpha model from it. Then, we conducted the first round of user acceptance testing (UAT) to gather real data from end users to the log, including the natural language question and possibly the expected SQL queries. This data is collected as our benchmark dataset for testing purpose. Instead of pouring this real data into the next round of training set directly, we firstly investigated those queries marked as "thumbs-down", and figured out what language pattern, observed in user's queries, can't be generated based on template language. For instance, there was only "customer (in) (san diego)" generated by template initially, but we've seen queries from the users like "customer **(from)** (san diego)", "**(san diego)** customer", and "**members** (in) (san diego)". Thus, we accommodate these language variations into the next round of data simulation by adding to templates.

Then, the model of next round will be trained entirely from the initial state but based on this new template. This process will be performed iteratively on certain cadence or on demand, which allows the model to evolve along with the utilization from users.

6 Experiment

6.1 Data

For the production model serving users, we build up our training dataset by the template-based data simulation process together with external data sources to improve the optimization to the focus domain for banking as well the model's robustness to linguistic variation.

We have two external data sources: WikiSQL [36] and Spider [33]. WikiSQL is limited to one-table scenarios so it can only be used to train the SELECT and WHERE parts of the entire model. Spider has more complex SQL queries like nested queries, which is not suitable for our model. In order to take advantage of the Spider train dataset, we only keep those queries that are compatible with our model. The criteria used to clean the external data sources include: 1) the total length of the concatenated input tokens to the BERT encoding layer tokenized by the WordPiece [31] tokenizer won't exceed the allowed maximum length, which is 512; 2) the where value index can be parsed through CoreNLP tokenizer [17]; 3) the SQL query can be represented in logical form [32]. After cleanup, there are only 2286 samples from Spider train and 205 from Spider dev satisfying our needs. We denote them as $Spider_{Select}$ train and $Spider_{Select}$ dev. As the Spider test set is not publicly accessible, we use the $Spider_{Select}$ dev as the test set. $Spider_{Select}$ dev has 90 easy, 93 medium, 20 hard and 2 extra-hard question, defined by Spider. We only select 2000 samples from WikiSQL as $WikiSQL_{Select}$.

We also collected our own benchmark dataset from 3 rounds of UAT consisting of 289 samples, which can represent a wide range of user's input questions.

We used the $Spider_{Select}$ dev and our collected benchmark dataset as the test sets. WikiSQL is not usable because it only contains one-table samples.

6.2   Experiment Settings

The pretrained language model that we employ is the uncased BERT-base from Huggingface's library [30]. The entire dataset is composed of the synthesis pairs of NL questions and queries from the data simulation and samples from $WikiSQL_{Select}$ and $Spider_{Select}$ train set. The total number of samples are 7886. It is further separated into the train set and dev set. We use mini-batch size 1 and an early stop criterion on the dev dataset. The Adam optimizer [12] was applied. Other settings are the same as Hwang et al. [8].

The entire NL2SQL model consists of a sequence of 9 successive modules. The downstream tasks often rely on the result of the upstream tasks (see Fig. 2). Thus, we employ teacher forcing [29] during the training phase for more efficiency.

We used a GPU to train the model, but CPU during inference. The UAT has confirmed the inference time for a request is acceptable to users, which is around 800ms on average. It allows us to deploy the system on clusters without GPU resources, which is more scalable.

6.3   Experiment Results

Table 2. Exact match accuracy comparison on $Spider_{Select}$ Dev

$Spider_{Select}$ Dev (205 samples)	Easy	Medium	Hard
Bridge v2 + BERT [16]	0.89	0.53	0.25
NL2SQL by data simulation + $WikiSQL_{Select}$ + $Spider_{Select}$ train	0.71	0.52	0.4
NL2SQL by $WikiSQL_{Select}$ + $Spider_{Select}$ train	0.56	0.28	0.2

We use the Bridge model by Lin et al. [16], which is the top model on the Spider leaderboard at the time of writing, as a comparison for the logical form exact match accuracy. In Table 2, our model trained on data simulation, $WikiSQL_{Select}$ and $Spider_{Select}$ train set underperform on the Easy queries of $Spider_{Select}$ Dev, compared to the Bridge model. However, our model catches up with the Bridge model on Medium and exceeds on Hard. Our model was exposed to a small portion of the original Spider train set while the Bridge was trained on the entire one. The Spider train and dev set shares the same databases and domains. When the expected SQL becomes more complex on Medium and Hard, the syntax-guided sketch of our model starts to show the advantage of composing longer and more difficult SQL queries.

We also trained a model on $WikiSQL_{Select}$ and $Spider_{Select}$ train set. It shows that a tool which can increase the size of training data like this simulation method can significantly improve the model performance on other domains even though the templates are not built for cross-domain data generation.

Table 3. Accuracy comparison on our banking benchmark

Banking benchmark (289 samples)	mt	rn	rt	sc	agg	wn	wc	wo	wv	Exact match
Bridge v2 + BERT										0.01
NL2SQL by data simulation + $WikiSQL_{Select}$ + $Spider_{Select}$ train	0.82	0.99	0.73	0.74	0.99	0.7	0.67	0.66	0.62	0.45
NL2SQL by $WikiSQL_{Select}$ + $Spider_{Select}$ train	0.8	0.35	0.28	0.2	0.85	0.15	0.07	0.06	0.03	0.01

In Table 3, the performance of our model is broken down into the different sub tasks introduced in Sect. 4.2. As the output of Bridge model is SQL, we parse it into the syntax-guided sketch for evaluation, which can prevent grammatical errors. The Bridge model and our model trained without data simulation underperform on our banking benchmark dataset collected from the UAT. Our model with the data simulation process gains significant improvement to 45% exact match accuracy because of its better adaptation to the linguistic patterns in the domain. Note that the Matching Process is applied to both the Bridge and our model in this comparison.

7   Conclusion

This paper presents an optimized system for the enterprise data mart, including auto completion, neural semantics parser, cell value matching process and data simulation method. With the feedback loop and data simulation, our system has the potential to be applied on any enterprise data mart in different domains. It can remove a significant barrier to entry for querying databases for many participants without SQL knowledge, enabling non-technical users to perform analyses and make decisions.

Our system has been deployed and opened to users on our analytical database [2] for the banking sector. We are also working on presenting data visualization to users by interpreting their intention in the questions in order to provide graph visualizations in response to natural language questions. We will continue to explore the syntax-guided sketch to extend the coverage of questions our system can answer.

References

1. Androutsopoulos, I., Ritchie, G.D., Thanisch, P.: Natural language interfaces to databases - an introduction. CoRR cmp-lg/9503016 (1995). http://arxiv.org/abs/cmp-lg/9503016
2. Aunalytics: Dayreak analytic database. https://www.aunalytics.com/products/daybreak/
3. Devlin, J., Chang, M., Lee, K., Toutanova, K.: BERT: pre-training of deep bidirectional transformers for language understanding. CoRR abs/1810.04805 (2018). http://arxiv.org/abs/1810.04805
4. Dhamdhere, K., McCurley, K.S., Nahmias, R., Sundararajan, M., Yan, Q.: Analyza: exploring data with conversation. In: Proceedings of the 22nd International Conference on Intelligent User Interfaces, pp. 493–504. IUI 2017. Association for Computing Machinery, New York, NY, USA (2017). https://doi.org/10.1145/3025171.3025227, https://doi.org/10.1145/3025171.3025227
5. Dong, L., Lapata, M.: Coarse-to-fine decoding for neural semantic parsing. In: Proceedings of the 56th Annual Meeting of the Association for Computational Linguistics (Volume 1: Long Papers), pp. 731–742. Association for Computational Linguistics, Melbourne, Australia. July 2018. https://doi.org/10.18653/v1/P18-1068, https://www.aclweb.org/anthology/P18-1068
6. Elastic: Elasticsearch. https://www.elastic.co/enterprise-search
7. Facebook: Duckling. https://duckling.wit.ai/
8. Hwang, W., Yim, J., Park, S., Seo, M.: A comprehensive exploration on WikiSQL with table-aware word contextualization. CoRR abs/1902.01069 (2019). http://arxiv.org/abs/1902.01069
9. Inmon, B.: Data mart does not equal data warehouse (1999)
10. Iyer, S., Konstas, I., Cheung, A., Krishnamurthy, J., Zettlemoyer, L.: Learning a neural semantic parser from user feedback. In: Proceedings of the 55th Annual Meeting of the Association for Computational Linguistics (Volume 1: Long Papers), pp. 963–973. Association for Computational Linguistics, Vancouver, Canada, July 2017. https://doi.org/10.18653/v1/P17-1089, https://www.aclweb.org/anthology/P17-1089
11. Janai, J., Güney, F., Behl, A., Geiger, A.: Computer vision for autonomous vehicles: problems, datasets and state of the art. Foundations Trends® Comput. Graph. Vis. **12**(1–3), 1–308 (2020). https://doi.org/10.1561/0600000079, http://dx.doi.org/10.1561/0600000079
12. Kingma, D.P., Ba, J.: Adam: a method for stochastic optimization. In: Bengio, Y., LeCun, Y. (eds.) 3rd International Conference on Learning Representations, ICLR 2015, 7–9 May 2015, San Diego, CA, USA, Conference Track Proceedings (2015). http://arxiv.org/abs/1412.6980
13. Kurita, K., Vyas, N., Pareek, A., Black, A.W., Tsvetkov, Y.: Measuring bias in contextualized word representations. In: Proceedings of the First Workshop on Gender Bias in Natural Language Processing, pp. 166–172. Association for Computational Linguistics, Florence, Italy, August 2019. https://doi.org/10.18653/v1/W19-3823, https://www.aclweb.org/anthology/W19-3823
14. Levenshtein, V.I.: Binary Codes Capable of Correcting Deletions. Insertions and Reversals. Soviet Physics Doklady **10**, 707 (1966)

15. Li, F., Jagadish, H.V.: NaLIR: an interactive natural language interface for query-ing relational databases. In: Proceedings of the 2014 ACM SIGMOD International Conference on Management of Data, pp. 709–712. SIGMOD 2014. Association for Computing Machinery, New York, NY, USA (2014). https://doi.org/10.1145/2588555.2594519, https://doi.org/10.1145/2588555.2594519

16. Lin, X.V., Socher, R., Xiong, C.: Bridging textual and tabular data for cross-domain text-to-SQL semantic parsing. In: Findings of the Association for Computational Linguistics: EMNLP 2020, pp. 4870–4888. Association for Computational Linguistics, Online, November 2020. https://doi.org/10.18653/v1/2020.findings-emnlp.438, https://www.aclweb.org/anthology/2020.findings-emnlp.438

17. Manning, C., Surdeanu, M., Bauer, J., Finkel, J., Bethard, S., McClosky, D.: The stanford CoreNLP natural language processing toolkit. In: Proceedings of 52nd Annual Meeting of the Association for Computational Linguistics: System Demonstrations, pp. 55–60. Association for Computational Linguistics, Baltimore, Maryland, June 2014. https://doi.org/10.3115/v1/P14-5010, https://www.aclweb.org/anthology/P14-5010

18. Navarro, G.: A guided tour to approximate string matching. ACM Comput. Surv. **33**(1), 31–88 (2001). https://doi.org/10.1145/375360.375365, https://doi.org/10.1145/375360.375365

19. Pennington, J., Socher, R., Manning, C.D.: GloVe: global vectors for word representation. In: Empirical Methods in Natural Language Processing (EMNLP), pp. 1532–1543 (2014). http://www.aclweb.org/anthology/D14-1162

20. Peterson, S.: Stars: A pattern language for query optimized schema (1994). http://c2.com/ppr/stars.html

21. Setlur, V., Battersby, S.E., Tory, M., Gossweiler, R., Chang, A.X.: Eviza: a natural language interface for visual analysis. In: Proceedings of the 29th Annual Symposium on User Interface Software and Technology, pp. 365–377. UIST 2016. Association for Computing Machinery, New York, NY, USA (2016). https://doi.org/10.1145/2984511.2984588, https://doi.org/10.1145/2984511.2984588

22. Setlur, V., Tory, M., Djalali, A.: Inferencing underspecified natural language utterances in visual analysis. In: Proceedings of the 24th International Conference on Intelligent User Interfaces, pp. 40–51. IUI 2019. Association for Computing Machinery, New York, NY, USA (2019). https://doi.org/10.1145/3301275.3302270, https://doi.org/10.1145/3301275.3302270

23. Shen, D., Wu, G., Suk, H.I.: Deep learning in medical image analysis. Ann. Rev. Biomed. Eng. **19**(1), 221–248 (2017). https://doi.org/10.1146/annurev-bioeng-071516-044442, https://doi.org/10.1146/annurev-bioeng-071516-044442, pMID: 28301734

24. Sun, T., et al.: Mitigating gender bias in natural language processing: literature review. In: Proceedings of the 57th Annual Meeting of the Association for Computational Linguistics, pp. 1630–1640. Association for Computational Linguistics, Florence, Italy, July 2019. https://doi.org/10.18653/v1/P19-1159, https://www.aclweb.org/anthology/P19-1159

25. Vaswani, A., et al.: Attention is all you need. CoRR abs/1706.03762 (2017). http://arxiv.org/abs/1706.03762

26. Wang, B., Shin, R., Liu, X., Polozov, O., Richardson, M.: RAT-SQL: relation-aware schema encoding and linking for text-to-sql parsers. CoRR abs/1911.04942 (2019). http://arxiv.org/abs/1911.04942

27. Wang, P., Shi, T., Reddy, C.K.: Text-to-SQL generation for question answering on electronic medical records. In: Huang, Y., King, I., Liu, T., van Steen, M. (eds.) WWW 2020: The Web Conference 2020, 20–24 April 2020, Taipei, Taiwan, pp. 350–361. ACM/IW3C2 (2020). https://doi.org/10.1145/3366423.3380120, https://doi.org/10.1145/3366423.3380120

28. Weir, N., et al.: DBPal: a fully pluggable NL2SQL training pipeline. In: Proceedings of the 2020 ACM SIGMOD International Conference on Management of Data, pp. 2347–2361. SIGMOD 2020, Association for Computing Machinery, New York, NY, USA (2020). https://doi.org/10.1145/3318464.3380589, https://doi.org/10.1145/3318464.3380589

29. Williams, R.J., Zipser, D.: A learning algorithm for continually running fully recurrent neural networks. Neural Comput. 1(2), 270–280 (1989). https://doi.org/10.1162/neco.1989.1.2.270

30. Wolf, T., et al.: Huggingface's transformers: state-of-the-art natural language processing. CoRR abs/1910.03771 (2019). http://arxiv.org/abs/1910.03771

31. Wu, Y., et al.: Google's neural machine translation system: bridging the gap between human and machine translation. CoRR abs/1609.08144 (2016). http://arxiv.org/abs/1609.08144

32. Xu, X., Liu, C., Song, D.: SQLNet: generating structured queries from natural language without reinforcement learning. CoRR abs/1711.04436 (2017). http://arxiv.org/abs/1711.04436

33. Yu, T., et al.: Spider: a large-scale human-labeled dataset for complex and cross-domain semantic parsing and text-to-sql task. CoRR abs/1809.08887 (2018). http://arxiv.org/abs/1809.08887

34. Zeng, J., et al.: Photon: A robust cross-domain Text-to-SQL system. In: Proceedings of the 58th Annual Meeting of the Association for Computational Linguistics: System Demonstrations, pp. 204–214. Association for Computational Linguistics, Online, July 2020. https://doi.org/10.18653/v1/2020.acl-demos.24, https://www.aclweb.org/anthology/2020.acl-demos.24

35. Zhong, V., Lewis, M., Wang, S.I., Zettlemoyer, L.: Grounded adaptation for zero-shot executable semantic parsing (2021)

36. Zhong, V., Xiong, C., Socher, R.: Seq2SQL: generating structured queries from natural language using reinforcement learning. CoRR abs/1709.00103 (2017). http://arxiv.org/abs/1709.00103

Time Aspect in Making an Actionable Prediction of a Conversation Breakdown

Piotr Janiszewski, Mateusz Lango(✉)(iD), and Jerzy Stefanowski(iD)

Faculty of Computing and Telecommunication, Institute of Computer Science,
Poznan University of Technology, ul. Piotrowo 2, 61-138 Poznan, Poland
1piotr.janiszewski@gmail.com
{mlango,jstefanowski}@cs.put.edu.pl

Abstract. Online harassment is an important problem of modern societies, usually mitigated by the manual work of website moderators, often supported by machine learning tools. The vast majority of previously developed methods enable only retrospective detection of online abuse, e.g., by automatic hate speech detection. Such methods fail to fully protect users as the potential harm related to the abuse has always to be inflicted. The recently proposed proactive approaches that allow detecting derailing online conversations can help the moderators to prevent conversation breakdown. However, they do not predict the time left to the breakdown, which hinders the practical possibility of prioritizing moderators' works. In this work, we propose a new method based on deep neural networks that both predict the possibility of conversation breakdown and the time left to conversation derailment. We also introduce three specialized loss functions and propose appropriate metrics. The conducted experiments demonstrate that the method, besides providing additional valuable time information, also improves on the standard breakdown classification task with respect to the current state-of-the-art method.

Keywords: Online abuse · Conversation breakdown prediction · Time aspects in online dialog · Hierarchical neural networks

1 Introduction

Cyberspace has a large potential for making constructive conversations, facilitating communication and cooperation of groups of people with similar interests, various areas of expertise. Unfortunately, some online discussions result in anti-social behaviors [16] since anonymity and an apparent sense of impunity limit the natural inhibitions interlocutors would have during a face-to-face conversation. A survey conducted in the US demonstrated that online harassment is a widespread phenomenon as approximately four-in-ten Americans were directly affected by some forms of it [8]. Online abuse can be a root cause of a wide range of mental problems, negatively affecting many aspects of victims' lives [2,18]. Even merely witnessing the harassment on the Internet can lead to a user's lower involvement in online service or even a complete refrain from using it [26,27].

© Springer Nature Switzerland AG 2021
Y. Dong et al. (Eds.): ECML PKDD 2021, LNAI 12979, pp. 351–364, 2021.
https://doi.org/10.1007/978-3-030-86517-7_22

Therefore, numerous websites leverage systems for hampering antisocial behavior. The most common methods include community moderation, up- and down-voting, the possibility to report comments, mute functionality, and banning users on the platform [5]. However, these simple approaches cannot successfully overcome the widespread problem, as a lot of hateful content can be overlooked by the moderators or simply not be reported by users. As a consequence, multiple machine learning techniques are used to support moderators by ranking unacceptable posts [9], automatically identifying cyberbulling [29] or detecting hate speech [11].

The majority of existing systems perform toxicity detection retrospectively. Even though such solutions mitigate the problem, they do not fully protect users as the potential harm has always to be inflicted to some extent, and only then the hostile comments can be filtered. These solutions do not make actionable classifications whether an online conversation is going to end in a personal attack or not, leaving no time for moderators to intervene before any harassment or conflict emerges.

A much more successful strategy would be to avert offensiveness when the discussion is still salvageable or at least hinder potential destructive effects. For instance, one could introduce to the conversation customized counter speech, which proved to be effective in combating offensiveness in various studies [15, 23]. Another solution would be to remind the interlocutors about the need for empathy and the rules of the service [17]. Even drawing moderators' attention to the derailing conversation can be beneficial as it reduces the response time and gives them an opportunity to intervene. Nevertheless, such solutions require a method for predicting conversation derailment in advance.

Moreover, just recognizing if the discussion is going to get out of hand may not be enough to obtain comprehensive and highly useful information about the potential derailment. Therefore, additional clues have to be provided. One of them is the *time to the breakdown*, which seems advantageous in many potential fields of application, especially when humans are in the loop and there is a need to prioritize actions to be performed. Such a forecast about the specific time of a breakdown may also help estimate the hostile tension in individual dialogs. This also can be a crucial hint for moderators who can recognize the most urgent cases and intervene on time. In addition, mistakes made on foreseeing how many utterances are left to the conversation derailment could be a valuable additional learning signal for the model and boost classification performance. This opens a new research and open problem since, to the best of our knowledge, such methods have not yet been proposed.

In this work, we propose a machine learning system based on deep neural networks that not only predicts whether the conservation will derail in the future but also estimates the number of utterances left to the derailment. We propose and explore three loss functions that allow for joint training of systems performing both the discussion breakdown prediction and time-to-the-breakdown estimation. We also introduce three valuable metrics for assessing the performance of models applied to foreseeing conversational breakdown with consideration of the time aspect. An experimental evaluation shows that the proposed approach, aside from providing additional and useful information about time to the derailment for moderators, also achieves better results on the standard classification task of discussion breakdown.

2 Related Works

A great deal of personal attacks in the cyberspace takes place during discussions when interlocutors disagree with each other, at least to some extent. Initially, a civil exchange may degenerate into a dispute resulting even in verbal aggression. Such "from within" derailments are potentially more dangerous and more troublesome to salvage than other types of toxicity (e.g., trolling or profanities), which a cybernaut can ignore more easily [28]. A conversation breakdown may have different faces and lead to distinct forms of antisocial behavior posing a considerable threat to the people involved.

Aside from causing emotional distress, failing conversations has also other negative impacts. For example, in online game industry, one of the main reasons leading users to stop playing the game is experiencing different kinds of toxicity during conversations with other players [26]. Therefore, it is crucial to forecast occurrences of offensiveness as a dialog develops and to make a correct prediction at the earliest possible moment, letting a moderator react appropriately. Even among Wikipedia editors community that is generally associated with well-educated people, abuse has proved to be a significant problem [27] that harms editors' willingness to further contribute.

Therefore, the problem of detecting various forms of toxicity in text data received recently considerable research attention. Methods for identifying cyberbullying [1,29], hate speech [11,14], doxing [24], or negative sentiment [12] proved to be useful to filter unacceptable content. Nevertheless, they focus on analyzing already posted, potentially harmful texts (so they work on historical recording).

Examining each text right before it is published creates an opportunity to identify abusive chunks on time [3,19]. For instance, the system can ask the user to modify the toxic comment. However, asking for changing a comment or proposing its corrected version [20,22] always requires an additional user's action, slows the exchange down, impede its natural flow and dynamics, and may discourage users from taking part in the discussion - especially when the prediction made by toxicity detector are too often incorrect. Another possibility is to remind the user about the need for empathy and rules of the service [17]; however, users who knowingly post hostile content might be completely unaffected by such a prompt. Therefore, more advanced solutions such as the introduction of customized counter-speech [15] are needed to solve the problem. Nevertheless, to apply techniques that prevent conversation failure, one needs to predict whether the conversation will derail first.

One method of foreseeing unacceptable content in online conversations was recently presented in [13]. The proposed approach determines whether any adverse utterance will be published below a post on Instagram basing on the set of initial comments. Another method was presented by Zhang et al. [28] who proposed an approach for forecasting whether a conversation is going to derail basing on the initial two utterances in a discussion. The approach uses the logistic regression classifier and bag-of-words features together with specially designed problem-specific features.

The current state-of-the-art method for predicting a discussion breakdown, called Conversational Recurrent Architecture for ForecasTing (CRAFT) [4] relies on a deep neural network. The approach models a conversation flow with Hierarchical Recurrent Encoder-Decoder [25] and performs forecasting in an online fashion. All the predictions are made as a dialog develops, i.e. the prediction is updated after seeing each new utterance. Although the presented solution outperforms previous approaches, there is still some room for improvement. In particular, this approach does not take into account the moment in which a first disruptive utterance comes, ignoring the time aspect that could be very useful in practice.

3   Time Aspect in Prediction of Conversation Breakdown

In this work, we propose a new method for detecting derailing conversations that provides additional information about the *time left to the conversation breakdown*, understood as the number of utterances left to the derailment. Note that all the previously proposed methods for this task do not provide such additional information.

Being able to predict when the dialog is going to fail would bring considerable benefits in practice. For instance, the websites would be able to manage their moderation resources more effectively by prioritizing the cases of abuse, paying most attention to the most urgent and most severe ones, and counteracting them more quickly.

3.1   Proposed Neural Network Architecture

An utterance context is a crucial factor to be considered when deciding if the utterance is abusive, as it can intensify or soften its overtone. Therefore, a breakdown should not be treated as a property of a single comment but rather as a property of a developing dialog. Following this idea, similarly to related works, the proposed method uses the hierarchical recurrent encoder-decoder (HRED) architecture [25] to model a developing dialog and to capture the conversation dynamics.

HRED consists of two recurrent neural networks called, utterance encoder and context encoder, respectively. The utterance encoder's goal is to construct a feature representation of a single user's utterance, which is then passed as an input to the context encoder. In our experiments, both networks are based on Gated Recurrent Units [6]. The input to utterance encoder is given as a sequence of words, previously processed by an embedding layer. The final hidden state of the encoder is forwarded as an input to the context encoder (Fig. 1).

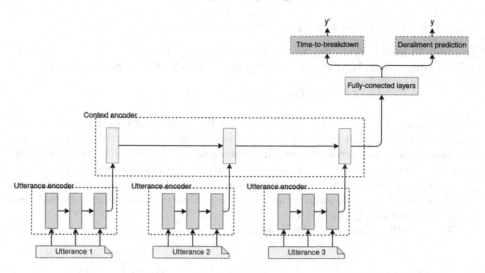

Fig. 1. The overview of the proposed neural network architecture, where y denotes the probability that the conversation will derail and y' is the prediction of the time left to the conversation breakdown.

In order to produce the useful feature representation for predicting both the probability of conversation failure and time-to-breakdown, the hidden state of the context encoder is passed through several fully-connected layers. Such constructed feature representation is processed by two separate output layers. The first one being the layer with only one sigmoid unit which predicts the probability that the conversation will derail in the future. The second output layer working on the same feature representation is a regression layer that predicts the time-to-breakdown (i.e., the number of utterances). The whole network architecture is trained jointly by back-propagation.

Note that when the sigmoid layer predicts that the dialog is not going to derail, the output of the regression layer can be discarded. However, the error related to the time-to-breakdown prediction provides an additional training signal to guide the model learning process. In related models without time-to-breakdown output, the error related to the derailment prediction is suffered usually only once, at the moment of conversation breakdown. Alternatively, to ensure that the model will predict possible derailment as soon as possible, one could enforce the derailment prediction after each utterance in the dialog. Nevertheless, such a solution will incorrectly introduce an association between usually the conversation beginning and the conversation breakdown class, adding unnecessary noise to the classifier training. By training the model with the additional output for time-to-breakdown we want to avoid these problems, at the same time providing a clear, additional training signal to the model.

3.2   Loss Functions Incorporating Time Aspect

In order to jointly train the network predicting the probability of conversation failure and time-to-breakdown, we propose to use the following loss function:

$$\min L(\theta) = \alpha L_{classification}(\theta) + (1 - \alpha)L_{time}(\theta)$$

This loss function has two components. The first one measures the error on the standard conversation failure prediction task, while the second term controls the time prediction error. These components are weighted with the parameter $\alpha \in (0,1)$ that controls the trade off between the model focus on the time-to-breakdown prediction and the classification task. In practice, this parameter could be tuned with the validation data, but in this work we treat both tasks as equally important, i.e. $\alpha = 0.5$.

The classification error is measured by the standard cross-entropy error:

$$L_{classification}(\theta) = \frac{1}{n} \sum_{i=1}^{n} [y_i \log h_\theta(x_i) + (1 - y_i) \log(1 - h_\theta(x_i))]$$

where $y_i \in \{0, 1\}$ is the label from the gold standard and $h_\theta(x_i)$ is the prediction of the classification layer.

The second term of the loss function is defined as

$$L_{time}(\theta) = \frac{1}{n} \sum_{i=1}^{n} f(g_\theta(x_i), y_i')$$

where y_i' is the gold standard for the time-to-breakdown task, $g_\theta(x_i)$ is the prediction of the regression layer and f is a function measuring error made on a particular example. Note that the time-to-breakdown y_i' is understood as the number of utterances left to the first uncivil utterance from i-th utterance. In this work, we will explore three possible ways of measuring the time-to-breakdown error: by a classical squared error for regression, casting the task to classification, and a time-dependent custom loss.

The classical squared error is defined as:

$$f_{MSE}(g_\theta(x), y') = (g_\theta(x) - y')^2$$

which is the squared difference between predicted time $g_\theta(x)$ and the time to derailment in the gold standard y'.

Yet another possibility is to treat time-to-breakdown prediction as a classification task by defining classes for specific ranges of time-to-breakdown. We use 11 classes, where each class j corresponds to a number of utterances left to the conversation failure and $j \in \{0..10\}$. The first class i.e. $j = 0$ represents a moment of an actual derailment and class $i = 10$ means that the dialog will break down in 10 comments or more. It is assumed that a discussion horizon longer than ten utterances is so distant and so uncertain that one can aggregate these cases into a single class. Adopting such a strategy should not have any strong negative impact on the quality of the prediction, nor on its usefulness for the potential action related to a possible failure. After casting the task to multi-class classification, we apply categorical cross-entropy error defined by

$$f_{CCE}(g_\theta(x), y') = -\sum_{j=0}^{10} y'_j \log g_\theta(x)_j$$

where $y'_j \in \{0, 1\}$ are binary variables indicating whether the time left to the breakdown belong to the class j. In this case, the activation of the respective output layer should be softmax.

We also explore the possibility of using a custom time-depended loss. The proposed loss follows the observation that the model makes more predictions for the utterances that are relatively far away from the discussion horizon in the case of long discourses. In such a case, the conversation outcome is difficult to foresee, not only because initially there can be no or little indicators that the conversation will fail but also because the prediction is based on a very small context. Therefore, in practice over- or under-estimating the time-to-breakdown about a constant value, e.g. 1, can be considered less severe if there is much time left to the derailment and considered more serious if the breakdown horizon is close.

We encompass this intuition in the following formulation:

$$f_{CTD}(g_\theta(x), y') = \begin{cases} \min\left\{\frac{|g_\theta(x) - y'|}{y'+1}; 1\right\}, & \text{for a failing conversation} \\ \max\left\{\frac{y' - g_\theta(x) + 1}{y'+1}; 0\right\}, & \text{for a civil conversation} \end{cases}$$

where for the civil conversation the y' is set to be the length of the conversation. If the discussion is derailing, the presented loss is computed as the minimum of one and the absolute prediction error divided by the actual number of remaining comments. Therefore, predictions with the high horizon do not generate higher cumulative losses and the loss value is always between 0 and 1. When the discourse stays civil, the loss is equal to zero when the model anticipates that the exchange will fail even later than it actually ends; thus, one is added in the numerator. Since the loss is computed as the maximum of 0 and the forecast error divided by the true number of remaining utterances, the higher the number of comments foreseen as remaining till the conversation breakdown, the smaller the loss.

3.3   Metrics Considering the Time-to-breakdown of Prediction

We propose three quality measures designed for the time-to-breakdown prediction. Each of them is bounded from 0 to 1 and expressed by an average of inverse errors, i.e.:

$$Q = \frac{1}{N} \sum_{i=1}^{N} \frac{1}{E_i + 1}$$

where N is the number of dialogues in a dataset, and E_i is the prediction error of a particular dialog, defined differently for each measure. Note that when the prediction error is equal to 0, our quality measures are equal to 1. Moreover, the measure values will not be dominated by predictions on long conversations since they are averaged over conversations and not over utterances.

The first measure, denominated as *average inverse prediction errors* (AIE), uses the classic definition of absolute error, so in that measure E_i is defined as

$$E_{AIE} = \frac{1}{K} \sum_{j=1}^{K} |\hat{y}'_j - y'_j|$$

where K is the length of the conversation, $\hat{y}'_j$ is the time-to-breakdown prediction at the time j while y'_j denotes corresponding gold standard value, i.e., the true time-to-breakdown.

The second measure, called *selected inverse prediction errors* (SIE_t) focuses on the quality of prediction at the specific time point before the possible conversation breakdown, and is defined as:

$$E_{SIE_t} = | \hat{y}'_t - y'_t |$$

where $\hat{y}'_t$ and y'_t are the predicted and the gold standard value at t utterances *before breakdown* or before the end of conversation. We hope that this measure could be important for practitioners, assuming that in practice one should have the information about the possible conversation failure at least e.g. $t = 5$ utterances before in order to have enough time to take action.

The third measure is *inverse prediction errors* at the highest probability point (IEH):

$$E_{IEH} = | \hat{y}'_{t*} - y'_{t*} |$$

where t^* denotes the moment in which the classifier predicted the breakdown with the highest probability. Such measure in a simplistic way takes into account that the output of the time-to-breakdown predictor will probably be used only when the classifier will assess the conversation as derailing.

Additionally, we measure the quality of time-to-breakdown prediction with standard macro-averaged F1-measure using the same eleven classes defined in the previous section for cross-entropy error. For the methods which return continuous prediction of time-to-breakdown, we round the predictions before calculating of F1-measure.

4   Experiments

The main aims of the experiments is to verify the usefulness of the new proposed approach and, in particular, to examine how the introduced loss functions influence the models' ability to predict the conversational breakdown and to approximate the time when it is going to happen. The method will be compared against the performance of the reference method – CRAFT, which is considered as the state of the art approach. The quality of inference will be estimated by using both standard classification measures, as well as the three proposed metrics.

4.1 Datasets

In our experiments, we use the same two datasets on which CRAFT's quality has been originally measured [4].

The first dataset consists of 4188 conversations retrieved from WikiConv [10]. It contains public discussions between Wikipedia contributors about the quality of entries and observance of the Wikipedia editing rules. Crowdworkers labelled them according to whether they contain a personal attack directed towards one of the interlocutors or not. Such an act of aggression should be committed by one of the contributors who took part in the dialog since its beginning.

The second dataset contains 6842 dialogues from the subreddit ChangeMyView. A conversation is considered as derailed if it contains a comment removed by a moderator due to a violation of Rule 2: "Don't be rude or hostile to other users". It means that there may exist discussions with abusive expressions without a correct label since they could go unnoticed by the moderators. The authors of the dataset additionally warrant that every deleted comment was written by a person previously involved in the conversation.

Additionally, every example which ended with a failure is paired with a civil one on the same topic in order not to let the model associate topic-specific information with individual labels (e.g. exchanges about politics are prone to fail). Significantly, in each derailing exchange all the utterances up to the toxic one are civil.

4.2 Experimental Setup

The setup of the method involves proper choosing of several architectural details in order to let the model learn effectively. In our experiments, HRED has two encoder networks (utterance and context encoder), each consisting of two GRU layers with the hidden layer of size of 500. The features for output layers are constructed by two fully-connected layers, the first one having 500 neurons and the second one with 250 units. As regularized we apply dropout with the rate of 0.1. Training batches contained 64 examples each and the process was optimized using Adam optimization algorithm with the learning rate of 10^{-5}. The end of training was determined using early-stopping in order to avoid overfitting.

Additionally, the HRED component was pre-trained on 1 million discussions from the Wikipedia Talk Page, using the generative pre-training technique proposed by the CRAFT's authors [4]. During such pre-training HRED component learns how to model the dynamics of a conversation in an unsupervised fashion.

The quality of forecasting whether a dialog will fail was measured using standard classification metrics, i.e., accuracy (Acc), precision (Prec), recall (Rec), false positive rate (FPR), and F1-score (F1). A conversation was deemed as failing, when at least one comment was identified as derailing before the dialog failed. Each forecast was based on the previous utterances from the same conversation, thus, for first utterances nothing was predicted. The metrics were computed on the test part of datasets as provided in [4] i.e. 20% of conversations were used for testing.

When foreseeing the number of utterances left to the derailment, the metrics described in Sect. 3.3 were used. For the SIE_t metric, we have used $t = 5$, i.e., the error was calculated looking at the prediction triggered by the fifth to last utterance in a discussion. If the conversation was shorter than 5, the prediction on the second utterance from the beginning was taken into account.

Note that the results achieved by CRAFT reported in this work are worse than those presented in [4]. During those experiments, predictions were triggered only for the last comments in each conversation. This gave CRAFT a special advantage, as each inference was drawn basing on the complete history of the conversation, providing the model with the best possible context for its forecast. It was serious facilitation, which would not happen in real-life setting, as the horizon of a dialog is unknown, and forecasts have to be made even if the available context is too short. Moreover, such an approach also makes it impossible to measure how the model works in the complete development of the conversation.

4.3   Results of Experiments

Table 1. Comparison of the proposed method with the three loss functions (MSE, CCE, CTD) and the state-of-the-art CRAFT model on the task of forecasting conversational derailment.

Approach	Wikipedia talk pages					Reddit CMV				
	Acc	Prec	Rec	FPR	F1	Acc	Prec	Rec	FPR	F1
CRAFT	0.606	0.573	0.776	0.574	0.660	0.524	0.522	0.572	0.523	0.546
MSE	**0.639**	**0.638**	0.641	**0.362**	0.640	0.546	**0.572**	0.364	**0.272**	0.445
CCE	0.616	0.597	0.710	0.479	0.649	**0.556**	0.546	**0.658**	0.547	**0.597**
CTD	0.614	0.591	**0.786**	0.554	**0.665**	0.534	0.529	0.626	0.557	0.573

The results of the experiments are presented in Table 1 and 2. In the classification task, one can observe that for both datasets CRAFT is outperformed by the methods proposed in this work on each of the metrics. Model which uses MSE time-to-breakdown error in the loss function achieved the best results on Wikipedia dataset, when it comes not only to accuracy, but also precision and false positive rate. These are significantly better scores compared to CRAFT. It also offered improvements on this measures on Reddit dataset, but it was CCE loss that provided the best accuracy, recall and F1-score on that dataset. The solution based on the Custom Time Dependent loss proposed in this work improves all the classification metric with respect to CRAFT on Wikipedia data and almost all (except FPR) on Reddit dataset. This demonstrates that the information about time-to-breakdown provides a useful additional learning signal to guide model training for this conversation breakdown prediction.

Table 2. Comparison between the performances of the proposed method with different loss functions on the task of predicting the number of comments left to a conversation breakdown.

Approach	Wikipedia talk pages				Reddit CMV			
	AIE	SIE$_5$	IEH	F1	AIE	SIE$_5$	IEH	F1
MSE	**0.480**	**0.400**	**0.572**	0.430	0.363	**0.398**	0.469	0.322
CCE	0.407	**0.400**	0.342	0.557	**0.428**	0.388	**0.686**	0.205
CTD	0.361	0.257	0.473	**0.602**	0.416	0.368	0.437	**0.370**

In the task of approximating time-to-breakdown on Wikipedia dataset, the proposed method with MSE achieved the best results on the inverse error metrics. The result on AIE close to 0.5 means that the model is wrong on average by only one comment. In our opinion, it should be sufficient to provide an effective support for online moderators. Surprisingly, our Custom Time Dependent Loss and not the standard cross-entropy provided better results on F1-score, i.e., while evaluating time-to-breakdown prediction as a multi-class classification task. On the Reddit dataset, CTD also gave the highest F1-score, but it was CCE that gave the highest values of AIE and IEH measures.

Note that the values of SIE$_5$ are generally lower than values of AIE and IEH for both datasets. This is because the prediction error taken into account when calculating SIE is calculated 5 comments before a personal attack, and the dialog context is often not sufficiently broad to make a good prediction. Nonetheless, this metric allows to check, what is the forecast quality, when the conversation is not completely developed and there is still much time to intervene. According to the definition of SIE the average number of conversations for which the best model was wrong is 1.5, which is a satisfactory result considering how early this prediction is made.

Furthermore, IEH values are usually higher than AIE and SIE$_5$ for most of approaches. This is due to the fact that the probability of derailment increases as the conversation develops in the failing direction and subsequent forecasts are made basing on wider contexts. This implies that as the model becomes more and more convinced that the exchange will eventually fail, it can more accurately foresee when it is going to happen. This is a good characteristics, as in case of dialogs with high tension (thus easier to detect) the final conflict should be potentially more serious, therefore it is especially important to identify how many comments are left to such conversation breakdown. For our best solution, the committed average error is only around 0.75 and 0.45 utterances for respective datasets.

5   Conclusions

This work introduces a new version of the online abuse conversation breakdown problem, which includes jointly predicting whether the conversation will derail and approximating time to the conversation breakdown. In particular, considering time aspects opens new research and application perspectives. Upon the current state-of-the-art, we presented a new approach to this problem by proposing three task-specific loss functions and extending the hierarchical recurrent neural network architecture.

The experiments with two datasets containing different real life online discussions have showed that the proposed methods (with these loss functions) achieve better results on the accuracy, F1-score, precision, and recall measures than the current state-of-the-art method for conversation breakdown prediction. Additionally, the proposed approach returns new type of information about time-to-breakdown, which could be very helpful in practice, for instance, to prioritize the cases handled by moderators.

Nevertheless, the approach described in this work could be still further developed. One possible option is to use a pre-trained architecture that models conversation dynamics in another way than HRED. In particular, recent experiments with the transformer-based models in many related natural language processing tasks may suggest that using the neural networks of this type may boost the results. Therefore, we also carried out some experiments by using contextual word embeddings produced by one light-weight transformer-based model, namely DistilBERT [21], but the results were not clear enough. Most importantly, the use of DistilBERT embeddings never produced better results than those obtained by any of the new proposed HRED-based methods, even though we have seen some improvements for some particular configurations of dataset and loss function. Nevertheless, we hypothesize that proposing a new transformer-based architecture dedicated to modeling conversations could be a topic of further research.

The other possible issue is that our model, similarly to related works, has been trained on a balanced dataset, even though online conversation derailments happen relatively less frequently. Therefore, while the system should be able to deal with a shifted class distribution. The question of how the low number of positive examples may influence the predictive performance of conversation breakdown predictors is still open.

Finally, the more advanced ways of dealing with the time aspect in predicting a conversation breakdown can be further explored. For instance, one can try to adopt early classification methods [7] that, instead of predicting time-to-breakdown, are directly trying to optimize the trade-off between the quality and earliness of event prediction, which can be useful in practice. Another possibility would be to explore ideas from the field of survival analysis or from the next event prediction problem in time series.

Acknowledgements. Mateusz Lango was supported by the Polish National Science Centre under grant No. 2016/22/E/ST6/00299. Moreover, the research of Jerzy Stefanowski was partially supported by the Polish Ministry of Education and Science, grant no. 0311/SBAD/0709. The authors also acknowledge the support from Google Cloud Platform research grant.

References

1. Agrawal, S., Awekar, A.: Deep learning for detecting cyberbullying across multiple social media platforms. In: Pasi, G., Piwowarski, B., Azzopardi, L., Hanbury, A. (eds.) ECIR 2018. LNCS, vol. 10772, pp. 141–153. Springer, Cham (2018). https://doi.org/10.1007/978-3-319-76941-7_11
2. Beran, T., Li, Q.: Cyber-harassment: a study of a new method for an old behavior. J. Educ. Comput. Res. **32**(3), 265 (2005)
3. Carton, S., Mei, Q., Resnick, P.: Extractive adversarial networks: high-recall explanations for identifying personal attacks in social media posts. In: Proceedings of the Conference on Empirical Methods in Natural Language Processing (2018)
4. Chang, J.P., Danescu-Niculescu-Mizil, C.: Trouble on the horizon: forecasting the derailment of online conversations as they develop. In: Proceedings of the Conference on Empirical Methods in Natural Language Processing (2019)
5. Cheng, J., Danescu-Niculescu-Mizil, C., Leskovec, J.: Antisocial behavior in online discussion communities. In: Proceedings of ICWSM (2015)
6. Cho, K., et al.: Learning phrase representations using RNN encoder-decoder for statistical machine translation. In: Proceedings of the Conference on Empirical Methods in Natural Language Processing (2014)
7. Dachraoui, A., Bondu, A., Cornuéjols, A.: Early classification of time series as a non myopic sequential decision making problem. In: Appice, A., Rodrigues, P.P., Santos Costa, V., Soares, C., Gama, J., Jorge, A. (eds.) ECML PKDD 2015. LNCS (LNAI), vol. 9284, pp. 433–447. Springer, Cham (2015). https://doi.org/10.1007/978-3-319-23528-8_27
8. Duggan, M.: Online harassment 2017. Pew Research Center (2017)
9. Hsu, C.F., Khabiri, E., Caverlee, J.: Ranking comments on the social web. In: 2009 International Conference on Computational Science and Engineering, vol. 4, pp. 90–97. IEEE (2009)
10. Hua, Y., Danescu-Niculescu-Mizil, C., Taraborelli, D., Thain, N., Sorensen, J., Dixon, L.: WikiConv: a corpus of the complete conversational history of a large online collaborative community. In: Proceedings of the 2018 Conference on Empirical Methods in Natural Language Processing, Brussels, Belgium, pp. 2818–2823. Association for Computational Linguistics (2018)
11. Janiszewski, P., Skiba, M., Walińska, U.: PUM at SemEval-2020 task 12: aggregation of transformer-based models' features for offensive language recognition. In: Proceedings of the International Workshop on Semantic Evaluation (SemEval) (2020)
12. Lango, M.: Tackling the problem of class imbalance in multi-class sentiment classification: an experimental study. Found. Comput. Decis. Sci. **44**, 151–178 (2019)
13. Liu, P., Guberman, J., Hemphill, L., Culotta, A.: Forecasting the presence and intensity of hostility on Instagram using linguistic and social features. In: Proceedings of AAAI Conference on Web and Social Media (ICWSM) (2018)

14. Malmasi, S., Zampieri, M.: Detecting hate speech in social media. In: Proceedings of the International Conference Recent Advances in Natural Language Processing (2017)
15. Mathew, B., et al.: Thou shalt not hate: countering online hate speech. In: Proceedings of the International AAAI Conference on Web and Social Media, vol. 13, pp. 369–380 (2019)
16. Mishra, P., Yannakoudakis, H., Shutova, E.: Tackling online abuse: a survey of automated abuse detection methods, CoRR (2019)
17. Munger, K.: Tweetment effects on the tweeted: experimentally reducing racist harassment. Polit. Behav. **39**(3), 629–649 (2017)
18. Munro, E.R.: The protection of children online: a brief scoping review to identify vulnerable groups. Childhood Wellbeing Research Centre (2011)
19. Noever, D.: Machine learning suites for online toxicity detection. arXiv preprint arXiv:1810.01869 (2018)
20. Prabhumoye, S., Tsvetkov, Y., Salakhutdinov, R., Black, A.W.: Style transfer through back-translation. In: Proceedings of the 56th Annual Meeting of the Association for Computational Linguistics (Volume 1: Long Papers) (2018)
21. Sanh, V., Debut, L., Chaumond, J., Wolf, T.: DistilBERT, a distilled version of BERT: smaller, faster, cheaper and lighter. In: Proceedings of the 5th Workshop on Energy Efficient Machine Learning and Cognitive Computing (co-located with NeurIPS) (2019)
22. Santos, C.N.d., Melnyk, I., Padhi, I.: Fighting offensive language on social media with unsupervised text style transfer. In: Proceedings of the 56th Annual Meeting of the Association for Computational Linguistics (2018)
23. Schieb, C., Preuss, M.: Governing hate speech by means of counterspeech on facebook. In: 66th ICA Annual Conference, pp. 1–23 (2016)
24. Snyder, P., Doerfler, P., Kanich, C., McCoy, D.: Fifteen minutes of unwanted fame: detecting and characterizing doxing. In: Proceedings of the 2017 Internet Measurement Conference, pp. 432–444 (2017)
25. Sordoni, A., Bengio, Y., Vahabi, H., Lioma, C., Simonsen, J.G., Nie, J.Y.: A hierarchical recurrent encoder-decoder for generative context-aware query suggestion. In: Proceedings of the 24th ACM International on Conference on Information and Knowledge Management (2015)
26. Stoop, W., Kunneman, F., van den Bosch, A., Miller, B.: Detecting harassment in real-time as conversations develop. In: Proceedings of the Third Workshop on Abusive Language Online, Florence, Italy, pp. 19–24. Association for Computational Linguistics, August 2019. https://doi.org/10.18653/v1/W19-3503
27. Wulczyn, E., Taraborelli, D., Thain, N., Dixon, L.: Ex machina: personal attacks seen at scale. In: International World Wide Web Conference (2017)
28. Zhang, J., Chang, J.P., Danescu-Niculescu-Mizil, C., Lucas Dixon, Y.H., Thain, N., Taraborelli, D.: Conversations gone awry: Detecting early signs of conversational failure. In: Proceedings of the Annual Meeting of the Association for Computational Linguistics (2018)
29. Zhao, R., Zhou, A., Mao, K.: Automatic detection of cyberbullying on social networks based on bullying features. In: Proceedings of the 17th International Conference on Distributed Computing and Networking. ICDCN 2016, Association for Computing Machinery, New York (2016). https://doi.org/10.1145/2833312.2849567

Feature Enhanced Capsule Networks for Robust Automatic Essay Scoring

Arushi Sharma[1,3] , Anubha Kabra[2,3] , and Rajiv Kapoor[3(✉)]

[1] Optum Global Advantage, Delhi, India
[2] Adobe Systems, Noida, India
[3] Delhi Technological University, Delhi, India
rajivkapoor@dce.ac.in

Abstract. Automatic Essay Scoring (AES) Engines have gained popularity amongst a multitude of institutions for scoring test-taker's responses and therefore witnessed rising demand in recent times. However, several studies have demonstrated that the adversarial attacks severely hamper existing state-of-the-art AES Engines' performance. As a result, we propose a robust architecture for AES systems that leverages Capsule Neural Networks, contextual BERT-based text representation, and key textually extracted features. This end-to-end pipeline captures semantics, coherence, and organizational structure along with fundamental rule-based features such as grammatical and spelling errors. The proposed method is validated by extensive experimentation and comparison with the state-of-the-art baseline models. Our results demonstrate that this approach performs significantly better on 6 out of 8 prompts on the Automated Student Assessment Prize (ASAP) dataset. In addition, it shows an overall best performance with a Quadratic Weighted Kappa (QWK) metric of 81%. Moreover, we empirically demonstrate that it is successful in identifying adversarial responses and scoring them lower.

Keywords: Automatic scoring · Capsule Neural Networks · Adversarial testing · BERT · Machine learning

1 Introduction

Writing compositions have been widely adopted by all language proficiency exams. The manual evaluation process is taxing and laborious; hence globally standardized exams such as GRE [2] resort to automatic scoring systems. However, many state-of-the-art AES tools suffer from adversarial attacks [10,19]. As a result, there is a dire need for computerized essay scoring systems that provide quick results while maintaining objectivity and accuracy in evaluations [17]. AES is a complex problem as these systems aim not just to point out grammatical or spelling errors but also to consider the semantics, identify the coherence in discourse, and ensure that the response is relevant to the question.

State-of-the-art AES systems can be broadly categorized into two types, feature engineered models and end-to-end deep learning models [14]. The feature-engineered

---

A. Sharma and A. Kabra—Equal Contribution - work done in Delhi Technological University.

© Springer Nature Switzerland AG 2021
Y. Dong et al. (Eds.): ECML PKDD 2021, LNAI 12979, pp. 365–380, 2021.
https://doi.org/10.1007/978-3-030-86517-7_23

models use handcrafted surface-level features, such as the length of the essay and grammatical errors, for scoring the responses [11]. While they are easily explainable and modifiable with scoring criteria, they lack the understanding of response as a whole. They cannot mimic pattern organization or coherence based on word and sentence level relations [21]. The latter type of AES system explores extracting semantic relationships within an essay. This method relies on word embeddings, which are used to express response essays in low dimensional vectors followed by dense, CNN or LSTM layers to represent the semantics in the text [1,23]. Models that incorporate deep learning techniques and handcrafted features aim to capture all the aspects required for essay scoring. However, they are vulnerable to adversarial attacks [15]. One of the challenges in such a task is identifying if the responses correctly answer the question. Consider an example of a question asking the test takers to describe an incident where they had to be patient. (*Prompt 7 Question: Write a story about a time when you were patient.*) If the test taker answers general sentences about patience and does not provide a personal story, the response should be scored lower as it does not answer the question. As shown in Table 1, the incorrect answer has all the features required for a good answer: correct grammar and spellings, topic relevance and coherence; yet this response does not qualify as the correct answer. While a human annotator can quickly identify this, an AES system requires learning each essay's underlying inconsistencies. Hence, understanding each essay response's composition, structure, and hierarchies need to be further explored to ensure correct scoring.

Table 1. For a given question: *Write a story about a time when you were patient*, the Table lists a correct and incorrect answer to illustrate that even a well-formed answer is marked incorrect simply because it does not answer the question.

Incorrect answer	Correct answer
Patience is the ability to endure difficult circumstances. Patience may involve perseverance in the face of delay. Patience is the level of endurance one can have before disrespect. It is also used to refer to the character trait of being steadfast...	I am not a patience person, as I can't sit in a room for more than five minutes, but there was one time I was patient, during my wife's operation. I was sitting quietly in the hospital. At that moment I felt that I needed that time to pray for her well being...

The problem above is addressed in this work, using the newly introduced Capsule Neural Network [24]. The CapsNet was introduced in image classification to overcome the drawbacks of a CNN network by allowing an effective combination of low-level features of images to high-level depiction using iterative routing. CNNs not only face problems in learning the transformational invariance of images but also are unable to ensure a local agreement between the features (due to the max-pooling layers). CapsNet understands the spatial correlation between a part and a whole and analyzes the current reference frame to generalize to new or unseen frames.

We aim to use this method to understand each response based on the intrinsic spatial relationship between the parts (sentences) and the whole (essay response). Moreover, this technique provides an understanding of the general structure of the response

[35]. Hence, the organization and transition amongst the lines in the response will be under scrutiny. We did not shy away from using key handcrafted features to enhance the model's capability to score the responses. We call the CapsNet architecture, CapsRater, and the feature engineering enhancement, FeatureCapture (Fig. 1). To provide a contextualized and semantic initial representation of essay responses, we use the BERT embeddings [9]. This architecture improves performance on six out of eight prompts and increases the overall average QWK score. Moreover, to check the robustness of this work, we extensively experimented with adversarially crafted essay responses [19]. CapsRater + FeatureCapture successfully detected these adversarial attacks and scored them relatively lower than the original essay responses.

Following are the significant contributions of this work[1]:

- We provide a novel architecture with two key components: BERT enhanced CapsNet model and the feature extraction model for Automatic Essay Scoring. This is the first work that leverages Capsule Neural Networks for Automatic Essay Scoring to the best of our knowledge.
- We develop critical features with emphasis on official rubrics for scoring. We empirically demonstrate the importance of these features in improving the QWK.
- The proposed method shows higher scores on the QWK metric than state-of-the-art models with a boosted overall average QWK of 81%.
- We analyze the proposed work's performance on multiple adversarial attacks proving that our model is considerably robust. It successfully scores most of the adversarial text lower.

The paper's organisation is as follows: we discuss related work in Sect. 2, and the proposed pipeline in Sect. 3. We illustrate substantial experimentation in Sect. 4, to evaluate performance with baseline models. In the same section, we also assess our model's sturdiness on the adversarially perturbed datasets.

2 Related Work

The previous work on essay scoring relies on human experts who build domain-specific features to check the lexical and grammatical errors. The systems then employ machine learning classifiers to predict the essay scores. For example, works such as [3,6,16,20,32], trained Naive Bayes, Linear Regression and Rank Support Vector Machine (RankSVM) models for essay scoring task. EASE [11], a popular AES engine, applies text analysis and feature engineering to several regression models. Using handcrafted features has had immense success on the AES task. While there are simple features such as sentence length, word count, there are also other features that have a convex mechanism of engineerings, such as readability [33], textual and discourse coherence [5,26]. Incorporating each of these requires specialized focus and a domain-specific approach.

More recent approaches have turned to neural networks, which encode an essay into richer representations of low-dimensional embeddings. For example, Alikaniotis et al.

---

[1] Our code is available at: https://github.com/ECMLPKDD/CapsRater-FeatureCapture.

[1] applied a deep LSTM layer, Taghipour et al. [27] employed the CNN layer. They map texts into sequences that can account for the variable input lengths. These strategies are popular because they provide an end-to-end solution. A fundamental limitation of deep learning approaches in this domain is that they are susceptible to adversarial attacks [10]. Researchers have conducted a study to showcase how simple tricks can deceive state-of-the-art AES tools [19]. In this light, Farag et al. [12] applied window-based local coherence to catch adversarial attacks. However, detecting and mitigating advanced attacks is still an open problem.

Previous works on Capsule Neural Networks showcased promise on the image, as well as text classification tasks [13,35]. The capsules utilize all the feature information, therefore, address the issues with information loss in CNNs. Variants of Capsule architectures have recently experimented with in text classification applications. For example, Zhao et al. [35] proposed two models to stabilize the routing mechanism between capsules from disturbances in the text, such as stopwords, as these do not contribute to the classification task at hand. Kim et al. [18] employ a static routing mechanism in place of the dynamic one, based on the observation that the document semantics can remain the same with a different order of sentences. Saha et al. [25] studied the joint optimization capabilities of BERT and capsule layer in their classification task.

Due to the success of Capsule Neural Networks to capture spatial inter-dependencies in text [36], we employed the CapsNet for the AES task. Our work leverages BERT embeddings to utilize the pre-trained contextual features. The capsule architecture captures the semantic inter-word relationships and spatial patterns of words and transfer learned parameters. It can effectively encode the information required for essay evaluation and learn the complex patterns in the data.

3   Methodology

Our architecture consists of two independent pipelines, called CapsRater and Feature-Capture. We report results on both of these models. However, the best performing model is their combination (Fig. 1), as shown in Sect. 5. To combine these models, we took the mean over their class-wise probabilities and passed the output through a final dense layer to get the resultant score vector.

3.1   CapsRater

Capsule Neural Networks were introduced to accurately identify hierarchical relationships between objects and the features that constitute those objects. A capsule is a structure that essentially contains information about the probability and orientation of these features. Following [24], we use the capsules to attain a vector output from the feature detectors in CNN. The capsule representation is used in place of the pooling layer's scalar output, which discards the text's positional information.

We employed BERT embeddings of essay text and fine-tuned them using the Capsule framework. BERT provides contextual word representations by applying bidirectional training of transformer to language modelling. During the pre-training on the Masked Language Model task, BERT's architecture enables it to capture the entire text

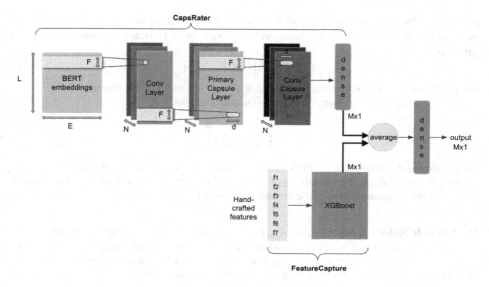

Fig. 1. The combined architecture of CapsRater + FeatureCapture model.

instead of sequential reading of directional models. As a result, BERT has a deeper understanding of language context and flow [9]. Thus, using BERT representation will provide the non-linear neural layers with the richer vector representations for the essay responses.

Consider the input sentence, S of length L, forms E dimensional BERT embeddings; then the input is represented as $S \in R^{L \times E}$.

Convolution Layer: We passed the input through the convolution layer to extract local textual features. In this step, the convolutional filter matrix, $W^a \in R^{F \times V}$ (F is the size of filter), slides across input of $s_{i:i+F-1}$ dimensions, multiplies with the input element-wise, and generates feature maps of size, $k^a \in R^{L-F+1}$. The resultant is added to a bias term, b_0, and a ReLU activation function f, is applied to it. Mathematically, this is denoted as:

$$k_i^a = f\left(\mathbf{s}_{i:i+F-1} \circ W^a + \mathbf{b}_0\right) \tag{1}$$

The above operation is iterated for N filters to widen the feature extraction process and concatenate the output [35]. Therefore there are N feature maps generated, represented by $K = [k_1, k_2, ..., k_N] \in R^{(L-F+1) \times N}$

Primary Capsule Layer: The feature maps are then passed through a convolution operation, with filter W^b, forming the first layer of capsules, denoted as $\Theta \in R^{(L-F+1) \times d}$, ($d$ is the dimension of capsule). This layer preserves the initial parameters belonging to each input feature instead of the scalar output from CNN's pooling operations. As a result, the capsules contain more information about the input, given as:

$$\Theta_i = (W^b)^T \times K_i + b_1 \tag{2}$$

where i denotes the size of the filter matrix: $1 \rightarrow L - F + 1$ and b_1 is the bias term.

A nonlinear squash function is applied on the above vectors.

$$\Theta_i = \frac{\|\Theta_i\|^2}{1 + \|\Theta_i\|^2} \frac{\Theta_i}{\|\Theta_i\|} \tag{3}$$

We performed the above steps with N number of filters and concatenated their corresponding outputs. For our domain, the capsule layer is aware of the semantics and ordering of sentences due to the vector representation of the instantiated parameters [35].

Part-Whole Relationship: Hinton et al. [13] defines this step as assigning part-to-whole.

We made use of two levels of capsules; each lower layer capsule assigns a vote vector $V_{low \rightarrow up}$ to each upper layer capsule. These vectors represent how much information is transferred from the different input capsules to the respective output capsule.

$$V_{low \rightarrow up} = W_{low \rightarrow up} \times \Theta_{low} \tag{4}$$

where $W_{low \rightarrow up}$ is a transformation matrix. These matrices solve exponentially taxing convolutions, and their insubstantial representation [24]. Moreover, they provide automated learning of part-whole relationship.

Dynamic Routing: This is an iterative process that builds a non-linear map to ascertain that each lower capsule's output is matched to the suitable upper capsule in the next layer. It controls the connection between the higher and lower layers' capsules. Following [24], there is an assignment probability $C_{low \rightarrow up}$ associated with input capsules, which measures the similarity between vote vector and output capsule. It is calculated as the multiplication of probability of the upper capsule for each lower capsule, $A'_{low \rightarrow up}$, with the softmax of logits of the assignment probability, $B_{low \rightarrow up}$.

$$C_{low \rightarrow up} = A'_{low \rightarrow up} \times \frac{\exp(B_{low \rightarrow up})}{\sum_{up=1}^{N} \exp(B_{low \rightarrow up})} \tag{5}$$

Here $B_{low \rightarrow up}$ is initialized as all 0s, and it measures the proportion of input capsule that makes the output capsule. It is updated according to the agreement between the upper layer capsule and the vote vector.

$$B_{low \rightarrow up} = B_{low \rightarrow up} + \Theta_{up}.V_{low \rightarrow up} \tag{6}$$

Output capsules are formed from the weighted sum of vote-vectors.

$$\Theta_{up} = squash(\frac{\sum_{low=1}^{H} (C_{low \rightarrow up} V_{low \rightarrow up})}{\sum_{low=1}^{H} (C_{low \rightarrow up})}) \tag{7}$$

$$A_{up} = |\Theta_{up}| \tag{8}$$

here, A_{up} is the activation probability of the upper layer capsule.

Convolutional Capsule Layer: The next layer gives the upper layer capsules formed from multiplying transformation matrices with the lower capsules, followed by dynamic routing. Using this routing mechanism, the capsules capture the importance and coherence of words while leaving out nonessential information. Finally, the upper layer capsules are flattened and passed through a dense layer.

3.2 FeatureCapture

Table 2. The key features extracted from the text to pass through the FeatureCapture model.

S. No.	Type	Feature
1	Prompt based	Number of words
2		Number of ! or ?
3		Correct POS tags
4		Number of spelling errors
5		Number of grammatical errors
6	Similarity based	Between prompt and responses
7		Between response sentences

Handcrafted features play an essential role in enhancing the performance of AES systems [21]. This can be observed in CapsRater + FeatureCapture model scores in Table 4, where adding features has illustrated best scores for the majority (5 out of 8) of prompts in the ASAP dataset. In this work, we used two types of handcrafted features: prompt-based and similarity-based (listed in Table 2). These were empirically decided by manual inspection of the responses and official rubrics given in the dataset. The prompt-based features are inspired by the EASE system [11].

- **Prompt Based Features**: We observed that shorter essays are scored lower by the annotators. Therefore, *Number of words* is an important length-based feature. We also took the frequency of *sentences ending with ! or ?* into account, as the essay response should not be primarily constituted of exclamatory or questioning sentences. For the next feature, *Correct POS tags*, we counted the number of erroneous unigrams and bigrams and subtracted those from total unigrams and bigrams. Using this, we get a statistical understanding of how coherent the essay response is. The above features were defined using the NLTK[2] library. Essays with many grammatical and spelling errors are penalized higher by the annotators; hence we included these in the feature space. *Number of spelling errors* were counted using pyspellchecker [4]. It relies on the Levenshtein Distance algorithm and compares all permutations in the frequency word list to correct an incorrect spelling. For identifying the *Grammatical Errors*, we use the popular grammar checking application called LanguageTool's python wrapper [29].
- **Similarity Based Features**: *The similarity score between the prompt and essay* shows if the response essay has borrowed text from the prompt by comparing the similarity between them. *Similarity between the sentences* gives an analysis of the amount of repetition within a response. Higher repetition reported by this feature should be scored lower. For calculating both, we use a fuzzy matching library, called rapidfuzz [22].

---

[2] https://www.nltk.org/.

All the above-extracted features are normalized to the 0–1 range, stacked and passed into the XGBoost [7], which is an ensemble learning method. Multiple individual base learners (models) are trained and combined for a final prediction. XGBoost has base learners that may generate average performance, but when sequentially added, they rectify the errors and lead to efficient predictions. Moreover, this algorithm is immensely scalable, and it relies on distributed computing which enhances fast learning [7]. The output of XGBoost is a probability vector for each target class. The final output vectors from CapsRater and FeatureCapture are averaged and passed through a dense layer for the final result.

4   Experimentation

4.1   Dataset

We perform our experiments on the widely used and accepted dataset for AES tasks, namely Automated Student Assessment Prize (ASAP). The dataset comprises 8 prompt questions, which students of grades 7 to 10 answer. The total number of answered essays are 12,976. More details about the dataset are shown below in Table 3.

Table 3. Description of the ASAP-AES Dataset used for evaluation of AES systems. Here RC refers to Reading Comprehension and # represents the count.

Prompt	# Responses	Type	Avg # words	Avg # sentences	Score range
1	1783	Argumentative	350	23	2–12
2	1800	Argumentative	350	20	1–6
3	1726	RC	150	6	0–3
4	1772	RC	150	4	0–3
5	1805	RC	150	7	0–4
6	1800	RC	150	8	0–4
7	1569	Narrative	250	12	0–30
8	723	Narrative	650	35	0–60

4.2   Evaluation Metric

We use the Quadratic Weighted Kappa (QWK) metrics for evaluation. It is a commonly used and accepted metric [21,27] which measures agreement between the AES scorer and the human annotators. QWK is calculated as:

$$k = 1 - \frac{\Sigma_{ij} w_{ij} Obs_{ij}}{\Sigma_{ij} w_{ij} Exp_{ij}} \tag{9}$$

Here, Obs and Exp are the observed and expected scores matrix respectively, while w denotes the weights. The scores assigned by the human and machine graders are i and j respectively. Given N is the number of possible scores, the weight matrix is defined as:

$$w_{ij} = \frac{(i-j)^2}{(N-1)^2} \tag{10}$$

The range of QWK score is from 0 to 1. The higher the score, the closer the machine-human agreement.

4.3   Baselines

We compared the proposed work with the recent state-of-the-art baselines: *EASE* [11], developed by *EdX*, is a feature-based model relying on n-grams and prompt word overlap. It performed third-best in the ASAP-AES competition. Taghipour et al. [27] build an architecture that applies CNN and LSTM with a mean over time layers. This model performs better than *EASE* on the ASAP dataset. *HISK+BOSWE* [8] is a statistical technique that captures text-based features using string kernels along with word embeddings. A reinforcement learning methodology, called *RL1*, was proposed by Wang et al. [30]. They used the QWK score as the reward function, which is governed by positive or negative feedback. *SkipFlow* [28] proposes the *neural coherence features* that capture semantics and coherence using deep neural networks. It uses the hidden states for extracting more information about the formation of the response. *MemoryNetworks* [34] takes one sample essay belonging to each score in the score range and saves it in the memory. This sample essay is used to calculate the similarity with new essays to score them. Using memory networks in grading helps them boost performance on 7 out of 8 prompts. *TSLF* [21] uses feature engineering along with neural networks for scoring. They form features for coherence, semantics and relevance using BERT embedding and employ the SVM classifier. They perform relatively higher than the state of the art methods. However, Kumar et al. [19] have shown that they lack robustness to tackle all kinds of adversarial attacks. R^2BERT [31] has multiple objective approaches where they explored two loss functions: the mean squared error and the batch-wise ListNet loss. They report improved results on baselines.

4.4   Implementation

We perform prompt-wise training on our models. The prompts in the dataset vary in terms of the genres (Argumentative, Reading Comprehension and Narrative), the scoring rubrics, and the grade of study of test-takers. We load the response essays to pass them through CapsRater and FeatureCapture, respectively. The first step in CapsRater involves transforming the data into the BERT-base model's embeddings ($V = 768$). It is then passed through the CapsRater model, where we use 32 filters, of size $F = 3$ words, in the first and second convolution layers while 16 filters in the convolution capsule layer. The capsules are also set as 16-dimensional vectors, and the dynamic routing process is iterated thrice for optimum loss convergence. In FeatureCapture, we use the XGBClassifier with the gbtree booster method. The max_depth parameter is set to 6, the objective function to multi:softprob, and n_estimators to 1000. The learning rate is set to 1e−4. For the CapsRater pipeline, we used the standard, cross-entropy loss function, Adam optimization, and the model is trained for 50 epochs with a 1e−4 learning rate.

5   Result and Analysis

This section compares CapsRater, FeatureCapture, and their combined pipeline with the prominent baseline works on the ASAP dataset. Table 4 shows the QWK scores on all 8 prompts available in the data. Moreover, we provide the average score to represent the overall performance. Table 4 reports that the proposed architecture boosts performance on six out of eight prompts, including both Argumentative, three Reading Comprehension and one Narrative prompt. Moreover, it leads to an overall increase of 2% in scores compared to the baselines.

Analysis: We had a thorough look at the essay responses and their scoring rubrics provided with the dataset. The rubrics can be categorized into three main parts[3]: Ideas + Content, Organization and Style. Ideas + Content focus on topic relevance and scrutinizes the main idea of

---

[3] https://www.kaggle.com/c/asap-aes.

Table 4. Performance comparison of proposed work with baseline models for each prompt in the dataset. Scores are calculated using the QWK metric. The best performance for each prompt is **emboldened**

PROMPT	1	2	3	4	5	6	7	8	Avg
EASE (SVR) [11]	0.781	0.621	0.630	0.749	0.782	0.771	0.727	0.534	0.699
EASE (BLRR) [11]	0.761	0.606	0.621	0.742	0.784	0.775	0.730	0.617	0.705
CNN [27]	0.804	0.656	0.637	0.762	0.752	0.765	0.750	0.680	0.726
LSTM [27]	0.808	0.697	0.689	0.805	0.818	0.827	0.811	0.598	0.756
CNN+LSTM [27]	0.821	0.688	0.694	0.805	0.807	0.819	0.808	0.644	0.761
HISK+BOSWE [8]	0.845	0.729	0.684	0.829	0.833	0.830	0.804	0.729	0.785
RL1 [30]	0.766	0.659	0.688	0.778	0.805	0.791	0.760	0.545	0.724
SkipFlow [28]	0.832	0.684	0.695	0.788	0.815	0.810	0.800	0.697	0.764
MemoryNets [34]	0.830	0.720	0.720	0.820	0.830	0.830	0.790	0.680	0.780
TSLF [21]	0.852	0.736	0.731	0.801	0.823	0.792	0.762	0.684	0.773
R^2BERT [31]	0.817	0.719	0.698	0.845	**0.841**	0.847	0.839	**0.744**	0.794
CapsRater (CR)	0.852	0.750	0.743	**0.847**	0.800	0.780	0.812	0.702	0.785
FeatureCapture (FC)	0.791	0.677	0.693	0.818	0.782	0.771	0.762	0.699	0.749
CR + FC	**0.866**	**0.764**	**0.751**	0.844	0.837	**0.852**	**0.843**	0.715	**0.809**

the response. Organization refers to the logical and structural organization of the response. Each response should give a meaningful, cohesive and complete meaning. Finally, the style includes features that emphasize spelling and grammatical errors. It castigates repetitive word and sentence usage. We also observed that poor-scored essays were generally shorter. While CapsRater has strong capabilities of demonstrating high performance, combining with FeatureCapture exhibits faster convergence, as it adheres to the prompt rubrics. This is evident in Fig. 2.

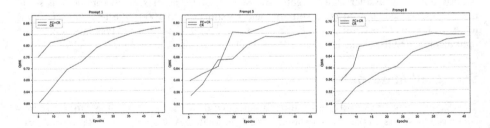

Fig. 2. Plot showing the increase in QWK metric performance with the increase in number of epochs on validation data during the training of prompt 1, 5 and 8 respectively.

The QWK score vs epoch plots in Fig. 2, shows that even at the beginning of the training process, CR + FC has significantly faster and greater convergence than CR on the validation data. Furthermore, as the number of epochs increases, there is an increase in the QWK scores. Introducing handcrafted features to CapsRater has penalized responses based on rubrics and has

Table 5. For prompt 1 question: *Write a letter to your local newspaper in which you state your opinion on the effects computers have on people*, the Table lists the original essay response along with adversarially perturbed responses. Green shows the addition of colored lines to original response, while red shows deletion of the colored lines from the response.

		Example
Original response		Dear Local Newspaper, I believe that computers are an extremely useful tool in society... Also, it lets you communicate with friends and family through the internet, for example, using facebook ... Each class teaches us to respect more and more of this culture... Also, is another website that helps us study for vocab. It also has many games that and learn our vocab words... I assume you will understand how much the computer has made a positive effect in society
Category	Attack	
ADD	AddSongs	Dear Local Newspaper, I believe that computers are an extremely useful tool in society. It helps people learn new things about different cultures . . . So shine bright, tonight you and I, We're beautiful like diamonds in the sky. . . I assume you will understand how much the computer has made a positive effect in society
ADD	RepeatSent	Dear Local Newspaper, I believe that computers are an extremely useful tool in society... Also, it lets you communicate with friends and family through the internet, for example, using facebook. Finally, it provides an accurate research tool for school projects, or interviews.Also, it lets you communicate with friends and family through the internet, for example, using facebook. Finally, it provides ...
DEL	DelRand	Dear Local Newspaper, I believe that computers are an extremely useful tool... It helps people learn new things about different cultures. Also, it lets you communicate with friends and family through the internet for example, using facebook... I assume you will understand how much the computer has made a positive effect in society
MOD	ModGrammar	Dear Local Newspaper, They believe this computer are the extremely useful tool is society. They helps this people should had learns ... Also, They lets u communicate with friend fam through the internets, 4 examples, using facebook. Finally, they provides the accurates researcher tools 4 schools project, or interview...
GEN	BabelGen	Computer with abandonment has not, and in all likelihood never will be boisterous, irreverent, and arrogant. Why is paper so accumulated to pondering? The reply to this query is that electronic computer is eternally and hastily incensed...

made a definite difference in the scoring. Interestingly, the average score of FeatureCapture (FC) by itself is 7% lower than the CR+FC model. This signifies that FeatureCapture, independently, does not have an excellent performance. Table 4 and Fig. 2 throw light on the importance of CapsRater, and the significance of its routing mechanism in modelling the hierarchical textual relationships. Moreover, using BERT embedding for initial text representation helps it grasp the contextual understanding of sentences in the response. Overall, this exhibits that the proposed work gains tremendously from utilizing both CapsNet architecture and extracted features.

5.1 Testing with Adversarial Essays

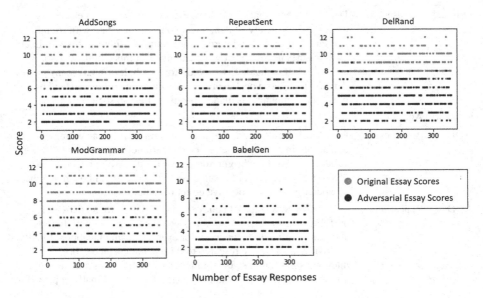

Fig. 3. CapsRater + FeatureCapture's scoring trends on Original and Adversarial essay responses for prompt 1

Recently, there has been thought-provoking work on the robustness of AES engines by conducting adversarial attacks on them [10, 19]. Kumar et al. [19] created a pipeline to generate various attacks on the ASAP-AES dataset. We employ their adversarially generated data to test the performance of our model. These are broadly divided into four categories: ADD (Text Addition), MOD (Text Modification), DEL (Text Deletion) and GEN (Random Text Generation). We experimented with two attacks from ADD and one from all other categories to check the robustness of the proposed model (CR + FC) and described the attacks with an example from prompt 1 response in Table 5. Here, *AddSongs* refers to the addition of lines of songs in the middle of the essay response. *RepeatSent* focused on repeating lines within the response essay to make it unreasonably wordy. *DelRand* created incoherent essays by removing random lines from the original response. *ModGrammar* introduced various grammatical errors in the original essay responses. *BabelGen* used a tool called the Babel Generator[4] to generate random sentences using prompt-specific keywords. *AddSongs*, *RepeatSent* and *DelRand* changed 25% of the original responses, while *ModGrammar* and *BabelGen* changed the entire original response. The training was done on the original training data, while the testing data was replaced with the perturbed testing data according to the five selected adversarial attacks. To keep the analysis concise, we have shown the difference between the grading of Prompt 1 and 8's original responses and their corresponding adversarially perturbed responses. The results are shown in Fig. 3 and 4.

---

[4] https://babel-generator.herokuapp.com/.

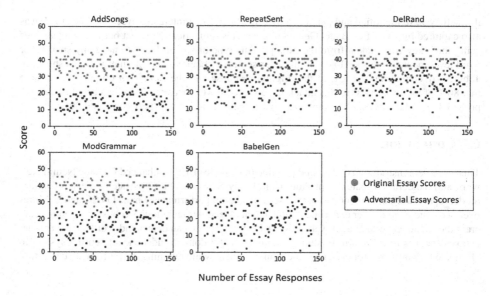

Fig. 4. CapsRater + FeatureCapture's scoring trends on Original and Adversarial essay responses for prompt 8.

Analysis: Original responses for prompt 1 are scored between the 2–12 range, and prompt 8 are scored between 0–60. We see that 90% of the original samples are scored between 8 and 12 for prompt 1. On similar lines, 88% of the original samples are scored between 30 and 60 for prompt 8. The overall trend is that adversarially perturbed responses have been scored consistently lower by CR+FC for all the attacks than the original essay responses. However, the maximum penalty was observed for the *ModGrammar* attack where 53% of responses of prompt 1 was scored the lowest possible score (2) and more than 70% of responses of prompt 8 was scored less than 30 (out of 60). *ModGrammar* made the responses grammatically incorrect and deformed the sentence structure such that it became semantically incorrect. For example, Table 5 shows a sentence where *ModGrammar* has introduced a singular demonstrative before plural noun: *"this people"*, a verb after a modal verb: *"should had"*, changed the semantics by introducing *"2"*, *"y"*, *"their"* incorrectly in the example. Hence, these responses were strongly penalized by CapsRater + FeatureCapture, due to excessive grammatical errors and semantic inconsistencies, leading to low scores. Another interesting observation is that while *RepeatSent* creates repetition of lines within the response, it still generates a relevant and coherent response. Even then, the attack was successfully identified, and most of the perturbed responses were scored lower for both prompt 1 and prompt 8. 45% of prompt 8 responses were scored between 20–30. We attribute this success to similarity-based features. *AddSongs* adds irrelevant lines to the responses, which tampers their relevance, structure and organization. We can see a clear drop in scores. Close to 65% of the perturbed responses for prompt 1 are scored between 2 and 4. Moreover, 40% of the responses for prompt 8 is scored between 15–20. This shows that the proposed work has grasped structural inconsistencies and penalized such responses. Similarly, for *DelRand*, most of the perturbed responses are scored lower. We noticed that the prompt 1 score bracket is inclined towards the range of 4–6, and that of prompt 8 is between 20–30. Deleting random lines from an essay disturbs the structure and transition within sentences. While these perturbed essays may lack feature-based flaws, they were identified and penalized by CapsRater. Most interestingly, *BabelGen* attack was also captured by CR + FC. Babel Generator generates semantically correct but random passages using three keywords taken from the prompt. This type of attack does not have any spelling or word level inconsistencies. However, the inter-sentence coherence is extremely low, as shown in Table 5, by comparing the text in the original response and the *BabelGen* output. We observe that 80% of the responses' in this attack were scored lower (between 2–5) than original data for prompt 1. Similarly, 85% of prompt 8 responses were scored lower (between 5 and 30).

6 Conclusion

This work demonstrates an end-to-end pipeline that applies BERT enhanced Capsule Neural Network and handcrafted features for Automatic Essay Scoring. The CapsRater + FeatureCapture model reports a significant increase in the state-of-the-art performance on the QWK metric. Moreover, we conducted deeper experimentation and analyzed our technique's robustness against several types of adversarial attacks. CapsRater + FeatureCapture can detect and score the adversarial essay responses low. We aim to study and implement a domain-independent scoring system to eliminate training costs for each new question statement while retaining the performance in future work.

References

1. Alikaniotis, D., Yannakoudakis, H., Rei, M.: Automatic text scoring using neural networks. arXiv preprint arXiv:1606.04289 (2016)
2. Attali, Y.: Automated essay scoring with e-rater®, vol 2.0. https://www.ets.org/Media/Research/pdf/RR-04-45.pdf
3. Attali, Y., Burstein, J.: Automated essay scoring with e-rater® v. 2. J. Technol. Learn. Assess. **4**(3) (2006)
4. Barrust: Pure Python spell checking. https://github.com/barrust/pyspellchecker
5. Chen, H., He, B.: Automated essay scoring by maximizing human-machine agreement. In: Proceedings of the 2013 Conference on Empirical Methods in Natural Language Processing, pp. 1741–1752 (2013)
6. Chen, H., Jungang, X., He, B.: Automated essay scoring by capturing relative writing quality. Comput. J. **57**(9), 1318–1330 (2014)
7. Chen, T., Guestrin, C.: XGBoost: a scalable tree boosting system. In: Proceedings of the 22nd ACM SIGKDD International Conference on Knowledge Discovery and Data Mining, pp. 785–794 (2016)
8. Cozma, M., Butnaru, A.M., Ionescu, R.T.: Automated essay scoring with string kernels and word embeddings. arXiv preprint arXiv:1804.07954 (2018)
9. Devlin, J., et al.: BERT: pre-training of deep bidirectional transformers for language understanding. arXiv:1810.04805 [cs.CL] (2019)
10. Ding, Y., Riordan, B., Horbach, A., Cahill, A., Zesch, T.: Don't take "nswvtnvakgxpm" for an answer-the surprising vulnerability of automatic content scoring systems to adversarial input. In: Proceedings of the 28th International Conference on Computational Linguistics, pp. 882–892 (2020)
11. Edx EASE: Ease (enhanced AI scoring engine) is a library that allows for machine learning based classification of textual content. this is useful for tasks such as scoring student essays. https://github.com/edx/ease
12. Farag, Y., Yannakoudakis, H., Briscoe, T.: Neural automated essay scoring and coherence modeling for adversarially crafted input. arXiv preprint arXiv:1804.06898 (2018)

13. Hinton, G.E., Krizhevsky, A., Wang, S.D.: Transforming auto-encoders. In: Honkela, T., Duch, W., Girolami, M., Kaski, S. (eds.) ICANN 2011. LNCS, vol. 6791, pp. 44–51. Springer, Heidelberg (2011). https://doi.org/10.1007/978-3-642-21735-7_6
14. Abdellatif Hussein, M., Hassan, H., Nassef, M.: Automated language essay scoring systems: a literature review. PeerJ Comput. Sci. **5**, e208 (2019)
15. Jia, R., Liang, P.: Adversarial examples for evaluating reading comprehension systems. arXiv preprint arXiv:1707.07328 (2017)
16. Jin, C., He, B., Hui, K., Sun, L.: TDNN: a two-stage deep neural network for prompt-independent automated essay scoring. In: Proceedings of the 56th Annual Meeting of the Association for Computational Linguistics (Volume 1: Long Papers), pp. 1088–1097 (2018)
17. Ke, Z., Ng, V.: Automated essay scoring: a survey of the state of the art. In: IJCAI, pp. 6300–6308 (2019)
18. Kim, J., Jang, S., Park, E., Choi, S.: Text classification using capsules. Neurocomputing **376**, 214–221 (2020)
19. Kumar, Y., Bhatia, M., Kabra, A., Junyi Li, J., Jin, D., Ratn Shah, R.: Calling out bluff: attacking the robustness of automatic scoring systems with simple adversarial testing. arXiv preprint arXiv:2007.06796 (2020)
20. Larkey, L.S.: Automatic essay grading using text categorization techniques. In: Proceedings of the 21st Annual International ACM SIGIR Conference on Research and Development in Information Retrieval, pp. 90–95 (1998)
21. Liu, J., Xu, Y., Zhu, Y.: Automated essay scoring based on two-stage learning. arXiv preprint arXiv:1901.07744 (2019)
22. Maxbachmann: Rapid fuzz similarity calculator. https://github.com/maxbachmann/rapidfuzz
23. Mikolov, T., Sutskever, I., Chen, K., Corrado, G., Dean, J.: Distributed representations of words and phrases and their compositionality. arXiv preprint arXiv:1310.4546 (2013)
24. Sabour, S., Frosst, N., Hinton, G.E.: Dynamic routing between capsules. arXiv preprint arXiv:1710.09829 (2017)
25. Saha, T., Jayashree, S.R., Saha, S., Bhattacharyya, P.: BERT-caps: a transformer-based capsule network for tweet act classification. IEEE Trans. Comput. Soc. Syst. **7**(5), 1168–1179 (2020)
26. Somasundaran, S., Burstein, J., Chodorow, M.: Lexical chaining for measuring discourse coherence quality in test-taker essays. In: Proceedings of COLING 2014, the 25th International Conference on Computational Linguistics: Technical Papers, pp. 950–961 (2014)
27. Taghipour, K., Tou Ng, H.: A neural approach to automated essay scoring. In: Proceedings of the 2016 Conference on Empirical Methods in Natural Language Processing, pp. 1882–1891 (2016)
28. Tay, Y., Phan, M., Tuan, L.A., Hui, S.C.: SkipFlow: incorporating neural coherence features for end-to-end automatic text scoring. In: Proceedings of the AAAI Conference on Artificial Intelligence, vol. 32 (2018)
29. Viraj: Python wrapper for grammar checking. https://pypi.org/project/grammar-check/1.3.1/
30. Wang, Y., Wei, Z., Zhou, Y., Huang, X.-J.: Automatic essay scoring incorporating rating schema via reinforcement learning. In: Proceedings of the 2018 Conference on Empirical Methods in Natural Language Processing, pp. 791–797 (2018)
31. Yang, R., Cao, J., Wen, Z., Wu, Y., He, X.: Enhancing automated essay scoring performance via cohesion measurement and combination of regression and ranking. In: Proceedings of the 2020 Conference on Empirical Methods in Natural Language Processing: Findings, pp. 1560–1569 (2020)
32. Yannakoudakis, H., Briscoe, T., Medlock, B.: A new dataset and method for automatically grading ESOL texts. In: Proceedings of the 49th Annual Meeting of the Association for Computational Linguistics: Human Language Technologies, pp. 180–189 (2011)

33. Zesch, T., Wojatzki, M., Scholten-Akoun, D.: Task-independent features for automated essay grading. In: Proceedings of the Tenth Workshop on Innovative Use of NLP for Building Educational Applications, pp. 224–232 (2015)
34. Zhao, S., Zhang, Y., Xiong, X., Botelho, A., Heffernan, N.: A memory-augmented neural model for automated grading. In: Proceedings of the Fourth (2017) ACM Conference on Learning@ Scale, pp. 189–192 (2017)
35. Zhao, W., Ye, J., Yang, M., Lei, Z., Zhang, S., Zhao, Z.: Investigating capsule networks with dynamic routing for text classification. arXiv preprint arXiv:1804.00538 (2018)
36. Zhao, W., Peng, H., Eger, S., Cambria, E., Yang, M.: Towards scalable and reliable capsule networks for challenging NLP applications. arXiv preprint arXiv:1906.02829 (2019)

TagRec: Automated Tagging of Questions with Hierarchical Learning Taxonomy

V. Venktesh$^{(\boxtimes)}$, Mukesh Mohania, and Vikram Goyal

Indraprastha Institute of Information Technology, New Delhi, India
{venkteshv,mukesh,vikram}@iiitd.ac.in

Abstract. Online educational platforms organize academic questions based on a hierarchical learning taxonomy (subject-chapter-topic). Automatically tagging new questions with existing taxonomy will help organize these questions into different classes of hierarchical taxonomy so that they can be searched based on the facets like chapter, topic. This task can be formulated as a flat multi-class classification problem. Usually, flat classification based methods ignore the semantic relatedness between the terms in the hierarchical taxonomy and the questions. Some traditional methods also suffer from the class imbalance issues as they consider only the leaf nodes ignoring the hierarchy. Hence, we formulate the problem as a similarity-based retrieval task where we optimize the semantic relatedness between the taxonomy and the questions. We demonstrate that our method helps to handle the unseen labels and hence can be used for taxonomy tagging in the wild, like the question-answer forums. In this method, we augment the question with its corresponding answer to capture more semantic information and then align the question-answer pair's contextualized embedding with the corresponding label (taxonomy) vector representations. The representations are aligned by fine-tuning a transformer based model with a loss function that is a combination of the cosine similarity and hinge rank loss. The loss function maximizes the similarity between the question-answer pair and the correct label representations and minimizes the similarity to unrelated labels. Finally, we perform extensive experiments on two real-world datasets. We empirically show that the proposed learning method outperforms representations learned using the multi-class classification method and other state of the art methods by **6%** as measured by Recall@k. We also demonstrate the performance of the proposed method on unseen but related learning content like the learning objectives without re-training the network.

Keywords: Hinge rank loss · Multi-class classification · Information retrieval

---

This work is supported by Extramarks Education (an education technology company) and TiH Anubhuti (IIITD).

Y. Dong et al. (Eds.): ECML PKDD 2021, LNAI 12979, pp. 381–396, 2021.
https://doi.org/10.1007/978-3-030-86517-7_24

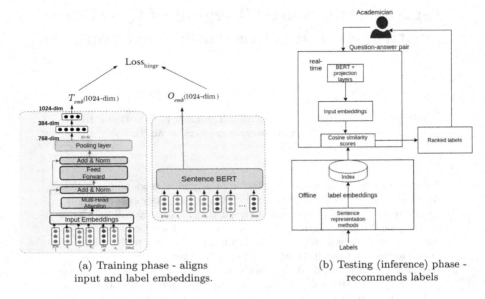

(a) Training phase - aligns input and label embeddings.

(b) Testing (inference) phase - recommends labels

Fig. 1. Training and testing phases for tagging questions with hierarchical labels

1   Introduction

Online learning platforms organize academic questions according to a hierarchical learning taxonomy (subject-chapter-topic). For instance a question about "electromotive force" is tagged with **"science - physics - electricity"**. This method of organization helps individuals navigate over large question banks. The taxonomy can also aid in *faceted* search. The *facets* could be topics, concepts, or chapters. However, manually tagging each question with the appropriate learning taxonomy is cumbersome. Hence there is a need for automated methods for tagging a question with the appropriate learning taxonomy. Automated tagging helps to organize acquired questions from third party vendors, which may be rarely linked to a learning taxonomy or are linked only at a "chapter" level. Also, the learning taxonomy is subject to change as the topic names or concept names could be replaced by synonyms or related concepts. Hence, the taxonomy tagging method should adapt to minor changes in the label (taxonomy) space without changes in the model architecture or re-training.

Automated categorization of content in online platforms is usually formulated as a multi-class classification problem [5,18]. However, there are some unique challenges when dealing with a hierarchical taxonomy and tagging short questions in the e-learning domain. *Firstly*, some of the traditional multi-class classification methods ignore the hierarchy and consider only leaf nodes of the hierarchical labels as labels. However, this formulation of the problem would suffer from class imbalance issues since a large number of contents may be tagged with a small number of leaf nodes leaving a smaller number of samples for other leaf nodes. The *second challenge* is that the labels are dynamic in nature as new topics could be added to the syllabus, and the old topics may no longer be valid or could be retired. This results in a change in the label space and thus gives rise to new labels. The new labels would have some similarity to some of the existing labels as the subject name and the chapter names could be semantically related to the existing chapter names. The traditional multi-class classification methods cannot exploit this semantic relatedness as they do not consider label representations. They require a change in architecture to incorporate the new labels and must be retrained. However, the hierarchical labels are an abstraction of their word descriptions and hence some of the terms in the hierarchical labels are semantically related to the words in the given questions. Hence, by learning a representation that captures the similarity between the labels and the related questions, the model can adapt to changes in label space.

To capture more semantic information from the given inputs, we augment the question with its answer as an auxiliary information. Hence, we refer to the augmented content as a *"question-answer"* pair and the hierarchical learning taxonomy is referred to as *"label"* or *"taxonomy"*. Our method, however would work even in cases where the answer is not given along with the question.

We propose a new method, named **TagRec**, for question-answer categorization in online learning platforms. In our method, the goal is to recommend relevant hierarchical learning taxonomy (label) for every question-answer pair to assist in organizing the learning content. Hence we adopt a similarity based retrieval method where hierarchical labels which are semantically related to the given question-answer pair. Figure 1 shows the basic architecture of the proposed method. Here, in the Fig. 1(a), the method projects the question-answer text and the corresponding label as inputs to a continuous vector space and aligns the input representations T_{emb} with the label representations O_{emb}. In the Fig. 1(b), during the recommendation (test time), when a new question arrives, the method projects the new question-answer pair to the vector space and computes the cosine similarity between the input representations and vector representations of all known labels. The labels are then ranked according to the similarity score, and the top-k labels are recommended for the given new question.

The proposed method can be used for tag recommendation in open source platforms like *StackExchange*. For example, a question about "Batch normalization" with tags "deep-learning" and "normalization" can be tagged with a hierarchical label $AI \rightarrow deep\ learning \rightarrow normalization \rightarrow Batch\ normalization$. The preprocessed data can then be fed to TagRec, which would be able to recommend hierarchical labels to new questions after the training.

The following are the key technical contributions of the paper:

- We propose a novel and efficient similarity based retrieval method to recommend a hierarchical taxonomy label to a given question-answer pair. The method decouples the computation of vector representations for the question input and the taxonomy labels, thus allowing label representations to be pre-computed and indexed for lookup.

- We propose a learning method to align the input and hierarchical label representations that involves a loss function combining the cosine similarity and the hinge rank loss [4].
- We employ a transformer based sentence representation method to represent the hierarchical labels. We conduct extensive experiments by varying the label representations in the architecture shown in Fig. 1(a) to empirically determine the effect of the label representations on the performance of the method. The proposed TagRec method outperforms the state of the art methods by upto **6%** with Recall@k as the metric.
- We demonstrate the ability of our method to adapt the changes in label space without any changes in architecture or retraining.
- We further demonstrate the ability of our method to categorize the unseen but related learning content like learning objectives. We extract 417 learning objectives from science textbooks and apply the proposed method to this data without any re-training. We observe that the proposed method is able to achieve high Recall@k at top-2 predictions and outperforms the existing state of the art methods by **7%**.

2   Related Work

In this section, we first provide an overview of multi-class classification methods that consider the hierarchical label structure and then briefly discuss the current state of the art sentence representation methods.

2.1   Multi-class Classification with Hierarchical Taxonomy

Many websites in the e-commerce and e-learning domains organize their content based on a hierarchical taxonomy [5,18]. The most common approaches for automatic categorization of the content to the hierarchical labels are flat multi-class single-step classification and hierarchical multi-step classifiers [17,19]. In multi-class single-step methods, the hierarchy is ignored and the leaf nodes are considered as labels. This leads to class imbalance issue, as discussed in Sect. 1. In the hierarchical multi-step approach, a classifier is trained to predict the top-level category and the process is repeated for predicting the sub-categories. However, the main problems associated with this approach are that the error from the classifiers at one level propagates to the next level and the number of classifiers increases at every step.

Several single-step classifiers have been proposed for the task of hierarchical classification. In [19], the word level features like n-grams were used with SVM as classifier to predict level 1 categories, whereas in [5] the authors have leveraged n-gram features and distributed representations from Word2Vec to obtain features and fed them to a linear classifier for multi-class classification. Several deep learning methods like CNN [6] and LSTM [17] have been proposed for the task of question classification. Since the pre-trained language models, like BERT [3], improve the performance, the authors in [18] propose a model BERT-QC, which fine tunes BERT on a sample of questions from science domain to classify them to a hierarchical taxonomy. The hierarchical multi-class classification problem has also been cast as a machine translation problem in [14] where the authors provide the product titles as input and use a seq2seq architecture to translate them to product categories that exhibit a hierarchy. However, all these above approaches do not consider the label representations. The hierarchical neural attention model [12] has been proposed, which leverages attention to obtain useful input sentence representation and uses an encoder-decoder architecture to predict each category in the hierarchical taxonomy. However, this approach may not scale to deep hierarchies.

In this paper, we take a similarity-based retrieval approach with the aim to recommend the relevant label (i.e., the hierarchical learning taxonomy) by aligning the input embeddings and the label embeddings. We do not explore the multi-level classifier approach owing to the shortcomings explained earlier in this section. The proposed method can also adapt to changes in the label space.

2.2 Sentence Representation Methods

Distributed representations that capture the semantic relationships [8] have helped to advance many NLP tasks like classification, retrieval. Methods like GloVe [10] learn vector representation of word by performing dimensionality reduction on a co-occurrence count matrix. Rather than averaging word represent ations to obtain sentence embeddings, an unsupervised method named Sent2Vec [9] for composing n-gram embeddings to learn sentence representations was proposed.

The Bidirectional Encoder Representation from Transformers (BERT) [3] is one of the current state of the art methods. However, one of the disadvantages of the BERT network structure is that no independent sentence embeddings are computed. The Sentence-BERT [11] model was proposed to generate useful sentence embeddings by fine-tuning BERT. Another transformer based sentence encoding model is the Universal Sentence Encoder (USE) [2] that has been specifically trained on semantic textual similarity task and generates useful sentence representations.

In this paper, we treat each label as a sentence and embed it using the sentence representation methods. For example, the label **Science - Physics - electricity** is treated as a sentence. In our experiments, we observe that USE embeddings and Sentence-BERT embeddings perform better than averaging word embeddings. These results are discussed in Sect. 4.

3 Methodology

In this section, we describe our method for classifying questions to hierarchical labels. The method consists of a training phase and testing phase, as shown in Fig. 1. The input to the method is a corpus of documents, $C = \{D_1, D_2...D_n\}$ where each document corresponds to a question-answer pair and the hierarchical

---

Algorithm 1. Tag Recommender

Input: Training set $T \leftarrow$ docs $\{D_1, ..D_n\}$, labels O of form (Subject-Chapter-Topic)
Output: Set of tags for test set , RO
    Training (batch mode)
1: Get input text embeddings , $T_{emb} \leftarrow BERT(D)$
2: Obtain label embeddings, $O_{emb} \leftarrow SENT\_BERT(O)$
3: $Index(labels) \leftarrow O_{emb}$
4: $loss \leftarrow \sum_{j \neq label} max(0, margin - cos(T_{emb}, O_{emb}(label)) + cos(T_{emb}, O_{emb}(j)))$
5: Fine-tune BERT to minimize $loss$ and align T_{emb} and O_{emb}
    Testing Phase
6: Compute embeddings for test set S using fine-tuned BERT $S_{emb} \leftarrow BERT(S)$
7: Rank set of unique labels $RO \leftarrow sorted(Sim(S_{emb}, O_{emb}))$
8: **return** Top-k labels from RO

---

labels $O = \{(S_1, Ch_1, T_1), (S_2, Ch_2, T_2)...\}$ where S_i, Ch_i and T_i denote subject, chapter, and topic respectively. The goal here is to learn an input representation that is close to the correct label in the vector space. We consider the label (S_i, Ch_i, T_i) as a sequence, $(S_i + Ch_i + T_i)$ and obtain a sentence representation for it using pre-trained models. We obtain contextualized representations for the inputs using BERT [3] followed by two projection layers. The linear projection layers are transformations that map the 768-D representation from BERT to the 1024-D or 512-D vector representation.

The steps of the proposed method are given in Algorithm 1. The details of the two phases in Algorithm 1 are as follows:

- In the *training* phase, the input question-answer pair is passed through a transformer based language model BERT followed by projection layers. The vector representations for the labels are obtained using a sentence representation method like USE [2] or Sentence-BERT [11]. The vector representations for all unique set of labels can be pre-computed and indexed for lookup. This saves computation cost and time during training and testing phases. The model is fine-tuned using a loss function that is a combination of cosine similarity and hinge rank loss [4]. This helps to align the contextualized input representations with the label representations.

- In the *testing* phase, as shown in Fig. 1b, the results are obtained in three steps. Firstly, the vector representations (embedding) for the input are computed using the fine-tuned BERT model. Secondly, the labels are ranked by computing cosine similarity between the input embeddings and the pre-computed label embeddings. Finally, top-k labels are chosen and metrics like Recall@k are computed for evaluating the performance of the model.

Our method is efficient as the label representations are pre-computed and indexed. Hence the time complexity at inference or testing time is $O(T_M N_{qa})$, where T_M is the time cost of the model (BERT + projection layers) and N_{qa} is the number of question-answer pairs at test time.

3.1   Contextualized Input Representations

The academic questions are mostly comprised of technical terms or concepts that are related with the "topic" component of the label. For example, a question that contains terms like *"ethyl alcohol"* is closely related with the topic *"alcohols and ethers"* and hence the question can be tagged with the label *"science - chemistry - alcohols and ethers"*. Academic questions also have terms that refer to different meanings depending on the context of their occurrence in the input sentence. For instance, the word "imaginary" in the sentence "Consider an imaginary situation" and its occurrence in the sentence "Given two imaginary numbers" has different meanings. This is an example of **polysemy** where the same word has different meanings in different contexts. Hence we need a method that can focus on important terms in the sequence and also tackle the problem of polysemy. To tackle the mentioned problems, we use a transformer based language model BERT for projecting the input text to the vector space. The BERT is a language model where the representations are learnt in two stages. In the first stage, the model is trained in an unsupervised manner. In the second stage, the model is fine-tuned on task specific labelled data to produce representations for downstream tasks. The "self-attention" mechanism in BERT helps in obtaining better vector representations and helps tackle the problem of polysemy.

Self-attention [15] is the core of transformer based language models, and BERT leverages it to obtain better representation for a word by attending other relevant words in the context; Thus, a word has different representations depending on the context it has been used in. Self-attention encodes each word in the sentence using Query (Q), Key(K) and Value(V) vectors to obtain attention scores which determines how much attention to pay to each word when generating an embedding for the current word. Mathematically,

$$Attention(Q, K, V) = \frac{Softmax(Q * K^T)}{\sqrt{d_k}} * V \tag{1}$$

$$Softmax(x_i) = \frac{exp(x_i)}{\sum_j^N exp(x_j)} \tag{2}$$

where d_k is the dimension of query, key, and value vectors and is used to scale the attention scores.

The self-attention mechanism helps to obtain contextualized representations that tackle the mentioned problems. We obtain contextualized representations of the input from BERT and pass them through the two projection layers, as shown in Fig. 1a. We fine-tune BERT and the projection layers to align the generated contextualized representations with label representations as given in Algorithm 1. We further explore the training phase in Sect. 3.3

3.2   Hierarchical Label Representations

Here, we describe how sentence representations are obtained for the labels. We consider the labels that have a hierarchical structure as a sequence of words and leverage sentence embedding methods to project them to vector space. We embed the labels this way to preserve the semantic relatedness between the labels. For instance, the label like **science - physics - electricity** must be closer to **science - physics - magnetism** than **science - biology - biomolecules** in the vector space. With simple vector arithmetic (cosine similarity), we observe that embedding the labels with sentence based representation methods like Sentence-BERT or Sent2Vec help to preserve the semantic relatedness when compared to averaging word embeddings from GLoVe [10]. The sentence representation methods also do not suffer from constituent words being out of vocabulary unlike traditional word embedding methods and are able to handle such words. Since the Sentence-BERT and the USE models have been explicitly trained on semantic textual similarity tasks they provide rich textual representations that can be used for similarity based retrieval tasks. Hence, in this paper, we extensively experiment with various sentence embeddings methods like Sent2Vec, Universal Sentence Encoder (USE), and Sentence-BERT. We also propose a method where the labels are represented using the mean of the GloVe vectors. We observe that sentence embedding methods significantly outperform the averaging of word vectors. The results are discussed in detail in the **Experiments and Results** section.

3.3 Loss Function

In the training phase in Algorithm 1, hinge rank loss is employed to maximize the similarity between contextualized input text embeddings and the vector representation of the correct label.

The hinge ranking loss is defined as :

$$loss(text, label) \leftarrow \sum_{j \neq label} max(0, margin - cos(T_{emb}, v(label)) + cos(T_{emb}, v(j)))$$

where T_{emb} denotes the input text embeddings from BERT, $v(label)$ denotes the vector representation of the correct label, $v(j)$ denotes the vector representation of an incorrect label, and cos denotes the cosine similarity function. The derivative of the loss function is propagated, and the linear projection layers are trained and the BERT layers are fine-tuned to minimize the loss as given in Algorithm 1. The margin was set to a value of 0.1, which is a fraction of the norm of the embedding vectors (1.0), and it yields the best performance.

4 Experiments

In this section, we discuss the experimental setup and the datasets on which the experiments were performed. All experiments are carried out on Google colab.

4.1 Datasets

To evaluate the effectiveness of the proposed method, we perform experiments on the following datasets:

Table 1. Some samples from the QC-Science dataset

Question	Answer	Taxonomy
The value of electron gain enthalpy of chlorine is more than that of fluorine. Give reasons	Fluorine atom is small so electron charge density on F atom is very high	Science→chemistry→classification of elements and periodicity in properties
What are artificial sweetening agents?	The chemical substances which are sweet in taste but do not add any calorie	Science→chemistry→chemistry in everyday life

- **QC-Science:** This dataset contains 47832 question-answer pairs belonging to the science domain tagged with labels of the form subject - chapter - topic. The dataset was collected with the help of a leading e-learning platform. The dataset consists of 40895 samples for training, 2153 samples for validation and 4784 samples for testing. Some samples are shown in Table 1. The average number of words per question is 37.14, and per answer, it is 32.01.
- **ARC** [18]: This dataset consists of 7775 science multiple choice exam questions with answer options and 406 hierarchical labels. The average number of words per question in the dataset is 20.5. The number of train, validation and test samples are 5597, 778 and 1400 respectively.
- **Learning Objectives:** This dataset consists of 417 learning objectives collected from the *"What you learnt"* section in class 8,9 and 10 science textbooks (K−12 system). The corresponding learning taxonomy was extracted from the "Table of contents" of the textbooks.

In our experiments we concatenate the question and the answer and it is considered as the input to the model (BERT), and the hierarchical taxonomy is considered as the label. Though BERT model has a context limit of 512 tokens, the length of each question-answer pair is within this range.

4.2 Analysis of Representation Methods for Encoding the Hierarchical Labels

In this section, we briefly provide an analysis of different vector representation methods for projecting the hierarchical labels (learning taxonomy) to a continuous vector space. We embed the hierarchical labels using sentence representations methods like Sent2Vec [9] and Sentence-BERT [11]. Additionally, we also average the word embeddings of individual terms in the hierarchical label using Glove to represent the label. We then compute the cosine similarity between the vectors of two different labels, and the results are as shown in Table 2. From Table 2, we observe that though "science→physics→electricity" and "science→chemistry→acids" are different, the representations obtained by averaging Glove embeddings output a high similarity score. This may be due to the loss

Table 2. Comparison of different representation methods for hierarchical labels

Method	Label1 (L1)	Label2 (L2)	cos (L1, L2)
Sentence-BERT	Science→physics→electricity	Science→chemistry→acids	0.3072
Sent2vec	Science→physics→electricity	Science→chemistry→acids	0.6242
GloVe	Science→physics→electricity	Science→chemistry→acids	0.6632

of information by averaging word vectors. Additionally here, the context of words like electricity is not taken into account when encoding the word physics. Additionally, "physics" and "chemistry" are co-hyponyms which may result in their vectors being close in the continuous vector space. We also observe that Sent2Vec is also unable to capture the semantics of the labels as it gives a similar high cosine similarity score. However, we observe that the vectors obtained using Sentence-BERT are not very similar, as indicated by the cosine similarity score. This indicates that Sentence-BERT is able to produce semantically meaningful sentence representations for the hierarchical labels. We also observe that Sentence-BERT outputs high similarity scores for semantically related hierarchical labels. Since this analysis is not exhaustive, we also provide a detailed comparison of methods using different vector representation methods in Sect. 5.

4.3   Methods and Experimental Setup

We compare TagRec with flat multi-class classification methods and other state of the art methods. In TagRec, the labels are represented using transformer based sentence representation methods like Sentence-BERT (Sent_BERT) [11] or Universal Sentence Encoder [2]. The methods we compare against are:

- **BERT+Sent2Vec:** In this method the training and testing phases are similar to TagRec. The labels representations are obtained using Sent2vec [9] instead of USE or Sent_BERT.
- **BERT+GloVE:** In this method, the labels are represented as the average of the word embeddings of their constituent words. The word embeddings are obtained from GloVe.

$$V(label) = mean((Gl(subject), Gl(chapter), Gl(topic)))$$

where, $V(label)$ denotes vector representation of the label, Gl denotes GloVe pre-trained model. The training and testing phases are same as TagRec.
- **Twin BERT:** This method is adapted from Twin BERT [7]. In this method, instead of using pre-trained sentence representation methods , we fine-tune a pre-trained BERT model to compute the label representations. The label representations correspond to the last layer hidden state of the first token. The first token is denoted as [CLS] in BERT, which is considered as the aggregate sequence representation. The BERT model that computes representations for the input and the BERT model for computing the label representations are fine-tuned simultaneously.

- **BERT multi-class** (label relation) [18]: In this method, we fine-tune a pre-trained BERT model to classify the input question-answer pairs to one of the labels. Here the labels are encoded using label encoder, and hence this is a flat **multi-class classification** method. At inference time, we compute the representations for the question-answer pairs and labels using the fine-tuned model. Then the labels are ranked according to the cosine similarity scores computed between the input text embeddings and the label embeddings.
- **BERT multi-class** (prototypical embeddings) [13]: To provide a fair comparison with TagRec, we propose another baseline that considers the similarity between samples rather than the samples and the label. A BERT model is fine-tuned in a flat multi-class classification setting similar to the previous baseline. Then for each class, we compute a prototype, which is the mean of the embeddings of randomly chosen samples for each class from the training set. The embedding for each chosen sample is computed as the concatenation of the [CLS] token of the last 4 layers of the fine-tuned BERT model. We observe that this combination provides the best result for this baseline. After the prototypes are formed for each class, at inference time, we obtain the embeddings for each test sample in the same way and compute cosine similarity with the prototype embeddings for each class. Then the classes are ranked using the cosine similarity and top-k classes are returned.
- **Pretrained Sent_BERT:** We implement a simple baseline where the vector representations of the input texts and the labels are obtained using a pre-trained Sentence-BERT model. There is no training involved in this baseline. For each input top closest matching labels are retrieved according to cosine similarity.

All the BERT models were fine-tuned for 30 epochs (with early stopping) with the ADAM optimizer, with learning rate of 2e−5 [3] and epsilon which is a hyperparameter to avoid division by zero errors is set to 1e−8. The random seed was set to a value of 42. The margin parameter in the hinge rank loss was set to a value of 0.1. All the implementations were done in Pytorch. The huggingface library [16] was used to fine-tune pre-trained BERT models.

Our code and datasets are publicly available at https://bit.ly/3jQpzEv.

5   Results and Discussion

The performance comparison of the methods described in the previous section is shown in Table 3. We use the Recall@k metric, which is a common metric for ranked retrieval tasks. From the results, we observe that the proposed method TagRec (BERT + USE and BERT + Sent_BERT) outperforms flat multi-class classification based baselines and other state of the art methods. We observe that representing the labels with transformer based sentence embedding methods perform the best. This is evident from the table as TagRec (BERT + USE) and TagRec (BERT + Sent_BERT) outperform BERT+Sent2Vec and BERT + GloVe methods. This is because Universal Sentence Encoder (USE)

Table 3. Performance comparison of TagRec with variants and baselines, † indicates TagRec's significant improvement at 0.001 level using *t-test*

Dataset	Method	R@5	R@10	R@15	R@20
QC-science	TagRec (BERT + USE) (proposed method)	**0.86**	0.92	**0.95**	0.96
	TagRec (BERT + Sent_BERT) (proposed method)	0.85†	**0.93†**	**0.95†**	**0.97†**
	BERT + sent2vec	0.79	0.89	0.93	0.95
	Twin BERT [7]	0.72	0.86	0.91	0.94
	BERT + GloVe	0.76	0.87	0.92	0.94
	BERT classification (label relation) [18]	0.39	0.50	0.57	0.63
	BERT classification (prototypical embeddings) [13]	0.83	0.91	0.93	0.95
	Pretrained Sent_BERT	0.30	0.40	0.47	0.52
ARC	TagRec (BERT + USE) (proposed method)	**0.67†**	**0.81†**	**0.86†**	**0.89†**
	TagRec (BERT + Sent_BERT) (proposed method)	0.65	0.77	0.84	0.88
	BERT + sent2vec	0.55	0.72	0.81	0.87
	Twin BERT [7]	0.46	0.63	0.72	0.78
	BERT + GloVe	0.56	0.73	0.82	0.86
	BERT classification (label relation) [18]	0.27	0.37	0.42	0.49
	BERT classification (prototypical embeddings) [13]	0.64	0.75	0.80	0.83
	Pretrained Sent_BERT	0.31	0.46	0.54	0.59

and Sentence-BERT use self-attention to produce better representations. This reinforces the hypothesis that averaging the word vectors to represent the labels does not preserve the required semantic relatedness between labels. The Twin BERT architecture does not perform well when compared with TagRec. This is because the label representations obtained through fine-tuned BERT may not preserve the semantic relatedness than the label representations obtained from pre-trained sentence embedding models.

Also both the Sentence-BERT and the Universal Sentence Encoder models are trained on semantic text similarity (STS) tasks thereby rendering them the ideal candidates for retrieval based tasks. Finally we observe that the TagRec method outperforms the flat classification based baselines confirming the hypothesis that the representations learnt by aligning the input text and label representations provide better performance. This is pivotal to the task of question-answer pair categorization as the technical terms in the short input text are strongly correlated with the words in the label. The first baseline (BERT label relation) performs poorly as it has not been explicitly trained to minimize the distance between the input and label representations. This implies that the representations learnt through flat classification has no notion of label similarity. But the prototypical embeddings based baseline performs better as the classification is done based on similarity between train and test sample representations. However this baseline also has no notion of label similarity. Hence does not perform well when compared to our proposed method, TagRec. We also observe that the

Table 4. Examples demonstrating the performance for unseen labels at test time.

Question text	Ground truth	Top 2 predictions	Method
A boy can see his face when he looks into a calm pond. Which physical property of the pond makes this happen? (A) flexibility (B) reflectiveness (C) temperature (D) volume	Matter→properties of material→reflect	Matter→properties of material→flex and **matter→properties of material→reflect**	TagRec (BERT + USE)
		Matter→properties of objects→mass and Matter→properties of objects→density	Twin BERT [7]
		Matter→states→solid and matter→properties of material→density	BERT + GloVe
		Matter→properties of material→specific heat and matter→properties of material	BERT + sent2vec
Which object best reflects light? (A) gray door (B) white floor (C) black sweater (D) brown carpet	Matter→ properties of material→reflect	Energy→light→reflect and **matter→properties of material→reflect**	TagRec (BERT + USE)
		Energy→thermal→ radiation and energy→light→generic properties	Twin BERT [7]
		Energy→light and energy→light→refract	BERT + GloVe
		Energy→light→reflect and energy→light→refract	BERT + sent2vec

simple baseline of performing semantic search using pretrained Sentence-BERT does not work well as the model is not fine-tuned to align the input and labels.

To further show the efficacy of our method, we perform statistical significance tests and observe that the predicted results are statistically significant. For instance, for Recall@20 we observe that the predicted outputs from TagRec are statistically significant (t-$test$) with p-values **0.000218** and **0.000816** for *QC-Science* and *ARC* respectively.

Table 5. Performance comparison for learning objective categorization

Method	R@1	R@2
TagRec (BERT + USE) (proposed method)	0.69	0.85
TagRec (BERT + Sent_BERT) (proposed method)	**0.77**	**0.91**
BERT+sent2vec	0.49	0.64
Twin BERT [7]	0.54	0.79
BERT+GloVe	0.62	0.84
BERT classification (label relation) [18]	0.46	0.59
BERT classification (prototypical embeddings) [13]	0.60	0.76
Pretrained Sent_BERT	0.39	0.54

The proposed method TagRec was also able to adapt to new labels. For instance, two samples in the test set of the **ARC** dataset were tagged with *"matter→properties of material→reflect"* unseen during the training phase as shown in Table 4. At test time, the label *"matter→properties of material→reflect"* appeared in top 2 predictions output by the proposed method (TagRec (BERT + USE)) for the two samples. We also observe that for the method (TagRec (BERT + Sent_BERT)) the label *"matter→properties of material→reflect"* appears in its top 5 predictions. We observe that for other methods shown in Table 4 the correct label does not appear even in top 10 predictions. The top 2 predictions from other methods for the samples are shown in Table 4. We also make similar observations for the BERT classification (label relation) and BERT classification (prototypical embeddings) baselines. We do not show them in Table 4 owing to space constraints. The top 2 predictions from BERT classification (prototypical embeddings) baseline for example 1 in Table 4 are *matter→properties of objects→temperature* and *matter→properties of objects→shape*.

For example 2, in Table 4, the top 2 predictions from BERT classification (prototypical embeddings) are *energy→light→reflect* and *matter→properties of material→color*.

The top 2 predictions from BERT classification (label relation) baseline for example 1 in Table 4 are *matter→properties of objects→ density* and *matter→properties of material→density*. For example 2, in Table 4, the top 2 predictions from BERT classification (label relation) are *energy→light→refract* and *matter→properties of material→luster*. This confirms our hypothesis that the proposed method can adapt to new labels without re-training or change in the model architecture unlike existing methods.

We also demonstrate the performance of TagRec on unseen but related learning content like the learning objectives. Learning objectives convey the learning goals and can be linked to learning content through the learning taxonomy.

We obtain the predictions for the given learning objectives using the models trained on the $QC - Science$ dataset. We do not fine-tune them on the given learning objectives dataset and directly use them as test set to obtain predictions.

The results of the learning objective categorization task are shown in Table 5. We show the recall at top 1 and top 2 predictions as the best results were obtained in top 2 predictions. We observe that the proposed method TagRec outperforms other methods. Particularly TagRec (BERT + Sent_BERT) which uses Sentence-BERT to represent the hierarchical labels gives the best performance. This demonstrates that the proposed method is able to generalize to unseen but related learning content without any re-training.

6 Conclusion

In this paper, we proposed a new method for learning to suggest hierarchical taxonomy (label) for short questions. We demonstrated that the representations learnt using the proposed similarity based learning method is better than flat classification methods and other state of the art methods [7]. Our method can easily adapt to unseen labels without a change in the architecture unlike flat classification based methods. We also demonstrated that the trained model can be used to categorize any related learning content like learning objectives without any retraining. The proposed method can also be used for taxonomy tagging in the forums like Quora and other discussion forums. The questions in Quora have a character limit of 50 words, but the answers could be longer than the context limit of the BERT model. To handle such long sequence lengths, we plan to explore new methods like Longformer [1]. Also in the future, we aim to explore the hyperbolic space to represent the hierarchical labels.

References

1. Beltagy, I., Peters, M.E., Cohan, A.: Longformer: The long-document transformer. arXiv preprint arXiv:2004.05150 (2020)
2. Cer, D., et al.: Universal sentence encoder for English. In: Proceedings of the 2018 Conference on Empirical Methods in Natural Language Processing: System Demonstrations, pp. 169–174. Association for Computational Linguistics, Brussels, Belgium, November 2018
3. Devlin, J., Chang, M., Lee, K., Toutanova, K.: BERT: pre-training of deep bidirectional transformers for language understanding. CoRR abs/1810.04805 (2018)
4. Frome, A., et al.: DeViSE: a deep visual-semantic embedding model. In: Advances in Neural Information Processing Systems, pp. 2121–2129 (2013)
5. Kozareva, Z.: Everyone likes shopping! multi-class product categorization for e-commerce. In: Proceedings of the 2015 Conference of the North American Chapter of the Association for Computational Linguistics: Human Language Technologies, pp. 1329–1333 (2015)
6. Lei, T., Shi, Z., Liu, D., Yang, L., Zhu, F.: A novel CNN-based method for question classification in intelligent question answering. In: Proceedings of the 2018 International Conference on Algorithms, Computing and Artificial Intelligence. ACAI 2018. Association for Computing Machinery, New York, NY, USA (2018)
7. Lu, W., Jiao, J., Zhang, R.: TwinBERT: distilling knowledge to twin-structured compressed BERT models for large-scale retrieval, pp. 2645–2652. CIKM 2020. Association for Computing Machinery, New York, NY, USA (2020)

8. Mikolov, T., Sutskever, I., Chen, K., Corrado, G., Dean, J.: Distributed representations of words and phrases and their compositionality. In: Proceedings of the 26th International Conference on Neural Information Processing Systems - Volume 2, pp. 3111–3119. NIPS 2013, Curran Associates Inc., Red Hook, NY, USA (2013)

9. Pagliardini, M., Gupta, P., Jaggi, M.: Unsupervised learning of sentence embeddings using compositional n-gram features. arXiv preprint arXiv:1703.02507 (2017)

10. Pennington, J., Socher, R., Manning, C.: GloVe: global vectors for word representation. In: Proceedings of the 2014 Conference on Empirical Methods in Natural Language Processing (EMNLP), pp. 1532–1543. Association for Computational Linguistics, Doha, Qatar, October 2014

11. Reimers, N., Gurevych, I.: Sentence-BERT: sentence embeddings using Siamese BERT-networks. In: Proceedings of the 2019 Conference on Empirical Methods in Natural Language Processing and the 9th International Joint Conference on Natural Language Processing (EMNLP-IJCNLP). Association for Computational Linguistics, Hong Kong, China, November 2019

12. Sinha, K., Dong, Y., Cheung, J.C.K., Ruths, D.: A hierarchical neural attention-based text classifier. In: Proceedings of the 2018 Conference on Empirical Methods in Natural Language Processing, pp. 817–823 (2018)

13. Snell, J., Swersky, K., Zemel, R.: Prototypical networks for few-shot learning. In: Advances in neural information processing systems, pp. 4077–4087 (2017)

14. Tan, L., Li, M.Y., Kok, S.: E-commerce product categorization via machine translation. ACM Trans. Manage. Inf. Syst. **11**(3), 1–14 (2020)

15. Vaswani, A., et al.: Attention is all you need. In: Advances in Neural Information Processing Systems, pp. 5998–6008 (2017)

16. Wolf, T., et al.: Transformers: state-of-the-art natural language processing. In: Proceedings of the 2020 Conference on Empirical Methods in Natural Language Processing: System Demonstrations, October 2020

17. Xia, W., Zhu, W., Liao, B., Chen, M., Cai, L., Huang, L.: Novel architecture for long short-term memory used in question classification. Neurocomputing **299**, 20–31 (2018)

18. Xu, D., et al.: Multi-class hierarchical question classification for multiple choice science exams. In: Proceedings of the 12th Language Resources and Evaluation Conference, pp. 5370–5382. European Language Resources Association, Marseille, France, May 2020

19. Yu, H.F., Ho, C.H., Arunachalam, P., Somaiya, M., Lin, C.J.: Product title classification versus text classification. Csie. Ntu. Edu. Tw, pp. 1–25 (2012)

Remote Sensing, Image and Video Processing

Checking Robustness of Representations Learned by Deep Neural Networks

Kamil Szyc⬛, Tomasz Walkowiak⬛, and Henryk Maciejewski(✉)⬛

Wroclaw University of Science and Technology, Wroclaw, Poland
{kamil.szyc,tomasz.walkowiak,henryk.maciejewski}@pwr.edu.pl

Abstract. Recent works have shown the vulnerability of deep neural networks to adversarial or out-of-distribution examples. This weakness may come from the fact that training deep models often leads to extracting spurious correlations between image classes and some characteristics of images used for training. As demonstrated, popular, ready-to-use models like the ResNet or the EfficientNet may rely on the non-obvious and counterintuitive features. Detection of these weaknesses is often difficult as classification accuracy is excellent and does not indicate that the model is non-robust. To address this problem, we propose a new method and a measure called robustness score. The method allows indicating which classes are recognized by the deep model using non-robust representations, i.e. representations based on spurious correlations. Since the root of this problem lies in the quality of the training data, our method allows us to analyze the training dataset in terms of the existence of these non-obvious spurious correlations. This knowledge can be used to attack the model by finding adversarial images. Consequently, our method can expose threats to the model's reliability, which should be addressed to increase the certainty of classification decisions. The method was verified using the ImageNet and Pascal VOC datasets, revealing many flaws that affect the final quality of deep models trained on these datasets.

Keywords: Deep neural networks · Robust representations · Spurious correlations · Explainable AI

1 Introduction

Recent developments in image classification using deep neural networks have led to remarkable improvement in classification accuracy. For instance, the top-1 results on the ImageNet benchmark are: 63.3% for AlexNet [16], 78.6% for ResNet-152 [12], 84.4% for EfficientNet-B7 [24], and 90.2% for one of the best currently models [21]. Some authors argue that deep models now surpass human-level performance [11]. However, recent works have shown that these optimistic figures do not fairly reflect performance of deep networks for real life recognition tasks, but are rather specific to the ImageNet benchmark. Deep networks

© Springer Nature Switzerland AG 2021
Y. Dong et al. (Eds.): ECML PKDD 2021, LNAI 12979, pp. 399–414, 2021.
https://doi.org/10.1007/978-3-030-86517-7_25

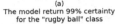

(a)	(b)	(c)	(d)
The model return 99% certainty for the "rugby ball" class	The saliency map	The part of the saliency map within bounding box	When cutting out bounding box the model is still return 99% certainty for the "rugby ball" class

Fig. 1. An image of one of the lowest robustness scores for n04118538 (rugby ball) class using the ResNet-152 as the CNN model and GradCam++ as the saliency map generation. There is spurious correlation of a ball with a player outfit. **(b)** shows saliency maps for this image, where the warm colors mark pixels with a more significant impact on the final prediction for this class, and cold colors indicate lower impact. **(c)** shows saliency map withing the bounding box (ROI). Robustness score is the ratio of saliency map summed within the bounding box **(c)** and saliency map summed over the entire image **(b)**. **(d)** shows the ROI (ball area) covered with white noise; the network response does not change - for both images **(a)** and **(d)**, the winning class's softmax output is above 0.99.

are vulnerable to adversarial or out-of-distribution examples that humans easily decipher [7,13,19,32]. This is a big concern in safety-critical applications of AI [1].

These weaknesses are primarily due to the way how deep networks learn representations of image classes: deep neural networks are excellent extractors of correlations between image categories and some characteristics of training data. Hence recognition of some image categories may rely on irrelevant correlations not perceived by human perception but strongly embedded in the training dataset. On the other hand, deep networks may also learn to overuse some relevant correlations (e.g., texture or color), which again leads to counter-intuitive behavior of deep models [29].

The problem of learning spurious correlations has been recently investigated by many authors, not only in the context of the ImageNet. [31] and [14] showed that deep models trained in the context of medical diagnostic tasks involving chest X-rays usually generalize poorly to new data or data from other sources (e.g., hospitals). This often results from the fact that models learn some undesirable features rather than clinically relevant features. [2] showed similar behavior in the context of recognition of animals in a new environment - a generalization of deep models to new locations was generally poor. [26] analyzed this in the context of the ImageNet. They showed that the unexpected behavior of deep models trained on the ImageNet might come from incorrect or ambiguous labels, which occur for some classes in this popular dataset. [3] analyzed the sources of label noise in ImageNet and proposed a new human annotation procedure that yields improved performance of models. The bias of image background was analyzed by [29]. The study proves that state-of-the-art models trained on the ImageNet are very sensitive to the change of background - adversarial background 'fools' recognition of up to 87.5% of images. [13] demonstrate a large collection of natural adversarial images, comprised of known as well as out-of-distribution examples, that lead to surprising yet high-confidence decisions of current models. Changes in network architectures do not improve recognition performance, as the root of the problem again lies in spurious, non-robust representations learned by the deep model.

In this work, we want to deal with this problem. We propose a method that allows us to quantify to what extend recognition of individual classes may rely on spurious, counter-intuitive correlations found in the training and validation data. The idea is to use saliency maps [8,20,27] to identify image areas/patterns with high impact on the classifier decision and to measure by how much these areas overlap with the bounding boxes surrounding the object (region-of-interest, ROI) corresponding to the image category. The rationale of the method is illustrated in Fig. 1 which shows a sample image for class 'rugby ball'. The saliency maps reveal that majority of the model's attention is focused outside the ROI (the rugby ball), and hence the recognition of this image relies primarily (solely?) on the background or surroundings of the object. We argue that recognition of image categories in which most of the model's attention lies outside the ROI should be considered non-robust, as the model either learns irrelevant spurious correlations present in the training data or learns to overuse relevant context (i.e. background or surroundings) in place of actual characteristics of the object of interest.

Our main contributions are the following. We propose a measure called robustness score which quantifies, broadly, the proportion of the attention (as expressed by saliency maps) a deep model tends to focus within the ROI while recognizing images of a particular class. *Small* values of the measure indicate classes with most likely non-robust representations. These image classes can be further analyzed by visualizing the training/validation images for the class of interest, as illustrated in Fig. 1. This allows us to explain the nature of these counter-intuitive, non-robust representations. In this way, we pinpoint image classes for which the training images should be improved by either providing more relevant labels, as postulated in [3], or by extending the collection of training examples with some form of 'background augmentation' in order to reduce the risk of learning spurious correlations by the training algorithm. We performed a series of computational experiments to demonstrate that the proposed method is effective for different deep neural network models (we tested ResNet-152, AlexNet, and some versions of the EfficientNet), different algorithms for saliency map generation (we tested the Grad-CAM++ [8] and Smooth Grad-Cam++ [20]), and for different image recognition datasets (we tested ImageNet and Pascal VOC).

In the feasibility studies of our method, we identified many categories recognized (by the ResNet and other models) with high accuracy which realize the robustness score below 0.3. We found that recognition of these categories relies on spurious correlations, and thus we showed that the train/validation data for these categories is biased by the existence of such counter-intuitive, not perceived correlations. Because these occur in both train and validation subsets, standard measures like accuracy cannot detect this problem. We used this knowledge to find some natural adversarial examples, which illustrates the low reliability of the models for these categories.

Illustrations of some of the findings are shown in Fig. 1 and 2. Continuing our example in Fig. 1, class "rugby ball" strongly correlates with player's suits, or the class pickelhaube (Fig. 2) correlates with a uniform in old-looking, monochromatic photograph. Figure 6 shows some natural adversarial images that are easy to find once we know the spurious correlations exist. It can be shown that for instance, the class "diaper" strongly correlates with babies, or class "miniskirt" strongly correlates with naked female legs. We can also use this method to identify classes for which the unreliability of models may result from ambiguous class labels, as illustrated in Fig. 5. As presented the classes like volleyball or basketball can mean both the game or the ball - however, models are unreliable if the latter meaning is assumed.

The proposed method allows us to pinpoint image categories recognized by state-of-the-art models using non-robust representations, and consequently to improve the trustworthiness of deep classifiers. Hence we believe that the method contributes towards trustworthy/explainable AI, which has recently become the filed of active research, see e.g. [9, 18, 22, 23] for some prominent directions.

All results are fully reproducible, and the code is available on the GitHub[1].

Finally, we want to refer the reader to other approaches to estimate neural model robustness. A commonly used way to evaluate model robustness is to measure the testing accuracy under some classes of powerful adversarial attacks, such as the PGD (projected gradient descent) [17], C&W [6], or recently proposed new class of gradient-based attacks [5]. The drawback of this approach is that it is not invariant from the adversarial attack type or defense method. Alternative methods propose attack-agnostic robustness metrics, e.g., [30] define model robustness as a measure of the stability of network prediction under input perturbations. For instance, [28] define robustness in terms of the minimum distortion from a given input example required to craft an adversarial example out of this input, or the method proposed in [10] allows for identifying safe regions in the input space where the network prediction is robust against adversarial input perturbations.

2   Method

We propose a simple and effective measure, we call robustness score, that allows us to detect which classes of images are recognized by a deep neural network using non-robust representations. Low values of the measure indicate a mismatch between the object of interest and areas in the image with a strong impact on classification.

---

[1] https://github.com/hmaciej/robustness_score.git.

Fig. 2. Chosen examples from the ImageNet with low class robustness score. For such classes the probabilities that the CNN model learned spurious correlations are higher. For the tested class images the (crs) refers to the mean(class) robustness score and the (acc) refers to the mean accuracy. There are strong spurious correlations for each example - e.g., the class n04264628 (space bar) correlates with keyboards and typewriter, the objects for class n03929855 (pickelhaube) are usually presented as old, monochromatic photographs with correlation with a military uniform, and for class n04228054 (ski) exist strong correlations with snow and winter suit.

Technically, to calculate the robustness score, we require that, in the validation data, bounding boxes are available that provide the location (ROI) of the object related to the image class, denoted $\ell(I)$. Here we assume that the bounding box is a binary indicator function $bbox(I, x, y)$ equal 1 if the pixel x,y of an image I is inside the box, and 0 otherwise.

Moreover, our method requires a saliency map generation which is applied to indicate the image areas with a big impact on final classification. Several methods have been proposed in literature, e.g. [8,20,27]. Originally these methods were based on occluding parts of the image. The current approaches rely on gradient backpropagation which is faster and more accurate. Here we assume that the saliency map, denoted as $\phi(I, x, y)$, returns values between 0 and 1, with larger values indicating pixels with higher impact.

Given the trained deep neural network, we propose the method that allows us to systematically verify the reliability of learned representations. The idea is to rank the learned image classes by the per-class robustness score and inspect the training/validation examples for classes with the smallest values of the score.

Technically, the proposed method is realized as follows.

1. For each image I we define the robustness score $(rs(I))$ as:

$$rs(I) = \frac{\sum_{(x,y)\in I} bbox(I,x,y) \cdot \phi(I,x,y)}{\sum_{(x,y)\in I} \phi(I,x,y)} \tag{1}$$

2. For each class c learned by the model calculate the class robustness score $(crs(c))$ as the average robustness scores over all validation set images for this class:

$$crs(c) = \frac{\sum_{I:\ell(I)=c} rs(I)}{|I:\ell(I)=c|} \tag{2}$$

3. Rank the classes by increasing the value of the crs score.

Given the ranking of image categories, we can pinpoint categories that most likely rely on spurious correlations, providing their class robustness score is *low*. It does not seem feasible to provide a threshold here. However, it can be noticed that the robustness score has clear interpretation as the percentage of the total 'attention' (as expressed by the saliency map) of the model which is focused on the object recognized (or, in other words, the portion of saliency map activation included within the ROI). For illustration: when analyzing the ResNet-152, we discovered 10 categories with the class robustness score $crs < 0.2$, 31 categories with $crs < 0.3$, and 59 categories with $crs < 0.4$. Interestingly, the ResNet consistently realizes high accuracy of recognition for these categories (over 0.9 in most cases), which makes identification of these suspicious classes with spurious correlations difficult, unless a robustness score is used. Some of the suspicious classes with high accuracy and small csr as shown in Fig. 2. The manual investigation of images from suspicious classes should start with these images with the smallest image robustness score, hence it is not required to inspect all images in the dataset to detect suspicious correlations or inconsistency in annotations.

An open question remains how much of the model's attention should be placed on the object (ROI), and how much on the surroundings (context). Clearly, models tend to learn some patterns in areas outside bounding boxes (commonly some saliency map activation is observed outside the ROI). Human recognition is similar: we focus not only on the analyzed object (marked by the bounding box, for example basket ball), but also on the surrounding areas (for example, playground floor). Surrounding areas provide humans with the context of images that clearly helps us to classify the image. The problem with model reliability starts when most of the attention is placed on the context rather than the object itself. Such models are prone to (natural) adversarial images, as shown in Sect. 3.5, as illustrated by the well-known 'Husky vs Wolf' recognition task, as reported in [22].

The application of the proposed method for a given dataset requires a saliency map generator and bounding boxes. The last one could be a crucial problem when lacking. However, we think that an object detection method like EfficientDet [25] or YOLO [4] could be useful to generate such boxes automatically. A more flexible procedure for the automatic generation of bounding boxes for unknown objects is proposed in [15].

3 Computational Experiments

In this section, we show the feasibility of the proposed method using different
deep network models, different algorithms of saliency map generation, and different image recognition datasets.

Firstly, we analyze ImageNet and one of the available deep network models
(ResNet-152 [12]). The ImageNet dataset includes bounding boxes labeled by
hand that is required by our method to work. Next in Sect. 3.2, we analyze
the performance of the method using different deep network models trained
on ImageNet, and different saliency map generation techniques. We compare
these settings in terms of the spurious correlations detected. Then in Sect. 3.3,
we illustrate the method using another image recognition dataset i.e. Pascal
VOC. Finally, we show some practical applications of our method: in Sect. 3.3,
we analyze the detected inconsistencies in the ImageNet annotations, and in
Sect. 3.5, we show the vulnerability of the networks to adversarial attacks, where
the (natural) adversarial images are suggested by detected spurious correlations.

3.1 ImageNet Feasibility Study

In the first experiment, we analyzed the robustness of representations of ResNet-
152 [12] classification model trained on the ImageNet, using the Grad-CAM++
[8] for generating saliency maps. In Figs. 1 and 2 we show the classes with the
smallest value of robustness score. Continuing our rugby ball example (Fig. 1),
the network classifies these images with high accuracy as n04118538 (rugby ball) -
correct classification, even though the model's attention is focused on the players
rather than the ball, as the saliency maps reveal. If we cover the ball with RGB
noise (image (d) in Fig. 1), we observe the same, high level of confidence, although
the ball is missing. This type of spurious correlations could be expected, and the
problem can be at least partly explained by the ambiguous labeling of training
images in this class (this is further discussed in Sect. 3.4).

In Fig. 2 we show selected examples of classes with a small class robustness score. Notice that the accuracy of recognition of these classes is generally
high, although the robustness score signals that this recognition is not reliable.
For instance, the class n04264628 (space bar) tends to be in strong association
with the keyboard or the typewriter (another ambiguous labeling-related issue).
Many classes with the small class robustness score refer to a sport where specific
objects, suits, or venues are required. When preparing the training data for such
classes, spurious correlations will likely occur. Examples of this are the classes
n04019541 (puck, hockey puck) or n03942813 (ping-pong ball). They refer to
specific objects while the network primarily learns elements specific to the game
rather than the object itself. Other interesting examples are n03929855 (pickelhaube) and n03770439 (miniskirt, mini). The pickelhaubes are usually presented
in old-looking, monochromatic photographs. CNN learns these features, mainly
because this style of photos is almost unique to this class. The miniskirts are
strongly correlated with women and women's legs, which lead to wrong classification (legs as miniskirt). The database lacks the miniskirt images with a

different background/context - hence the learned bias. There are 1300 miniskirt images in the training set - only 55 are without woman context (miniskirt on a hanger or a dummy, or white background), and almost all of the rest are with clear women context (mostly with naked legs). Therefore it is not surprising that the network strongly correlates mini skirts with a female body.

Since small values of the robustness score often involve classes with very good accuracy of recognition, we conclude that the ImageNet dataset is poorly prepared for these classes and, as both training and validation partitions include the same spurious correlation, which leads to poor generalization to images that do not include this context.

Fig. 3. Comparison of different CNN models and techniques to generate saliency maps. Saliency maps generated for the newer models tend to be more detailed and match the object more closely. Despite this, the proposed method gives similar results for all tested variants.

3.2 Sensitivity Study for Different Deep Models and Saliency Map Generators

In the previous section, we showed that the proposed methods works for the ResNet-152 and Grad-CAM++ on ImageNet and detects non-robust classes. Here, we investigate how the change of the model architecture and saliency map generation method affect the operation.

Therefore, in the next experiments we used the AlexNet [16] and different versions of the EfficientNet (B0, B3, and B7) [24]. The idea was to choose an old model, and one of new, well-known CNN architectures; we wanted to analyze models with different top-1/top-5 accuracy. The ResNet-152 (baseline) achieves 78.6%/94.3%, AlexNet 63.3%/84.6%, EfficientNet-B0 76.3%/93.2%, EfficientNet-B3 81.1%/95.5%, and EfficientNet-B7 84.4%/97.1% on the ImageNet.

For generating saliency maps, we tested another recent algorithms: the Smooth Grad-Cam++ [20].

In Fig. 3 we demonstrate the difference in saliency maps generated by different techniques using different networks. We can notice that in newer models the areas highlighted by saliency maps tend to be more detailed and match the class objects more closely. However, this does not significantly change the ranking of classes by the robustness scores that we do in step 3 of the method. The main idea of the proposed method is to verify if a spurious correlation exists and to analyze the correlations in detail. Too accurate or sensitive techniques (see the example of EfficentNet-B7 in Fig. 3) may seem less comfortable for the user who wants to investigate the nature of the problem signaled by the robustness score in a particular class. Therefore bigger and more readable saliency maps seem more comfortable for human analysis.

To verify how changes in the used methods affect the detection of spurious correlations, in Table 1 we list the top-20 classes with the smallest robustness score selected by the 'baseline' method (i.e. ResNet-152, GradCam++). In the following columns of the table, we show the ranking positions of these classes returned by the analyzed methods (different model and/or saliency map generator).

It could be noticed that the methods identify mostly the same classes in top-20 lists. A few exceptions, e.g., n03379051 (football helmet), may indicate that each technique handles correlations for these classes differently. We also notice that bigger differences in rankings occur for less accurate models, like AlexNet and EfficientNet-B0.

Finally, we used Fisher's exact test to compare the sets of top-20 classes returned by different methods shown in Table 1. Technically, we verified the null hypothesis that the indicator variables 'in-top-20 list' returned by each of the methods and the baseline method are not related. All the tests returned the p-value very close to 0, hence the lists of classes with the lowest robustness scores returned by different methods are strongly related. It confirms that the model architecture and saliency map detection algorithm do not significantly influence the final results. Similar results come from the top-100 lowest robustness score ranking lists: the number of co-ocurrences of the same class in different top-100 ranking lists is (notation as in Table 1 caption): $|Bs \cap Sm| = 93$, $|Bs \cap Al| = 91$, $|Bs \cap Ef0| = 83$, $|Bs \cap Ef3| = 87$, $|Bs \cap Ef7| = 84$, which yield p-values of the Fisher's exact tests equal 0.

These analyses confirm that the root of spurious correlations lies in the training data, and not in the models themselves. We identify the most problematic classes of images in the ImageNet, which require improvements in training data to obtain more trustworthy models.

Table 1. Comparison of the class rankings with 20 smallest robustness score, returned by our method (step 3) using different models and saliency map generators. **Bs** denotes the 'baseline' method, i.e. ResNet-152 + GradCam++, other settings are: ResNet-152 + SmoothGradCAM++ (**Sm**), AlexNet + GradCam++ (**Al**), and EfficientNet + GradCam++ in different versions B0 (**Ef0**), B3 (**Ef3**) and B7 (**Ef7**). The table shows the ranking position based on the class robustness scores determined by each of the analyzed techniques. Results for all techniques are very similar, with the number of classes co-occuring in two different ranking equal to: $|Bs \cap Sm| = 20$, $|Bs \cap Al| = 17$, $|Bs \cap Ef0| = 17$, $|Bs \cap Ef3| = 18$, $|Bs \cap Ef7| = 17$. Fisher's exact tests prove that there is no chance that this is the random co-occurrence. This shows that the proposed method effectively detects spurious correlations and non-robuts representations using different model architectures and saliency map algorithms.

Name	Bs	Sm	Al	Ef0	Ef3	Ef7	Name	Bs	Sm	Al	Ef0	Ef3	Ef7
basketball	1	1	1	1	1	2	sunglasses	11	9	22	18	9	11
ping-pong ball	2	2	2	2	2	1	rugby ball	12	13	11	16	15	13
volleyball	3	3	3	3	3	3	croquet ball	13	12	18	14	17	12
pickelhaube	4	4	6	4	4	4	horizon..bar	14	14	26	26	13	10
swim..trunks	5	6	4	5	5	6	switch	15	17	12	11	20	21
bathing cap	6	8	5	6	7	9	snorkel	16	15	20	17	16	16
space bar	7	5	8	13	6	5	racket	17	19	15	10	11	14
bearskin	8	7	10	8	8	7	diaper	18	18	17	23	23	27
miniskirt	9	11	7	7	12	17	nail	19	20	14	12	19	26
balance beam	10	10	9	9	10	15	flagpole	20	16	25	25	21	19

3.3   Pascal VOC Feasibility Study

To show the feasibility of the method we apply it to the Pascal VOC 2007 dataset. This dataset contains 20 classes where each image has assigned both labels for a classification problem and bounding boxes for the object detection problem. We trained the ResNet-101 model using the transfer-learning technique and we achieved 81.14% accuracy for the classification task. Applying our method allows us to detect classes where the model learns some spurious correlations. As shown in Table 2 the classes with the lowest robustness score are 'bottle', 'chair', and 'boat'. We analyzed these classes and found that the bottle is strongly correlated with humans and tables; similar correlations occur for the class 'chairs'; and the boats are strongly correlated with water background (sea or lake), as demonstrated in Fig. 4. As shown the accuracy for the "dining table" class is very low, we looked closer at it, and the most examples from this class were assigned by the network to the "bottle" (36%) or the "chair" (33%) classes. That shows how significant impact the spurious correlations have on the final classification.

bottle   **bottle**   **chair**   **chair**   **boat**

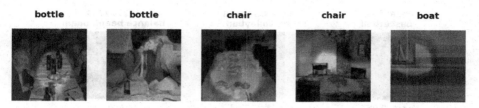

Fig. 4. Images from the Pascal VOC where the trained CNN network could not focus on target objects. Our method identifies some classes (i.a. 'bottle', 'chair', 'boat') where the model learned spurious correlations. Closer analysis shows that in this dataset, images in category 'bottle' also present people and/or tables.

Table 2. Accuracy (Acc) and Class robustness score (Crs) for all Pascal VOC classes. Our method shows some problems especially with the "bottle" class (lowest robustness score). Closer analysis reveals that this class is strongly correlated with the class tables. This also accounts for low accuracy of recognition of the class "dining table".

Class	Acc	Crs	Class	Acc	Crs
aeroplane	0.92	0.59	dining table	0.17	0.64
bicycle	0.81	0.49	dog	0.93	0.64
bird	0.90	0.49	horse	0.92	0.60
boat	0.84	0.40	motorbike	0.90	0.71
bottle	0.59	0.18	person	0.73	0.64
bus	0.72	0.51	potted plant	0.58	0.44
car	0.89	0.55	sheep	0.76	0.54
cat	0.94	0.68	sofa	0.44	0.70
chair	0.71	0.34	train	0.92	0.62
cow	0.87	0.59	tv monitor	0.68	0.47

3.4   Inconsistency in the ImageNet Annotations

Our experiments with ImageNet-based models show that recognition of classes with the smallest scores most likely relies on spurious correlations. Training images from these classes are worth investigating to analyze the nature of these correlations.

The ImageNet's authors delivered bounding boxes. However, there are some lapses in annotated boxes, which negatively influence the performance of the proposed method. There are images where not all objects are marked. It often happens in classes where presented objects are small as n01440764 (bee), n01443537 (golf ball), or n01484850 (nail). In these cases, the robustness score is smaller than it should be, which should not happen when bounding boxes are correctly marked.

The other important issue is the ambiguous meaning of some classes. For example, in the case of n04540053 (volleyball) or n02802426 (basketball), the

Fig. 5. The classes with ambiguous meaning like n02802426 (basketball), n04540053 (volleyball) or n02777292 (balance beam, beam) can achieve a small class robustness score. It happens because the training images relate to the general meaning of the word - e.g., to a whole sport or event while bounding boxes have been marked for the specific meaning, e.g., for a ball or a beam. The proposed method detects such situations, which should help the database author decide which meaning is correct and consequently modify labels or training images.

class name (due to the polysemy) can refer to a sport discipline or only to a ball. The network tends to treat these classes as sports disciplines (see saliency maps on Fig. 5). However, the bonding boxes cover only balls. Looking at the ImageNet tree hierarchy for the n04540053 (volleyball) class, i.e.: ImageNet → Instrumentality, instrumentation → Equipment Game → Equipment → Ball, we can state that bounding boxes are correct. That the volleyball class is defined as a ball, not as a sport discipline. However, looking at the ImageNet training set, there is a lack of images focusing directly on the ball. And even worse there are images without balls.

The above remarks show that the proposed method allows detecting inconsistencies between images and annotations (class assignments or bounding boxes). It allows the database authors to fix them, by modifying database content or by changing bounding boxes and therefore the semantic meaning of the class. Such cleaning of the database will result in an improvement of the deep models' reliability.

3.5   Adversarial Attacks

It is possible to perform an attack on the network using existing spurious correlations in the image set. An example of attacks on ResNet-152 model is shown in Fig. 6. As discussed in Sect. 3.1, the n03770439 (miniskirt, mini) class strongly correlates with women context. Hence it is enough to show only female legs, and the CNN will indicate miniskirt as the first answer (Fig. 6, image 5).

We can distinguish two types of attacks. The first one relies on forcing the network to misclassify an image by showing the object without typical spurious correlations - as shown in the first two examples in Fig. 6. For instance, to attack the class n04019541 (puck, hockey puck), it is enough to show an image with a puck on grass or placed side by side. To attack the class n04264628 (space bar) we can show an image where a space key was pulled out from a keyboard. In these cases, a valid label is unseen in the top-5 network answers.

| n04019541 | n04264628 | n03929855 | n04039381 | n03770439 | n03134739 | n03188531 |
| puck | space bar | pickelhaube | racket | miniskirt | croquet ball | diaper |

face powder	buckle	pickelhaube	racket	miniskirt	croquet ball	diaper
bottlecap	spatula	military uniform	volleyball	clog	golf ball	bassinet
Petri dish	cleaver	bearskin	Italian greyhound	sandal	crutch	scale
barrel	letter opener	ballplayer	unicycle	Loafer	matchstick	swimming trunks
pill bottle	scabbard	prison	knee pad	cowboy boot	pinwheel	crib

Fig. 6. Example attacks of the specific classes from ImageNet. New images - outside the ImageNet set - are wrongly classified because of existing spurious correlations. There are captions below each image containing information with the top-5 response from the baseline model (ResNet-152). The first two examples are wrongly classified because of a lack of usually existing correlations. The rest examples make CNN choose attacked classes based only on spurious correlations, even when there is no class object in images.

The second type of attack forces the network to point out a specific class by showing only spurious correlations without a real object - see the last five examples in Fig. 6. For example, to attack the class n03188531 (diaper, nappy, napkin) it is enough to show an image with a baby but without a diaper, and the network responds "diaper" as the first answer. To attack the class n03134739 (croquet ball), we can show an image containing other pieces of the croquet equipment, even if the ball is missing. Interestingly, the confidence of recognition (the network's output after at the softmax layer) of these adversarial images was usually excellent.

4 Conclusion

In this paper, we proposed a simple method that is useful in identifying, which of the image categories learned by a deep neural network are likely to be recognized by the network using spurious, counter-intuitive representations. The method relies on a measure we call robustness score. The score signals discrepancies between objects specific to the category and image areas with a high impact on the classifier, as marked by saliency maps.

We applied the method using the ResNet and some other models trained on the ImageNet and discovered several classes recognized by these models by spurious correlations. This leads to low reliability of prediction for these classes, as the models generalize poorly to images other than ImageNet examples. Additionally, the models are vulnerable to natural adversarial images that are easy to find once these spurious representations are analyzed.

Many state-of-the-art deep models trained on the ImageNet published today are affected by this problem. Low reliability of models for these classes is generally not signaled by high accuracy of prediction, calculated on the ImageNet test benchmark. This comes from the observation that the spurious correlations tend to occur both in ImageNet training and validation examples.

Our method relies on the availability of bounding boxes that identify objects of interest in training images. If the binding boxes are not available in training data, we believe that object detectors like EfficientDet, YOLO, or recently proposed method [15] could be used. We showed that the proposed method is a useful tool for the analysis of different deep neural networks, trained on different image recognition datasets (ImageNet and Pascal VOC), and with different saliency map generators. In our analyses, we discovered similar spurious correlations in different CNN models, as clearly the root of the problem lies in the quality of the train and validation subsets.

The method has some limitations. Since we analyze discrepancies between the region of interest (ROI) and the saliency maps, the method is blind to spurious correlations related to overusing some relevant correlations, such as texture or color, that occur within the ROI. Further research is required to mitigate this limitation.

References

1. Amodei, D., Olah, C., Steinhardt, J., Christiano, P., Schulman, J., Mané, D.: Concrete problems in AI safety. arXiv preprint arXiv:1606.06565 (2016)
2. Beery, S., Van Horn, G., Perona, P.: Recognition in terra incognita. In: Ferrari, V., Hebert, M., Sminchisescu, C., Weiss, Y. (eds.) ECCV 2018. LNCS, vol. 11220, pp. 472–489. Springer, Cham (2018). https://doi.org/10.1007/978-3-030-01270-0_28
3. Beyer, L., Hénaff, O.J., Kolesnikov, A., Zhai, X., Oord, A.V.D.: Are we done with imagenet? arXiv preprint arXiv:2006.07159 (2020)
4. Bochkovskiy, A., Wang, C.Y., Liao, H.Y.M.: Yolov4: optimal speed and accuracy of object detection. arXiv preprint arXiv:2004.10934 (2020)
5. Brendel, W., Rauber, J., Kümmerer, M., Ustyuzhaninov, I., Bethge, M.: Accurate, reliable and fast robustness evaluation. arXiv preprint arXiv:1907.01003 (2019)
6. Carlini, N., Wagner, D.: Towards evaluating the robustness of neural networks. In: 2017 IEEE Symposium on Security and Privacy (SP), pp. 39–57. IEEE (2017)
7. Chakraborty, A., Alam, M., Dey, V., Chattopadhyay, A., Mukhopadhyay, D.: Adversarial attacks and defences: a survey. arXiv preprint arXiv:1810.00069 (2018)
8. Chattopadhay, A., Sarkar, A., Howlader, P., Balasubramanian, V.N.: Grad-CAM++: Generalized gradient-based visual explanations for deep convolutional networks. In: 2018 IEEE Winter Conference on Applications of Computer Vision (WACV), pp. 839–847. IEEE (2018)
9. Das, A., Rad, P.: Opportunities and challenges in explainable artificial intelligence (XAI): a survey. arXiv preprint arXiv:2006.11371 (2020)
10. Gopinath, D., Katz, G., Pasareanu, C.S., Barrett, C.: DeepSafe: a data-driven approach for checking adversarial robustness in neural networks. arXiv preprint arXiv:1710.00486 (2017)
11. He, K., Zhang, X., Ren, S., Sun, J.: Delving deep into rectifiers: surpassing human-level performance on ImageNet classification. In: Proceedings of the IEEE International Conference on Computer Vision, pp. 1026–1034 (2015)
12. He, K., Zhang, X., Ren, S., Sun, J.: Deep residual learning for image recognition. In: Proceedings of the IEEE Conference on Computer Vision and Pattern Recognition, pp. 770–778 (2016)

13. Hendrycks, D., Zhao, K., Basart, S., Steinhardt, J., Song, D.: Natural adversarial examples. arXiv preprint arXiv:1907.07174 (2019)
14. Jabbour, S., Fouhey, D., Kazerooni, E., Sjoding, M.W., Wiens, J.: Deep learning applied to chest x-rays: exploiting and preventing shortcuts. In: Machine Learning for Healthcare Conference, pp. 750–782. PMLR (2020)
15. Joseph, K., Khan, S., Khan, F.S., Balasubramanian, V.N.: Towards open world object detection. arXiv preprint arXiv:2103.02603 (2021)
16. Krizhevsky, A., Sutskever, I., Hinton, G.E.: ImageNet classification with deep convolutional neural networks. In: Proceedings of the 25th International Conference on Neural Information Processing Systems, NIPS 2012, Red Hook, NY, USA, vol. 1, pp. 1097–1105 (2012)
17. Madry, A., Makelov, A., Schmidt, L., Tsipras, D., Vladu, A.: Towards deep learning models resistant to adversarial attacks. arXiv preprint arXiv:1706.06083 (2017)
18. Mothilal, R.K., Sharma, A., Tan, C.: Explaining machine learning classifiers through diverse counterfactual explanations. In: Proceedings of the 2020 Conference on Fairness, Accountability, and Transparency, pp. 607–617 (2020)
19. Nguyen, A., Yosinski, J., Clune, J.: Deep neural networks are easily fooled: high confidence predictions for unrecognizable images. In: Proceedings of the IEEE Conference on Computer Vision and Pattern Recognition, pp. 427–436 (2015)
20. Omeiza, D., Speakman, S., Cintas, C., Weldermariam, K.: Smooth Grad-CAM++: an enhanced inference level visualization technique for deep convolutional neural network models. arXiv preprint arXiv:1908.01224 (2019)
21. Pham, H., Xie, Q., Dai, Z., Le, Q.V.: Meta pseudo labels. arXiv preprint arXiv:2003.10580 (2020)
22. Ribeiro, M.T., Singh, S., Guestrin, C.: "why should i trust you?" explaining the predictions of any classifier. In: Proceedings of the 22nd ACM SIGKDD International Conference on Knowledge Discovery and Data Mining, pp. 1135–1144 (2016)
23. Samek, W., Wiegand, T., Müller, K.R.: Explainable artificial intelligence: Understanding, visualizing and interpreting deep learning models. arXiv preprint arXiv:1708.08296 (2017)
24. Tan, M., Le, Q.V.: EfficientNet: rethinking model scaling for convolutional neural networks. arXiv preprint arXiv:1905.11946 (2019)
25. Tan, M., Pang, R., Le, Q.V.: EfficientDet: scalable and efficient object detection. In: Proceedings of the IEEE/CVF Conference on Computer Vision and Pattern Recognition, pp. 10781–10790 (2020)
26. Tsipras, D., Santurkar, S., Engstrom, L., Ilyas, A., Madry, A.: From ImageNet to image classification: contextualizing progress on benchmarks. arXiv preprint arXiv:2005.11295 (2020)
27. Wang, H., et al.: Score-CAM: score-weighted visual explanations for convolutional neural networks. In: Proceedings of the IEEE/CVF Conference on Computer Vision and Pattern Recognition Workshops, pp. 24–25 (2020)
28. Weng, T.W., et al.: Evaluating the robustness of neural networks: an extreme value theory approach. arXiv preprint arXiv:1801.10578 (2018)
29. Xiao, K., Engstrom, L., Ilyas, A., Madry, A.: Noise or signal: the role of image backgrounds in object recognition. arXiv preprint arXiv:2006.09994 (2020)
30. Yu, F., Qin, Z., Liu, C., Zhao, L., Wang, Y., Chen, X.: Interpreting and evaluating neural network robustness. arXiv preprint arXiv:1905.04270 (2019)

31. Zech, J.R., Badgeley, M.A., Liu, M., Costa, A.B., Titano, J.J., Oermann, E.K.: Variable generalization performance of a deep learning model to detect pneumonia in chest radiographs: a cross-sectional study. PLoS Med. **15**(11), e1002683 (2018)
32. Zhou, Z., Firestone, C.: Humans can decipher adversarial images. Nature Commun. **10**(1), 1–9 (2019)

CHECKER: Detecting Clickbait Thumbnails with Weak Supervision and Co-teaching

Tianyi Xie[1]([✉]), Thai Le[2], and Dongwon Lee[2]

[1] Shanghai Jiao Tong University, Shanghai, China
lsyhxty@sjtu.edu.cn
[2] The Pennsylvania State University, University Park, USA
{tql3,dongwon}@psu.edu

Abstract. Clickbait thumbnails on video-sharing platforms (e.g., YouTube, Dailymotion) are small catchy images that are designed to entice users to click to view the linked videos. Despite their usefulness, the landing videos after click are often inconsistent with what the thumbnails have advertised, causing poor user experience and undermining the reputation of the platforms. In this work, therefore, we aim to develop a computational solution, named as CHECKER, to detect clickbait thumbnails with high accuracy. Due to the fuzziness in the definition of clickbait thumbnails and subsequent challenges in creating high-quality labeled samples, the industry has not coped with clickbait thumbnails adequately. To address this challenge, CHECKER shares a novel clickbait thumbnail dataset and codebase with the industry, and exploits: (1) the *weak supervision* framework to generate many noisy-but-useful labels, and (2) the *co-teaching* framework to learn robustly using such noisy labels. Moreover, we also investigate how to detect clickbaits on video-sharing platforms with both thumbnails and titles, and exploit recent advances in vision-language models. In the empirical validation, CHECKER outperforms five baselines by at least 6.4% in F1-score and 4.2% in AUC-ROC. The codebase and dataset from our paper are available at: https://github.com/XPandora/CHECKER.

Keywords: Clickbait thumbnail · Weak supervision · Co-teaching · Learning with noisy labels

1 Introduction

In recent a few years, the popularity of video-sharing platforms (e.g., YouTube, Dailymotion, and Vimeo) has dramatically increased. According to the recent survey by Pew Research[1], for instance, around three-quarters of U.S. adults

---

[1] http://tiny.cc/3jkvtz.

Part of the work was done while the author visited Penn State during the summer of 2019 as an intern.

© Springer Nature Switzerland AG 2021
Y. Dong et al. (Eds.): ECML PKDD 2021, LNAI 12979, pp. 415–430, 2021.
https://doi.org/10.1007/978-3-030-86517-7_26

(73%) use YouTube, surpassing 69% of U.S. adults using Facebook. As such, it is a critically important problem for such platforms to maintain a clean ecosystem and provide pleasant experience to users. However, one phenomenon severely polluting this ecosystem is the prevalence of the so-called **clickbait thumbnails**, small catchy images that are designed to entice users to click to view the linked videos (e.g., several examples shown in Fig. 1). Such clickbait thumbnails are often deceptive, sensationalized, exaggerating, or misleading, sometimes accompanied by eye-catching titles. The emergence of thumbnails is partially due to the desire of content creators to increase the view counts for diverse reasons (e.g., monetary gain). Despite their attractiveness at first glance, however, the landing videos may have the contents different from what the thumbnails have advertised. Such inconsistency then leads to users' unpleasant online experience and deteriorates the reputation of video-sharing platforms.

One trivial solution to combat clickbait thumbnails is to employ human annotators to review and tag clickbait thumbnails. However, not only it is costly, but also it cannot scale well to match the sheer volume of videos uploaded on popular video-sharing platforms, calling for computational and scalable solutions. Therefore, to mitigate this phenomenon of clickbait thumbnails on video-sharing platforms, the aim of this work to develop a machine learning based solution that can detect clickbait thumbnails with a high accuracy. Despite the closely related problem of detecting (text-based) clickbait news headlines has been well studied (e.g., [6,9,24]), the detection of clickbait thumbnails has been relatively less explored and existing solutions (e.g., [23,28]) are based on impractical settings or show unsatisfactory accuracies. Moreover, solving the problem of detecting clickbait thumbnails using machine learning framework needs to cope with a few inherent challenges:

- Due to the subjective and ambiguous nature in the definition of clickbait thumbnails, it is non-trivial to build a clean supervised learning environment with ample labeled samples. As the tolerance levels of people often differ, a clearly annoying clickbait thumbnail to A can be perfectly entertaining thumbnail to B. Even if one uses human annotators to tag clickbait thumbnails, it is unclear what specific instruction one has to give to the annotators.
- As such, achieving consensus on a single clickbait thumbnail among multiple human annotators is challenging (and costly). Further, even after consensus, human annotated labels for clickbait thumbnails can be noisy.
- Finally, achieving high detection accuracy using rich features found in various meta-data of landing videos may not be a practical solution (e.g., [23,28]). This is because in real settings, users are often given only a pair of information (i.e., thumbnail and title) to determine to click or not. Therefore, an ideal solution is to mimic the situation and detect clickbait thumbnails using multimodal features from the pair of thumbnail and title.

In an attempt to address the aforementioned challenges in detecting clickbait thumbnails, this paper presents CHECKER (Clickbait tHumbnail dEtection with Co-teaching and weaK supERvision), which leverages weak supervision to generate noisy-but-useful labels and adopts co-teaching [13] to learn robustly from such noisy labels. In addition, different from prior works [23,28], we are interested in detecting clickbait thumbnails using only the pair of a thumbnail

Fig. 1. Examples of clickbait thumbnails. Though they are eye-catching at first glance, the content of the linked videos is inconsistent with what these thumbnails have advertised.

and title, which simulates the real users' experience while browsing video-sharing platforms and avoids the cold start problem when statistics of a new video is not available. To this end, we first collect 8,987 videos along with their metadata from YouTube, including the thumbnail and title. Note that the collected metadata of video are used to generate noisy labels, but will not be used in either training or inference. Then, we collect the initial labels for a small subset of these thumbnails via crowdsourcing on the Amazon Mechanical Turk platform. Note that most of the thumbnails remain unlabeled. To make a full use of these unlabeled thumbnails, then, we adopt the weak supervision framework and generate noisy-but-useful labels for them. Then, to prevent the powerful neural networks (NNs) from memorizing these noisy labels (thus degrading accuracy), we furthermore adopt the co-teaching strategy [13] to filter out thumbnails with wrong labels while training. By and large, our main contributions are as follows:

- We release a clickbait thumbnail detection dataset, which consists of 8,987 videos with their metadata from YouTube, and 787 of them get labeled through crowdsourcing.
- We propose CHECKER for clickbait thumbnail detection, which leverages weak supervision to generate labels for thumbnails with over 80% accuracy. Specifically, based on the characteristics of clickbait thumbnails, we design several useful labeling functions as weak supervision sources and then combine them to generate labels. Furthermore, co-teaching strategy is also applied in the training to cope with the noise among generated labels.
- We exploit recent advances in vision-language models and make a comprehensive comparison. Moreover, extensive experiments are conducted to show that our method effectively alleviates the issue of high-quality labeled training data shortage in training clickbait thumbnail detectors.

2   Related Work

2.1   Clickbait Headline Detection

There is a growing interest in studying misinformation on social media. One line of research focuses on the detection of clickbait headlines. Online content creators use these clickbait titles to attract attention and lure visitors to click on a hyperlink of a target landing web page [9], which may contain misinformation. Thus, clickbait headlines have become a popular medium for mass propagation of false news. To explore what makes a headline "clickbaity", [17] conduct three clickbait studies. To effectively detect clickbait headlines, most of existing approaches train machine learning (ML) detectors with features that are either carefully engineered [5,6,10] or automatically learnt via deep NNs [1,22]. Moreover, [15,24,26] further improves those detectors by augmenting their training dataset with synthetic clickbait headlines. In this work, we turn to study another type of clickbait but deserve more attention in the current literature: *clickbait thumbnail.*

2.2   Clickbait Thumbnail Detection

Clickbait thumbnails are small catchy images that are designed to entice users to click to view a particular video, with a defining characteristic of being deceptive, sensationalized, exaggerating, or misleading. Compared to clickbait headlines, only a few pioneering works start to study these misleading thumbnails. To the best of our knowledge, [28] first studies the clickbait problem on Youtube and builds a VAE-based model for automatic detection. [23] proposes a content-agnostic approach to detect clickbait videos, which mainly makes use of the comments of videos. In spite of their progress, both of them suffer from the shortage of a reliable training corpus. [19] also indicates that automatic clickbait detection on YouTube is still far out of reach due to the paucity of training data. To deal with the lack of available datasets, [28] retrieves videos from clickbait and non-clickbait channels, and obtain labels for videos based on the label (clickbait or non-clickbait) of the channels they belong to. However, this approach is not convincing since even non-clickbait channels may publish clickbait videos. [23] also constructs a dataset of 625 videos, but such size is usually too small to train a robust deep neural network. Hence, in this paper, we make further efforts to tackle the shortage of training samples in clickbait thumbnail detection.

2.3   Vision-Language Model

Various vision-language tasks have attracted the attention of the research community in recent years, such as Image Captioning and Visual Question Answer, which require the capability to understand and fuse multimodal features. Early works in vision and language understanding usually design separate models for each modality followed by a multi-modal fusion layer. In this case, bi-linear fusion is thought to be more expressive but tends to result in an excess of parameters. Subsequent work address this issue through low-rank decomposition [3,4,12]. In addition, more recent works show that a joint pre-training over both modalities enables the model to easily adapt to downstream tasks. Some work therefore train a holistic network on a large training corpus, which is able to give a joint embedding of vision and language, such as VisualBERT [16], LXMERT [25] and UNITER [8]. In this work, we apply and compare these state-of-the-art methods and models to the clickbait thumbnail detection task.

Table 1. The overview of our clickbait thumbnail dataset. As we can see, even clickbait channels may use non-clickbait thumbnails, and the same is with non-clickbait channels.

		Clickbait channel	Non-clickbait channel	total
Train	# Clickbait thumbnail	146	38	184
	# Non-clickbait thumbnail	150	256	406
	# Unlabeled thumbnail	3851	4349	8200
Test	# Clickbait thumbnail	49	15	64
	# Non-clickbait thumbnail	45	88	133
	# Unlabeled thumbnail	–	–	–

3   Building Dataset

In this work, we aim to study the clickbait thumbnail detection problem on YouTube. Since there is not any reliable dataset of clickbait thumbnails in the literature, we first need to collect a high-quality labeled dataset for our study. Our data collection process includes two steps: (i) data acquisition and (ii) label collection.

3.1   Data Acquisition

There are many more videos with benign than with clickbait thumbnails. Due to this imbalanced nature between clickbaits and non-clickbaits, collecting data points randomly from video-sharing platforms will result in a dataset with a highly skewed class distribution. Thanks to prior work [28], we first retrieve a list of clickbait and non-clickbait channels on YouTube. By leveraging YouTube Data API[2], we crawl 8,987 videos as well as the metadata from these channels, which are published between May and July of 2019. Note that here we use the video's source as an approximation for its clickbaitness and we also try to collect the same amount of videos from each channel to prevent uneven data distribution.

Generally, the metadata can be categorized into four groups: (1) title and description; (2) thumbnail; (3) statistics (e.g., like and dislike count, etc.); (4) comments. Particularly, assuming that popular comments represent the opinion of the majority, we select only the top 10 comments with the highest like count for each video.

3.2   Label Collection

Though we have collected a large number of data from YouTube, all of them are still unlabeled. For the sake of model evaluation, ground truth labels are

---

[2] https://developers.google.com/youtube/v3.

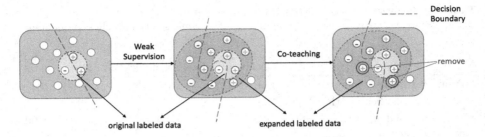

Fig. 2. The overview of data flow in CHECKER. The decision boundary will change along with the distribution of training data.

indispensable. However, due to the vague and ambiguous definition of clickbait thumbnails, it is impossible to annotate all of them in a short time. To collect high-quality labels, we first define clickbait thumbnail as follows:

Definition: Clickbait Thumbnail. Clickbait thumbnail is a thumbnail that is inconsistent with the gist of the corresponding video that it represents.

Based on this definition, we publish labeling tasks on the Amazon Mechanical Turk platform and utilize crowdsourcing to label parts of samples in the dataset. To simulate the experience when users are browsing video-sharing portals, we ask workers to first inspect the thumbnail and title of a video. Then workers are required to watch the video for at least one minute to grab the gist. By comparing the content of the video to the meaning conveyed by the thumbnail and title, workers should be able to tell whether the thumbnail is a clickbait.

To ensure the quality of labels, for each sample, we invite 5 workers to label and use the majority vote to determine the final label. Finally, 787 samples get labeled through crowdsourcing. For experiments, we take 197 of them as the test set while others as a part of the training set. Table 1 provides an overview of our collected dataset.

4   The Proposed Method: CHECKER

Our objective is to train a discriminative model with partially labeled training data. In this paper, we present our framework CHECKER, which takes the advantage of both weak supervision and co-teaching. Basically, our framework can be split into two stages: **generating noisy labels** and **learning from noisy labels**. Specifically, we leverage weak supervision to generate noisy labels while adopt the co-teaching algorithm to remove samples with wrong labels in learning from noisy labels. Figure 2 presents the basic data flow in our framework.

4.1   Generating Noisy Labels

Though we have collected some labels through crowdsourcing, a large number of samples are still unlabeled. To make these unlabeled thumbnails available for

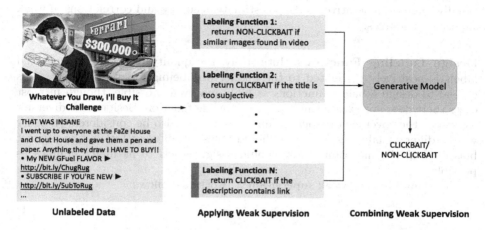

Fig. 3. The overview of generating labels. Specifically, we first design labeling functions as weak supervision sources and then use a generative model to combine them to produce the final label.

training models, we leverage the weak supervision to generate labels for them. Specifically, weak supervision means noisy, limited, or imprecise sources are used to provide supervision signals for labeling large amounts of training data. These cheap labels can be obtained through a set of simple rules instead of manual annotation. This approach, to a great extent, releases researchers from spending too much time in acquiring high-quality labels. In the clickbait thumbnail detection task, weak supervision sources can be various labeling functions based on the characteristics of the thumbnail. For instance, the presence of the word 'clickbait' in the comments of a video on Youtube indicates that this video's thumbnail may be a clickbait.

Here we explain why weak supervision is suitable for generating labels for clickbait thumbnail detection. First, though there is no explicit definition for clickbait thumbnail that enables us to label data quickly, we can easily speak out several rules to roughly judge whether the thumbnail is a clickbait. One simple rule can be that if the thumbnail is one frame of the video, then it should be a non-clickbait thumbnail since it does truthfully reflect the content of this video. Such rules can be regarded as weak supervision sources and are easy to implement. Second, correctly identifying clickbait thumbnails requires people to fully understand the video content and then compare it with the thumbnail, which is extremely time-consuming, while utilizing weak supervision can prevent such heavy work. Third, since we can get various weak supervision sources by designing different labeling rules, combining them as an ensemble enables us to obtain high-quality labels.

Once proper labeling functions are designed, the critical problem becomes how to regulate and utilize these results. Recent advances [11,20] in weak supervision have already made some breakthroughs with regard to this problem, which usually builds a generative model to estimate accuracy and correlations of weak supervision sources.

Design Labeling Functions. Intuitively, the quality of the final generated labels is positively correlated to the quality of labeling functions. Hence, it is crucial to design labeling functions as high quality as possible, though in most cases there does not exist a single perfect labeling function. Besides, the diversity as well as the coverage of labeling functions should also be considered. In other words, different labeling functions should focus on different features to prevent bias, and in the meantime, they should assign labels to as many samples as possible.

To formalize, each weak supervision λ_j works as follows:

$$\tilde{y}_{ij} = \lambda_j(x_i), \tag{1}$$

where x_i denotes the feature of i-th data sample, including title x_i^{ti}, thumbnail x_i^{th}, description x_i^d, video x_i^v, statistics x_i^s and comment x_i^c, and $\tilde{y}_{ij} \in \{-1, 0, 1\}$ denotes the labeling result given by j-th labeling function. Note that '-1' refers to abstain, '0' refers to non-clickbait while '1' refers to clickbait.

Based on the characteristic of clickbait thumbnails, we design labeling functions according to the following aspects:

- *Channel.* [28] once used the label of the channel for the videos inside. Though this is actually not corrected, the label of channels indeed indicates the general property of thumbnails. As shown in Table 1, most of clickbait thumbnails are from clickbait channels while non-clickbait channels seldom upload clickbait thumbnails. Hence, we adopt the label of the channel as one labeling function.
- *Thumbnail.* As shown in Fig. 1, One main critical feature of clickbait thumbnail is the presence of those striking texts, which are artificially added by video uploaders. To draw the attention of users, such text usually occupies a large space of a thumbnail. We therefore employ the optical character recognition (OCR) service[3] to measure the ratio of the text area to the whole image. With a proper threshold, a thumbnail whose text area exceeds the threshold value can be categorized as clickbait. In addition, since telling whether a thumbnail is a clickbait needs comparison with video content, we also adopt dHash algorithm h[4] to calculate the similarity between the thumbnail and frames of the video. Specifically, we calculate the L1 distance between the dHash code of the thumbnail and that of each frame, and the similarity score is the minimum value among all the distances. To formulize, the similarity score is calculated as follows:

$$d_{in} = \left\| h(x_i^{th}) - h(x_{in}^v) \right\|_1, \tag{2}$$

$$s_i = \min\{d_{i1}, d_{i2}, ..., d_{iN}\}, \tag{3}$$

---

[3] https://cloud.google.com/vision/docs/ocr.

[4] http://www.hackerfactor.com/blog/index.php?/archives/529-Kind-of-Like-That.html.

Table 2. Statistics of each labeling function on labeled data. Note that polarity represents the set of labels that labeling functions will output and '1' refers to clickbait while '0' refers to non-clickbait.

Labeling function	Polarity	Coverage	Overlaps	Conflicts	Correct	Incorrect	Acc.
Channel&thumbnail-based	1	0.202	0.108	0.089	110	49	0.692
Channel&statistics-based	1	0.088	0.067	0.048	45	24	0.652
Channel-based	0	0.495	0.348	0	364	26	0.933
Title-based	0	0.492	0.411	0.105	295	93	0.760
Thumbnail-based	0	0.131	0.119	0.016	95	8	0.922
Description-based	0	0.084	0.079	0.002	59	7	0.894

where x_{in}^v denotes the n-th frame of the video and N is the frame number. d_{in} denotes the L1 distance between the thumbnail and n-th frame while s_i denotes the similarity score between the thumbnail and the video. A high similarly score means that the thumbnail indeed reflect the video content, which indicates a benign thumbnail.

- *Title.* Clickbait thumbnails are usually presented with eye-catching titles. Generally, to catch users' attention, exaggerated titles tend to exhibit strong subjectivity. On the other hand, a title with high subjectivity indicates a great possibility of clickbait. Therefore we also use TextBlob[5] to mine the deep semantics behind the title and consider those with high subjective scores as clickbait.
- *Description.* Clickbait on video-sharing platforms usually displays links to other websites in the description for the purpose of advertising. Thus, according to whether the link exists in the description, we can judge the class of thumbnails.
- *Statistics.* Statistics includes like count, dislike count, view count and comment count. Generally, users tend to close the video webpage without leaving comments once they discover it's a clickbait. Hence, we consider the video with a low comment to view ratio as a potential clickbait.

Based on the above observation, we write 6 labeling functions $(\lambda_1, ..., \lambda_N,$ where $N = 6)$. Performance of labeling functions on labeled data is provided in Table 2 and their detailed implementation can be found in the provided codebase.

Combining Labeling Results. By applying all labeling functions to all unlabeled data, we obtain a label matrix Λ, where $\Lambda_{i,j} = \lambda_j(x_i)$. To combine the different labeling results, we are essentially aiming to build a generative model G that functions as follows:

$$\tilde{y}_i = G(\Lambda_i), \tag{4}$$

where Λ_i refers to the labeling result of all weak supervision sources for the unlabeled data sample x_i.

---

[5] https://github.com/sloria/textblob.

Table 3. Comparison of different generative model.

Method	Accuracy	F1 score	Precision	Recall
Majority voter	0.836	0.635	0.717	0.570
Epoxy [7]	0.784	0.637	0.638	0.635
Snorkel [20]	0.808	0.667	0.670	0.663

We evaluate and compare three different generative models on the labeled data. The comparison results is presented in Table 3. Note that for Snorkel [20] and Epoxy [7], we train them with unlabeled data before evaluation. Based on the comparison result, we adopt the majority voter for further experiments for two reasons: (1) the accuracy of majority voter is higher so that there is less noise among the generated labels, and (2) considering that clickbait training samples are more important due to its paucity in our dataset, a higher precision means more high-quality 'clickbait' labels, which enables the model to learn a better decision boundary. Using this generative model, we generate labels for 7,039 unlabeled data in total while 1,061 samples remain unlabeled since none of labeling functions assigns labels for them. After label generation, the size of our labeled training samples has increased to 7,630.

4.2   Learning from Noisy Labels

The objective of this stage is to train a robust vision-language classifier with the generated labels. Specifically, given a thumbnail x_i^{th} and title x_i^{ti}, the task of this classifier is to predict a label $\hat{y}_i$ indicating whether it is a clickbait. Plus, though we have obtained a large number of labels with weak supervision, these generated labels are noisy. Note that noisy labels mean that not all labels are correct. It is known that the strong fitting capability of machine learning models such as neural networks may lead itself to overfit the noise, which would finally result in a poor generalization. Hence, for robust learning, it is also critical to combat noisy labels during training.

Model Architecture. Since the clickbait thumbnail detection is a vision-language task, we exploit recent advances in vision-language areas to build a clickbait detector, as shown on the left side of Fig. 4. Specifically, we use the ResNet-50 model [14] pre-trained on ImageNet to extract the image embedding while adopt GloVe [18] to capture the sentence embedding. The image embedding, a 2048-dimension vector, is the output of the final pooling layer. As for the sentence embedding, we adopt the GloVe of 100-dimension version pre-trained on Wikipedia and Gigaword. By feeding both image and sentence embeddings to a following fusion layer and a fully connected layer, the model will output the predicted result. In regard to the design of fusion, we investigate and compare several recent works, such as MCB [12], Mutan [3] and so on. Comparison results among different fusion layers are presented in the experiment part.

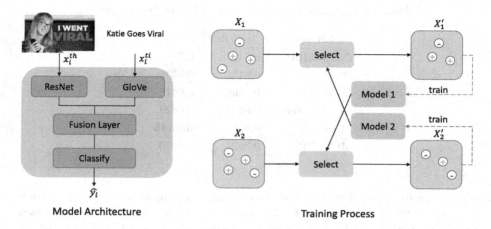

Model Architecture   Training Process

Fig. 4. The architecture of our proposed model and its training process. To tackle noisy labels, we adopt co-teaching to filter out data with wrong labels during training. Note that X_1, X_2, X_1', X_2' refer to the batch of training samples, and Model 1 and Model 2 share the same model architecture but with different initialized parameters.

Learning Strategy. As for robust learning with noisy labels, following the idea of [13], we exploit the co-teaching method, which filters out wrong labels while training. Concretely, we set up two identical networks to teach each other. In each training batch, each network selects instances with small loss as useful knowledge and teaches these instances to the peer network for further training. The basic assumption behind this strategy is that, on a noisy dataset, deep networks tend to first learn easy and clean patterns in initial epochs. Note that when applying the co-teaching, we oversample the clickbait samples in each batch to make labels of training samples balanced. The reason why we do this is that co-teaching tends to drop positive samples when the number of negative samples is much more. With such configuration, wrongly labeled instances that are out of normal pattern and usually lead to high loss can be removed.

5   Experimental Validation

5.1   Set-Up

After labeling generation, our dataset consists of 787 labeled data and 7,039 weakly labeled data. For evaluation, we select 197 labeled data as the test set while the other 590 data as a part of the training set. Besides, for a fair comparison, we use '5-fold validation' to evaluate each method. Specifically, we conduct 5 experiments for each method and, in each experiment, we select one-fifth from 590 labeled training data as the validation set to pick the best model. The averaged result of 5 experiments is models' final performance.

For our method, we fine-tuned the ResNet while training, and use the average of word embeddings to represent the sentence embedding. For neural network

Table 4. Performance comparison of different fusion layer.

Fusion layer	AUC-ROC	F1 score
ConcatMLP	0.8427	0.6404
Block [4]	0.8452	0.6538
Mutan [3]	**0.8659**	0.6415
BlockTucker [4]	0.8392	0.6329
MFH [27]	0.8603	**0.6585**
MCB [12]	0.8626	0.6170

models, we fix batch size as 32 and set the learning rate as 1e−4. We train each method for 20 epochs and select the one that performs best in the validation set for evaluation. As for the optimizer, we use Adam with the default hyperparameters in Pytorch.

Since the label of our test set is not balanced, which is the same case with real data distribution in video-sharing platforms nowadays, we employ F1 score and AUC-ROC as the evaluation metric. The F1 score is the harmonic mean of the precision and recall, which is usually better than accuracy when evaluating with imbalanced labels. AUC-ROC curve is a performance measurement for the classification problems at various threshold settings. ROC is a probability curve and AUC represents the degree of separability. In other words, it represents the capability of the model to distinguish between classes. Note that we also use the AUC-ROC to pick the best model during training.

5.2   Performance Comparison

Fusion Layer Comparison. We first compare several recent works on multimodal fusion using our built model architecture. For this comparison, we only use the labeled data for training models with different fusion Layers. The comparison result is reported in Table 4. Note that ConcatMLP simply concatenates the image embedding and sentence embedding for fusion. For subsequent experiments, we select the Block, Mutan and MFH for further comparison, which perform best among all the fusion layers.

Comparison with Baselines. We then compare our models with several representative and state-of-the-art vision-languages models, including SVM, Logistic Regression, VisualBERT, LXMERT and UNITER. For SVM and logistic Regression, we concatenate the features including the outputs of the pre-trained ResNet-50 and GloVe as their input, which is identical with our model. Note RBF kernel function is used for SVM. As for VisualBERT, LXMERT and UNITER, we use their pre-trained models in the VQA task and fine-tune them to our task.

As shown in Table 5, our methods consistently outperform the baselines. For SVM and Logistic Regression, though they share the same embedding format

Table 5. Comparison results using different models and with/without generated labels.

Method	w/o generated labels		w/ generated labels	
	AUC-ROC	F1 score	AUC-ROC	F1 score
SVM	0.7149	0.3830	0.7355	0.4000
Logistic regression	0.7144	0.4912	0.7629	0.5986
VisualBERT [16]	–	–	0.8460	0.6722
LXMERT [25]	–	–	0.8458	0.6640
UNITER [8]	–	–	0.8196	0.6554
Ours + Block [4]	0.8452	0.6538	0.8644	0.6831
Ours + Mutan [3]	**0.8659**	0.6415	**0.8666**	**0.6933**
Ours + MFH [27]	0.8603	**0.6585**	0.8603	0.6884

"–": Does not converge due to a lack of data

with our proposed model, their performance is not satisfactory. On one hand, since they are not end-to-end models, they are unable to fine-tune the ResNet during training, which may result in inappropriate image feature representation. In contrast, our end-to-end model does not have such constraint and can fine-tune the ResNet to get a better image feature representation for our task. On the other hand, the fitting and generalization capability of classical machine learning models is not as great as neural networks. As for the current SOTA vision-language models which are based on the transformer, they usually take the object detection results as the input. In this context, they greatly rely on the object detection networks like Faster R-CNN [21], and these networks would not be fine-tuned while training the vision-language models. However, the images used for training objection detection networks are usually different from thumbnails exhibited on video-sharing platforms. In short, there exists a data distribution discrepancy. As a result, the objection results may beyond our expectation and are not ideal for the clickbait thumbnail detection task. That's the possible reason for the limited performance of these BERT-like vision-language models. To improve their performance, an object recognition dataset specific to thumbnails on video websites may be required, which is unavailable currently. Moreover, compared to these transformer-based networks, our model is more light-weight and can adapt to a new domain with much less training data.

Effectiveness of Generated Labels. With weak supervision and majority voter, we generate 7039 labels for unlabeled data with 83.6% accuracy on labeled data. To access the impacts of these generated labels toward models' performance, we make a comparison of models trained with and without generated labels. Note we only have 591 samples for training without generated labels while 7620 samples with generated labels. As reported in Table 5, all the methods benefit from these additional training samples. Experimentally, we found that, without generated labels, transformer-based models are very hard to converge

Table 6. Performance comparison with different forget rate τ.

Forget rate	F1 socre			AUC-ROC		
	Block	Mutan	MFH	Block	Mutan	MFH
$\tau = 0.00$	0.6831	0.6933	0.6884	0.8644	0.8666	0.8603
$\tau = 0.05$	0.6941	0.6877	0.6759	**0.8714**	0.8663	0.8469
$\tau = 0.15$	0.7102	**0.7122**	**0.7127**	0.8680	**0.8712**	**0.8805**
$\tau = 0.30$	**0.7153**	0.7039	0.7100	0.8672	0.8692	0.8695

and their training loss barely falls down. This also demonstrates the effectiveness of our generated labels. As for the different improvements in AUC-ROC and F1-score, we think that adding generated labels enables models to hold a better decision boundary when the threshold is 0.5, but with a similar ability to distinguish two classes.

5.3   Understanding Co-teaching

Despite the improvement we obtain with generated labels, the performance of models is still limited by the noise in them. In this section, we conduct experiments to verify the effectiveness of co-teaching in combating noisy data, where the choosing of forget rate is critical. Generally, at the initial learning epochs, we can safely update the parameters of the network using all entire noisy data since the network will not memorize the noise in the early stage of training [2]. But as the learning proceeds, the network has to 'forget' some noisy data to prevent fitting them. In other words, we will drop some instances that are considered as noise. And the forget rate means how many instances should be considered as noise and would be dropped in every training batch. To understand how forget rate τ affects the co-teaching, we vary $\tau = \{0, 0.05, 0.15, 0.3\}$ and make a comparison.

Table 6 presents the comparison results of using different forget rate τ. We can observe that all three models benefit from co-teaching, which verifies its effectiveness to tackle noise. Note that $\tau = 0$ means co-teaching is not employed for training. Besides, co-teaching with $\tau = 0.15$ performs better than other forget rate setting. Considering that the accuracy of generated labels is 83.6% in the evaluation, the $\tau = 0.15$ setting helps remove most of the samples with wrong labels at the meanwhile of reserving as many valid training samples as possible, which accounts for the good performance of models in this setting.

5.4   Limitation and Future Work

Our proposed framework CHECKER detects clickbait thumbnails using their visual features in conjunction with their titles without the need to comprehensively process the target videos' contents. This is because CHECKER aims to stimulate the users' experience where ones can detect a clickbait thumbnail even before watching its video. In the future, we hope to explore if utilizing different video comprehension techniques can further improve our model.

6   Conclusion

In this paper, we propose to leverage weak supervision to address the training data shortage in clickbait thumbnail detection. To this end, we first construct a dataset consisted of Youtube videos and invite workers to manually annotate some of them. To make use of unlabeled data, based on characteristics of clickbait thumbnails, we design several high-quality labeling functions as weak supervision sources to generate labels for them. Then, with recent advances in multimodal fusion, we build a multimodal model that takes the thumbnail and title as input to identify clickbait. Furthermore, to deal with noise in generated labels, we adopt co-teaching to filter out samples with wrong labels to train a robust classifier. The experiment results demonstrate the effectiveness of our proposed method.

Acknowledgement. The works of Thai Le and Dongwon Lee were in part supported by NSF awards #1742702, #1820609, #1909702, #1915801, #1934782, and #2114824.

References

1. Agrawal, A.: Clickbait detection using deep learning. In: NGCT (2016)
2. Arpit, D., et al.: A closer look at memorization in deep networks. In: ICML (2017)
3. Ben-Younes, H., Cadene, R., Cord, M., Thome, N.: MUTAN: multimodal tucker fusion for visual question answering. In: ICCV (2017)
4. Ben-Younes, H., Cadene, R., Thome, N., Cord, M.: BLOCK: bilinear superdiagonal fusion for visual question answering and visual relationship detection. In: AAAI (2019)
5. Biyani, P., Tsioutsiouliklis, K., Blackmer, J.: 8 amazing secrets for getting more clicks: detecting clickbaits in news streams using article informality. In: AAAI (2016)
6. Chakraborty, A., Paranjape, B., Kakarla, S., Ganguly, N.: Stop clickbait: detecting and preventing clickbaits in online news media. In: ASONAM (2016)
7. Chen, M.F., et al.: Train and you'll miss it: interactive model iteration with weak supervision and pre-trained embeddings. arXiv:2006.15168 (2020)
8. Chen, Y.-C., et al.: UNITER: UNiversal image-TExt representation learning. In: Vedaldi, Andrea, Bischof, Horst, Brox, Thomas, Frahm, Jan-Michael. (eds.) ECCV 2020. LNCS, vol. 12375, pp. 104–120. Springer, Cham (2020). https://doi.org/10.1007/978-3-030-58577-8_7
9. Chen, Y., Conroym, N.J., Rubin, V.L.: Misleading online content: recognizing clickbait as false news. In: ACM on Workshop on Multimodal Deception Detection (2015)
10. Elyashar, A., Bendahan, J., Puzis, R.: Detecting clickbait in online social media: you won't believe how we did it'. arXiv:1710.06699 (2017)
11. Fu, D., Chen, M., Sala, F., Hooper, S., Fatahalian, K., Ré, C.: Fast and three-rious: speeding up weak supervision with triplet methods. In: ICML (2020)

12. Fukui, A., Park, D.H., Yang, D., Rohrbach, A., Darrell, T., Rohrbach, M.: Multimodal compact bilinear pooling for visual question answering and visual grounding. arXiv:1606.01847 (2016)
13. Han, B., et al.: Co-teaching: Robust training of deep neural networks with extremely noisy labels. In: NIPS (2018)
14. He, K., Zhang, X., Ren, S., Sun, J.: Deep residual learning for image recognition. In: CVPR (2016)
15. Le, T., Shu, K., Molina, M.D., Lee, D., Sundar, S.S., Liu, H.: 5 sources of clickbaits you should know! using synthetic clickbaits to improve prediction and distinguish between bot-generated and human-written headlines. In: ASONAM (2019)
16. Li, L.H., Yatskar, M., Yin, D., Hsieh, C.-J., Chang, K.-W.: VisualBERT: a simple and performant baseline for vision and language. arXiv:1908.03557 (2019)
17. Molina, M., Sundar, S.S., Roy, M.M.U., Hassan, N., Le, T., Lee, D.: Does clickbait actually attract more clicks? Three clickbait studies you must read. In: CHI (2021)
18. Pennington, J., Socher, R., Manning, C.D.: GloVe: global vectors for word representation. In: EMNLP (2014)
19. Qu, J., Hißbach, A.M., Gollub, T., Potthast, M.: Towards crowdsourcing clickbait labels for YouTube videos. In: HCOMP (2018)
20. Ratner, A., Bach, S.H., Ehrenberg, H., Fries, J., Wu, S., Ré, C.: Snorkel: rapid training data creation with weak supervision. In: VLDB (2017)
21. Ren, S., He, K., Girshick, R., Sun, J.: Faster R-CNN: towards real-time object detection with region proposal networks. In: NIPS (2015)
22. Rony, M.M.U., Hassan, N., Yousuf, M.: Diving deep into clickbaits: who use them to what extents in which topics with what effects?. In: ASONAM (2017)
23. Shang, L., Zhang, D.Y., Wang, M., Lai, S., Wang, D.: Towards reliable online clickbait video detection: a content-agnostic approach. Knowl. Based Syst. **182**, 104851 (2019)
24. Shu, K., Wang, S., Le, T., Lee, D., Liu, H.: Deep headline generation for clickbait detection. In: ICDM (2018)
25. Tan, H., Mohit B.: LXMERT: learning cross-modality encoder representations from transformers. In: EMNLP (2019)
26. Xu, P., et al.: Clickbait? Sensational headline generation with auto-tuned reinforcement learning. In: EMNLP (2019)
27. Yu, Z., Yu, J., Xiang, C., Fan, J., Tao, D.: Beyond bilinear: generalized multimodal factorized high-order pooling for visual question answering. IEEE Trans Neural Netw. Learn. Syst. **29**, 5947–5959 (2018)
28. Zannettou, S., Chatzis, S., Papadamou, K., Sirivianos, M.: The good, the bad and the bait: detecting and characterizing clickbait on Youtube. In: IEEE Security and Privacy Workshops (SPW) (2018)

Crowdsourcing Evaluation of Saliency-Based XAI Methods

Xiaotian Lu[1]([✉])(iD), Arseny Tolmachev[2], Tatsuya Yamamoto[2], Koh Takeuchi[1],
Seiji Okajima[2], Tomoyoshi Takebayashi[2], Koji Maruhashi[2], and Hisashi Kashima[1]

[1] Kyoto University, Kyoto, Japan
lu@ml.ist.i.kyoto-u.ac.jp, takeuchi@kyoto-u.ac.jp,
kashima@i.kyoto-u.ac.jp
[2] Fujitsu Research, Fujitsu Ltd., Tokyo, Japan
{t.arseny,tyamamo,okajima.seiji,takebayashi.tom,
maruhashi.koji}@fujitsu.com

Abstract. Understanding the reasons behind the predictions made by deep neural networks is critical for gaining human trust in many important applications, which is reflected in the increasing demand for explainability in AI (XAI) in recent years. Saliency-based feature attribution methods, which highlight important parts of images that contribute to decisions by classifiers, are often used as XAI methods, especially in the field of computer vision. In order to compare various saliency-based XAI methods quantitatively, several approaches for automated evaluation schemes have been proposed; however, there is no guarantee that such automated evaluation metrics correctly evaluate explainability, and a high rating by an automated evaluation scheme does not necessarily mean a high explainability for humans. In this study, instead of the automated evaluation, we propose a new human-based evaluation scheme using crowdsourcing to evaluate XAI methods. Our method is inspired by a human computation game, "Peek-a-boom", and can efficiently compare different XAI methods by exploiting the power of crowds. We evaluate the saliency maps of various XAI methods on two datasets with automated and crowd-based evaluation schemes. Our experiments show that the result of our crowd-based evaluation scheme is different from those of automated evaluation schemes. In addition, we regard the crowd-based evaluation results as ground truths and provide a quantitative performance measure to compare different automated evaluation schemes. We also discuss the impact of crowd workers on the results and show that the varying ability of crowd workers does not significantly impact the results.

Keywords: Explainable AI · Interpretability · Evaluation · Crowdsourcing

1 Introduction

Recent significant advances in AI technologies have introduced innovations in various fields. In particular, deep neural networks (DNNs) exhibit remarkable performance in a wide range of real-world applications, such as natural language processing [8,27], image classification [6,7,16,29], and human action recognition [15,19]. DNNs can

© Springer Nature Switzerland AG 2021
Y. Dong et al. (Eds.): ECML PKDD 2021, LNAI 12979, pp. 431–446, 2021.
https://doi.org/10.1007/978-3-030-86517-7_27

extract intricate underlying patterns from large and high-dimensional datasets and have reduced the demand for feature engineering. However, the internal mechanism of DNNs is a black box, i.e., it is difficult to understand the relationships between their inputs and outputs. In low-risk environments, errors made by DNNs do not have severe impacts; for example, in movie recommendation systems, the impact of making a recommendation error is relatively low. However, in other fields such as healthcare, a single misdiagnosis can be fatal; therefore, it is essential to explain the predictions. In regulated industries such as the judicial system and financial markets, a mandate for explanations in addition to model predictions is emerging in legal norms. However, most current DNN models are opaque and provide no information about their decision-making process, which has been a significant obstacle to the implementation of AI in essential applications. Understanding the reasons behind their predictions is critical for gaining human trust in many important applications [22], which is reflected in the increasing demand for explainability in AI (XAI) in recent years.

To satisfy the requirements of XAI, various explanation and interpretation methods have emerged, especially for black-box predictions made by already-trained neural networks. One of the major approaches to this problem involves the estimation of the influence of a subset of input features on the predictions of a model. By understanding the important features, the model can be improved, model predictions can be trusted, and undesirable behaviors can be isolated [12]. For example, in image classification tasks, the generation of saliency maps, which assign an importance measure for each part (or pixel) of an input image, is a major research direction; the representative methods include Vanilla Gradients [3, 10, 25], SmoothGrad [26], Guided-Backpropagation [28], and Grad-CAM [24, 30] (Fig. 1).

While various XAI methods have been proposed, their evaluation strategy has not been established well, and there is an urgent demand for quantitative measures to answer the question "Given several XAI methods of a black-box prediction model, which one yields the best interpretations?" Several automated evaluation schemes have been proposed [12, 21, 23]; they usually delete or replace pixels that are said important by an XAI method, and check the deterioration in the prediction performance. However, as pointed out by a recent research [20], high interpretability for machines does not imply the same for humans. A machine may recognize an object based on its relation to the background rather than the object itself. For example, the background of an image of an airplane is often the sky. This is a reasonable strategy for machines to make decisions based on statistical information, but it is probably different from how humans recognize objects. After all, interpretability for humans can ultimately only be evaluated by humans.

In this study, we perform human-based evaluation of XAI methods by using crowdsourcing. Our XAI evaluation approach is based on a human computation game. In human computation [18], some approaches embed human intelligence tasks into games, usually referred to as game with a purpose (GWAP). Peek-a-boom is a type of GWAP with two players named Peek and Boom; Boom reveals a part of a given image and Peek guesses the image object from the revealed part. We use a Peek-a-boom-based XAI evaluation in which a human plays the Peek, and an XAI method plays the Boom instead of another human.

(a) Original image   (b) Vanilla Gradients   (c) SmoothGrad

(d) Guided-Backpropagation   (e) Grad-CAM

Fig. 1. Different saliency maps produced by different XAI methods. The bright areas indicate important areas.

We implement a crowd-based evaluation interface (Fig. 2)[1], recruit crowd workers for executing evaluation tasks of four popular XAI methods on two real datasets, and compare the results with those by four automated evaluation schemes.

The results show that the proposed scheme gives different evaluations from the automated evaluation schemes. Subsequently, We consider the proposed crowd-based scheme as ground truths and evaluate the automated evaluation schemes in terms of interpretability for humans. Finally, we analyze the ability of crowd workers and find that, even if their ability may vary considerably, the final results are not significantly affected.

The contributions of this study are summarized as follows:

1) We propose a new crowd-based evaluation scheme for XAI methods.
2) We experimentally investigate the difference between automated and crowd-based evaluation schemes.
3) We provide a performance measure for automated evaluation schemes based on their similarity to the proposed crowd-based evaluation scheme.
4) We examine the impact of the number and ability of crowd workers on the results.

---

[1] Our crowdsourcing evaluation interface can be tried at https://17bit.github.io/crowddemo/index.html .

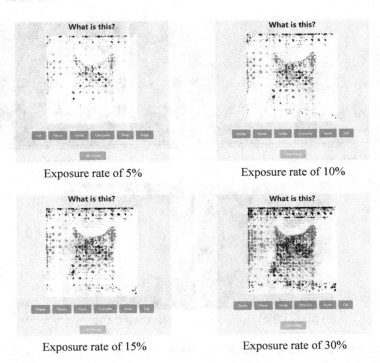

Fig. 2. Example of how the proposed evaluation interface gradually reveals an image to crowd workers at different exposure rates (5%. 10%, 15%, and 30%).

2 Related Work

We briefly review the XAI methods used in this study, automated evaluation schemes that can automatically evaluate the XAI methods, and existing crowd-based evaluation schemes.

2.1 XAI Methods

Most of the existing XAI methods attribute the output of a pre-trained neural network to a part of its input. For a multi-class classification problem with C classes, let $f : \mathbb{R}^D \to \mathbb{R}^C$ be a pre-trained neural network which takes an input feature vector $x \in \mathbb{R}^D$ and output a vector representing the degree of classification into each class. A typical XAI method provides a *saliency map* $s : \mathbb{R}^D \to \mathbb{R}^D$ that maps an input feature vector to a vector whose d-th element indicates the importance of the d-th feature of input x [1].

Vanilla gradient [3,10,25] is the most basic method to create a saliency map as $s^{\text{Vanilla}}(x) = \frac{\partial f}{\partial x}$, which quantifies the influence of a small change in each input dimension on the output of the network. One of the drawbacks of this method is that the results are sensitive to the noise in the original image (Fig. 1b).

SmoothGrad [26] is an improved version of the vanilla gradient which estimates the gradient more robustly. It takes the average of the vanilla gradient over perturbed

Table 1. Dataset split modification of automated evaluation schemes. "top" indicates that the top-ranked pixels are modified first, "bottom" implies that the bottom-ranked pixels are removed first, and '–' means that the dataset split is used without modification.

XAI method	Test set	Training set
ROAR [12]	–	Top
KAR [12]	–	Bottom
ROAE [21,23]	Top	–
KAE [21,23]	Bottom	–

inputs, $s^{\text{Smooth}}(x) = \frac{1}{N} \sum_{i=1}^{N} s^{\text{Vanilla}}(x + \Delta_i)$, where the perturbation of the input Δ_i is sampled by a Gaussian distribution, which results in clearer saliency maps than the vanilla gradient (Fig. 1c).

Guided-Backpropagation [28] gives the contribution of an input dimension of a neuron to the output by distributing the output back to the input. For better interpretability, it back-propagates the output of "active" ReLU units, which highlights important edges in images (Fig. 1d).

Grad-CAM [24] focuses on the last convolution layer of a CNN (Convolutional Neural Network) and visualizes the globally-average-pooled gradients In contrast with the previously mentioned XAI methods, only Grad-CAM relies on both gradients and feature maps of the convolutional layer. The results focus more on important "areas" rather than edges as shown in Fig. 1e. A recent research shows that these saliency maps based on only the gradients will not change greatly even if the parameters of DNN model are randomized [1]. However, saliency map of GradCAM is different from saliency maps based on gradients. Our experiments also show that GradCAM performs the best in our crowd-based evaluation scheme.

2.2 Automated Evaluation Schemes for XAI Methods

Most automated evaluation schemes work by modifying (either train or test) data, and compare the differences in the prediction performance of models. Hooker et al. [12] proposed several automated evaluation schemes for XAI methods. In their studies, they argued that by removing data features from the training set, better evaluation robustness can be archived in comparison to schemes that modify the test set.

In the Remove and Retrain (ROAR) scheme, the top-ranked pixels given by XAI methods are removed from the images in the training dataset, a new model is trained on the modified training set, and the resulting model is evaluated on the non-modified test dataset. The ROAR scheme was mainly compared to the Keep and Retrain (KAR) scheme, in which the bottom-ranked pixels were removed from the training set. It should be noted that Hooker et al. did not perform comparisons to crowd-based evaluation schemes.

In our experiments (Sect. 4), we also test two schemes that change the test set while leaving the training set unchanged, namely, Remove and Evaluate (ROAE) and Keep

and Evaluate (KAE) [21,23]. Table 1 summarizes the comparison between the automated evaluation schemes.

2.3   Crowd-Based Evaluation Schemes for XAI Methods

Several crowd-based evaluation schemes have been proposed to measure the ability
of XAI methods. Hutton et al. [13] used crowdsourcing to assess the explanations for
supervised text classification. Crowd workers were asked to compare human-generated
and XAI method-generated explanations and indicate which they preferred and why.
Selvaraju et al. [24] and Jeyakumar et al. [14] asked crowd workers to choose better
explanations directly. Similarly, Can et al. [5] asked crowd workers to rate the saliency
maps of Grad-CAM on the visual characteristics of venues.

Doshi-Velez and Kim [9] concluded that there are three different types of crowd-
based evaluation schemes: 1) binary forced choices in which humans are presented
with pairs of explanations and choose better ones, 2) forward simulation/prediction in
which humans are presented with an explanation and an input, and simulate the output
of the model, and 3) counterfactual simulation in which humans are presented with
an explanation, an input, and an output, and tell what must be changed to change the
prediction to a desired output.

Our proposed scheme is similar to none of the aforementioned schemes. We do not
directly show the saliency maps to workers nor force them to make a binary choice;
instead, we transform into a simpler task, which makes it easier for workers to make
objective choices that are less dependent on subjective judgments of workers.

3   Proposed Crowd-Based Evaluation Scheme for XAI Methods

We propose a crowd-based evaluation scheme for XAI methods based on Peek-a-
boom [2] which is an online human computation game as shown in Fig. 3. As suggested
by the name, this cooperative game has two players, namely, "Peek" and "Boom."

Peek starts with a blank screen, while Boom starts with an image and a word related
to it. At each round of the game, Boom can specify a small area in the image and
reveals the area to Peek, and Peek enters a guess of the word on the basis of the revealed
parts. The both players get more points when Peek correctly answers the word earlier;
therefore, Boom has an incentive to reveal only the areas of the image necessary for
Peek to guess the correct word.

We use a Peek-a-boom style Web interface, in which a crowd worker plays the Peek,
and an XAI method plays the Boom instead of another human. In the web interface,
crowd workers are asked to perform an image classification task, i.e., assigning a label
to an image from a set of labels. First, we reveal a small percentage of image pixels
with an option to reveal more if it is impossible to assign a label with confidence. We
show a correct label, several randomly selected wrong labels, and an "I don't know"
button. If the worker cannot provide a confident answer, they can select "I don't know";
then, more parts of the image will be revealed. In the case of an incorrect answer, more
parts will be revealed as well. Once the worker selects the right answer, or "I don't
know" is selected with a fully-shown image, we give the worker a new image. For a

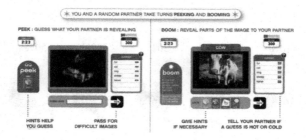

Fig. 3. Interface of Peek-a-boom human computation game [2]. Peek sees the left screen, and Boom sees the right one. Boom determines which parts are exposed to Peek so that Peek correctly guess the image content (that is a cow in this example.)

given image, XAI methods rank the pixels in descending order of their importance. For each crowd worker, our crowdsourcing interface starts from an almost blank image (i.e., exposure rate $= 0.05$); gradually, the pixels are revealed in order of importance (i.e., increase the exposure rate) (Fig. 2). At each exposure rate $r \in [0, 1]$, the crowd worker is asked to guess the object in the image, typically in terms of multiple-choice questions. We consider that if an XAI method is "interpretable" enough, the crowd worker can correctly answer the question at a small exposure rate r.

Specifically, the evaluation procedure consists of the following steps:

1) Prepare a pre-trained prediction model, a dataset, and several XAI methods to be evaluated.
2) Apply all XAI methods and a random baseline to each image.
3) Get a saliency map from each XAI method and the random baseline.
4) A series of images with a part of top importance pixel features are generated from the saliency map (Fig. 4).
5) Start with the smallest percentage of pixels (e.g., 5%) and ask the crowd workers about the class of object. If they do not know, show more pixels and record the percentage of images when the worker answers correctly.

4 Results

We conduct experiments to answer the following four questions:
Q1. Are human and automated evaluations really different?
Q2. Which XAI methods are deemed better by humans?
Q3. Which automated evaluation scheme is closer to humans?
Q4. How does the number and ability of crowd evaluators affect the evaluation results?

Exposure Rate

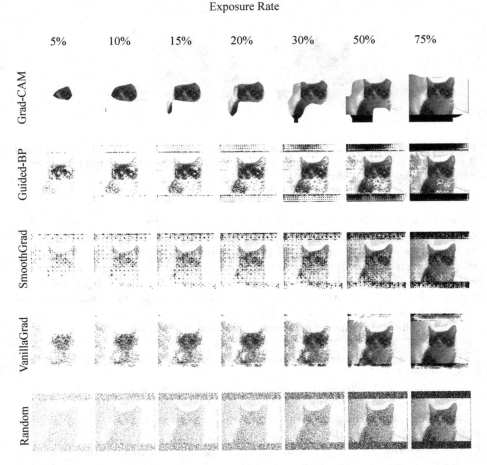

Fig. 4. "Explanations" for a cat image provided by five XAI methods (Grad-CAM, Guided-BP, SmoothGrad, VanillaGrad, and Random) at different exposure rates (5%, 10%, 15%, 20%, 30%, 50%, and 75%).

4.1 Experimental Settings

We use two datasets for the evaluation; namely, Food101 [4] and Animal95. Animal95 is a subset of OpenImages v6 dataset [17]. We use bounding box data to filter the dataset to extract object areas. Subsequently, we select only images with either single animals or multiple animals of the same class. Images with multiple classes (multiple types of animals) are not included, but those with several animals of a single class (e.g., three dogs) or non-animal classes are included.

We shortlist the 30 most common food classes from all the classes in Food101, and randomly select ten images in each class. Similarly, we select 95 most frequent animals from OpenImages for the Animal95 dataset.

The XAI methods used in our experiments include GradCAM, Guided-backpropagation, SmoothGrad, and Vanilla Gradient. We implement the four XAI methods as well as the random baseline with a pre-trained ResNet50 model [11]. In total, 1500 pairs and 4750 of (image, XAI method) are generated for Food101 and Animal95, respectively[2].

We use Amazon Mechanical Turk (AMT)[3] and Lancers[4] as the crowdsourcing platforms for crowd-based evaluation. Each (image, XAI method)-pair is evaluated ten times. Each crowd worker is required to evaluate 20 pairs for a reward of USD 0.5 in AMT, or JPY 40 in Lancers. To avoid biases, we randomly sample (image, XAI method)-pairs assigned to each worker. Approximately 3200 crowd workers participate in the evaluation tasks.

We compare ROAR, KAR, ROAE, and KAE schemes (introduced in Sect. 2.2) as the representatives of automated evaluation schemes. The performance of each XAI method is evaluated by an accuracy-exposure curve. For the crowd-based evaluation, We calculate the average accuracy of crowd answers at each exposure rate, while the human evaluators are replaced by a machine classifier in the automated evaluation.

In addition, we also provide the area under curve (AUC) of each accuracy-exposure curve. Let a series of exposure rates be $r_1 = 0, r_2, r_3, \ldots, r_n = 1$, where $r_i < r_j$ for $i < j$. Let a_i^k denote the accuracy at exposure rate r_i for XAI method k both in crowd-based and automated evaluations. The value of AUC in XAI method k denoted by AUC^k is defined as

$$\text{AUC}^k = \sum_{i=2}^{n} \frac{1}{2}(r_i - r_{i-1})(a_i^k + a_{i-1}^k).$$

4.2   Results

Q1. Are Human and Automated Evaluations Really Different?

The first question we investigate is the difference between automated and crowd-based evaluation schemes, because the latter requires higher time and financial costs, and there is no reason to resort to human evaluation if they both give the same results.

Figure 5 shows the accuracy-exposure curves of different XAI methods at different exposure rates by different evaluation methods. For the crowd-based evaluation scheme (denoted by Crowd) and two automated evaluation schemes (KAR and KAE), higher curves indicate better performance. In contrast, for the other two automated evaluation schemes, ROAR and ROAE, lower curves indicate better performance.

Table 2 shows the AUCs of each scheme; each row and column correspond to an evaluation scheme and an XAI method, respectively. The numbers in the brackets show the ranks. The bold numbers show the best results. Although we can see some consistency between the ranking of AUC by the crowd-based ranking and those by the

---

[2] 30 classes $\times$ 10 images $\times$ 5 XAI methods $=$ 1500 for Food101, and 95 classes $\times$ 10 images $\times$ 5 XAI methods $=$ 4750 for Animal95.

[3] https://www.mturk.com/.

[4] https://www.lancers.jp/.

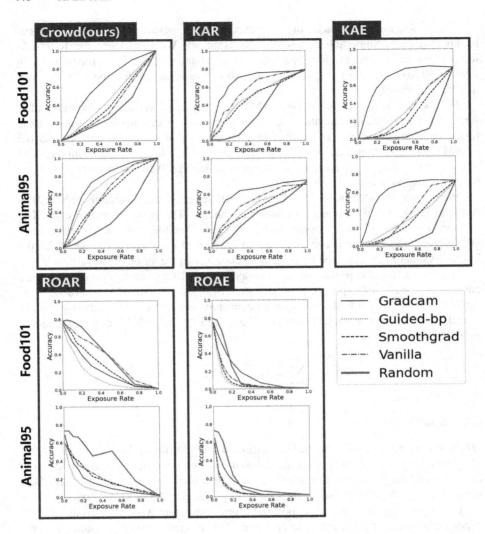

Fig. 5. Performance curves of different XAI methods on the two datasets (Food101 and Animal95). The horizontal and vertical axis indicates the exposure rate and accuracy, respectively. For the crowd-based evaluation (Crowd), KAR, and KAE, upper curves indicate better performance. In contrast, for ROAR and ROAE, lower curves are better.

automatic evaluation schemes, no automated evaluation scheme obtains the same ranking of AUC as the crowd-based evaluation. (We will see more detailed comparisons later.)

Now, we discuss the impact of different datasets on the results. It is evident from Table 2 that the values of AUC are different among the two datasets, Food101 and Animal95. The AUCs for Animal95 datasets are generally better than those for Food101 in

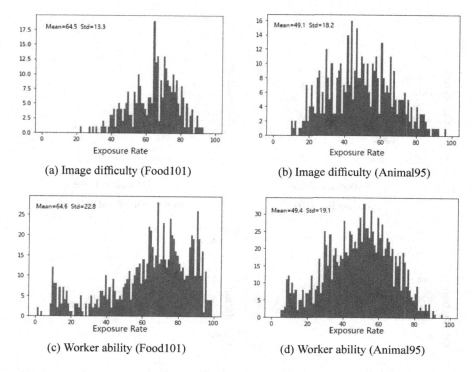

(a) Image difficulty (Food101)

(b) Image difficulty (Animal95)

(c) Worker ability (Food101)

(d) Worker ability (Animal95)

Fig. 6. Histograms of average exposure rates of correct answers. The horizontal and vertical axes indicate the exposure rate and frequency, respectively; the top and bottom rows indicate the frequencies of images and workers, which show the distributions of "image difficulty" and "worker ability", respectively. Comparing (a) and (b), the mean image difficulty of Animal95 dataset is higher than that of Food101, indicating the Animal95 dataset is relatively easier than Food101. In the bottom row ((c) and (d)), the variance of the worker ability in the Food 101 dataset is higher than that of Animal95, which is probably because the difficulty of recognizing food can be significantly affected by cultural differences. In spite of the large variations in the worker ability, Table 3 shows they have no significant impacts on the results.

all the schemes. This is probably because it is rather easier to recognize animals than foods; this is also suggested by Fig. 6 showing the distribution of the "difficulty" of the images.

In contrast, Fig. 5 shows that the ranking of XAI methods is not entirely different among the datasets, except for the slight difference in SmoothGrad and Vanilla Gradients. This shows that the difference of datasets does not significantly impact the relative superiority or inferiority of the different schemes. However, this conclusion is drawn from only two datasets and needs to be validated with more datasets in the future.

Q2. Which XAI Methods Are Deemed Better by Humans?

Because different XAI methods provide different pixel rankings, the next question we investigate is which XAI method is more reliable. It can be observed from Table 2 that

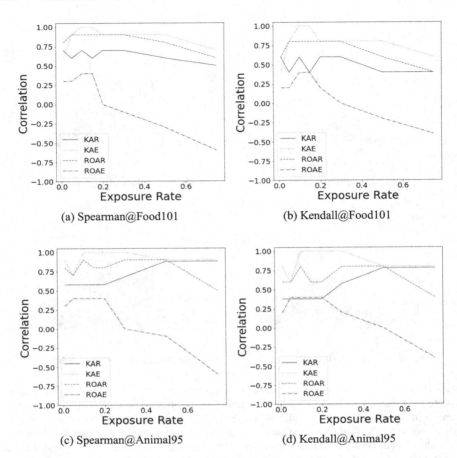

(a) Spearman@Food101

(b) Kendall@Food101

(c) Spearman@Animal95

(d) Kendall@Animal95

Fig. 7. Correlations of XAI methods ranking between the crowd-based evaluation and the four automated evaluations at different exposure rates in the Food101 and Animal95 datasets (the higher, the better). The correlations are given both in the Spearman ranking correlation and the Kendall ranking correlation. The horizontal and vertical axes indicate the exposure rate and the ranking correlation value, respectively.

GradCAM exhibits the best performance in the crowd-based evaluation scheme. This is probably because GradCAM produces low resolution feature maps (e.g., 7×7 for ResNet family when the input size is 224×224), which are then linearly interpolated to the resolution of the input image, thereby producing mostly connected regions rather than distributed regions.

In addition, this result is also consistent with the conclusion of previous work on sanity checking [1].

Q3. Which Automated Evaluation Scheme Is Closer to Humans?

Once we assume that the human assessments are the ground truths, automated eval-

Table 2. AUCs of different methods by different evaluation schemes. The numbers in the bracket show the rank of the XAI method. The bold numbers show the best results. For the crowd-based evaluation (Crowd), KAR, and KAE, larger AUC values show better performance, that is, important areas of images are shown earlier. On the other hand, for ROAR and ROAE, smaller AUCs indicate better performance, that is, important areas are removed earlier. (Also see Fig. 5.)

Dataset	Scheme	GradCAM	Guided-bp	SmoothGrad	Vanilla Gradients	Random
Food101	Crowd (ours)	**0.639** (1)	0.469 (2)	0.425 (3)	0.396 (4)	0.334 (5)
	KAR	**0.667** (1)	0.494 (3)	0.478 (4)	0.570 (2)	0.340 (5)
	KAE	**0.669** (1)	0.340 (2)	0.265 (4)	0.316 (3)	0.136 (5)
	ROAR	0.211 (2)	**0.140** (1)	0.258 (3)	0.346 (4)	0.366 (5)
	ROAE	0.159 (5)	**0.060** (1)	0.072 (2)	0.087 (3)	0.140 (4)
Animal95	Crowd (ours)	**0.752** (1)	0.696 (2)	0.592 (4)	0.608 (3)	0.354 (5)
	KAR	**0.627** (1)	0.456 (3)	0.445 (4)	0.515 (2)	0.365 (5)
	KAE	**0.619** (1)	0.311 (3)	0.294 (4)	0.354 (2)	0.137 (5)
	ROAR	0.142 (2)	**0.088** (1)	0.194 (3)	0.200 (4)	0.385 (5)
	ROAE	0.115 (4)	**0.048** (1)	0.054 (2)	0.059 (3)	0.137 (5)

uation scheme which is closer to the crowd-based evaluation scheme indicates better performance. Quantifying the goodness of automated evaluation schemes is not only useful for evaluating them but will also help improving themselves.

We investigate the ranking similarity between the four automated evaluations and the crowd-based evaluations. Figure 7 shows the correlations of XAI methods ranking between the crowd-based evaluation and the four automated evaluations at different exposure rates in the Food101 and Animal95 datasets.

Most of the automated evaluation schemes show positive correlations with crowd-based evaluation, but only ROAE shows the lowest correlations, and even gives negative correlations for high exposure rates; this is probably because the mechanism of ROAE equals adding white noises to the images at high exposure rates, which is also known as an approach for generating adversarial examples.

ROAR shows the better performance than ROAE, which is consistent with the report in the previous study [12], which implies the importance of re-training. KAE consistently performs well independent of the change of datasets, correlation types, and exposure rates. KAR performs sub-optimally, but it maintains almost the same performances at high exposure rates, while other automated evaluation schemes tend to decrease the performance at high exposure rates.

Q4. How Does the Number and Ability of Crowd Evaluators Affect the Evaluation Results?

Finally, we investigate the stability of the proposed crowd-based evaluation scheme in terms of the number of crowd workers participating in the evaluation. Crowd workers have different abilities and diligence; for example, some crowd workers do not work seriously on tasks. Figure 6c and Fig. 6d show the histograms of the average exposure rate at which each crowd worker made a correct answer, which can be considered as the distribution of the worker ability; large variations are observed in the ability of the workers.

Table 3. AUCs for different average numbers of crowd workers per (image, XAI method)-pair. The numbers in the bracket show the rank of the XAI method. Although the performance of crowd workers varies greatly due to their different ability and diligence (Fig. 6 (c)(d)), the ranking of XAI methods does not change according to the number of workers.

Dataset	Workers per image	GradCAM	Guided-bp	SmoothGrad	Vanilla Gradients	Random
Food101	0.3	0.647 (1)	0.492 (2)	0.474 (3)	0.422 (4)	0.337 (5)
	0.5	0.618 (1)	0.468 (2)	0.450 (3)	0.419 (4)	0.348 (5)
	1	0.610 (1)	0.465 (2)	0.446 (3)	0.422 (4)	0.331 (5)
	3	0.632 (1)	0.462 (2)	0.421 (3)	0.395 (4)	0.330 (5)
	5	0.636 (1)	0.470 (2)	0.431 (3)	0.399 (4)	0.332 (5)
	7	0.638 (1)	0.469 (2)	0.428 (3)	0.397 (4)	0.333 (5)
	10	0.639 (1)	0.469 (2)	0.425 (3)	0.396 (4)	0.334 (5)
Animal95	0.3	0.742 (1)	0.675 (2)	0.624 (3)	0.622 (4)	0.367 (5)
	0.5	0.752 (1)	0.667 (2)	0.591 (4)	0.616 (3)	0.347 (5)
	1	0.745 (1)	0.683 (2)	0.603 (4)	0.622 (3)	0.339 (5)
	3	0.758 (1)	0.695 (2)	0.604 (4)	0.608 (3)	0.365 (5)
	5	0.755 (1)	0.696 (2)	0.596 (4)	0.608 (3)	0.357 (5)
	7	0.752 (1)	0.696 (2)	0.593 (4)	0.609 (3)	0.358 (5)
	10	0.752 (1)	0.696 (2)	0.592 (4)	0.608 (3)	0.354 (5)

Table 3 summarizes the AUC values of the performance curves when the average number of workers per (XAI method, image)-pair is changed. Some variations are observed in the results when the average number of workers was changed; however, no significant change was found in the qualitative results, which shows the stability and efficiency of the proposed crowd evaluation scheme.

5   Conclusion

In this study, we investigated schemes for evaluation of XAI methods. Based on the hypothesis that interpretability for humans can ultimately only be assessed by humans, We proposed a new human-based evaluation scheme using crowdsourcing and compared it with existing automated evaluation schemes. We convened a total of 3,200 crowd workers to conduct experiments using four XAI methods and two datasets. The results showed that there are differences between the crowd-based evaluation and automatic evaluation. Among the various automatic evaluation schemes, KAE gave the most similar XAI evaluations to human evaluation.

In the report by Hooker et al. [12], ROAR performed better than KAR, but the results of our experiment indicate the opposite, which can be further investigated in the future.

In addition, among the four XAI methods, Grad-CAM was found to be the XAI method closest to human evaluation. This is rather counter intuitive if we focus only on saliency maps because Guided-Backpropagation and SmoothGrad highlight the outline of objects more accurately (as shown in Fig. 1); however, in our scheme, we present a combination of the original image and the saliency map so that Grad-CAM can convey more information with the fewer pixels.

We also confirmed that the number of crowd workers and datasets did not significantly impact the results; however, larger-scale experiments using more datasets will be desirable in the future.

References

1. Adebayo, J., Gilmer, J., Muelly, M., Goodfellow, I., Hardt, M., Kim, B.: Sanity checks for saliency maps. In: Advances in Neural Information Processing Systems, vol. 31, pp. 9505–9515 (2018)
2. von Ahn, L., Liu, R., Blum, M.: Peekaboom: a game for locating objects in images. In: Proceedings of the SIGCHI Conference on Human Factors in Computing Systems (CHI), pp. 55–64 (2006)
3. Baehrens, D., Schroeter, T., Harmeling, S., Kawanabe, M., Hansen, K., Müller, K.R.: How to explain individual classification decisions. J. Mach. Learn. Res. **11**, 1803–1831 (2010)
4. Bossard, L., Guillaumin, M., Van Gool, L.: Food-101 – mining discriminative components with random forests. In: Fleet, D., Pajdla, T., Schiele, B., Tuytelaars, T. (eds.) ECCV 2014. LNCS, vol. 8694, pp. 446–461. Springer, Cham (2014). https://doi.org/10.1007/978-3-319-10599-4_29
5. Can, G., Benkhedda, Y., Gatica-Perez, D.: Ambiance in social media venues: visual cue interpretation by machines and crowds. In: Proceedings of the 2018 IEEE/CVF Conference on Computer Vision and Pattern Recognition Workshops (CVPRW), pp. 2363–2372 (2018)
6. Ciregan, D., Meier, U., Schmidhuber, J.: Multi-column deep neural networks for image classification. In: Proceedings of the IEEE Conference on Computer Vision and Pattern Recognition (CVPR), pp. 3642–3649 (2012)
7. Ciresan, D., Giusti, A., Gambardella, L., Schmidhuber, J.: Deep neural networks segment neuronal membranes in electron microscopy images. In: Advances in Neural Information Processing Systems, vol. 25, pp. 2843–2851 (2012)
8. Collobert, R., Weston, J., Bottou, L., Karlen, M., Kavukcuoglu, K., Kuksa, P.: Natural language processing (almost) from scratch. J. Mach. Learn. Res. **12**(76), 2493–2537 (2011)
9. Doshi-Velez, F., Kim, B.: Towards a rigorous science of interpretable machine learning. arXiv preprint arXiv:1702.08608 (2017)
10. Erhan, D., Bengio, Y., Courville, A., Vincent, P.: Visualizing higher-layer features of a deep network. Univ. Montreal **1341**(3), 1 (2009)
11. He, K., Zhang, X., Ren, S., Sun, J.: Deep residual learning for image recognition. In: Proceedings of the IEEE Conference on Computer Vision and Pattern Recognition (CVPR), pp. 770–778 (2016)
12. Hooker, S., Erhan, D., Kindermans, P.J., Kim, B.: A benchmark for interpretability methods in deep neural networks. In: Advances in Neural Information Processing Systems, vol. 32, pp. 9737–9748 (2019)
13. Hutton, A., Liu, A., Martin, C.: Crowdsourcing evaluations of classifier interpretability. In: 2012 AAAI Spring Symposium Series (2012)
14. Jeyakumar, J.V., Noor, J., Cheng, Y.H., Garcia, L., Srivastava, M.: How can i explain this to you? An empirical study of deep neural network explanation methods. In: Advances in Neural Information Processing Systems, vol. 33, pp. 4211–4222 (2020)
15. Ji, S., Xu, W., Yang, M., Yu, K.: 3D convolutional neural networks for human action recognition. IEEE Trans. Pattern Anal. Mach. Intell. **35**(1), 221–231 (2012)
16. Krizhevsky, A., Sutskever, I., Hinton, G.E.: ImageNet classification with deep convolutional neural networks. In: Advances in Neural Information Processing Systems, vol. 25, pp. 1097–1105 (2012)

17. Kuznetsova, A., et al.: The open images Dataset V4: unified image classification, object detection, and visual relationship detection at scale. Int. J. Comput. Vis. **128**, 1956–1981 (2020)

18. Law, E., Ahn, L.V.: Human Computation. Morgan & Claypool Publishers (2011)

19. Le, Q.V., Zou, W.Y., Yeung, S.Y., Ng, A.Y.: Learning hierarchical invariant spatio-temporal features for action recognition with independent subspace analysis. In: Proceedings of the IEEE Conference on Computer Vision and Pattern Recognition (CVPR), pp. 3361–3368 (2011)

20. Narayanan, M., Chen, E., He, J., Kim, B., Gershman, S., Doshi-Velez, F.: How do humans understand explanations from machine learning systems? An evaluation of the human-interpretability of explanation. arXiv preprint arXiv:1802.00682 (2018)

21. Nguyen, T.T., Le Nguyen, T., Ifrim, G.: A model-agnostic approach to quantifying the informativeness of explanation methods for time series classification. In: Lemaire, V., Malinowski, S., Bagnall, A., Guyet, T., Tavenard, R., Ifrim, G. (eds.) AALTD 2020. LNCS (LNAI), vol. 12588, pp. 77–94. Springer, Cham (2020). https://doi.org/10.1007/978-3-030-65742-0_6

22. Ribeiro, M.T., Singh, S., Guestrin, C.: "Why should I trust you?" explaining the predictions of any classifier. In: Proceedings of the 22nd ACM SIGKDD International Conference on Knowledge Discovery and Data Mining (KDD), pp. 1135–1144 (2016)

23. Samek, W., Binder, A., Montavon, G., Lapuschkin, S., Müller, K.R.: Evaluating the visualization of what a deep neural network has learned. IEEE Trans. Neural Netw. Learn. Syst. **28**(11), 2660–2673 (2016)

24. Selvaraju, R.R., Cogswell, M., Das, A., Vedantam, R., Parikh, D., Batra, D.: Grad-CAM: Visual explanations from deep networks via gradient-based localization. In: Proceedings of the IEEE International Conference on Computer Vision (ICCV), pp. 618–626 (2017)

25. Simonyan, K., Vedaldi, A., Zisserman, A.: Deep inside convolutional networks: visualising image classification models and saliency maps. arXiv preprint arXiv:1312.6034 (2013)

26. Smilkov, D., Thorat, N., Kim, B., Viégas, F., Wattenberg, M.: SmoothGrad: removing noise by adding noise. arXiv preprint arXiv:1706.03825 (2017)

27. Socher, R., et al.: Recursive deep models for semantic compositionality over a sentiment treebank. In: Proceedings of the Conference on Empirical Methods in Natural Language Processing (EMNLP), pp. 1631–1642 (2013)

28. Springenberg, J.T., Dosovitskiy, A., Brox, T., Riedmiller, M.: Striving for simplicity: the all convolutional net. arXiv preprint arXiv:1412.6806 (2014)

29. Szegedy, C., et al.: Going deeper with convolutions. In: Proceedings of the IEEE Conference on Computer Vision and Pattern Recognition (CVPR), pp. 1–9 (2015)

30. Zhou, B., Khosla, A., Lapedriza, A., Oliva, A., Torralba, A.: Learning deep features for discriminative localization. In: Proceedings of the IEEE Conference on Computer Vision and Pattern Recognition (CVPR), pp. 2921–2929 (2016)

Automated Machine Learning for Satellite Data: Integrating Remote Sensing Pre-trained Models into AutoML Systems

Nelly Rosaura Palacios Salinas[1]([✉]), Mitra Baratchi[1][iD], Jan N. van Rijn[1][iD], and Andreas Vollrath[2]

[1] Leiden Institute of Advanced Computer Science,
Leiden University, Leiden, The Netherlands
`n.r.palacios.salinas@umail.leidenuniv.nl`
[2] Phi-Lab, ESA/ESRIN, Frascati, Italy

Abstract. Current AutoML systems have been benchmarked with traditional natural image datasets. Differences between satellite images and natural images (e.g., bit-wise resolution, the number, and type of spectral bands) and lack of labeled satellite images for training models, pose open questions about the applicability of current AutoML systems on satellite data. In this paper, we demonstrate how AutoML can be leveraged for classification tasks on satellite data. Specifically, we deploy the Auto-Keras system for image classification tasks and create two new variants, IMG-AK and RS-AK, for satellite image classification that respectively incorporate transfer learning using models pre-trained with (i) natural images (using ImageNet) and (ii) remote sensing datasets. For evaluation, we compared the performance of these variants against manually designed architectures on a benchmark set of 7 satellite datasets. Our results show that in 71% of the cases the AutoML systems outperformed the best previously proposed model, highlighting the usefulness of a customized satellite data search space in AutoML systems. Our RS-AK variant performed better than IMG-AK for small datasets with a limited amount of training data. Furthermore, it found the best automated model for the datasets composed of near-infrared, green, and red bands.

Keywords: Remote sensing · AutoML · Transfer learning · Classification

1 Introduction

Remote sensing satellites continuously monitor the Earth's surface and collect data representing the state and health of the planet. The range of applications that can benefit from such data varies from environmental mapping to urban planning, emergency response, and many more [3]. To make use of such data, remote sensing practitioners commonly adopt methods of computer vision and

---

A. Vollrath—Affiliated with FAO since 09/2020.

The original version of this chapter was revised: The given name and surname of the author Nelly Rosaura Palacios Salinas have been wrongly attributed in some parts of the original publication. This has now been corrected. The correction to this chapter is available at https://doi.org/10.1007/978-3-030-86517-7_32

Y. Dong et al. (Eds.): ECML PKDD 2021, LNAI 12979, pp. 447–462, 2021.
https://doi.org/10.1007/978-3-030-86517-7_28

machine learning. Classical machine learning approaches benefit from domain-specific, hand-crafted features to account for dependencies in time or space, but rarely exploit spatio-temporal dependencies exhaustively. Modern deep learning methods can automatically extract such spatio-temporal features. However, currently, two obstacles are limiting the use of deep learning for satellite data. The first one is the lack of sufficient labeled data and the difficulty of getting labels considering that satellite images are not as interpretable as natural images for the human eye [3]. The second obstacle lies in the difficulty of designing appropriate architectures that take the characteristics of satellite images into account. Satellite images are different from natural images due to their additional spectral information content. Natural color images always include the same three channels (RGB) but for satellite images, the number and type of channels are variable, depending on the satellite instrument. A multi-spectral satellite image captures information of the electromagnetic spectrum related to different processes on Earth (e.g., land, ocean, atmosphere). The images from the most common satellites can have up to 13 spectral bands that each could be relevant for observing a different process. For instance, examples of channels related to vegetation features are near-infrared and short-wave infrared bands.

Furthermore, natural images have an 8 bits precision, while remote sensing input data usually comes at higher precision (16 or 32 bits). Creating new high-performing models for satellite data requires designing new architectures while taking into account these characteristics. Furthermore, the hyperparameters need to be set properly. These tasks can be complex for remote sensing experts.

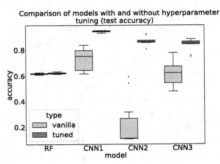

Fig. 1. Preliminary experiments using the EuroSAT dataset [9]. A random forest and three different CNNs built from scratch based on machine learning (a simple CNN with 3 convolutional layers (CNN1)) and remote sensing literature (CNN2 [1], CNN3 [15]) are compared. For each model, two versions are shown: a vanilla model performance using default configurations and a tuned model. The tuned models show the performance after applying hyperparameter tuning for the optimizer, learning rate, batch size, and the number of epochs in the case of the CNNs and the number of features for the random forest.

To overcome these obstacles, we propose to systematically leverage recent developments in two different machine learning fields: (i) transfer learning [22] and (ii) automated machine learning (AutoML) [11]. Transfer learning addresses the lack of labeled data by re-using the knowledge gained from previously seen tasks and transferring it to a newly created model in another task (e.g., through using pre-trained models). AutoML [11] aims to automatically design high-performing models for each dataset in a data-driven manner and thus making machine learning accessible to non-machine learning experts. Hyperparameter optimization and Neural Architecture Search (NAS) are both exemplars of techniques that are scoped in this field. Specifically, with the increased interest in using deep learning algorithms, NAS has become an important area that aims at finding the best neural network architecture given a task and a dataset by automatically tuning various hyperparameters.

As far as we are aware, NAS research systems have been benchmarked with natural image datasets but not with satellite images. This brings us to the questions: *what is the performance of current AutoML systems for satellite data?* and *how can we further improve their performance for satellite data by transferring the knowledge gained from previous research in the field of remote sensing?* Fig. 1 shows the results of one of our preliminary experiments, demonstrating the potential of applying the most recent advances in AutoML regarding hyperparameter optimization to a remote sensing dataset. We know that positive results in specific applications are based on human priors. By incorporating domain expert prior knowledge into machine learning systems the performance of resultant models can significantly improve. Therefore, in this paper, we propose to tailor the neural architecture search space of Auto-Keras [12] (a popular AutoML system) by integrating findings of the remote sensing field in form of pre-trained models on ImageNet and remote sensing datasets.

To the best of our knowledge, this is the first work considering the design of AutoML systems for machine learning tasks based on remote sensing datasets. More specifically, to achieve this goal our contributions in this paper are as follows: (i) composing a diverse benchmark of already available satellite datasets using a standardized format, (ii) evaluating the performance of the deployed AutoML NAS system on these datasets, and finally, (iii) enriching this system by incorporating pre-trained models on remote sensing datasets in a new block called RS-AK.

2   Related Work

In this section, we review the most popular deep learning approaches applied to satellite data and the current status of AutoML in remote sensing.

Deep Learning in Remote Sensing: The remote sensing research community increasingly relies on the use of deep learning models. The authors of [3] indicate that CNN-based methods have obtained impressive results when numerous annotated samples to fine-tune or train a network from scratch are available. Due to the difficulty of acquiring labeled data, researchers typically rely on techniques from transfer learning, with models pre-trained on natural image datasets (e.g., ImageNet) but also remote sensing benchmark datasets (e.g., EuroSat [9], BigEarthNet [25]). Some works that rely on this technique are [9,16,18,23]. The authors of [19] analyze three different transfer learning strategies to improve the performance of CNNs for satellite image scene classification, i.e., full training, fine-tuning, and using CNNs as feature extractors. They conclude that the fine-tuning approach tends to be a good option in various scenarios. The authors of [9] evaluated various CNN architectures on the EuroSAT dataset, achieving the best accuracy using a fine-tuned ResNet-50 pre-trained on ImageNet for the RGB data. The authors of [19,30] reported high-performance results using CNNs too. The authors of [15] suggest that an ensemble of Inception and ResNet modules is an effective architecture for land cover classification. Current remote sensing research does not fully exploit hyperparameter tuning to further improve these models; researchers have mainly considered optimizing a subset of hyperparameters using a parameter sweep approach [6,18]. The authors of [13] have considered AutoML for a specific application of high-throughput image-based plant phenotyping. They use Auto-Keras and compare its results with human-designed ImageNet pre-trained CNN architectures, finding the best performance while using the pre-trained network. However, they did not use all the potential of Auto-Keras. In this paper, we consider more general applicability by performing a systematic analysis on a diverse benchmark of problems and we propose the customization of Auto-Keras for satellite tasks. Moreover, we see that many architectures have been applied to remote sensing problems, but no clear consensus has been reached about which one works best. This makes a compelling argument for using AutoML, which can explore and select the best option in a data-driven way.

AutoML and Neural Architecture Search: AutoML aims to automate the different stages of a machine learning pipeline. These steps typically are data collection, data preparation, feature engineering, preprocessing, algorithm selection, hyperparameter optimization, model training, and deployment. Current AutoML systems commonly cover stages from data preparation to model training [11]. Auto-Sklearn [7], Auto-WEKA [26] and T-POT [20] are examples of AutoML systems focusing on traditional machine learning (such as SVM, random forest, K-nearest neighbors). So far, only a few open-source AutoML systems focus on deep learning. One of the biggest challenges of NAS compared to previously mentioned AutoML systems is maintaining computational efficiency. The time required to successfully solve the NAS problem is linked to the time needed to train a candidate network and the number of candidates existing in the search space. Two popular AutoML systems that focus on deep learning are Auto-Keras [12] and Auto-Pytorch [17], both supporting image classification tasks. Auto-Pytorch uses multi-fidelity optimization and Bayesian optimization (BOHB) [5] while Auto-Keras uses a Bayesian optimization with a neural network kernel and a tree-structured acquisition function to search for the best settings. The search space of Auto-Keras is defined based on network morphism, it encloses all architectures that can be created by morphing the initial architecture. Auto-Pytorch is delimited to multi-layer perceptron networks and funnel-shaped residual networks. To deal with the memory limitations, Auto-Pytorch asks the user to choose between small, medium, and full configuration spaces, whereas Auto-Keras adapts the configuration space automatically based on a memory estimation function. Both systems have focused on solving traditional machine learning tasks, in the case of image classification the attention is only on natural images. Our goal in this paper is to focus on Earth observation data. We propose to customize AutoML systems for satellite data tasks. The challenge of adapting NAS for specific problems falls into a right delimitation of the search space. By doing this, the remote sensing practitioners can reduce the amount of time needed to find a suitable model for their data and instead focus on other major tasks.

3   Methodology

To discover automatically generated high-performance architectures for satellite data classification tasks, we integrate the deep learning solutions for remote sensing in an AutoML framework. We propose to increase the efficiency of AutoML systems by reducing the complexity of the search space focusing on the most likely well-performance architectures for satellite data tasks.

We selected one of the deployed AutoML systems to build upon. We select Auto-Keras [12], an efficient NAS with network morphism, where Bayesian optimization is used to guide the exploration of the search space. The search space of Auto-Keras is based on network morphism, enclosing all architectures that can be created by morphing the initial architecture. The generation of the candidate architectures depends on the acquisition function of the Bayesian optimization. As the NAS space is not Euclidean, Auto-Keras uses an edit-distance neural network kernel for the Gaussian Process. This kernel measures the number of operations needed to morph one network into another one. It considers morphing the layers and the skip-connections. Different from fixed layer width methods [14], the morphism operations can change the number of filters in a convolutional layer and then make the search space larger. Therefore, finding a good architecture could take more time. By focusing on the most likely well-performance architectures for specific tasks, the searching time would be reduced.

To measure the benefits of the development of specific tasks for satellite data, we decided to gradually enhance the search space of the system and proposed three different settings for our experiments. Those settings and the motivation behind them are explained in the following subsections.

3.1   Original Auto-Keras System (V-AK)

Auto-Keras search space is built upon network morphism where the search space of NAS is created using morphism operations. An initial network architecture G is given and, with the use of morphism operations, new networks are created. Auto-Keras' authors use a three-layer convolutional network as starting architecture for the experiments presented in their paper to test the efficiency of their approach compared with other methods. However, the deployed Auto-Keras system has a task-oriented API, in which 3 different initial architectures are applied for the image classification task: first, it tries a vanilla network with 2 layers, second a ResNet50 model without pre-training, and thirdly an Efficientb7 network pre-trained with ImageNet. This change influences the possible architectures to select and outperforms the system initialized with a three-layer convolutional network. To the best of our knowledge, the selection of the initial architectures was based on human expert knowledge and state-of-the-art architectures for specific tasks based on natural image data.

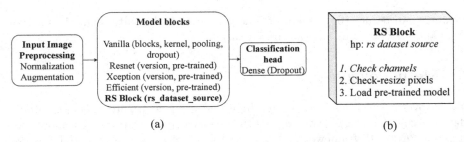

(a)   (b)

Fig. 2. (a) An abstract illustration of how the final architecture can be build based on pre-defined blocks. A network consists of one preprocessing block, several model blocks, and one classification head. V-AK and IMG-AK compose the model by using Vanilla, Resnet, Xception and/or Efficient blocks. In addition to these, RS-AK can make use of the RS Block as well. (b) The RS Block, only available to RS-AK. It can be extracted from various different remote sensing datasets, which is controlled by the hyperparameter *rs_dataset_source*

3.2 Models Pre-trained Using ImageNet Dataset (IMG-AK)

Based on remote sensing research, we know that models pre-trained with ImageNet can lead to promising results for satellite data classification tasks [3]. The Auto-Keras search space already includes blocks with weights acquired by pre-training on ImageNet. However, the decision to use such blocks depends on the process of selecting new candidate architectures. It could be the case that, due to trials budget and the vast search space, these pre-trained architectures are not considered. Figure 2 provides an abstract illustration of how the final architecture for the image classification task can be build based on pre-defined blocks existing in Auto-Keras. The model blocks in which the ImageNet weights are available have a hyperparameter called *pretrained*, which defines whether or not a pre-trained version of the model will be used.

Therefore, in this approach, although we make use of the available pre-trained models in the current systems, we still modify the configuration of G by defining an initial architecture for the new specific task: satellite image classification. We expect to improve the classification results by starting the neural architecture search with a block pre-trained with ImageNet. The model block can be selected based on the remote sensing literature findings. As reviewed in Sect. 2, ResNet architectures have shown promising results in classification of satellite images in the literature (see, e.g., [9,15]). Thus, we configure the initial G with a ResNet block and we set the parameter *pretrained* to *true*.

3.3 Models Pre-trained Using Remote Sensing Datasets (RS-AK)

Transfer learning can be most successful when the source and target domain are similar [4,18,27]. Within the remote sensing community, there are models pre-trained with remote sensing datasets [18,25] but none of these are available yet in AutoML systems. Therefore, we proposed to incorporate this type of pre-trained models and customize the Auto-Keras image classification task for satellite data.

We need to initially decide what needs to be changed in Auto-Keras to be able to add this feature. The Auto-Keras task-oriented NAS approach can be inferred from the open-source deployed system. The image classification task builds an architecture based on pre-defined cells or blocks. These blocks can be divided into three categories: preprocessing, model, and classification head. For the preprocessing category, two blocks are considered: (i) normalization, which performs a feature-wise normalization on the data; and (ii) an image augmentation block, which can apply various methods including flipping, rotation, and translation. The addition of such blocks to the final architecture in Auto-Keras is treated as a hyperparameter. The model blocks represent all the possible cells that will shape the hidden layers of the network. Each block consists of parameterized modules of well-known CNNs with various hyperparameters to be tuned. The third category is the classification head block, which creates the output layer of the network based on the number of classes and the classification type. The only hyperparameter to tune in this block is a dropout value. The preprocessing steps correspond to the ones applied by the authors of our satellite datasets, and the classification head block does not need to be changed because the nature of the classification is the same as any image classification task. We only need to change the model blocks and how our new block (which we refer to as *RS Block*) will interact with the classification head block. Figure 2 is an abstract illustration of this. The *RS Block* first checks the shape of the input and resizes the pixels if necessary. It chooses between different pre-trained module versions (trained with satellite data). This choice is considered as another hyperparameter to tune. Hence, it uses the same hyperparameter tuner that is used for all the other blocks. The optimization method is explained next.

Hyperparameter Tuning. Different tuners can be used to determine which combination of hyperparameters will be sent for training in each trial during NAS. We used an oracle combining random search and greedy algorithm [12,21] presented inside of Auto-Keras. The hyperparameters are arranged by grouping them into several categories according to the level or functionality. The oracle tunes each category separately using random search. In each trial, it uses a greedy strategy to generate new values for one of the categories of hyperparameters and use the best trial so far for the rest.

Remote Sensing Pre-trained Models. Our *RS Block* is composed of modules of different satellite learning representations acquired from different pre-trained models. These pre-trained models were trained with 5 different satellite datasets (BigEarthNet [25], EuroSAT [9], RESISC-45 [2], So2Sat [30], and UC Merced [28]). Based on the number of spectral bands of the collected datasets we considered two types of pre-trained models: (i) 3-channels and (ii) 13-channels. Figure 3 shows the architectures and datasets used for pre-training. The 3-channel pre-trained models were taken from the publicly available models posted by [18]. Inspired by the findings in [15] and the selected architecture in [18], we decided to create in-domain representations for 13-channels datasets using ResNet architectures and training with the EuroSAT dataset [9].

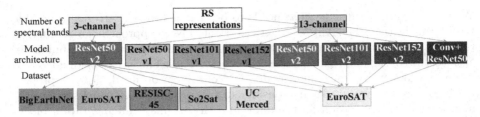

Fig. 3. Remote sensing pre-trained models considered for the RS Block. The first layer indicates the number of channels, the second layer the architecture used, and, the third layer the remote sensing dataset used. 3-channels models were created by Google Research [18], 13-channels were created by us.

To rapidly test the performance of our new block, we made two changes in the Auto-Keras search space. We first added the proposed *RS Block* to the model blocks structure. Secondly, we adapted the initial architecture G to start with our new remote sensing block. We would like to be able to study which of the remote sensing representations (pre-trained blocks) are used more often and, thus are more promising. We can inspect this, by studying the *rs_dataset_source* parameter of the *RS Block*, which indicates the source dataset used for pre-training in the case of the 3-channel datasets.

4   Experiments

In our evaluation we aim to address the following research questions:

- **Q1.** Can we achieve a performance similar to the non-automated deep learning research in remote sensing by using AutoML systems?
- **Q2.** How do different Auto-Keras variants perform for datasets with different characteristics (different number of spectral bands, sizes, and class distributions)?
- **Q3.** Which of the remote sensing pre-trained modules used in the RS-AK shows more promising results for developing NAS systems for remote sensing?

4.1   Datasets

To have a broader idea of the applicability of this framework in the remote sensing field, we have composed a benchmark of 7 diverse and well-known multi-spectral satellite datasets. Furthermore, this selection shows a variety of classification tasks with presumably different degrees of difficulty and complexity. Table 1 presents the characteristics of these datasets and summarizes the approach taken by their corresponding authors, as well as its performance. Except for the EuroSAT, So2Sat, and UC Merced datasets, the performance and approach showed in this table is the state-of-the-art (SOTA) considered for our experiments. For the case of these 3 datasets, better results are reported by the Google research

Table 1. Overview of available labelled datasets and the presented approach and performance from the paper in which the dataset was introduced.

Dataset	Satellite (Bands)	Resolution	Images	Labels	No.	Perf (%)	Approach
BigEarthNet	Sentinel-2 (3/12)	Med-high	590k (L)	Land	43	67.59	CNN 3-Conv[25]
BrazilDam	Sentinel-2 (13)	High	1.92k (S)	Dam?	2	94.1	DenseNet [6]
Brazilian Coffee	SPOT (3)	High	2.87k (S)	Coffee?	2	83.04	2 OverFeat networks [23]
Cerrado-Savanna	RapidEye (3)	High	1.31k (S)	Veg.	4	90.5	Fine-tuning AlexNet [19]
EuroSAT	Sentinel-2 (3/13)	High	27k (L)	Land	10	98.57	Pre-trained ResNet [9]
So2Sat	Sentinel-2 (3)	High	376k (L)	Land	17	61	ResNet [30]
UC Merced	USGS(3)	Very high	2.1k (S)	Land	21	NA	BoVW [28]

team in [18] using ResNet models pre-trained with remote sensing datasets; thus their results are the SOTA in Table 2.

The use of bands different from the RGB spectrum is a common practice in remote sensing applications due to the additional information that can be extracted from other spectral bands. A clear example is the creation of vegetation indexes for different applications; such indexes involve non-RGB channels like near-infrared. The number of samples available for training in remote sensing real-world problems is usually small. The Coffee scenes, BrazilDam, and Cerrado-Savanna datasets meet these characteristics. The Cerrado-Savanna scenes [19] is one of the most challenging datasets for classification. As explained by the authors, this is due to the high intraclass variance of the dataset, caused by different spatial configurations and densities of the same vegetation type, as well as its high inter-class similarity, caused by the similar appearance of different vegetation species [19]. Moreover, from 1,311 samples included in this dataset, 73% correspond to the Arboreal vegetation.

4.2 Experimental Setup

For all our experiments, the datasets were first randomly divided into train and test sets. The test set was created by reserving 20% of all the available data from Eurosat, BigEarthNet, So2Sat, and UC Merced datasets. In the case of the Brazil-Dam dataset, only the Sentinel fold from 2019 was extracted to study. The Coffee scenes and Savanna datasets are originally divided into five folds. The first four were used for training and the last fold is reserved for testing. Next, another split of 80-20 was applied to the training set, assigning 20% of it for validation, which is used for the AutoML system to tune hyperparameters and select the best model. As most of the datasets are also used as a source for creating pre-trained models, when evaluating RS-AK, we should be careful not to include pre-trained blocks from the dataset that we want to use to test on, to avoid being exposed to labels from the test set. As such, when evaluating on a given dataset, we remove the pre-trained blocks coming from this dataset from the search space. To exclude the corresponding dataset, before running the task, we keep out this option from the set of pre-trained models available for the $rs\_dataset\_source$ hyperparameter in the *RS Block*.

To be able to show the significance of the results we performed a Wilcoxon signed-rank test, first ensuring that the data was not normally distributed and considering a p value of 0.05. The outcomes presented in this paper are based on the 10 trials experiments. Each trial, varying per dataset, ranges from few minutes to around 6 h. All the experiments were run on a compute cluster using nodes with 4 GPUs (PNY GeForce RTX 2080TI). We delimited the memory to 32 and 64 GB for the experiments. For better reproducibility, we have made the source code of our experiments available in a public repository.[1]

5   Results

In this section we will answer the research questions that were stated in Sect. 4.

5.1   AutoML vs Non-automated Models

Table 2 summarizes the performance of the three different AutoML approaches on the test set for the different datasets. The performance metric shown here, same as the baseline papers, is the overall classification accuracy. For the BigEarthNet-rgb dataset, we decided to change the performance metric to be able to compare with the baseline. We achieved an F1-score of 67.84% using an ImageNet pre-trained module, while the result presented in [25] is 67.59%. There is no benchmark performance available for the full spectral version of EuroSAT. Resultant of our experiments, we established one with 97.8% overall accuracy.

To answer **Q1** we grouped the results of the three variants (V-AK, IMG-AK, and RS-AK) and we took the maximum performance. In this way, we can analyze the AutoML competency against the non-automated architectures. We outperformed the literature in 5 out of 7 datasets, improving the state-of-the-art result for So2Sat by a rate of 34.5%. Therefore, we can conclude that the performance found by using AutoML systems can be competitive and even better for some of these datasets.

5.2   AutoML Variants and the Different Type of Datasets

To address **Q2**, we group our datasets based on size, number, and type of spectral bands (channels). We consider four small datasets. We have 2 datasets with 13 channels (BrazilDam and EuroSAT-all) and 6 with 3 channels. The 3-channels are either RGB bands or near-infrared, green, and red bands. Note that the EuroSAT-all dataset has an empty entry for the SOTA and RS-AK approach. Since the pre-trained blocks from the 13-band dataset come all from the EuroSAT-all dataset, we could not fairly deploy this model (see experimental setup). To facilitate comparisons with the SOTA found in literature and among our experiments, in Table 2 the boldfaced entries indicate the best approach among the 3 Auto-Keras variants.

---

[1] https://github.com/palaciosnrps/automl-rs-project.

Table 2. Performance on test dataset considering 10 runs (Except for BigEarthNet which had 3 runs) of each of our experiments and the state-of-the-art (SOTA) found in literature for each dataset. BigEarthNet performance metric is F1-score, all the other datasets use overall accuracy. An asterisk (*) represents statistically significant results.

Dataset	Type	SOTA	V-AK	IMG-AK	RS-AK
BrazilDam	Small-13	94.1 [6]	**89.09 ± .05**	76.54 ± .13	85.57 ± .01
Coffee scenes	Small-3	83.4 [23]	86.18 ± .02	82.96 ± .04	**88.84 ± .00***
Cerrado-Savanna	Small-3	90.5 [19]	85.79 ± .01	84.33 ± .03	**89.92 ± .01**
UCMerced	Small-rgb	99.61 [18]	**99.62 ± .00**	76.43 ± .13	91.19 ± .06
EuroSAT-all	Large-13	–	95.38 ± .02	**97.82 ± .00***	–
EuroSAT-rgb	Large-rgb	99.2 [18]	99.18 ± .00	**99.54 ± .00***	95.90 ± .01
So2Sat-rgb	Large-rgb	63.25 [18]	95.47 ± .00	**97.80 ± .00***	76.92± .00
BigEarthNet-rgb	Large-rgb	67.59 [25]	50.62 ± .00	**67.84 ± .00**	65.29 ± .00

If the results are statistically significant to both other approaches according to the Wilcoxon Signed rank test the entry is marked with an asterisk (*). Please note that the paired comparison of second-best approaches is not shown in the table.

The original Auto-Keras V-AK and the IMG-AK version performed well on the EuroSAT-all dataset. In this case, IMG-AK performs better than V-AK. For the case of the BrazilDam dataset, the initialization with a pre-trained ImageNet model did not benefit the performance (see IMG-AK Table 2) and it even decreased the average accuracy. This can be explained considering the difference in the number of input channels (increasing the complexity) and the size of the dataset. BrazilDam dataset has 13 channels; therefore, the direct use of pre-trained models from ImageNet (3-channel) does not apply. Different from EuroSAT, the number of labeled samples of BrazilDam is small. We can notice an improvement using RS-AK but this is not enough to beat the baselines.

We can see that for the RGB channel datasets either V-AK or IMG-AK approaches lead to the best performance. We achieved a large improvement for the So2Sat-rgb dataset, compared to the work presented in [18]. Even though the authors of [18] also used pre-trained models, the variety of model versions and the more sophisticated hyperparameter tuning method provided by the AutoML systems played an important role in achieving better performance for this dataset. Conversely, the RS-AK variant obtained the best results for the Coffee scenes and Cerrado-Savanna datasets. These two datasets are composed of near-infrared, green, and red bands and the classification task differs from land cover identification. Based on that, we can infer that the 3-channel remote sensing representations are an option for transfer learning when the target dataset is different from the well-known RGB channel datasets. In the case of the 13-channel representations used for the BrazilDam dataset, the results were not as successful as what was obtained by manually designed architectures. The best-automated model generated using the original Auto-Keras consists of convolutional blocks without pre-trained modules, suggesting that for this dataset training from scratch rather than using the available pre-trained models is a better approach. Based on the results of the non-RGB datasets, we can expect that improving the 13-channel representations could lead us to better performance.

Considering the dataset size, we notice that comparing the initialization of G with ImageNet pre-trained models (IMG-AK) versus the implementation of remote sensing pre-trained models (RS-AK), RS-AK gives better performance for the small datasets. Meanwhile, IMG-AK consistently results in better performance for large datasets. This could be explained by (i) the amount of data available for pre-training and (ii) the degree of similarity between the target and source domains that both determine the quality of the transfer-learning technique [24,27,29]. Bigger datasets should produce better representations. But data similarity also needs to be taken into account. It is possible that for the classes represented in the small datasets the current remote sensing representations are enough and the best performance is acquired, as the domain source is similar. However, in the case of the large datasets the quality of the representations generated with the ImageNet dataset (being over 2 times bigger than the BigEarthNet dataset) gain over the domain similarity. To improve the performance of classification for the bigger datasets using RS-AK, more studies are needed and some of those should investigate different fine-tuning strategies and improving the performance of the BigEarthNet representation, which so far is the most promising one.

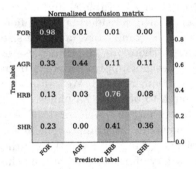

(a) Confusion matrix for the Cerrado-Savanna dataset using a pre-trained remote sensing block.

(b) Confusion matrix, Cerrado-Savanna dataset using only convolutional blocks (no pre-trained versions).

	FOR	AGR	HRB	SHR
Precision	0.94	0.57	0.71	0.67
Recall	0.98	0.44	0.76	0.36

	FOR	AGR	HRB	SHR
Precision	0.90	0.00	0.61	0.83
Recall	0.99	0.00	0.71	0.23

Fig. 4. Comparison of confusion matrices for Cerrado-Savanna dataset. Classes are Agriculture (AGR), Arboreal Vegetation (FOR), Herbaceous Vegetation (HRB) and Shrubby Vegetation (SHR).

The overall accuracy only gives a general idea of the performance, for datasets in which the samples per class are not balanced we need to look with more detail into the performance achieved for each class to know if there is still any room for improvement. We generate confusion matrices to inspect the performance in more detail. Figure 4a is the confusion matrix of the best model found for the Cerrado-Savanna dataset by using RS-AK. The classes with originally more samples (FOR, HRB) are the classes with better performance. For the SHR and AGR classes, the misclassification is still high. However, while comparing with the results given by using a non-pre-trained model obtained with V-AK (Fig. 4b), we can appreciate a big improvement of 13% and 44% in the less representative classes (SHR, AGR) acquired by the use of pre-trained blocks.

Table 3 summarizes the findings of the confusion matrices for datasets with a major difference in the distribution of class samples. To measure the impact of pre-trained blocks, in this table, we compare the performance achieved for the minority and majority classes, with and without pre-training. We notice that while using pre-trained blocks, the recall of the least representative classes in all datasets increases between 7% and 44 % while the values for majority classes slightly decrease between 1% and 9%. However, the overall accuracy is impacted more by the majority class, ignoring the large improvements on the minority classes. For remote sensing applications in which the class distribution is non-balanced, this improvement for the minority class is important.

5.3   The Remote Sensing Block RS-AK

In this section, we aim to address **Q3**. Figure 5 shows the frequency at which each source model was selected as part of the customized block for each dataset. For the Savanna Cerrado, Coffee scenes, and So2Sat datasets the most chosen pre-trained model was BigEarthNet. So2Sat was the most selected model in the case of UC Merced dataset and it tied with BigEarthNet for the EuroSAT dataset. These results are expected due to the big size of the datasets but differ from the findings of [18] who conclude that the RESISC-45 representation achieves the highest performance. We found the RESISC-45 representation to achieve the best results only when used for classification on the BigEarthNet dataset. Our experiments differ in the way we are using a more efficient framework for tuning a large set of possible hyperparameters (including learning rate, optimizer, regularization, preprocessing) and selecting the design choices using an oracle combining random search and a greedy algorithm (explained in Sect. 3.3) while the authors of [18] optimize by sweeping only a fixed set of hyperparameters (learning rate, weight decay, training schedules, preprocessing). The authors of [18] utilized the same ResNet50V2 architecture [8] to fine-tune the remote sensing datasets using SGD with momentum set to 0.9, in our approach the pre-trained model is only a block that is part of the full architecture (see Fig. 2). In [18], the comparison of the different pre-trained models was made after finishing the fine-tuning using partial (100, 1000) and full training samples; in our study, the selection of the best-performed model was based on the validation set inside the Auto-Keras framework. Considering that, we believe that our experiments have exploited the potential of each

Table 3. Recall value of the classes with most and least samples for the non-balanced datasets.

Dataset	Class	Non-pre	Pre-trained
Cerrado	Majority (73.6%)	0.99	0.98
savanna	Minority (3.4%)	0.00	0.44
So2Sat	Majority (12.3%)	0.95	0.99
	Minority (0.6%)	0.76	0.94
BrazilDam	Majority (57.9%)	0.95	0.86
	Minority (42.1%)	0.78	0.85

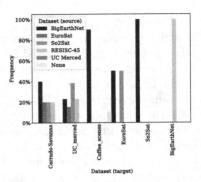

Fig. 5. Remote sensing pre-trained models selected for the 3-channels datasets during the 3rd experiment.

dataset representation by using a more sophisticated framework for the design of the architecture and the hyperparameter tuning; moreover, our results are consistent with the expectations of the remote sensing community about the promising applications of BigEarthNet on remote sensing tasks [25].

6   Conclusions and Future Work

We demonstrated how AutoML can be used to leverage the implementation of deep learning models for satellite data tasks, outperforming some state-of-the-art research results. We focused on classification tasks for multi-spectral satellite datasets. We assessed the performance of the original Auto-Keras [12] (V-AK) and modified its search space to create two different variants of its image classification task: (i) initializing the architecture to morph with a model pre-trained on ImageNet (IMG-AK) and (ii) adding models pre-trained on well-known remote sensing datasets (RS-AK) such as BigEarthNet and UC Merced. Our experimental results on a varied selection of satellite datasets showed that for 3-channel datasets, current AutoML systems can beat state-of-the-art results for land cover classification tasks. Analyzing the performance of the two Auto-Keras variants initialized with pre-trained blocks (IMG-AK and RS-AK), we noticed that RS-AK performed better for small datasets meanwhile IMG-AK was best for relatively large datasets. Moreover, we showed that these pre-trained versions exhibit superior performance on minority classes. The use of bands different from RGB is a common practice in remote sensing due to the extra spectral information that can be extracted from such bands. Besides, the amount of samples available for training in remote sensing real-world problems is often small. Our remote sensing block achieved the best results in such situations. This highlights the usefulness of a customized satellite data search space in AutoML systems for real-world datasets. The 13-channel pre-trained models can be downloaded and used for other remote sensing tasks; due to the number of channels these models are useful when working with Sentinel-2 satellite images. There is still room for improvement in such remote sensing representations. In future work, we will first aim at improving the transferability of the remote sensing pre-trained models and work on covering the widely used image segmentation task. A more sophisticated transfer learning method, deep meta-learning [10], or customized techniques per dataset & task (based on [24,29]) integrated into AutoML systems could improve the usage of remote sensing data representations. Based on our experiments, we recommend the remote sensing practitioners to make use of the existing open-source AutoML tools. By making this framework publicly available, we enable the community to further experiment with relevant remote sensing datasets and expect to expand the use of AutoML for different applications.

Acknowledgement. This work is partially supported by PAME, a Dutch Research Council (NWO) project under grant number OCENW.KLEIN.425 and it has been performed using the ALICE & GRACE compute resources provided by Leiden University.

References

1. Basu, S., et al.: DeepSat: a learning framework for satellite imagery. In: 23rd SIGSPATIAL International Conference on Advances in Geographic Information Systems (2015)
2. Cheng, G., et al.: Remote sensing image scene classification: benchmark and state of the art. Proc. IEEE **105**(10), 1865–1883 (2017)
3. Cheng, G., et al.: Remote sensing image scene classification meets deep learning: challenges, methods, benchmarks, and opportunities. IEEE J. Sel. Topics Appl. Earth Observations Remote Sens. **13**, 3735–3756 (2020)
4. Cui, Y., et al.: Large scale fine-grained categorization and domain-specific transfer learning. In: Proceedings of the IEEE Conference on Computer Vision and Pattern Recognition, pp. 4109–4118 (2018)
5. Falkner, S., et al.: BOHB: robust and efficient hyperparameter optimization at scale. In: 35th International Conference on Machine Learning, pp. 2323–2341 (2018)
6. Ferreira, E., et al.: BrazilDam: a benchmark dataset for tailings dam detection. In: Latin American GRSS & ISPRS Remote Sensing Conference, pp. 339–344 (2020)
7. Feurer, M., et al.: Efficient and robust automated machine learning. In: Advances in Neural Information Processing Systems, vol. 28, pp. 2962–2970 (2015)
8. He, K., Zhang, X., Ren, S., Sun, J.: Identity mappings in deep residual networks. In: Leibe, B., Matas, J., Sebe, N., Welling, M. (eds.) ECCV 2016. LNCS, vol. 9908, pp. 630–645. Springer, Cham (2016). https://doi.org/10.1007/978-3-319-46493-0_38
9. Helber, P., et al.: EuroSat: a novel dataset and deep learning benchmark for land use and land cover classification. IEEE J. Sel. Topics Appl. Earth Observations Remote Sens. **12**(7), 2217–2226 (2019)
10. Huisman, M., van Rijn, J.N., Plaat, A.: A survey of deep meta-learning. Artif. Intell. Rev. 1–59 (2021). https://doi.org/10.1007/s10462-021-10004-4
11. Hutter, F., et al.: Automated machine learning: Methods, systems, challenges. Springer, Cham (2018). https://doi.org/10.1007/978-3-030-05318-5
12. Jin, H., et al.: Auto-keras: an efficient neural architecture search system. In: Proceedings of the 25th ACM SIGKDD International Conference on Knowledge Discovery & Data Mining, pp. 1946–1956 (2019)

13. Koh, J.C., et al.: Automated machine learning for high-throughput image-based plant phenotyping. Remote Sens. **13**, 858 (2021)
14. Liu, H., et al.: DARTS: differentiable architecture search. In: Proceedings of ICLR (2019)
15. Mahdianpari, M., et al.: Very deep convolutional neural networks for complex land cover mapping using multispectral remote sensing imagery. Remote Sens. **10**(7), 1119 (2018)
16. Marmanis, D., et al.: Deep learning earth observation classification using ImageNet pretrained networks. IEEE Geosci. Remote Sens. Lett. **13**(1), 105–109 (2016)
17. Mendoza, H., et al.: Towards automatically-tuned deep neural networks. In: AutoML: Methods, Systems, Challenges, vol. 7, pp. 141–156 (2018)
18. Neumann, M., et al.: Training general representations for remote sensing using in-domain knowledge. In: 2020 IEEE International Geoscience and Remote Sensing Symposium (2020)
19. Nogueira, K., et al.: Towards vegetation species discrimination by using data-driven descriptors. In: 9th IAPR Workshop on Pattern Recogniton in Remote Sensing, pp. 1–6(2016)
20. Olson, R., et al.: Evaluation of a tree-based pipeline optimization tool for automating data science. In: Proceedings of GECCO 2016, pp. 485–492 (2016)
21. O'Malley, T., et al.: Keras Tuner (2019). https://github.com/keras-team/keras-tuner
22. Pan, S.J., Yang, Q.: A survey on transfer learning. IEEE Trans. Knowl. Data Eng. **22**(10), 1345–1359 (2010)
23. Penatti, O.A., et al.: Do deep features generalize from everyday objects to remote sensing and aerial scenes domains? In: IEEE Conference on Computer Vision and Pattern Recognition Workshops, pp. 44–51 (2015)
24. Soekhoe, D., van der Putten, P., Plaat, A.: On the impact of data set size in transfer learning using deep neural networks. In: Boström, H., Knobbe, A., Soares, C., Papapetrou, P. (eds.) IDA 2016. LNCS, vol. 9897, pp. 50–60. Springer, Cham (2016). https://doi.org/10.1007/978-3-319-46349-0_5
25. Sumbul, G., et al.: BigEarthNet: a large-scale benchmark archive for remote sensing image understanding. In: IEEE International Geoscience and Remote Sensing Symposium, pp. 5901–5904 (2019)
26. Thornton, C., et al.: Auto-WEKA: combined selection and hyperparameter optimization of classification algorithms. In: Proceedings of the 19th ACM SIGKDD International Conference on Knowledge Discovery & Data Mining, pp. 847–855 (2013)
27. Yang, F., et al.: Transfer learning strategies for deep learning-based PHM algorithms. Appl. Sci. **10**(7), 2361 (2020)
28. Yang, Y., Newsam, S.: Bag-of-visual-words and spatial extensions for land-use classification. In: 18th SIGSPATIAL International Conference on Advances in Geographic Information Systems, pp. 270–279 (2010)
29. Yosinski, J., et al.: How transferable are features in deep neural networks? In: Advances in Neural Information Processing Systems, vol. 27, pp. 3320–3328 (2014)
30. Zhu, X., et al.: So2Sat LCZ42: a benchmark data set for the classification of global local climate zones. IEEE Geosci. Remote Sens. Mag. **8**(3), 76–89 (2018)

Multi-task Learning for User Engagement and Adoption in Live Video Streaming Events

Stefanos Antaris[1,2](✉), Dimitrios Rafailidis[3], and Romina Arriaza[2]

[1] KTH Royal Institute of Technology, Stockholm, Sweden
antaris@kth.se
[2] Hive Streaming AB, Stockholm, Sweden
{stefanos.antaris,romina.arriaza}@hivestreaming.com
[3] University of Thessaly, Volos, Greece
draf@uth.gr

Abstract. Nowadays, live video streaming events have become a mainstay in viewer's communication in large international enterprises. Provided that viewers are distributed worldwide, the main challenge resides on how to schedule the optimal event's time so as to improve both the viewer's engagement and adoption. In this paper we present a multi-task deep reinforcement learning model to select the time of a live video streaming event, aiming to optimize the viewer's engagement and adoption at the same time. We consider the engagement and adoption of the viewers as independent tasks and formulate a unified loss function to learn a common policy. In addition, we account for the fact that each task might have different contribution to the training strategy of the agent. Therefore, to determine the contribution of each task to the agent's training, we design a Transformer's architecture for the state-action transitions of each task. We evaluate our proposed model on four real-world datasets, generated by the live video streaming events of four large enterprises spanning from January 2019 until March 2021. Our experiments demonstrate the effectiveness of the proposed model when compared with several state-of-the-art strategies. For reproduction purposes, our evaluation datasets and implementation are publicly available at https://github.com/stefanosantaris/merlin.

Keywords: Multi-task learning · Reinforcement learning · Live video streaming

1 Introduction

Over the last years, video streaming technologies have been widely exploited by large international enterprises as the main internal communication medium [3]. The enterprises schedule several live video streaming events to communicate with thousands of their employees, who are spread around the world. To ensure that

© Springer Nature Switzerland AG 2021
Y. Dong et al. (Eds.): ECML PKDD 2021, LNAI 12979, pp. 463–478, 2021.
https://doi.org/10.1007/978-3-030-86517-7_29

every employee/viewer attends the event without experiencing poor network performance, the enterprises exploit distributed live video streaming solutions. Such solutions account for each office's internal bandwidth to overcome network congestion and distribute the streaming video to viewers [4]. Although distributed solutions ensure that every viewer can attend the event, an erroneously scheduled time of an event negatively affects the viewer's engagement, that is the percentage of the event's duration that a viewer attends [1]. In practice, the viewers partially attend the entire duration of an event, when an event is erroneously scheduled on a non-preferred time e.g., day and hour, resulting in a low viewer's engagement. Moreover, the erroneously scheduled time impacts the number of enterprise's events that each viewer participates, reflecting on the viewer's adoption. In particular, the viewers with several time zones have low adoption, when organizing the events and ignoring the viewer's availability. Instead of manually organizing the events, it is important for the enterprises to develop a mechanism to learn how to schedule an event on the day and hour that optimizes both the viewer's engagement and adoption.

To organize an event, enterprises interact with a centralized agent that is located in a company offering the live video streaming solution. However, current streaming solutions do not account for the optimal selection of the time of the next event. To overcome the shortcomings of current live video streaming solutions, in this study we follow a reinforcement learning strategy and design an agent that receives the viewer's engagement and adoption as two different reward signals for the selection of the event's time. Reinforcement learning has been proven an efficient means for optimizing a reward signal in various domains such as robotics [18,28], games [19,27], recommendation systems [14,26], and so on. However, such approaches train an agent on a single task, where the learned policy maximizes a single cumulative reward. Nonetheless, the goal of the agent in our case of the event's time selection problem is to optimize both the viewer's engagement and adoption rewards. Recently, multi-task reinforcement learning approaches have been proposed to generate a single agent that learns a policy which optimizes multiple tasks, with each task corresponding to a different reward signal [8,11,23]. State-of-the-art approaches train an agent by sharing knowledge among similar tasks [25]. For example, the attentive multi-task deep reinforcement learning (AMT) model [5] exploits a soft-attention mechanism to train a single agent on tasks that follow different distributions in the reward signal. However, AMT transfers knowledge among similar tasks, while isolating dissimilar tasks during the agent's training. This means that AMT achieves sub-optimal performance when tasks have completely different characteristics, as it happens in the case of live video streaming events. For instance, as we will demonstrate in Sect. 2 the viewers have a low engagement behavior over time, whereas the viewer's adoption increases among consecutive events.

In addition, to efficiently select the event's time, the agent has to capture the evolution of the viewer's engagement and adoption. Towards this aim, the Transformer's architecture has been emerged as a state-of-the-art learning model across a wide variety of evolving tasks [24]. For example, in [17] the Transformer's architecture has been exploited in a reinforcement learning strategy to provide memory to the agent by preserving the sequence of the past observations. However, baseline approaches based on the Transformer's architecture have not been studied for multi-task reinforcement learning problems.

To address the shortcomings of state-of-the-art strategies, in this study we propose a **M**ulti-task l**EaR**ning model for user engagement and adoption in **Li**ve v**I**deo streami**N**g events (MERLIN), making the following contributions:

- We formulate the viewer's engagement and adoption tasks as different Markov Decision Processes (MDPs) and propose a multi-task reinforcement learning strategy to train an agent that selects the optimal time, that is day and hour of the enterprise's next event aiming to maximize both tasks.
- We design a Transformer's architecture to weigh the importance of each task during the training of the agent, that is to determine the contribution of each task to the learning strategy of the agent's policy.
- We transfer knowledge among tasks through a joint loss function in a multi-task learner component and compute a common policy that optimizes both the viewer's engagement and adoption in a live video streaming event.

Our experimental evaluation on four real-world datasets with live video streaming events show the superiority of the proposed MERLIN model over baseline multi-task reinforcement learning strategies. The remainder of this paper is organized as follows, in Sect. 2 we present the main characteristics of the live video streaming events as well as the evolution of the viewer's engagement and adoption. In Sect. 3 we formally define the multi-task problem of scheduling live video streaming events, and detail the proposed MERLIN model. Then, in Sect. 4 we present the experimental evaluation of our model against baseline strategies, and conclude the study in Sect. 5.

2   Live Video Streaming Events

Table 1. Statistics of the datasets with all the live video streaming events that took place in four international enterprises from January 2019 until March 2021.

	Enterprise 1 (E1)	Enterprise 2 (E2)	Enterprise 3 (E3)	Enterprise 4 (E4)
#Events	833	1, 303	3, 025	7, 249
#Viewers	98, 296	59, 090	194, 026	508, 654
#Countries	63	97	167	150
#Time zones	5	12	19	22
Avg. Engagement (u_t)	0.455	0.422	0.383	0.409
Avg. Adoption (v_t)	1.275	6.905	8.528	6.375

We collected four real-world datasets with all the events that occurred in four large enterprises worldwide from January 2019 until March 2021. The video

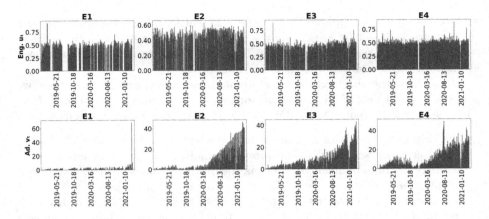

Fig. 1. Evolution of viewer's engagement u_t and adoption v_t in the events.

streaming solution of the events was supported by our company. We monitored a set $\mathcal{E}$ of live video streaming events, where for each event $e_t \in \mathcal{E}$ on date t the viewers reported to a backend server of our company the timezones, as well as their joining and leaving times during the event. The datasets were anonymized and made publicly available. In Table 1, we summarize the statistics of the four evaluation datasets. Each enterprise has a different number of viewers, located in several countries around the world with different time zones. We observe that the viewers in Enterprise 1 are distributed to less time zones than the other enterprises, whereas Enterprise 4 hosts the largest number of live video streaming events with approximately 0.5M viewers in total. In Fig. 1, we present the average viewer's engagement to the live video streaming events throughout the time span. We define the average engagement u_t of the viewers that participated in the event $e_t \in \mathcal{E}$ on the date t as follows:

$$u_t = \frac{1}{n} \sum_{i=1}^{n} \frac{k_i}{m} \qquad (1)$$

where n is the number of viewers that participated in the event e_t, k_i is each viewer's attendance time and m is the duration of the event. In all enterprises the viewers have low engagement, that is in all enterprises the viewers attended less than the half duration of each live video streaming event with average viewer's engagement $u_t < 0.5$ (Table 1). In addition, the average viewer's adoption expresses how many events the viewers attended until a date t, where large adoption scores indicate that viewers were willing to participate in the enterprise's previous events. We formally define the average adoption v_t as follows:

$$v_t = \frac{\sum_{i=1}^{n} c_i}{n} \qquad (2)$$

where c_i is the number of events that each viewer i attended prior to the event e_t. We observe that the viewers in Enterprise 1 adopted less events than the other enterprises with average adoption $v_t = 1.275$. On one of the last dates Enterprise 1 organized an all-hands event where all the viewers were invited, which explains the pick of the adoption score for Enterprise 1 in Fig. 1. The adoption scores for Enterprises 2, 3 and 4 increase over time in the last year, as enterprises started to organize more events than the previous years for viewers who most of them worked from home due to the COVID'19 pandemic.

3 Proposed Model

An enterprise organizes $T = |\mathcal{E}|$ events, where each event on a date/step t is defined as $e_t = (h, n, m, u_t, v_t, \mathbf{z})$, with h being a timestamp that corresponds to the event's day and hour. Notice that a date/step t has 24 different timestamps h and an event e_t has a duration of m minutes with n viewers. The viewers attend the event with different time zones which is represented as an one-hot vector $\mathbf{z} \in \mathbb{R}^{d_z}$, where d_z is the number of different time zones of the viewers. The goal of the enterprise is to organize each event $e_t \in \mathcal{E}$ on the timestamp h, to maximize the average engagement u_t and adoption v_t of the viewers. We formulate the scheduling of the next event as a Markov Decision Process (MDP), where the agent interacts with the environment/enterprise by selecting the timestamp h of the next event e_{t+1} and maximizing the cumulative rewards. In particular, we define the MDP of the live video streaming event as follows [21]:

Definition 1. *Live Video Streaming Event MDP. At each step $t = 1, \ldots, T$, the agent interacts with the environment and selects an action $\mathbf{a}_t \in \mathcal{A}$. An action $\mathbf{a}_t$ corresponds to the selection of the timestamp h of the next event e_{t+1} based on the state $\mathbf{s}_t \in \mathcal{S}$ of the enterprise. We define the state $\mathbf{s}_t$ of the enterprise as a sequence of the l previous events $\mathbf{s}_t = \{e_{t-l}, \ldots, e_t\}^1$. The agent receives a reward $r(\mathbf{s}_t, \mathbf{a}_t, \mathbf{s}_{t+1}) \in \mathcal{R}$ for selecting the action $\mathbf{a}_t \in \mathcal{A}$ in state $\mathbf{s}_t \in \mathcal{S}$, considering the enterprise transitions to state $\mathbf{s}_{t+1}$ with a probability $p(\mathbf{s}_{t+1}|\mathbf{s}_t, \mathbf{a}_t) \in \mathcal{P}$. The goal of the agent is to find the optimal policy $\pi_\theta : \mathcal{S} \times \mathcal{A} \to \mathcal{R}$, where θ is the set of policy parameters, assigning a probability $\pi_\theta(\mathbf{a}_t|\mathbf{s}_t)$ of selecting an action $\mathbf{a}_t \in \mathcal{A}$ provided a state $\mathbf{s}_t \in \mathcal{S}$. Having computed the policy π_θ, the agent maximizes the expectation of the discounted cumulative reward $\max \mathbb{E}[\sum_{t=0}^{T} \gamma^t r(\mathbf{s}_t, \mathbf{a}_t, \mathbf{s}_{t+1})|\pi_\theta]$, with $\gamma \in [0, 1]$ being the discount factor.*

In our model, we focus on training a common agent that optimizes both the viewer's engagement u_t and adoption v_t. As mentioned in Sect. 2, the viewer's engagement and adoption behavior vary over time. Therefore, we first consider the viewer's engagement and adoption as independent tasks, and then train a common agent to optimize the cumulative rewards of both tasks at the same time. We define the multi-task Reinforcement Learning (RL) problem in live video streaming events as follows [5,6,8,11]:

---

[1] We consider only the l previous events to capture the most recent viewers behavior. As we will demonstrate in Sect. 4, considering large values of l does not necessarily improve the model's performance.

Definition 2. *Multi-task RL in Live Video Streaming. In the multi-task RL problem for live video streaming events, we consider a set of tasks $\mathcal{T}$, that is the engagement and adoption tasks with $|\mathcal{T}| = 2$. We formulate each task $\tau \in \mathcal{T}$ as a different MDP, where the tasks have the same state $\mathcal{S}$ and action space $\mathcal{A}$ with a different set of rewards $\mathcal{R}$. For the engagement task we compute reward $r(\mathbf{s}_t, \mathbf{a}_t)$ as the average engagement u_t in Eq. 1, and for the adoption task the reward corresponds to the average adoption v_t in Eq. 2 at the t-th step. The goal of the agent is to learn a common policy π_θ that solves each task $\tau \in \mathcal{T}$, by maximizing the expected return $\max \mathbb{E}_{\tau \sim \mathcal{T}} [[\sum_{t=0}^{T} \gamma^t r(\mathbf{s}_t^\tau, \mathbf{a}_t^\tau, \mathbf{s}_{t+1}^\tau) | \pi_\theta]]$ for both tasks. $\mathbf{s}_t^\tau$ is the state of the agent and $\mathbf{a}_t^\tau$ is the action taken by the agent for the task τ at the t-th step.*

3.1   MERLIN's Architecture

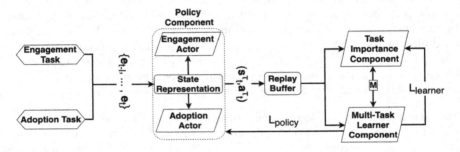

Fig. 2. The architecture of the proposed MERLIN model for the viewer's engagement and adoption tasks. MERLIN consists of: (i) the policy (ii) task importance and (iii) multi-task learner components.

As illustrated in Fig. 2, the proposed MERLIN model consists of three main components: the policy, task importance and multi-task learner components. The goal of MERLIN is to compute a common policy π_θ that maximizes the future rewards for the viewer's engagement and adoption tasks $\tau \in \mathcal{T}$.

- Policy Component. The role of the policy component is to compute the action $\mathbf{a}_t^\tau$ of both tasks. During training, the agent interacts with two environments in the enterprise, that is the different two tasks $\tau \in \mathcal{T}$. The input of the policy component is the l previous events $\{e_{t-l}^\tau, \ldots, e_t^\tau\}$ of each task. We implement a shared state representation module to compute the state $\mathbf{s}_t^\tau$ of task τ. In our architecture, we design the respective two actors to generate the actions $\mathbf{a}_t^\tau$ for the engagement and adoption tasks [8]. Then, the generated state-action transitions by both actors are stored in the replay buffer with size l_b to train the common agent.

- Task Importance Component. The task importance component determines the contribution of each task to the learning process of the agent. Notice that state-of-the-art RL strategies are designed to learn a policy of a single agent that optimizes similar tasks, ignoring the information of each task's state-action transition [25]. Instead, in the proposed MERLIN model to account for the impact of each state-action transition on the policy π_θ, we consider the encoder model of the Transformer's architecture for the state-action transition sequences. In doing so, we capture the information of the state-action transitions of both the engagement and adoption actors over time [15,17]. In addition, the task importance component computes a weight matrix $\mathbf{M} \in \mathbb{R}^{l_b \times |\mathcal{T}|}$ which reflects on the contribution of each actor to the learning process of the policy π_θ.

- **Multi-task Learner Component.** The role of the multi-task learner component is to optimize the policy π_θ based on the l_b state-action transitions stored in the replay buffer. Provided the stored state-action transitions in the replay buffer and the weight matrix $\mathbf{M}$ of the task importance component, the multi-task learner updates the policy parameters through a joint loss function $\mathcal{L}_{policy}$ and the parameters of the task importance component via the $\mathcal{L}_{learner}$ function, following the temporal-difference learning strategy [21]. In particular, matrix $\mathbf{M}$ first weighs the state-action transitions in the replay buffer, and then the multi-task learner optimizes the joint loss function $\mathcal{L}_{policy}$ to compute the parameters of the policy component. In addition, the multi-task learner learns its parameters via the joint loss function $\mathcal{L}_{learner}$, and updates the parameters of the task importance component accordingly.

3.2   Policy Component

At each step $t = 1, \ldots, T$, the policy component takes as an input the l previous events $\{e_{t-l}^\tau, \ldots, e_t^\tau\}$ of each task $\tau \in \mathcal{T}$. The goal of the policy component is to learn a policy π_θ that solves each task τ. Provided that the engagement and adoption tasks have the same state space $\mathcal{S}$ and action space $\mathcal{A}$, the policy component consists of a shared state representation module and two actors, that is the engagement and adoption actors.

- **State Representation Module.** The state representation module takes as an input the l previous events $\{e_{t-l}^\tau, \ldots, e_t^\tau\}$, and generates the state $\mathbf{s}_t^\tau$ of each task τ at the t-th step. We represent each event e_t^τ as a d_x-dimensional vector $\mathbf{x}_t^\tau \in \mathbb{R}^k$ concatenating the event's features $\mathbf{x}_t^\tau = Concat(h, n, m, g, o, \mathbf{z})$. Given the representations $\{\mathbf{x}_{t-l}^\tau, \ldots, \mathbf{x}_t^\tau\}$ of the l previous events, we compute the d_s-dimensional state representation vector $\mathbf{s}_t^\tau \in \mathbb{R}^{d_s}$ as follows: [29,30]:

$$\mathbf{s}_t^\tau = \xi_w(\mathbf{x}_t^\tau, \Delta(t)) = \text{Time-LSTM}(\mathbf{x}_t^\tau, \Delta(t)) \tag{3}$$

where w are the trainable parameters of the Time-LSTM function $\xi(\cdot)$ [29]. Notice that Time-LSTM models the time difference $\Delta(t)$ of the event e_t^τ and the previous event e_{t-1}^τ as follows:

$$\mathbf{g}_t = \sigma\left(\mathbf{x}_t^\tau \mathbf{W}_{xg} + \sigma(\Delta(t)\mathbf{W}_g + b_g)\right)$$
$$\mathbf{q}_t = \mathbf{f}_t \odot \mathbf{q}_{t-1} + \mathbf{i}_t \odot \mathbf{g}_t \odot \sigma\left(\mathbf{x}_t^\tau \mathbf{W}_{xq} + \mathbf{s}_{t-1}\mathbf{W}_{sq} + b_q\right) \qquad (4)$$
$$\mathbf{o}_t = \sigma(\mathbf{x}_t^\tau \mathbf{W}_{xo} + \Delta(t)\mathbf{W}_o + \mathbf{s}_{t-1}^\tau \mathbf{W}_{so} + \mathbf{q}_t \odot \mathbf{W}_{qo} + b_o)$$
$$\mathbf{s}_t^\tau = \mathbf{o}_t \odot \sigma(\mathbf{q}_t)$$

where $\mathbf{g}_t$ is the time dependent gate influencing the memory cell and the output gate $\mathbf{o}_t$, $\mathbf{q}_t$ is the memory cell of LSTM, and $\mathbf{f}_t$ and $\mathbf{i}_t$ are the forget and input gates, respectively [10, 30]. The symbol $\odot$ represents the Hadamard element-wise product and $\sigma(\cdot)$ is the sigmoid function. The different weight matrices $\mathbf{W}_*$ in Eq. 4 transform the event embedding $\mathbf{x}_t^\tau$ and the time difference $\Delta(t)$ to the d_s-dimensional latent space, and b_* are the respective bias terms. Notice that the time difference $\Delta(t)$ is important to capture the similarity among consecutive events in the state $\mathbf{s}_t^\tau$. Provided that the engagement and adoption of the viewers vary over time, our goal is to capture the most recent viewer's behaviour in the state space $\mathbf{s}_t^\tau$. Therefore, the Time-LSTM in Eq. 4 tends to forget events with high time difference, and focuses on the recent events.

- **Engagement and Adoption Actors.** The engagement and adoption actors take as input the state $\mathbf{s}_t^\tau$ of each task $\tau \in \mathcal{T}$. The state representation $\mathbf{s}_t^\tau$ captures the evolution of the enterprise over time. Given the state $\mathbf{s}_t^\tau$ and a policy π_θ, each actor computes a d_a-dimensional action vector $\mathbf{a}_t^\tau \in \mathbb{R}^{d_a}$, where d_a is the number of all the possible timestamps. Each dimension of the action vector $\mathbf{a}_t^\tau$ corresponds to the probability of selecting the timestamp h for the next event e_{t+1}. We implement a two-layer perceptron (MLP) to transform the state vector $\mathbf{s}_t^\tau \in \mathbb{R}^b$ to the action vector $\mathbf{a}_t^\tau \in \mathbb{R}^u$ as follows:

$$\mathbf{a}_t^\tau = \pi_\theta(\mathbf{s}_t^\tau) = MLP(\mathbf{s}_t^\tau) \qquad (5)$$

where θ are the trainable parameters of the MLP, that is the policy parameters of the agent. Given the action vector $\mathbf{a}_t^\tau$ of each actor, we normalize the action vector $\mathbf{a}_t^\tau$ based on the softmax function and select the action with the highest value using the ϵ-greedy exploration technique [21]. The generated state-action transitions are stored in the replay buffer to learn the optimal policy π_θ based on the past experiences of each task.

3.3 Task Importance Component

The goal of the task importance component is to determine the contribution of each task to the learning strategy of the policy π_θ. The input of the task importance component is the set of state-action transitions stored in the replay buffer by the engagement and adoption actors. At each step $t = 1, \ldots, T$, the engagement and adoption actors store in the replay buffer the respective state-action transition $(\mathbf{s}_t^\tau, \mathbf{a}_t^\tau)$ of the task $\tau \in \mathcal{T}$. Having stored the l_b state-action transitions of each task τ in the replay buffer, the task importance component computes the similarity among the tasks. As the replay buffer contains a sequence of state-action transitions, we employ the encoder of the Transformer's model to capture the information of the l_b states to d_y-dimensional vectors $\mathbf{Y}^\tau \in \mathbb{R}^{l_b \times d_y}$ [24]. To overcome any stability problems that might occur at the early stages of the training, we implement the Gated Transformer(-XL) (GTrXL) model of the Transformer's architecture as follows [17]:

$$\mathbf{Y}^\tau = \psi_\eta(\{\mathbf{s}_{t-l_b}^\tau, \ldots, \mathbf{s}_t^\tau\}) = GTrXL(\{\mathbf{s}_{t-l_b}^\tau, \ldots, \mathbf{s}_t^\tau\}) \qquad (6)$$

where $\{\mathbf{s}_{t-l_b}^\tau, \ldots, \mathbf{s}_t^\tau\}$ is the states sequence of the task τ stored in the replay buffer. Parameters η denote the trainable weights of the GTRrXL function $\psi(\cdot)$ [17].

By computing the d_y-dimensional vectors, that is the rows of matrix $\mathbf{Y}^\tau$ of each task τ, we deduce the importance of each state $\mathbf{s}_t^\tau$ in the actions selected by the actor over time for task τ. Therefore, we can compute a weight matrix $\mathbf{M} \in \mathbb{R}^{l_b \times |\mathcal{T}|}$ of each state $\mathbf{s}_t^\tau$ during the training of the agent's policy π_θ. To calculate the weight matrix $\mathbf{M}$, we employ a two-layer MLP with softmax activation:

$$\mathbf{M} = \lambda_\omega(\mathbf{Y}^\tau) = softmax\left(MLP(\mathbf{Y}^\tau)\right) \qquad (7)$$

where ω are the parameters of the MLP transformation function $\lambda(\cdot)$. Intuitively, we give stronger preference to the states $\mathbf{s}_t^\tau$ that contribute more to the learning strategy of the agent than the rest of the states. This means that our agent learns the policy π_θ based on the most important states $\mathbf{s}_t^\tau$.

3.4 Multi-task Learner Component

According to our architecture in Sect. 3.1 the multi-task learner optimizes the joint loss function $\mathcal{L}_{policy}$ to compute the parameters w and θ of the policy component of Eqs. 3 and 5. In addition, based on the joint loss function $\mathcal{L}_{learner}$ we calculate the parameters ζ of the multi-task learner component, and update the parameters η and ω of the task importance component of Eqs. 6 and 7.

The input of the multi-task learner component is the l_b state-action transitions, of each task τ, stored in the replay buffer, and the weight matrix $\mathbf{M}$ generated by the task importance component. The multi-task learner component calculates the state-action value $Q(\mathbf{s}_t^\tau, \mathbf{a}_t^\tau)$, which is an approximation of the expected cumulative rewards of the agent, given the state $\mathbf{s}_t^\tau$ and action $\mathbf{a}_t^\tau$. We compute the state-action value $Q(\mathbf{s}_t^\tau, \mathbf{a}_t^\tau)$, as follows:

$$Q(\mathbf{s}_t^\tau, \mathbf{a}_t^\tau) = \phi_\zeta(\mathbf{s}_t^\tau, \mathbf{a}_t^\tau) = MLP(\mathbf{s}_t^\tau \oplus \mathbf{a}_t^\tau) \qquad (8)$$

where ζ are the trainable parameters of the MLP function $\phi(\cdot)$, and $\oplus$ denotes the concatenation of the state $\mathbf{s}_t^\tau$ and action $\mathbf{a}_t^\tau$ vectors. Intuitively, the value $Q(\mathbf{s}_t^\tau, \mathbf{a}_t^\tau)$ corresponds to the benefit of the agent in terms of the expected reward for each task τ, when taking the action $\mathbf{a}_t^\tau$ given the state $\mathbf{s}_t^\tau$ and following the policy π_θ. By computing the value $Q(\mathbf{s}_t^\tau, \mathbf{a}_t^\tau)$ based on Eq. 8, we can optimize the joint loss function $\mathcal{L}_{policy}$ with respect to the parameters w and θ as follows [13, 20]:

$$w \leftarrow w - \alpha \nabla_w \mathcal{L}_{policy}(\pi_\theta)$$
$$\theta \leftarrow \theta - \alpha \nabla_\theta \mathcal{L}_{policy}(\pi_\theta) \tag{9}$$
$$\min_{w,\theta} \mathcal{L}_{policy} = -\tfrac{1}{|\mathcal{T}|l_b} \sum_{\tau \in \mathcal{T}} \sum_{k=0}^{l_b} log \pi_\theta(\mathbf{a}_k^\tau, \mathbf{s}_k^\tau)[r(\mathbf{s}_k^\tau, \mathbf{a}_k^\tau) - M_{\tau,k} Q(\mathbf{s}_k^\tau, \mathbf{a}_k^\tau)]$$

where α is the learning rate. The term $[r(\mathbf{s}_k^\tau, \mathbf{a}_k^\tau) - M_{\tau,k} Q(\mathbf{s}_k^\tau, \mathbf{a}_k^\tau)]$ corresponds to the benefit of taking the action $\mathbf{a}_k^\tau$ given the state $\mathbf{s}_k^\tau$. The expected value $Q(\mathbf{s}_k^\tau, \mathbf{a}_k^\tau)$ is weighted by $M_{\tau,k}$ so as to strengthen/weaken the contribution of the state $\mathbf{s}_k^\tau$ when learning the policy π_θ, accordingly.

The joint loss function $\mathcal{L}_{learner}$ is formulated as a minimization mean squared error function with respect to parameters η ω and ζ as follows:

$$\eta \leftarrow \eta - \alpha \nabla_\eta \mathcal{L}_{learner}(\pi_\theta)$$
$$\omega \leftarrow \omega - \alpha \nabla_\omega \mathcal{L}_{learner}(\pi_\theta)$$
$$\zeta \leftarrow \zeta - \alpha \nabla_\zeta \mathcal{L}_{learner}(\pi_\theta) \tag{10}$$
$$\min_{\eta,\omega,\zeta} \mathcal{L}_{learner} = \tfrac{1}{|\mathcal{T}|l_b} \sum_{\tau \in \mathcal{T}} \sum_{k=0}^{l_b} \left(r(\mathbf{s}_k^\tau, \mathbf{a}_k^\tau) - M_{\tau,k} Q(\mathbf{s}_k^\tau, \mathbf{a}_k^\tau) \right)^2$$

Overall, to train our model we consider that the agent interacts with the environment in an episodic manner [21]. This means that the agent interacts with the environment within a finite horizon of T interactions/events. We train our model for multiple episodes and optimize the joint loss functions $\mathcal{L}_{policy}$ and $\mathcal{L}_{learner}$ in Eqs. 9 and 10 with respect to the parameters w, θ, η ω and ζ through backpropagation with the Adam optimizer [12].

4   Experiments

4.1   Setup

- **Environment.** In our experiments, we evaluate the performance of the proposed model to select the timestamp h of each event that maximizes the viewer's engagement u_t and adoption v_t. For each dataset we order the events according to the timestamps, and consider the first 70% of the events as training set $\mathcal{E}^{train}$, 10% for validation $\mathcal{E}^{val}$ and 20% for testing $\mathcal{E}^{test}$. The agent interacts with an emulated environment[2] which models the behavioural policy π_β of the events of each dataset. Following [7,9,30], to emulate the behavioural policy π_β we train a multi-head neural network on each dataset, which takes as input a sequence of events and outputs the average engagement and adoption of the next event. During the agent's training, we initialize the reinforcement learning environment with the events of the training set $\mathcal{E}^{train}$. To initialize the state $\mathbf{s}_t^\tau$ of the agent, we randomly select an event $e_t \in \mathcal{E}^{train}$ of the training set.

---

[2] Provided the high risk that might hinder when evaluating the learned policy π_θ directly to the enterprises, in our study we perform off-line A/B testing based on the events of each dataset [9,30].

At each step $t = 1, \ldots, T$, the agent takes an action $\mathbf{a}_t^\tau$ for each task τ. Then, the agent receives the average engagement u_t and adoption v_t generated by the behavioural policy π_β as a reward of each task. To evaluate the learned policy π_θ, we initialize the reinforcement learning environment with the events of the test set $\mathcal{E}^{test}$. Similar to the training strategy, the state $\mathbf{s}_t^\tau$ of the agent is initialized by randomly selecting an event $e_t \in \mathcal{E}^{test}$ from the test set. The agent takes an action $\mathbf{a}_t^\tau$ and receives the reward by the multi-head network which models the behaviour policy π_β of the test set $\mathcal{E}^{test}$.

- **Evaluation Metrics.** We evaluate the performance of our proposed model in terms of the step-wise variant of Normalized Capping Importance Sampling (NCIS) for each task as follows: [22,30]:

$$
\begin{aligned}
NCIS &= \sum_{t=1}^{T} \frac{\bar{\rho} r(\mathbf{s}_t^\tau, \mathbf{a}_t^\tau)}{\sum_{k=1}^{T} \bar{\rho}} \\
\bar{\rho} &= \min\{\delta, \prod_{t=1}^{T} \frac{\pi_\theta(\mathbf{a}_t^\tau | \mathbf{s}_t^\tau)}{\pi_\beta(\mathbf{a}_t^\tau | \mathbf{s}_t^\tau)}\}
\end{aligned}
\tag{11}
$$

where $\bar{\rho}$ is the max capping of the importance ratio, and δ is a threshold to ensure small variance and control the bias of the policy π_θ towards the behavioural policy π_β. The term $\bar{\rho} r(\mathbf{s}_t^\tau, \mathbf{a}_t^\tau)$ is the capped importance weighted reward of a task τ. Intuitively, by adopting different rewards in the term $\bar{\rho} r(\mathbf{s}_t^\tau, \mathbf{a}_t^\tau)$, we can measure the performance of the policy π_θ to approximate the behavioural policy π_β. By setting each reward $r(\mathbf{s}_t^\tau, \mathbf{a}_t^\tau)$ equal to the viewer's engagement and adoption as in Sect. 3, we can evaluate the performance of the proposed model based on the respective metrics Eng. NCIS and Ad. NCIS for both tasks. As the emulated environment is initialized randomly, we repeated our experiments five times and report average Eng. NCIS and Ad. NCIS in our experiments.

- **Baselines.** We compare the proposed MERLIN model against the following strategies: FeedRec [30], AMT[3] [5], IMPALA[4] [8] and PopART [11]. As there are no publicly available implementations of FeedRec and PopART, we implemented both from scratch and published our source codes[5].

- **Parameter Configuration.** For each examined model, we tuned the hyperparameters on the validation set, following a grid-selection strategy. In FeedRec, we set the state representation dimensionality $d_s = 256$ for Enterprises 1 and 3, and $d_s = 128$ for Enterprises 2 and 4. At the t-th step, the FeedRec model takes as an input all the events occurred prior to the current step with $l = 0$. In AMT we fix a $d_s = 128$ dimensional state representation for all datasets, with a time window $l = 30$ previous events. In IMPALA and PopART the state representation's dimensionality is fixed to $d_s = 64$ for all Enterprises. The window length l in IMPALA and PopART is set to 20 and 23, respectively. In the proposed MERLIN model we use a $d_s = 128$ dimensional state representation for Enterprises 1 and 4, and 256 and 64 for Enterprises 2 and 3, respectively. The window length l is fixed to 10 for Enterprise 1, and 15 for Enterprises 2, 3 and 4. In addition, the size of the replay buffer l_b is set to 128 for all Enterprises. In all the examined

---

[3] https://github.com/braemt/attentive-multi-task-deep-reinforcemt-learning.
[4] https://github.com/deepmind/scalable_agent.
[5] https://github.com/stefanosantaris/merlin.

models, we follow an ϵ-greedy exploration-exploitation strategy and set $\epsilon = 0.1$. The discount factor γ is fixed to 0.92 and the learning rate is set to $\alpha = 0.001$. In the emulated environment, we set the number of interactions/events to 200 and the number of episodes to 300.

All our experiments were conducted on a single server with an Intel Xeon Bronze 3106, 1.70 GHz CPU. The operating system of the server was Ubuntu 18.04 LTS. We accelerated the training of the model using the GPU Geforce RTX 2080 Ti graph card. Our proposed MERLIN model was implemented in Pytorch 1.7.1 and we created the reinforcement learning environment with the OpenAI Gym 0.17.3 library.

4.2   Performance Evaluation

Table 2. Performance comparison of the examined models on the engagement and adoption tasks in terms of average Eng. NCIS and Ad. NCIS. Bold values indicate the best method using a statistical significance t-test with $p < 0.01$.

Task	Model	Datasets			
		E1	E2	E3	E4
Avg. Eng. NCIS	**FeedRec**	0.553	0.591	0.423	0.467
	AMT	0.462	0.513	0.371	0.380
	IMPALA	0.452	0.493	0.352	0.314
	PopART	0.421	0.460	0.432	0.401
	MERLIN	**0.622**	**0.663**	**0.512**	**0.552**
Avg. Ad. NCIS	**FeedRec**	8.122	15.271	14.393	27.292
	AMT	6.284	12.781	11.842	20.962
	IMPALA	5.023	10.523	9.232	18.284
	PopART	4.891	9.362	9.013	16.642
	MERLIN	**10.112**	**17.292**	**16.961**	**29.554**

In Table 2, we evaluate the performance of the examined models in terms of average Eng. NCIS and Ad. NCIS over the five trials in the emulated environment for the engagement and adoption tasks, respectively. The proposed MERLIN model significantly outperforms the baselines in all datasets. This indicates that MERLIN can efficiently learn a common policy π_θ that optimizes both tasks concurrently. Compared with the second best method FeedRec, MERLIN achieves relative improvements of 15.76 and 15.96% in terms of Eng. NCIS and Ad. NCIS, respectively. FeedRec performs better than the other baseline approaches because FeedRec formulates a joint loss function for training the agent on the different tasks. However, each task in FeedRec contributes equally when learning the policy π_θ, and therefore the agent ignores the evolutionary patterns and the

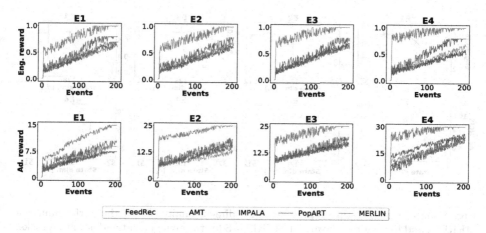

Fig. 3. The Eng. reward and Ad.reward based on Eqs. 1 and 2 of the examined models for the engagement and adoption tasks, when the interactions/events evolve in the emulated environment.

importance of the state-action transitions for each task. The proposed MER-LIN model overcomes this problem by integrating the training parameters of the task importance component in the common learning strategy of the policy and multi-task learner components. In doing so, MERLIN balances the contribution of each task to the generated policy.

In Fig. 3 we report the Eng. reward and Ad. reward based on Eqs. 1 and 2 for the engagement and adoption tasks, respectively, when the interactions/events evolve in the emulated environment. We observe that MERLIN constantly achieves higher rewards than the other baseline approaches at the first interactions. This demonstrates the effectiveness of MERLIN to weigh the importance of each task during training and learn a policy that optimizes both tasks. In addition, we observe that the Ad. reward in the adoption task of MERLIN converges faster in Enterprises 2, 3 and 4 than in Enterprise 1. As discussed in Sect. 2, the viewer's adoption in Enterprises 2, 3 and 4 increase over time. Therefore, the task importance component promotes the adoption task during the training of the policy, thus achieving high reward in Enterprises 2, 3 and 4 at the beginning of the interactions.

4.3   Multi-task Vs Single-Task Learning in Parameter Configuration

In the next set of experiments we compare the proposed MERLIN model with its variant MERLIN-S. In particular, the agent of the variant MERLIN-S is trained on a single task, ignoring the multi-task learning strategy of MERLIN. In Fig. 4, we study the impact of the state representation's dimensionality d_s on the performances of MERLIN and MERLIN-S in terms of Eng. NCIS and Ad. NCIS for the engagement and adoption tasks, when varying d_s in $\{32, 64, 128, 256, 512\}$. We observe that MERLIN achieves the best performance when setting 128 dimen-

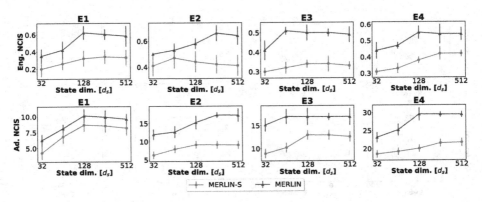

Fig. 4. Impact of the state representation's dimensionality d_s on the performance of MERLIN and its single-task variant MERLIN-S for the engagement and adoption tasks.

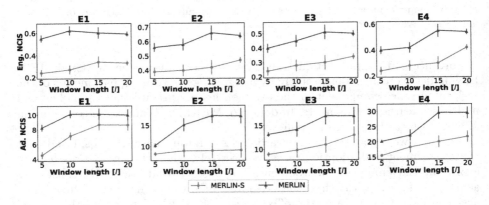

Fig. 5. Impact of the window length l on MERLIN and MERLIN-S.

sions for Enterprises 1 and 4, 256 for Enterprise 2, and 64 for Enterprise 3. By increasing the dimensionality d_s of the state representation, the agent of MER-LIN achieves similar performances in both tasks. We observe that MERLIN significantly outperforms the MERLIN-S model in both tasks, indicating the importance of the multi-task learning strategy to efficiently extract knowledge from both tasks. In Fig. 5, we present the impact of the window length l on MER-LIN and MERLIN-S. We vary the window length l from 5 to 20 by a step of 5. MERLIN requires 10 past events in Enterprise 1, and 15 events in Enterprises 2, 3 and 4. Moreover, we observe that MERLIN constantly outperforms the single task variant MERLIN-S. Notice that MERLIN-S achieves the best performance when the window length l is set to 15 past events for Enterprise 1, and 20 for Enterprises 2, 3 and 4. Therefore, MERLIN-S requires a higher window length l than MERLIN in all Enterprises, as MERLIN-S omits the auxiliary information of the other task when training the agent.

5   Conclusions

In this study, we presented a multi-task reinforcement learning strategy to train an agent so as to select the optimal time of a live video streaming event in large enterprises, aiming to improve the viewer's engagement and adoption. In the proposed MERLIN model, we formulate the engagement and adoption tasks as different MDPs and design a joint loss function to extract knowledge from both tasks. To determine the contribution of each task to the training strategy of the agent, we implement a task importance learner component that extracts the most important information, that is the most important state-action transitions from the replay buffer based on the Transformer's architecture. Having weighted the transitions, the agent of MERLIN learns a common policy for both tasks. Our experiments with four real-world datasets demonstrate the superiority of our model against several baseline approaches in terms of viewer's engagement and adoption. The proposed MERLIN model can significantly help enterprises in selecting the optimal time of an event. Provided that nowadays the majority of the events are online, the enterprises want to ensure that their employees/viewers adopt the video streaming events with high engagement. This means that with the help of MERLIN in scheduling the live video streaming events, the enterprises can communicate with their employees efficiently, which as a consequence reflects on significant productivity gains [2]. An interesting future direction is to study the influence of distillation strategies on the proposed MERLIN model [16].

References

1. Break up your big virtual meetings. https://hbr.org/2020/04/break-up-your-big-virtual-meetings (2020). Accessed 19 Mar 2021
2. Gauging demand for enterprise streaming - 2020 - investment trends in times of global change. https://www.ibm.com/downloads/cas/DEAKXQ5P (2020). Accessed 29 Jan 2021
3. Using video for internal corporate communications, training & compliance. https://www.ibm.com/downloads/cas/M0R85GDQ (2021). Accessed 30 Mar 2021
4. Antaris, S., Rafailidis, D.: Vstreamdrls: Dynamic graph representation learning with self-attention for enterprise distributed video streaming solutions. In: ASONAM, pp. 486–493 (2020)
5. Bräm, T., Brunner, G., Richter, O., Wattenhofer, R.: Attentive multi-task deep reinforcement learning. In: Brefeld, U., Fromont, E., Hotho, A., Knobbe, A., Maathuis, M., Robardet, C. (eds.) ECML, pp. 134–149 (2020)
6. Calandriello, D., Lazaric, A., Restelli, M.: Sparse multi-task reinforcement learning. In: NIPS (2014)
7. Chen, M., Beutel, A., Covington, P., Jain, S., Belletti, F., Chi, E.H.: Top-k off-policy correction for a reinforce recommender system. In: WSDM, pp. 456–464 (2019)
8. Espeholt, L., et al.: IMPALA: Scalable distributed deep-RL with importance weighted actor-learner architectures. In: ICML, pp. 1407–1416 (2018)
9. Gilotte, A., Calauzènes, C., Nedelec, T., Abraham, A., Dollé, S.: Offline a/b testing for recommender systems. In: WSDM, pp. 198–206 (2018)

10. Graves, A.: Generating sequences with recurrent neural networks. arXiv preprint arXiv:1308.0850 (2013)
11. Hessel, M., Soyer, H., Espeholt, L., Czarnecki, W., Schmitt, S., van Hasselt, H.: Multi-task deep reinforcement learning with popart. In: AAAI, pp. 3796–3803. AAAI Press (2019)
12. Kingma, D.P., Ba, J.: Adam: a method for stochastic optimization. In: ICLR (2015)
13. Lillicrap, T.P., et al.: Continuous control with deep reinforcement learning (2019)
14. Liu, F., Guo, H., Li, X., Tang, R., Ye, Y., He, X.: End-to-end deep reinforcement learning based recommendation with supervised embedding. In: WSDM, pp. 384–392 (2020)
15. Loynd, R., Fernandez, R., Celikyilmaz, A., Swaminathan, A., Hausknecht, M.: Working memory graphs. In: ICML (2020)
16. Parisotto, E., Salakhutdinov, R.: Efficient transformers in reinforcement learning using actor-learner distillation (2021)
17. Parisotto, E., et al.: Stabilizing transformers for reinforcement learning. In: ICML, pp. 7487–7498 (2020)
18. Polydoros, A.S., Nalpantidis, L.: Survey of model-based reinforcement learning: applications on robotics. J. Intell. Robot. Syst. **86**(2), 153–173 (2017)
19. Silver, D., et al.: Mastering chess and shogi by self-play with a general reinforcement learning algorithm (2017)
20. Silver, D., Lever, G., Heess, N., Degris, T., Wierstra, D., Riedmiller, M.: Deterministic policy gradient algorithms. In: ICML, pp. 387–395 (2014)
21. Sutton, R.S., Barto, A.G.: Reinforcement learning: An introduction. MIT Press, Cambridge (2018)
22. Swaminathan, A., Joachims, T.: The self-normalized estimator for counterfactual learning. In: Cortes, C., Lawrence, N., Lee, D., Sugiyama, M., Garnett, R. (eds.) NeurIPS, vol. 28 (2015)
23. Teh, Y.W., et al.: Distral: robust multitask reinforcement learning. In: NIPS, pp. 4499–4509 (2017)
24. Vaswani, A., et al.: Attention is all you need. In: NIPS, pp. 5998–6008 (2017)
25. Vithayathil Varghese, N., Mahmoud, Q.H.: A survey of multi-task deep reinforcement learning. Electronics **9**(9), 1363 (2020)
26. Xin, X., Karatzoglou, A., Arapakis, I., Jose, J.M.: Self-supervised reinforcement learning for recommender systems. In: SIGIR, pp. 931–940 (2020)
27. Ye, D., et al.: Mastering complex control in moba games with deep reinforcement learning. AAAI **34**(04), 6672–6679 (2020)
28. Zhu, H., et al.: The ingredients of real world robotic reinforcement learning. In: ICLR (2020)
29. Zhu, Y., et al.: What to do next: Modeling user behaviors by time-lstm. In: IJCAI-17, pp. 3602–3608 (2017)
30. Zou, L., Xia, L., Ding, Z., Song, J., Liu, W., Yin, D.: Reinforcement learning to optimize long-term user engagement in recommender systems. In: KDD, pp. 2810–2818 (2019)

Social Media

Explainable Abusive Language Classification Leveraging User and Network Data

Maximilian Wich[(✉)](ID), Edoardo Mosca(ID), Adrian Gorniak(ID),
Johannes Hingerl(ID), and Georg Groh(ID)

Technical University of Munich, Munich, Germany
{maximilian.wich,edoardo.mosca,adrian.gorniak,
johannes.hingerl}@tum.de, grohg@in.tum.de

Abstract. Online hate speech is a phenomenon with considerable consequences for our society. Its automatic detection using machine learning is a promising approach to contain its spread. However, classifying abusive language with a model that purely relies on text data is limited in performance due to the complexity and diversity of speech (e.g., irony, sarcasm). Moreover, studies have shown that a significant amount of hate on social media platforms stems from online hate communities. Therefore, we develop an abusive language detection model leveraging user and network data to improve the classification performance. We integrate the explainable AI framework SHAP (SHapley Additive exPlanations) to alleviate the general issue of missing transparency associated with deep learning models, allowing us to assess the model's vulnerability toward bias and systematic discrimination reliably. Furthermore, we evaluate our multimodel architecture on three datasets in two languages (i.e., English and German). Our results show that user-specific timeline and network data can improve the classification, while the additional explanations resulting from SHAP make the predictions of the model interpretable to humans.

Keywords: Hate speech · Abusive language · Classification model · Social network · Deep learning · Explainable AI

1 Introduction

Hate speech is a severe challenge that social media platforms such as Twitter and Facebook face nowadays. However, it is not purely an online phenomenon and can spill over to the offline world resulting in physical violence [36]. The Capitol riots in the US at the beginning of the year are a tragic yet prime example. Therefore, the fight against hate speech is a crucial societal challenge.

The enormous amount of user-generated content excludes manual monitoring as a viable solution. Hence, automatic detection of hate speech becomes the

Warning: This paper contains content that may be abusive or offensive.

© Springer Nature Switzerland AG 2021
Y. Dong et al. (Eds.): ECML PKDD 2021, LNAI 12979, pp. 481–496, 2021.
https://doi.org/10.1007/978-3-030-86517-7_30

key component of this challenge. A technology to facilitate the identification is *Machine Learning*. Especially in recent years, *Natural Language Processing* (NLP) has made significant progress. Even if these advances also enhanced hate speech classification models, there is room for improvement [29].

However, gaining the last points of the F1 score is a massive challenge in the context of hate speech. Firstly, abusive language has various forms, types, and targets [32]. Secondly, language itself is a complex and evolving construct; e.g., a word can have multiple meanings, people create new words or use them differently [29]. This complexity exacerbates classifying abusive language purely based on textual data. Therefore, researchers have started to look beyond pure text-driven classification and discovered the relevance of social network data [10]. Kreißel et al. [11], for example, showed that small subnetworks cause a significant portion of offensive and hateful content on social media platforms. Thus, it is beneficial to integrate network data into the model [3,5,6,15,22]. However, to the best of our knowledge, no one has investigated the impact of combining the text data of the post that is meant to be classified, the user's previous posts, and their social network data.

An issue with such an approach is its vulnerability to bias, meaning that a system "systematically and unfairly discriminate[s] against certain individuals or groups of individuals in favor of others" [7, p. 332]. *Deep Learning* (DL) models often used in NLP are particularly prone to this issue because of their black-box nature [17]. Conversely, a system combining various data sources and leveraging user-related data has a more considerable potential of discriminating individuals or groups. Consequently, such systems should integrate *eXplainable AI* (XAI) techniques to address this issue and increase trustworthiness.

We address the following two research questions in our paper concerning the two discussed aspects:

RQ1 Can abusive language classification be improved by leveraging users' previous posts and their social network data?

RQ2 Can explainable AI be used to make predictions of a multimodal hate speech classification model more understandable?

To answer the research questions, we develop an explainable multimodal classification model for abusive language using the mentioned data sources[1]. We evaluate our model on three different datasets—WASEEM [33], DAVIDSON [4], and WICH [35]. Furthermore, we report findings of integrating user and social network data that are relevant for future work.

2   Related Work

Most work in the abusive language detection domain has focused on developing models that only use the text data of the document to be classified [16,24,29].

---

[1] Code available on https://github.com/mawic/multimodal-abusive-language-detection.

Other works, however, have started to integrate context-related data into abusive language detection [18,24,29]. One promising data source is the users' social network because it has been shown that hater networks on social media platforms cause a considerable amount of online hate [8,11]. Combining network and text data from Twitter was already successfully applied to predict whether an account is verified [2] or to identify extremist accounts [38]. In the case of abusive language, Papegnies et al. [19] built a classification model using local and global topological measures from graphs as features for cyberbullying detection (e.g., average distance, betweenness centrality). A similar approach has been applied by Chatzakou et al. [3], but they also integrated user-related data (e.g., number of posts, account age) and textual data (e.g., number of hashtags). This approach was picked up and extended by other researchers [5,6] (e.g., integrating users' gender, geolocation) who confirmed the usefulness of additional context-related data sources. They all have in common that the network features are only topological measures and do not contain any information about the relations. Mishra et al. [15] addressed this downside and modeled the users' follower network with a node2vec embedding that serves as an additional input for the classification model. Ribeiro et al. [22] developed a similar model; they, however, used the graph embedding GraphSAGE to model the retweet network and combined it with a document embedding for the text data [9]. For this purpose, they collected a dataset that has a fully connected network. Unfortunately, they released only the network data and the document embeddings but not the raw text. Recently, Li et al. [12] refined this approach.

Another data source that supports abusive language detection is the user's history of previous posts. Qian et al. [20] improved a hate speech classifier for tweets by adding the previous tweets of the author. Raisi and Huang [21] proposed a model that leverages the user's history of posts and the post directed to the user to calculate a bully and victim score for each user. However, to the best of our knowledge, no one has integrated user's previous posts and social networks into abusive language detection.

Besides multimodality, XAI in abusive language detection is another topic that we have to consider in this section. Since XAI is a relatively new field, it has not been frequently applied to abusive language detection with some exceptions [14,18,27,30,31,34]. All models use only the text as input, except [30]. Their model also relies on network data. But the network submodel is very simple; it is only a binary vector encoding whether the user follows pre-defined hater accounts. Furthermore, the explanations for this submodel are not detailed. Hence, the explainable model that we propose is an advancement.

3   Data

For our experiment, we use three abusive language datasets that are from Twitter. Table 1 provides an overview of the datasets' characteristics. Figure 1 visualizes the social network graph of the datasets.

Table 1. Overview of the datasets' statistics

	Davidson			Waseem			Wich	
Number of tweets	14,939			16,907			68.443	
Number of users	6,725			2,024			939	
Avg. number of tweets per user	2.22			8.35			72.9	
Class	Hate	Offensive	Neither	Sexism	Racsim	None	Offensive	Non-offensive
Class distribution	814	11,800	2,325	3,430	1,976	11,501	26,205	42,238
Network: avg. degree	1.85			3.44			1.63	
Network: graph density	0.0005			0.0034			0.0002	

DAVIDSON Davidson et al. [4] released an English abusive language dataset containing 24,783 tweets annotated as hate, offensive, or neither. Unfortunately, the dataset does not contain any data about the user or the network. Therefore, we used the Twitter API to get the original tweets and the related user and network data. Since not all tweets are still available on Twitter, our dataset has shrunk to 14,939 tweets.

WASEEM Waseem et al. [33] published an English abusive language dataset containing 16,907 tweets annotated as sexist, racist, or none. Similar to DAVIDSON, the dataset does not provide any user- or network-related data. The authors of [15] shared their enriched WASEEM dataset with us containing the user and network data.

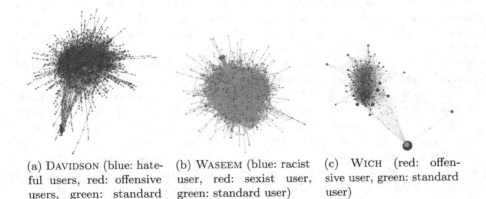

(a) DAVIDSON (blue: hateful users, red: offensive users, green: standard user)

(b) WASEEM (blue: racist user, red: sexist user, green: standard user)

(c) WICH (red: offensive user, green: standard user)

Fig. 1. Visual comparison of the network topologies. Standalone nodes or very small subnetworks that do not connect to the main graph for DAVIDSON and WASEEM are excluded. (Color figure online)

WICH Wich et al. [35] released a German offensive language dataset containing 4,647,200 tweets annotated as offensive or non-offensive. Most of the tweets are pseudo-labeled with a BERT-based classifier; a smaller portion of the dataset is also manually annotated. The difference between this dataset and the other two is the way it was collected. Wich et al. applied a snowball sampling strategy focusing on users. Starting from seed users, the authors collected the connected users and their tweets based on their offensiveness. Hence, the network graph has a star-shaped network topology contrary to the other two, as depicted in Fig. 1c. We select only 68,443 tweets and the related user and network information to better handle the data. The manually annotated tweets are used as a test set.

4   Methodology

The section is split into two subsections. The first one deals with the model architecture and training of the multimodal classification model. The second one considers the XAI technique that we use to explain the predictions of our multimodal model.

4.1   Multimodal Classification Model

Architecture. The multimodal classification model for abusive language consists of three submodels that process the different inputs:

1. **Text model**: It processes the text data of the tweet that is meant to be classified. For this purpose, we use DistilBERT with a classification head.
2. **History model**: It processes the tweet history of the user.
3. **Network model**: It processes the social network data of the tweet's user. To model the network data, we use the vector embedding framework Graph-SAGE.

The three models' outputs are combined in a linear layer, which outputs the prediction for the tweet to be classified.

Text Model. The text data of the tweet is fed into a pre-trained DistilBERT model with a classification head. DistilBERT is a lighter and faster version of the transformer-based model BERT [23]. Despite the parameter reduction, its performance is comparable to BERT in general [23] and in the context of abusive language detection [28]. In order to implement the model, we use the Transformers library from Hugging Face[2] and its implementation of DistilBERT [37]. As pre-trained models, we use `distilbert-base-uncased` for the English datasets and `distilbert-base-german-cased` for the German one. Before tokenizing the text data, we remove username mentions from the tweets, but we keep the "@" from the mention[3]. The purpose of this procedure is to avoid the classifier memorizing the username and associating it with one of the classes. But the classifier should recognize that the tweet addresses another user.

---

[2] https://huggingface.co/transformers/.

[3] If a user is mentioned in a tweet, an "@" symbol appears before the user name.

History Model. We use a bag-of-words model to model the user's tweet history, comprising the 500 most common terms from the dataset based on term frequency-inverse document frequency (tf-idf). For each user, it is a 500-dimensional binary vector that reflects which of the most common terms appear in the user's tweet history.

Network Model. In order to model the user's social network, we apply the inductive representation learning framework GraphSAGE [9]. The advantage of an inductive learning framework is that it can be applied to previously unseen data, meaning the model can generate an embedding for a new user in a network, which is a desirable property for our use case. Our GraphSAGE model is trained on the undirected network graph of the social relations. Furthermore, we assign to each user/node a class derived from the labels of their tweets. The output of the model is a 32-dimensional graph embedding for each user. The graphs are modeled as follows:

- DAVIDSON: An edge between two users exists if at least one follows the other. A user is labeled as hater, if he or she has at least one hate tweet; as offensive, if he or she has at least one offensive tweet, but no hate tweet; as neither, if he or she has only neither tweets.
- WASEEM: An edge between two users exists if at least one follows the other. A user is labeled as racist, if he or she has at least one tweet labeled as racist; same for sexist; as none, if he or she is neither racist nor sexist.
- WICH: An edge between two users exists if at least one has retweeted the other. A user is labeled as offensive, if he or she has at least three offensive tweets.

Users without network connections in their respective dataset, so-called solitary users, do not receive a GraphSAGE embedding; their embedding vector only contains zeros.

The output of the three models is concatenated to a 534 or 535 respectively dimensional vector (DistilBERT: 2 or 3 dimensions depending on the output speech classes; GraphSAGE: 32 dimensions; bag-of-words: 500 dimensions) and fed into a hidden linear layer. This final layer with softmax activation reduces the output to the number of classes according to the selected dataset.

Training. Several challenges have to be faced when it comes to training the model. In terms of sampling, we cannot randomly split the dataset: We have to ensure that tweets of any user do not appear in the train and test set; otherwise, we would have a data leakage. Therefore, sampling is done on the user level. Users are categorized into groups based on their class and the existence of a network. We gather six different categories for WASEEM and DAVIDSON and four categories for WICH. The train, validation, and test set all contain users from different classes by sampling these categories to prevent bias toward certain user groups. Due to the different tweet counts per user, the train set size varies between 60–70% depending on the dataset.

We under- and oversample the classes during training since all datasets are unbalanced. Moreover, we have to train the three submodels separately because the unsupervised training process of GraphSAGE cannot be combined with the supervised training of DistilBERT. DistilBERT is fine-tuned for two epochs with a batch size of 64 and an Adam optimizer (initial learning rate of 5×10^{-5} and a weight decay of 0.01). We train our GraphSAGE model, consisting of three hidden layers with 32 channels each, for 50 epochs with an Adam optimzer (initial learning rate of 5×10^{-3}). The bag-of-words model does not require training. After training the submodels, we freeze them and train the hidden layer (10 epochs; Adam optimizer with an initial learning rate of 1×10^{-3}).

4.2 Explainable AI Technique

We set model interpretability as a core objective of our work. To this end, we produce Shapley-values-based explanations at different levels of granularity. Shapley values are an established technique to estimate the contribution of input features w.r.t. the model's output [13,25]. Their suitability for this task has been proven both on a theoretical as well as on an empirical level [13].

As computing exact Shapley values is exponentially complex w.r.t. the input size and hence not feasible, accurate approximations are fundamental for their estimation [13]. As shown in Algorithm 1, we compute them by iteratively averaging each feature's marginal contribution to a specific output class. We find that 15 iterations are sufficient for Shapley values to converge. A random sampling of features was used for reasons of simplicity. Finally, we can assign each feature a Shapley value, representing its relative impact score. A similar approximation approach has been used in [26].

There are two different granularity levels in terms of features: For instance, we can treat each model component (tweet, network, history) as a single feature and derive impact scores (Shapley values) for these components. Alternatively, each model component input or feature (e.g., each token of a tweet) can be treated separately on a more fine-grained level. As Shapley values are additive, they can be aggregated to represent component-level Shapley values. The way feature and components are excluded in order to compute their respective Shapley value changes based on these two levels listed in Table 2. Thus, our multimodal model can be explained on a single instance, and the role played by each model can always be retrieved.

Additionally, we partition the network graph into communities using the Louvain algorithm to derive Shapley values for individual network connections [1]. All user edges in that community with the target user are disabled to obtain the impact of a specific community, resulting in a new GraphSAGE generated user embedding as input for the multimodal model. The embedding vectors of solitary users that only contain zeros result in Shapley values equal to zero for the network component of all these users.

Result: Shapley value $\{\phi_t\}_{t=1}^M$ for every feature $\{x_t\}_{t=1}^M$
Input: p sample probability, x instance, f model, I number of iterations
for $i = 0, ..., I$ **do**
    for $t = 1, ..., M$ **do**
        sample a Bernoulli vector $P = \{0,1\}^M$ with probability p
        pick S a subset of the features $\{x_t\}_{t=1}^M \setminus \{x_t\}$ according to P
        build x_S alteration of x with only features in S
        $\phi_t \leftarrow \phi_t \frac{i-1}{i} + \frac{f(x_{S \cup \{x_t\}}) - f(x_S)}{i}$
    end
end

Algorithm 1: Shapley value approximation algorithm. In our experiments, $p = 0.7$ and $I = 15$ were used as parameters.

Table 2. Masking strategies for SHAP on component and feature level

	Text	Network	History
Component wise	Masking BERT output with 0s	Setting GraphSAGE embedding to 0	Setting all vocabulary counts to 0
Feature wise	Masking each token individually	Disabling edges to user based on community and generating new embedding	Setting each vocabulary token count to 0 individually

5   Results

In the first subsection, we deal with answering RQ1 based on the classification performance of our architecture. The second subsection addresses the explainability of the models and related findings to answer RQ2.

Table 3. Classification models' performance by different architectures and datasets

Model	Davidson			Waseem			Wich		
	P	R	F1	P	R	F1	P	R	F1
Text	75.3	77.1	76.1	77.5	84.1	80.3	89.8	91.7	90.7
Text + History	73.7	77.8	75.5	79.3	87.8	**82.7**	89.8	91.7	90.7
Text + Network	75.3	77.2	76.2	77.5	84.4	80.4	89.9	91.7	**90.8**
All	74.5	78.9	**76.5**	79.2	88.1	**82.7**	90.0	91.7	**90.8**

5.1   Classification Performance

Table 3 displays the different model architecture performance metrics for the three datasets. We find that combining text, history, and network increases the macro F1 score of WASEEM by 2.4 pp and of DAVIDSON by 0.4 pp. In the case of WICH, we observe only a minor increase of the precision by 0.1 pp. We ascribe these diverging increases to two aspects: Firstly, the network of WASEEM is the densest one of all three, followed by DAVIDSON and WICH, as depicted in Table 1. Secondly, WICH's text model has a high F1 score, meaning that this submodel presumably drives the predictions of the multimodal model. Our impact analysis using SHAP to identify each submodel's relevance confirms this hypothesis, as depicted in Fig. 2. It shows that the network and history data are less relevant for WICH's multimodal model than for the other two models.

In order to answer RQ1, these results signify that leveraging a user's previous posts and their social network data does improve abusive language classification. Additionally, the improvement of the F1 score is proportional to the network's density – the higher the density, the higher the improvement.

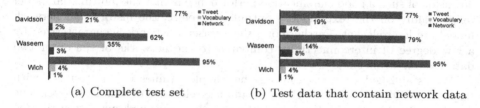

(a) Complete test set   (b) Test data that contain network data

Fig. 2. Avg. impact of each classifier's submodels on the respective test set based on shapley values

5.2   Explainability

In this subsection, we present the results of the XAI technique, SHAP, that we applied to our multimodal model. Firstly, we further investigate the impact of the network and history data added to the text model. Secondly, we show the explanation of a single tweet.

Impact Analysis of the Submodels. Figure 2 visualizes the impact of the submodels on the multimodal model. We calculate the impact by aggregating the Shapley values for each submodel based on the tweets in the test set. Figure 2a displays the impact on the complete test set of each dataset, while Fig. 2b shows the impact on test data that contains network data[4].

Our first observation is that all classifiers are mainly driven by the text model, followed by the history and network model. Comparing Fig. 2a and 2b, we see that network data, if available, contributes to the predictions of WASEEM's and DAVIDSON's multimodal models. If we compare the network model's impact of both datasets in the context of network density (DAVIDSON: 5×10^{-4}; WASEEM: 3.4×10^{-3}), we can conclude that the denser the network is, the more relevant it is for the classification. These findings confirm our answer to RQ1.

In the case of WASEEM, we observe a large contribution of the history model (35%) for the complete test set. We can trace it back to four users that produced

---

[4] Network data is not avaiable for all users.

a large portion of the dataset and mainly produced all abusive tweets. In general, the number of tweets in the user's history correlates positively with the Shapley value for the history model, reflecting the impact of the history model on the prediction. While the correlation within WICH's dataset is only weak ($r_{Wich} = 0.172$), we observe a moderate correlation for the other two datasets ($r_{Davidson} = 0.500$ and $r_{Waseem} = 0.501$).

Regarding WICH's dataset, the Shapley values indicate that the text model dominates (95%) the multimodal model's prediction, while the other two (4% and 1%) play only a minor role. There are two reasons for this: First, the tweets are pseudo-labeled by a BERT model. Since we use a DistilBERT model similar to BERT, we achieve an outstanding F1 score of the text model (90.7%). The downside of such a good classification performance is that the multimodal model relies mainly on the text model's output. Therefore, the history and network model are less relevant. Furthermore, the dataset's network is characterized by a low degree of interconnectivity compared to the networks of the other two datasets (cf. Table 1).

We established that aggregating the Shapley values of the test set with respect to RQ2 helps us better understand the relevance of each submodel. The insights gained by the applied XAI technique also confirmed our answer to RQ1 that user's network and history data contribute to abusive language detection.

Explaining a Single Tweet Classification. After investigating the model on an aggregated level, we focus on explaining the prediction of a single tweet. To do so, we select the following tweet from the DAVIDSON dataset that is labeled and correctly predicted as hateful by our multimodal model:

> @user i can guarantee a few things: you're white. you've never been any-where NEAR a real ghetto. you, or a relative is a pig. 100%

In the following, we demonstrate the explainable capabilities of our multimodal model based on the selected tweet. Figure 3 plots the Shapley values of the tweet's tokens and the user's history and network (last two rows). These Shapley values indicate the relevance of the feature on the multimodal model's prediction as hateful. A positive value (red-colored) represents a contribution favoring the classification as hateful, a negative value (blue-colored) that favors the classification as non-hateful.

We see that the most relevant word for the classification as hateful is "white", which should not be surprising because of the racist context. Furthermore, the @-symbol (representing a user mention) and "you(')re" are relevant for the classification model, indicating that directly addressing someone is recognized as a sign of hate for the classifier. In contrast, the punctuation of the tweet negatively influences the classification as hateful. A possible explanation is that correct spelling and punctuation are often disregarded in the context of abusive language. Beyond the textual perspective, we observe that the history and network submodels favor the classification as hateful. These inputs are relevant for our multimodal model to classify the tweet correctly. Considering Fig. 4a

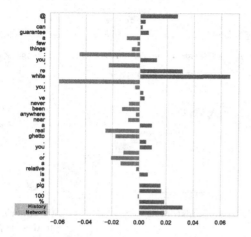

Fig. 3. Relevance of the different features in the form of Shapely values; positive, red values represent favoring a classification as hateful; negative, blue ones the opposite; Shapley values for history and network submodel are aggregated (Color figure online)

(an alternative visualization of the Shapely values), we see that the text model slightly favors the classification as non-hateful, represented by the negative sum of Shapley values. Due to the input from the other two submodel, however, the multimodal model classifies the tweet correctly, making this an excellent example of how abusive language detection can profit from additional data.

Figures 4b and 4c break down the contribution of the history and network model, where Fig. 4b is a waterfall chart displaying the most relevant terms that the user used in their previous posts—less relevant terms are summarized in the column named REST. As in the previous charts, red represents a positive contribution to the classification as hateful and blue vice versa. The last column, called OVERALL, is the sum of all terms' Shapley values. In this case, the previous tweets of the user contain words words that are primarily associated with hateful tweets; consequently, the history model favors a classification as hateful. Figure 4c shows the user's ego network and its impact on the classification. The nodes connected to the user represent communities identified by the Louvain algorithm. The first number of a node's label is an identifier; the second number is the number of haters in the community; the third number is the community's total number of users. The color of the nodes and edges have the same meaning as in the other visualizations. In our case, two connected communities contribute to a hateful classification, while the left-pointing community counteracts this.

The presented explanations of the complete model and its submodels provide meaningful and reasonable information to understand better how the model decides to make predictions. These findings extend our answer to RQ2 from the previous section. Our explainable model provides explanations on an aggregated level and a single prediction level to make the classification more understandable.

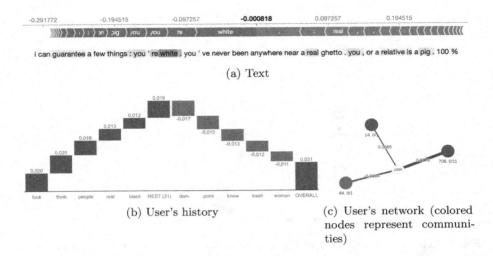

(a) Text

(b) User's history

(c) User's network (colored nodes represent communities)

Fig. 4. Explanations for predictions of test, history, and network submodel in the form of Shapely values (red, positive values favor a classification as hateful; blue, negative values favor a classification as non-hateful) (Color figure online)

6 Discussion

We demonstrated that leveraging a user's history and ego network can improve abusive language detection regarding RQ1, consistent with the findings from other researchers [15,20,22]. Our multimodal approach is novel because we combine text, users' previous tweets, and their social relations in one model. The additional data sources provide further indications for the classification model to detect abusive language better. That can be helpful, especially when the classifier struggles with a precise prediction, as in our example in Sect. 5.2. Other examples are implicit language, irony, or sarcasm, which are hard to detect from a textual perspective. The improvement, however, varies between the datasets. We trace this back to the network density of the available data. WASEEM has the network with the highest density and exhibits the best improvement if we integrate history and network data. In contrast, the classification model based on WICH, the dataset with the least dense network, could be improved only slightly. A further difficulty concerning WICH's dataset is that the tweets are pseudo-labeled with a BERT model, and our text submodel uses DistilBERT. Hence, our text submodel performs so well that the multimodal model nearly ignores the outputs of the history and network submodels. Therefore, it was hard to identify any improvement. Relating to DAVIDSON, we had the problem of data degradation. Since the dataset does not contain any user or network data, we used the Twitter API to obtain them. But not all tweets were still available, causing us to use only 60% of the original dataset for our experiment. We require more appropriate datasets to investigate the integration of additional data sources in abusive language detection and refine this approach.

For example, Riberio et al. [22] have released a comprehensive dataset containing 4,972 labeled users. Unfortunately, they have not published the tweets of the users. We are aware that releasing a dataset containing social relations and text might violate the users' privacy. Therefore, we suggest anonymizing the data by replacing all user names with anonymous identifiers.

We proved that our multimodal model combined with the SHAP framework provides reasonable and meaningful explanations of its predictions associated with RQ2. These explanations allow us to gain a better understanding with respect of the models in two different ways: (1) the influence of the different submodels on the final predictions on an aggregated level; (2) the relevance of individual features (e.g., word, social relationship) for a single prediction. These explainable capabilities of our multimodal model are a further novelty. To our best knowledge, no one has developed such an explainable model for abusive language detection.

Even though the SHAP explanations are only an approximation, they are necessary for the reliable application of a hate speech detection model, as we have developed. It should be humanly interpretable how each of the three models influences predictions since we combine various data sources, which is especially true when one data source, such as the social network, is not fully transparent for the user. The reason for the missing transparency is that our network submodel learns patterns from social relations, which are more challenging to understand without any additional information than the ones from the text model. Therefore, these explainable capabilities are indispensable for such a system to provide a certain degree of transparency and build trustworthiness.

After focusing on the individual research questions, we have to add an ethical consideration regarding our developed model for various reasons. One may criticize that we integrate social network data, which is personal data, into our model and that the benefit gained by it bears no relation to the invasion of the user's privacy. However, we argue against it based on the following reasons: (1) We use social network data to train embeddings and identify patterns that do not contain any personal data. (2) The user's history and network are shown to enhance the detection rate, even if the used datasets are not the most appropriate ones for this experiment because of the limited density. Furthermore, detecting abusive language can be challenging if the author uses irony, sarcasm, or implicit wording. Therefore, context information (e.g., user's history or network) should be included because its benefit outweighs the damage caused by abusive language.

Another point of criticism could be the possible vulnerability to bias and systematic discrimination of users. In general, DL models are vulnerable to bias due to their black-box nature. In the case of a multimodal model, however, the issue is more aggravated because one submodel can dominate the prediction without any transparency for the user. For example, a model that classifies a user's tweet only because of their social relations discriminates the user with a high probability. We address this challenge by adding explainable capabilities with SHAP. Therefore, we claim that our multimodal model is less vulnerable to bias than classical abusive language detection models applying DL techniques without XAI integration.

7 Conclusion and Outlook

This paper investigated whether users' previous posts and social network data can be leveraged to achieve good, humanly interpretable classification results in the context of abusive language. Concerning the classification performance (RQ1), we showed that the additional data improves the performance depending on the dataset and its network density. For WASEEM, we increased the macro F1 score by 2.4 pp, for DAVIDSON by 0.4 pp, and WICH by 0.1 pp. We found that the denser the network, the higher the gain. Nevertheless, the availability of appropriate datasets is a remaining challenge.

The model's interpretability (RQ2) demonstrated that our multimodal model using the SHAP framework produces meaningful and understandable explanations for its predictions. The explanations are provided both on a word level and connections to social communities in the user's ego network. The explanations help better understand a single prediction and the complete model if relevance scores are aggregated on a submodel level. Furthermore, explainability is a necessary feature of such a multimodal model to prevent bias and discrimination.

Integrating a user's previous posts and social network to enhance abusive language detection produced promising results. Therefore, the research community should continue exploring this approach because it might be a feasible way to address the challenge of detecting implicit hate, irony, or sarcasm. Concrete aspects that have to be addressed by future work are the following: (1) collecting appropriate data (in terms of size and network density) to refine our approach, (2) improving our model's architecture.

Acknowledgments. We would like to thank Anika Apel and Mariam Khuchua for their contribution to this project. The research has been partially funded by a scholarship from the Hanns Seidel Foundation financed by the German Federal Ministry of Education and Research.

References

1. Blondel, V.D., Guillaume, J.L., Lambiotte, R., Lefebvre, E.: Fast unfolding of communities in large networks. J. Stat. Mech: Theory Exp. **2008**(10), P10008 (2008)
2. Campbell, W., Baseman, E., Greenfield, K.: Content + context networks for user classification in twitter. In: Frontiers of Network Analysis, NIPS Workshop, 9 December 2013 (2013)
3. Chatzakou, D., Kourtellis, N., Blackburn, J., De Cristofaro, E., Stringhini, G., Vakali, A.: Mean birds: Detecting aggression and bullying on twitter. In: WebSci, pp. 13–22 (2017)
4. Davidson, T., Warmsley, D., Macy, M., Weber, I.: Automated hate speech detection and the problem of offensive language. In: Proceedings of 11th ICWSM Conference (2017)

5. Fehn Unsvåg, E., Gambäck, B.: The effects of user features on Twitter hate speech detection. In: Proceedings of 2nd Workshop on Abusive Language Online (ALW2), pp. 75–85. ACL (2018)
6. Founta, A.M., Chatzakou, D., Kourtellis, N., Blackburn, J., Vakali, A., Leontiadis, I.: A unified deep learning architecture for abuse detection. In: WebSci, pp. 105–114. ACM (2019)
7. Friedman, B., Nissenbaum, H.: Bias in computer systems. ACM Trans. Inf. Syst. 14(3), 330–347 (1996)
8. Garland, J., Ghazi-Zahedi, K., Young, J.G., Hébert-Dufresne, L., Galesic, M.: Countering hate on social media: large scale classification of hate and counter speech. In: Proceedings of 4th Workshop on Online Abuse and Harms, pp. 102–112 (2020)
9. Hamilton, W.L., Ying, Z., Leskovec, J.: Inductive representation learning on large graphs. In: NIPS, pp. 1024–1034 (2017)
10. Hennig, M., Brandes, U., Pfeffer, J., Mergel, I.: Studying Social Networks. A Guide to Empirical Research, Campus Verlag, New York (2012)
11. Kreißel, P., Ebner, J., Urban, A., Guhl, J.: Hass auf Knopfdruck. Rechtsextreme Trollfabriken und das Ökosystem koordinierter Hasskampagnen im Netz, Institute for Strategic Dialogue (2018)
12. Li, S., Zaidi, N.A., Liu, Q., Li, G.: Neighbours and kinsmen: hateful users detection with graph neural network. In: Karlapalem, K., Cheng, H., Ramakrishnan, N., Agrawal, R.K., Reddy, P.K., Srivastava, J., Chakraborty, T. (eds.) PAKDD 2021. LNCS (LNAI), vol. 12712, pp. 434–446. Springer, Cham (2021). https://doi.org/10.1007/978-3-030-75762-5_35
13. Lundberg, S.M., Lee, S.I.: A unified approach to interpreting model predictions. In: NeurIPS (2017)
14. Mathew, B., Saha, P., Yimam, S.M., Biemann, C., Goyal, P., Mukherjee, A.: Hatexplain: A benchmark dataset for explainable hate speech detection. arXiv preprint arXiv:2012.10289 (2020)
15. Mishra, P., Del Tredici, M., Yannakoudakis, H., Shutova, E.: Author profiling for abuse detection. In: COLING, pp. 1088–1098. ACL (2018)
16. Mishra, P., Yannakoudakis, H., Shutova, E.: Tackling online abuse: A survey of automated abuse detection methods. arXiv preprint arXiv:1908.06024 (2019)
17. Molnar, C.: Interpretable Machine Learning (2019). https://christophm.github.io/interpretable-ml-book/
18. Mosca, E., Wich, M., Groh, G.: Understanding and interpreting the impact of user context in hate speech detection. In: Proceedings of 9th International Workshop on Natural Language Processing for Social Media, pp. 91–102. ACL (2021)
19. Papegnies, E., Labatut, V., Dufour, R., Linarès, G.: Graph-based features for automatic online abuse detection. In: Camelin, N., Estève, Y., Martín-Vide, C. (eds.) SLSP 2017. LNCS (LNAI), vol. 10583, pp. 70–81. Springer, Cham (2017). https://doi.org/10.1007/978-3-319-68456-7_6
20. Qian, J., ElSherief, M., Belding, E., Wang, W.Y.: Leveraging intra-user and inter-user representation learning for automated hate speech detection. In: NAACL 2018 (Short Papers), pp. 118–123. ACL (2018)
21. Raisi, E., Huang, B.: Cyberbullying detection with weakly supervised machine learning, ASONAM 2017, pp. 409–416. Association for Computing Machinery, New York (2017)
22. Ribeiro, M., Calais, P., Santos, Y., Almeida, V., Meira Jr, W.: Characterizing and detecting hateful users on twitter. In: Proceedings of International AAAI Conference on Web and Social Media, vol. 12 (2018)

23. Sanh, V., Debut, L., Chaumond, J., Wolf, T.: DistilBERT, a distilled version of BERT: smaller, faster, cheaper and lighter. In: 2019 5th Workshop on Energy Efficient Machine Learning and Cognitive Computing - NeurIPS 2019 (2019)
24. Schmidt, A., Wiegand, M.: A survey on hate speech detection using natural language processing. In: Proceedings 5th International Workshop on Natural Language Processing for Social Media, pp. 1–10. ACL (2017)
25. Shapley, L.: Quota solutions of n-person games. Contrib. Theor. Games **2**, 343–359 (1953)
26. Štrumbelj, E., Kononenko, I.: Explaining prediction models and individual predictions with feature contributions. Knowl. and Inf. Syst. **41**(3), 647–665 (2013). https://doi.org/10.1007/s10115-013-0679-x
27. Švec, A., Pikuliak, M., Šimko, M., Bieliková, M.: Improving moderation of online discussions via interpretable neural models. In: Proceedings of 2nd Workshop on Abusive Language Online (ALW2), pp. 60–65. ACL (2018)
28. Vidgen, B., et al.: Detecting East Asian prejudice on social media. In: Proceedings of 4th Workshop on Online Abuse and Harms, pp. 162–172. ACL (2020)
29. Vidgen, B., Harris, A., Nguyen, D., Tromble, R., Hale, S., Margetts, H.: Challenges and frontiers in abusive content detection. In: Proceedings of 3rd Workshop on Abusive Language Online, pp. 80–93. ACL (2019)
30. Vijayaraghavan, P., Larochelle, H., Roy, D.: Interpretable multi-modal hate speech detection. In: Proceedings of International Conference on Machine Learning AI for Social Good Workshop (2019)
31. Wang, C.: Interpreting neural network hate speech classifiers. In: Proceedings of 2nd Workshop on Abusive Language Online (ALW2), pp. 86–92. ACL (2018)
32. Waseem, Z., Davidson, T., Warmsley, D., Weber, I.: Understanding abuse: A typology of abusive language detection subtasks. In: Proceedings of 1st Workshop on Abusive Language Online, pp. 78–84. ACL (2017)
33. Waseem, Z., Hovy, D.: Hateful symbols or hateful people? predictive features for hate speech detection on Twitter. In: Proceedings of NAACL Student Research Workshop, pp. 88–93. ACL (2016)
34. Wich, M., Bauer, J., Groh, G.: Impact of politically biased data on hate speech classification. In: Proceedings of 4th Workshop on Online Abuse and Harms, pp. 54–64. ACL (2020)
35. Wich, M., Breitinger, M., Strathern, W., Naimarevic, M., Groh, G., Pfeffer, J.: Are your friends also haters? identification of hater networks on social media: data paper. In: Companion Proceedings of Web Conference 2021, ACM (2021)
36. Williams, M.L., Burnap, P., Javed, A., Liu, H., Ozalp, S.: Hate in the machine: Anti-black and anti-muslim social media posts as predictors of offline racially and religiously aggravated crime. Br. J. Criminol. **60**(1), 93–117 (2020)
37. Wolf, T., et al.: Transformers: State-of-the-art natural language processing. In: Proceedings of 2020 Conference on Empirical Methods in Natural Language Processing: System Demonstrations, pp. 38–45. ACL (2020)
38. Xu, J., Lu, T.C., et al.: Automated classification of extremist twitter accounts using content-based and network-based features. In: 2016 IEEE International Conference on Big Data, pp. 2545–2549. IEEE (2016)

Calling to CNN-LSTM for Rumor Detection: A Deep Multi-channel Model for Message Veracity Classification in Microblogs

Abderrazek Azri[1](✉), Cécile Favre[1](✉), Nouria Harbi[1](✉),
Jérôme Darmont[1](✉) (iD), and Camille Noûs[2]

[1] Université de Lyon, Lyon 2, UR ERIC 5 Avenue Pierre Mendès France,
69676 Bron Cedex, France
{a.azri,cecile.favre,nouria.harbi,jerome.darmont}@univ-lyon2.fr
[2] Laboratoire Cogitamus, Université de Lyon, Lyon 2, Lyon, France
camille.nous@cogitamus.fr

Abstract. Reputed by their low-cost, easy-access, real-time and valuable information, social media also wildly spread unverified or fake news. Rumors can notably cause severe damage on individuals and the society. Therefore, rumor detection on social media has recently attracted tremendous attention. Most rumor detection approaches focus on rumor feature analysis and social features, i.e., metadata in social media. Unfortunately, these features are data-specific and may not always be available. In contrast, post contents (including images or videos) play an important role and can indicate the diffusion purpose of a rumor. Furthermore, rumor classification is also closely related to opinion mining and sentiment analysis. Yet, to the best of our knowledge, exploiting images and sentiments is little investigated. Considering the available multimodal features from microblogs, notably, we propose in this paper an end-to-end model called deepMONITOR that is based on deep neural networks, by utilizing all three characteristics: post textual *and* image contents, as well as sentiment. deepMONITOR concatenates image features with the joint text and sentiment features to produce a reliable, fused classification. We conduct extensive experiments on two large-scale, real-world datasets. The results show that deepMONITOR achieves a higher accuracy than state-of-the-art methods.

Keywords: Social networks · Rumor detection · Deep neural networks

1 Introduction

Nowadays, more and more people consume news from social media rather than traditional news organizations, thanks to social media features such as information sharing, real time, interactivity, diversity of content and virtual identities. However, conveniently publishing news also fosters the emergence of various

© Springer Nature Switzerland AG 2021
Y. Dong et al. (Eds.): ECML PKDD 2021, LNAI 12979, pp. 497–513, 2021.
https://doi.org/10.1007/978-3-030-86517-7_31

rumors and fake news that can spread promptly through social networks and result in serious consequences.

To detect rumors on microblogs, which we particularly target in this paper, most existing studies focus on the social features available in social media. Such features are post metadata, including the information on how post propagate, e.g., the number of retweets, followers, hashtags (#), user information, etc. To exploit such features, many innovative solutions [4,23] have been proposed. Unfortunately, these features are not always available, e.g., in case the rumor has just been published and not yet propagated, and do not indicate the purpose of a rumor, which is one of its most important aspects. Moreover, although social features are useful in rumor analysis, contents reveal more relevant in expressing the diffusion purpose of rumors [17]. Hence, in this paper, we analyse message contents from three aspects to automatically detect rumors in microblogs.

First, social media messages have rich textual contents. Therefore, understanding the semantics of a post is important for rumor detection. Attempts to automate the classification of posts as true or false usually exploit natural language processing and machine learning techniques that rely on hand-crafted and data-specific textual features [4,16]. These approaches are limited because the linguistic characteristics of fake news vary across different types of fake news, topics and media platforms. Second, images and videos have gained popularity on microblogs recently and attract great attention. Rich visual information can also be helpful in classifying rumors [10]. Yet, taking images into account for verifying post veracity is not sufficiently explored, with only a few recent studies exploiting multimedia content [10,11]. Third, liars can be detected, as they tend to frequently use words carrying negative emotions out of unconscious guilt [20]. Since emotion is closely related to fake news [1], analyzing emotions with opinion mining and sentiment analysis methods may help classifying rumors.

Automating rumor detection with respect to one of the three characteristics mentioned above is already challenging. Hand-crafted textual features are data-specific and time consuming to produce; and linguistic characteristics are not fully understood. Image features and emotions, which are a significant indicators for fake news detection in microblogs, are still insufficiently investigated.

To address these limitations, we propose an end-to-end model called deep-MONITOR, based on deep neural network that are efficient in learning textual or visual representations and that jointly exploits textual contents, sentiment and images. To the best of our knowledge, we are the first to do this. Hence, deep-MONITOR can leverage information from different modalities and capture the underlying dependencies between the context, emotions and visual information of a rumour.

More precisely, deepMONITOR is a multi-channel deep model where we first employ a Long-term Recurrent Convolutional Network (LRCN) to capture and represent text semantics and sentiments through emotional lexicons. This architecture combines the advantages of Convolutional Neural Network (CNN) for extracting local features and the memory capacity of Long Short-Term Memory Networks (LSTM) to connect the extracted features well. Second, we employ the pretrained VGG19 model [26] to extract salient visual features from post images. Image features are then fused with the joint representations of text and sentiment to classify messages. Eventually, we experimentally show that deep-MONITOR outperforms state-of-the-art rumor detection models on two large multimedia datasets collected from Twitter.

The remainder of this paper is organized as follows. In Sect. 2, we survey and discuss related works. In Sect. 3, we thoroughly details the deepMONITOR framework. In Sect. 4, we experimentally validate deepMONITOR with respect to the state of the art. Finally, in Sect. 5, we conclude this paper and hint at future research.

2   Related Works

Most studies in the literature address the automatic rumor detection task as feature-based. Features can be extracted from text, social context, sentiment and even attached images. Thus, we review existing work from the following two categories: single modality-based rumor detection and multimodal-based rumor detection.

2.1   Monomodal-Based Rumor Detection

Textual Features are extracted from textual post contents. They are derived from the linguistics of a text, such as lexical and syntactic features. In the literature, there is a wide range of textual features [4,25]. Unfortunately, linguistic patterns are highly dependent on specific events and the corresponding domain knowledge. Thus, it is difficult to manually design textual features for traditional machine learning-based rumor detection models. To overcome this limitation, a Recurrent Neural Network (RNN) can learn the representations of posts in time series as textual features [18].

Social Context Features represent user engagements in news on social media, such as the number of mentions(@), hashtags(#) and URLs [25]. Graph structures can capture message propagation patterns [27]. However, as textual features, social context features are very noisy, unstructured and require intensive labor to collect. Moreover, it is difficult to detect rumors using social context-based methods when the rumor has just popped up and not yet propagated, i.e., there is no social context information.

Visual Features are typically extracted from images and videos. Very few studies address the verification of multimedia content credibility on social media. Basic message features are characterized [8,27] and various visual features are extracted [11]. Visual features include clarity, coherence, diversity and clustering scores, as well as similarity distribution histogram. However, these features remain hand-crafted and can hardly represent complex distributions of visual contents.

Sentiment Features are emotional signals. There exists a relationship between rumors and sentiments in messages and an emotion feature, i.e., the ratio of the count of negative and positive words, can be built [1]. Besides, emotion features can also be extracted with respect to emotional lexicons from news contents [6].

2.2   Multimodal Rumor Detection

To learn feature representations from multiple aspects, deep neural networks, and especially CNNs and RNNs, are successfully applied to various tasks, including visual question answering [2], image captioning [12] and rumor detection [10,28]. In [10] authors propose a deep model uses attention mechanisms to fuse and capture the relations between visual features and joint textual/social features. Yet, it is very hard to identify high-level visual semantics in rumor detection, compared with object-level semantics in traditional visual recognition tasks. As a result, there is no mechanism that explicitly guarantees the learning of this matching relation in the attention model.

Zhou et al. [28] propose a neural-network-based method named SAFE that utilizes news multimodal information for fake news detection, where news representation is learned jointly by news textual and visual information along with their relationship (similarity). Assessing the similarity between text and image helps classify rumors where objects in the image are not mentioned in the text. Yet, other types of rumors escape this rule, e.g., caricatures widely used by journalists, where the text might be very different from the image, while it does not necessarily mean that the article is fake.

3   deepMONITOR Model

In this section, we formally define the problem and introduce some key notations, then introduce the components of deepMONITOR.

3.1   Problem Definition and Model Overview

We define a message instance as $M = \{T, S, V\}$ consisting of textual information T, Sentiment information S, and visual information V. We denote C_T, C_S and C_V the corresponding representations. Our goal is to learn a discriminable feature representation C_M as the aggregation of T, S and V for a given message M, to predict whether M is a fake ($\hat{y} = 1$) or a real message ($\hat{y} = 0$). First, we learn text with a CNN, then we merge the output with a sentiment vector with two stacked LSTMs, which generates a joint representation C_{TS} for these two modalities. Visual feature C_V is obtained with a pretrained deep CNN model. Finally, C_{TS} and C_V are concatenated to form the final multimodal feature representation C_M of message M. C_M is the input of a binary classifier that predicts whether the message instance is fake or real. A global overview of deepMONITOR is presented in Fig. 1.

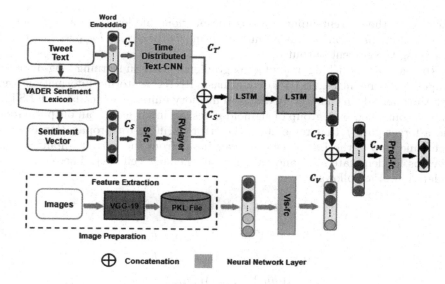

Fig. 1. Overview of deepMONITOR

3.2   LSTM Networks

For completeness, we present a brief introduction of the sequential LSTM model. LSTM is a special type of feed-forward RNN that can be used to model variable-length sequential information. Its structure is shown in Fig. 2.

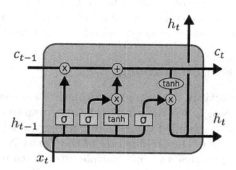

Fig. 2. Structure of an LSTM cell

Given an input sequence $\{x_1, x_2..., x_T\}$, a basic RNN model generates the output sequence $\{y_1, y_2..., y_T\}$, where T depends on the length of the input. Between the input layer and the output layer, there is a hidden layer, and the current hidden state h_t is estimated using a recurrent unit:

$$h_t = f(h_{t-1}, x_t) \tag{1}$$

where x_t is the current input, h_{t-1} is the previous hidden state and f can be an activation function or other unit accepting both x_t and h_{t-1} as input and producing the current output h_t.

To deal with vanishing or exploding gradients [3, 21] in learning long-distance temporal dependencies, LSTMs extend basic RNNs by storing information over long time periods in elaborately designed memory units. Specifically, each LSTM cell c is controlled by a group of sigmoid gates: an input gate i, an output gate o and a forget gate f that remembers the error during error propagation [9]. For each time step t, the LSTM cell receives input from the current input x_t, the previous hidden state h_{t-1} and the previous memory cell c_{t-1}. These gates are updated [5, 9] as follows:

$$i_t = \sigma(W_x^i x_t + W_h^i h_{t-1} + b_i) \tag{2}$$

$$f_t = \sigma(W_x^f x_t + W_h^f h_{t-1} + b_f) \tag{3}$$

$$o_t = \sigma(W_x^o x_t + W_h^o h_{t-1} + b_o) \tag{4}$$

$$\tilde{c}_t = tanh(W_x^c x_t + W_h^c h_{t-1} + b_c) \tag{5}$$

$$c_t = f_t \odot c_{t-1} + i_t \odot \tilde{c}_t \tag{6}$$

$$h_t = o_t \odot tanh(c_t) \tag{7}$$

where W^i, W^f, W^o are weight matrices for corresponding gates, and b are bias terms that are learned from the network. $\odot$ denotes the element-wise multiplication between two vectors. σ is the logistic sigmoid function. $tanh$ is the hyperbolic tangent function. The input gate i decides the degree to which new memory is added to the memory cell. The forget gate f determines the degree to which the existing memory is forgotten. The memory cell c is updated by forgetting part of the existing memory and adding new memory $\tilde{c}$.

3.3   Multimodal Feature Learning

Text Feature Extraction. To extract informative features from textual contents, we employ a CNN. CNNs have indeed been proven to be effective in many fields. We incorporate a modified CNN model, namely a Text-CNN [15], in our textual feature extraction. The architecture of the Text-CNN is shown in Fig. 3.

The Text-CNN takes advantage of multiple filters with various window sizes to capture different granularities of features. Specifically, each word in the message is first represented as a word embedding vector that, for each word, is initialized with a pretrained word embedding model. Given a piece of message with n words, we denote as $T_i \in R^k$ the corresponding k dimensional word embedding vector for the ith word in the message. Thus, the message can be represented as:

$$T_{1:n} = T_1 \oplus T_2 \oplus ... \oplus T_n \tag{8}$$

where $\oplus$ is the concatenation operator. To produce a new feature, a convolution filter with window size h takes the contiguous sequence of h words in the message

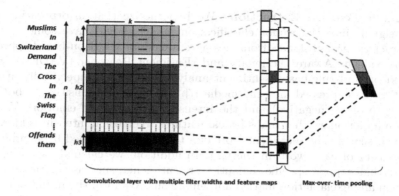

Convolutional layer with multiple filter widths and feature maps   Max-over-time pooling

Fig. 3. Text-CNN architecture

as input. For example, the feature t_i generated from a window size h starting with the ith word, can be represented as:

$$t_i = \sigma(W_c.T_{i:i+h-1} + b_c) \tag{9}$$

where, $W_c \in R^{hk}$ and $b_c \in R$ are the weight and bias of the filter, respectively, and σ is the rectified linear activation function (ReLU). This filter is applied to each possible window of h words in the message to produce a feature map:

$$t = [t_1, t_2, ..., t_{n-h+1}] \tag{10}$$

For every feature vector $t \in R^{n-h+1}$, we then apply a max-pooling operation to capture the most important information. Now, we get the corresponding feature for one particular filter. The process is repeated until we get the features of all filters. In order to extract textual features with different granularities, various window sizes are applied. For a specific window size, we have d different filters. Thus, assuming there are c possible window sizes, we have $c \times d$ filters in total. Following the max-pooling operations, a flatten layer is needed to ensure that the representation of the textual features $C_{T\prime} \in R^{c \times d}$ is fed back as input to the LSTM network.

Note that the Text-CNN above is only capable of handling a single message, transforming it from input words into an internal vector representation. We want to apply the Text-CNN model to each input message and pass on the output of each input message to the LSTM as a single time step. Thus, We need to repeat this operation across multiple messages and allow the next layer (LSTM) to build up internal state and update weights across a sequence of the internal vector representations of input messages. Thus, we wrap each layer in the Text-CNN in a Time-Distributed layer [14]. This layer achieves the desired outcome of applying the same layers multiple times and providing a sequence of message features to the LSTM to work on.

Sentiment Feature Extraction. We hypothesize that incorporating emotional signals into the rumor classification model should have some benefits. To extract emotional signals from messages, we adopt a lexicon-based approach, i.e., the Valence Aware Dictionary and sEntiment Reasoner (VADER), which is a lexicon and rule-based sentiment analysis tool that is specifically attuned to sentiments expressed in social media [7]. This model is sensitive to both the polarity (positive/negative) and the intensity (strength) of emotion. VADER relies on a dictionary that maps lexical features to emotion intensities known as sentiment scores. The sentiment score of a text can be obtained by summing up the intensity of each word in the text. In addition, we calculate some textual features that express specific semantics or sentiments, such as emotional marks (question and exclamation marks) and emoticons. We form the initial sentiment representation $C_S = [s_1, s_2, ..., s_l]^T$, where l is the dimension of sentiment features and s_i is the scalar value of the ith dimension. We first use a fully connected layer (S-fc in Fig. 1) to output a proper representation of sentiment vector $C_{S'}$:

$$C_{S'} = W_{sf} C_S \tag{11}$$

where W_{sf} are weights in the fully-connected layer. Then, we use a Repeat Vector layer [13] to ensure that $C_{S'}$ has the same dimension (3D) as the representation of the textual features $C_{T'}$. To connect the extracted features well, the representations of sentiment and those of textual features are then concatenated and fed as input to a two-stacked LSTM. Stacking LSTM hidden layers makes the model deeper, enables a more complex representation of our sequence data, and captures information at different scales. At each time step i, the LSTM takes as input $[C_{T_i'}, C_{S'}]$, i.e., the concatenation of the ith message $C_{T_i'}$ and the transformed sentiment feature $C_{S'}$. The resultant joint representation of text and sentiment features, denoted as $C_{TS} \in R^p$, has the same dimension (denoted as p) as the visual feature representation that is addressed in the next subsection. The whole process is illustrated in Fig. 4.

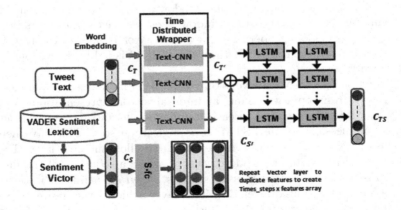

Fig. 4. Fusion process of text and sentiment features with Text-CNN and LSTM

Image Feature Extraction. The images attached to messages form the input of the visual sub-network (the bottom branch in Fig. 1). We employ the pre-trained VGG-19 model [26] to generate visual neurons as image features. We retain all front layers of the VGG-19 model and remove the last dense output layer, as well as the classification output layer. We extract the features from all images and store them into files. The benefit is that the very large pretrained VGG-19 does not need to be loaded, held in memory and used to process each image while training the textual submodel. For each loaded visual feature, we add a fully connected layer (Vis-fc in Fig. 1) to adjust the dimension of the final visual feature representation $C_V \in R^p$, as follows:

$$C_V = \psi(W_{vf}C_{V_{vgg}}) \tag{12}$$

where $C_{V_{vgg}}$ is the visual feature representation obtained from pretrained VGG-19, W_{vf} is the weight matrix of the fully connected layer and ψ denotes the ReLU activation function. The resultant joint representation of textual and sentiment features C_{TS} and the visual feature representation C_V are then concatenated to form the final multimodal feature representation of a given message, denoted as $C_M = C_{TS} \oplus C_V \in R^{2p}$.

3.4   Model Learning

Till now, we have obtained the joint multimodal feature representation C_M of a given message M, which is fed into a first fully connected layer with ReLu activation function, and a second fully connected layer with sigmoid activation function to predict whether the messages are fake. The output of the sigmoid layer for the ith message, denoted as $p(C_{M^i})$, is the probability of this post being fake:

$$p(C_{M^i}) = \sigma(W_{df2}\psi(W_{df1}C_{M^i})) \tag{13}$$

where W_{df1} and W_{df2} are weights in the two fully-connected layers, C_{M^i} is the multimodal representation of the ith message instance and σ and ψ are the sigmoid and ReLu functions, respectively. We employ the cross-entropy to define the detection loss of ith message:

$$L(M^i) = -y^i \log p(C_{M^i}) - (1 - y^i) \log (1 - p(C_{M^i})) \tag{14}$$

where y^i represents the ground truth label of the ith message instance with 1 representing false messages and 0 representing real messages. To minimize the loss function, the whole model is trained end-to-end with batched Stochastic Gradient Descent:

$$L = -\frac{1}{N} \sum_{i=1}^{N} [y^i \log p(C_{M^i}) + (1 - y^i) \log (1 - p(C_{M^i}))] \tag{15}$$

where N is the total number of message instances.

4   Experimental Validation

In this section, we first detail two real-world social media datasets used in our experiments. Then, we present the state-of-the-art rumor detection approaches, followed by the details of our experimental setup. We finally analyze the performance of deepMONITOR with respect to existing methods.

4.1   Datasets

To provide a fair evaluation deepMONITOR's performance, we conduct experiments on two real-world social media datasets collected from Twitter. Let us first detail both datasets.

FakeNewsNet [24] is one of the most comprehensive fake news detection benchmark. Fake and real news articles are collected from the fact-checking websites PolitiFact and GossipCop. Ground truth labels (fake or true) of news articles in both datasets are provided by human experts, which guarantees the quality of labels. We consider that all the tweets that discuss a particular news article bear the truth value, i.e., the label of the article, because it contributes to the diffusion of a rumor (true or false), even if the tweet denies or remains skeptical regarding the veracity of the rumor.

Since we are particularly interested in images in this work, we extract and exploit the image information of all tweets. We first remove duplicated and low-quality images. We also remove duplicated tweets and tweets without images, finally obtaining 207,768 tweets with 212,774 attached images. We carefully split the training and testing datasets so that tweets concerning the same events are not contained in both the training and testing sets.

DAT@Z20 is a novel dataset we collected from Twitter. More concretely, we retrieve all statements and reports of various nature verified by human experts from a fact-checking website; specifically contents published on June 1st, 2020. To guarantee a high quality ground truth, we retain only the data and metadata from 8,999 news articles explicitly labeled as fake or real. To extract tweets that discuss news articles, we create queries with the most representative keywords from the articles' abstracts and titles. Then, we refine keywords by adding, deleting or replacing words manually with respect to each article's context. We use the Twitter API to obtain the searched tweets by sending, as arguments, the queries prepared previously. Moreover, we employ Twitter Get status API to retrieve the available surrounding social context (retweets, reposts, replies, etc.) of each tweet.

Since we aim to build a multimedia dataset with images, we collect both the tweets' textual contents and attached images. Thus, from the 2,496,980 collected tweets, we remove text-only tweets and duplicated images to obtain 249,076 tweets with attached images. Finally, we split the whole dataset into training and testing sets and ensure that they do not contain any common event. Tweets take the label of the news articles they refer to, for the same reason as above. The detailed statistics of the two datasets are shown in Table 1.

Table 1. Datasets statistics

Statistics	Dataset					
	FakeNewsNet			DAT@Z20		
	True	Fake	Overall	True	Fake	Overall
News articles	17,441	5,755	23,196	2,503	6,496	8,999
News articles with images	17,214	1,986	19,200	455	858	1313
All Tweets	1,042,446	565,314	1,607,760	875,205	1,621,775	2,496,980
Tweets with images	161,743	46,025	207,768	81,452	167,624	249,076
Images	163,192	49,582	212,774	93,147	202,651	295,798

4.2 Experimental Settings

To learn a textual representation of tweets, we use the pretrained GloVe word embedding model [22] after standard text preprocessing. We obtain a $k = 50$-dimensional word embedding vector for each word in both datasets. One reason to choose the GloVe model is that the embedding is trained on tweets. We set the Text-CNN network's filters number to $d = 32$ and the window size of filters to $\{4, 6, 8\}$. We extract 14 sentiment features from both datasets (Table 2). The hidden size of the fully connected layer of sentiment features is 32. The joint representation of text and sentiment uses a first LSTM with hidden size 64 and a second LSTM with hidden size 32.

Table 2. Sentiment features' details

Feature
Vader negative/Positive/Neutral/Compound score
Positive/Negative words, Fraction of positive/negative words
Sad/Happy emoticons, # Exclamation/question mark
Uppercase characters, Words/Characters

Image features come from the output of the antepenultimate layer of the pretrained VGG-19 model, to generate a 4096-dimensional vector. This vector is fed to a fully connected layer with hidden size 32. The final multimodal feature representation is fed into a fully connected layer with hidden size 10. deepMONITOR uses a batch size of 64 instances. In our experiments, each dataset was separated into 70% for training and 30% for testing. The number of iterations is 100 in the training stage with an early stopping strategy on both datasets. The learning rate is 10^{-2}.

4.3 Baselines

We compare deepMONITOR with three groups of baseline methods: monomodal methods, multimodal methods, and a variant of deepMONITOR.

Monomodal Methods. We propose three baselines, where text, sentiment and image information are used separately for rumor classification.

– **Text**: deepMONITOR using textual information only.
– **Image**: deepMONITOR using visual information only.
– **Sent**: deepMONITOR using sentiment information only.

Multimodal Methods. We compare deepMONITOR with two state-of-the-art methods for multi-modal rumor detection.

– **att-RNN** [10] is a deep model that employs LSTM and VGG-19 with attention mechanism to fuse textual, visual and social-context features of news articles. We set the hyper-parameters as in [10] and exclude the social context features for a fair comparison.
– **SAFE** [28] is a neural-network-based method that explores the relationships (similarities) between the textual and visual features in news articles. We set the hyper-parameters as in [28].

Eventually, we also include a variant **deepMONITOR-** of deepMONITOR, where sentiment information is removed.

4.4 Performance Analysis

We first present the general performance of deepMONITOR by comparing it with baselines. Then, we conduct a component analysis by comparing deepMONITOR with its variants. Finally, we analyze the LRCN part. We use accuracy, precision, recall, and F_1 score as evaluation metrics.

General Performance Analysis. Table 3 shows the experimental results of baselines and deepMONITOR on FakeNewsNet and DAT@Z20. We can observe that the overall performance of deepMONITOR is significantly better than the baselines in terms of accuracy, recall and F_1 score. Moreover, the general performance of multimodal methods is deepMONITOR > SAFE > att-RNN. Deep-MONITOR indeed achieves an overall accuracy of 94.3% on FakeNewsNet set and 92.2% on DAT@Z20, which indicates it can learn effectively the joint features of multiple modalities. Compared to the state-of-the-art methods, deepMONI-TOR achieves an accuracy improvement of more than 6% and 8% with respect to SAFE; and 15% and 18% with respect to att-RNN, on FakeNewsNet and DAT@Z20, respectively.

Table 3. Performance comparison

		Text	Image	Sent	deep MONITOR-	att-RNN	SAFE	deep MONITOR
FakeNews Net	**Acc.**	0.865	0.776	0.650	0.874	0.799	0.888	**0.943**
	Prec.	0.875	0.775	0.638	0.932	0.787	0.866	**0.934**
	Rec.	0.852	0.778	0.698	0.808	0.823	0.943	**0.955**
	F_1	0.863	0.777	0.667	0.865	0.805	0903	**0.944**
DAT@Z20	**Acc.**	0.840	0.714	0.568	0.885	0.742	0.842	**0.922**
	Prec.	0.847	0.728	0.574	0.928	0.774	0.843	**0.938**
	Rec.	0.829	0.684	0.532	0.836	0.582	0.903	**0.905**
	F_1	0.838	0.705	0.552	0.880	0.665	0.872	**0.921**

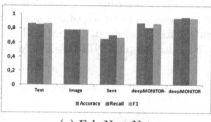 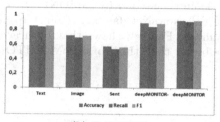

(a) FakeNewsNet   (b) DAT@Z20

Fig. 5. Component analysis results

Component Analysis. The performance of deepMONITOR and its variants are presented in Table 3 and Fig. 5. Results hint at the following insights.

1. Integrating tweets' textual information, sentiment and image information performs best among all variants. This confirms that integrating multiple modalities works better for rumor detection.
2. Combining textual and visual modalities (deepMONITOR-) performs better than monomodal variants because, when learning textual information, our model employs a CNN with multiple filters and different word window sizes. Since the length of each message is relatively short (smaller than 240 characters), the CNN may capture more local representative features, which are then fed to LSTM networks to deeply and well connect the extracted features.
3. The performance achieved with textual information is better than that of visual information. Textual features are indeed more transferable and help capture the more shareable patterns contained in texts to assess the veracity of messages. The reason is probably that both dataset have sufficient data diversity. Thus, useful linguistic patterns can be extracted for rumor detection.
4. Visual information is more important than sentiment information. Although images are challenging in terms of semantics, the use of the powerful tool VGG19 allows extracting useful features representations.

5. The performance achieved with sentiment information is the worst among multimodal variants, because without textual and visual contents, the actual meaning of tweets is lost. However, its contribution is non-negligible since the use of sentiment features (deepMONITOR- vs. deepMONITOR) can improve accuracy by 6% and 4% on FakeNewsNet and DAT@Z20, respectively.

LRCN Analysis. In this subsection, we analyze the importance of the LRCN component from the quantitative and qualitative perspectives.

Quantitative Analysis. From deepMONITOR, we design two new models, removing the text-CNN in the first (deepMONITOR-CNN), and the two LSTM networks in the second (deepMONITOR-LSTM). Then, we run the two models on the FakeNewsNet dataset. Figure 6 displays the results in terms of F_1 score and accuracy. Figure 6 shows that both accuracy and F_1 score of deepMONITOR are better than those of deepMONITOR-CNN and deepMONITOR-LSTM.

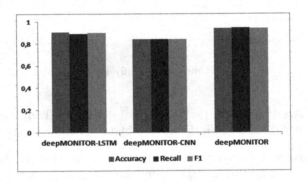

Fig. 6. Performance comparison of the LRCN component

Qualitative Analysis. To further analyze the importance of the LRCN component in deepMONITOR, we qualitatively visualize the feature representation C_{TS} learned by deepMONITOR, deepMONITOR-CNN and deepMONITOR-LSTM on the testing data of FakeNewsNet with t-SNE [19] (Fig. 7). The label of each post is fake (orange color) or real (blue color). We can observe that deepMONITOR-CNN and deepMONITOR-LSTM can learn discriminable features, but the learned features are intertwined. In contrast, the feature representations learned by deepMONITOR are more discriminable and there are bigger segregated areas among samples with different labels. This is because, in the training stage, the Text-CNN can effectively extract local features and the LSTM networks connect and interpret the features across time steps. Thus, we can draw the conclusion that incorporating the LRCN component is essential and effective for the task of rumor detection.

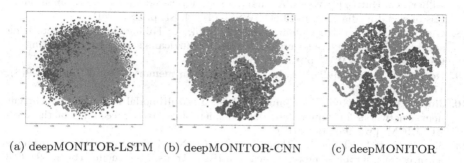

(a) deepMONITOR-LSTM   (b) deepMONITOR-CNN   (c) deepMONITOR

Fig. 7. Visualizations of learned latent text and sentiment feature representations on the testing data of FakeNewsNet (the orange colored points are fake tweets and the blue ones are real) (Color figure online)

5 Conclusion

In this paper, we propose deepMONITOR, a deep hybrid model for rumour classification in microblogs. The model extracts and concatenates textual, visual and sentiment information altogether. For a given message, we first fuse text and emotional signals with an LRCN network, which is an appropriate architecture for problems that have a 1-dimension structure of words in a sentence, such as microblog posts. This joint representation is then fused with image features extracted from a pretrained deep CNN. Extensive experiments on two large-scale dataset collected from Twitter show that deepMONITOR outperforms state-of-the-art methods.

A future line of research is to further investigate the contribution of sentiment features in the detection of rumors. Dedicating a deep submodel for learning such features instead of using our current, lexicon-based approach could indeed further improve the performance of deepMONITOR.

References

1. Ajao, O., Bhowmik, D., Zargari, S.: Sentiment aware fake news detection on online social networks. In: ICASSP 2019, pp. 2507–2511. IEEE (2019)
2. Antol, S., et al.: Vqa: Visual question answering. In: Proceedings of the IEEE ICCV, pp. 2425–2433 (2015)
3. Bengio, Y., Simard, P., Frasconi, P.: Learning long-term dependencies with gradient descent is difficult. IEEE Trans. Neural Netw. **5**(2), 157–166 (1994)
4. Castillo, C., Mendoza, M., Poblete, B.: Information credibility on twitter. In: Proceedings of the 20th IC on WWW, pp. 675–684 (2011)
5. Gers, F.A., Schraudolph, N.N., Schmidhuber, J.: Learning precise timing with lstm recurrent networks. J. Mach. Learn. Res. **3**, 115–143 (2002)
6. Giachanou, A., Rosso, P., Crestani, F.: Leveraging emotional signals for credibility detection. In: Proceedings of the 42nd ACM SIGIR, pp. 877–880 (2019)

7. Gilbert, C., Hutto, E.: Vader: A parsimonious rule-based model for sentiment analysis of social media text. In: 8th ICWSM-14, vol. 81, p. 82 (2014)

8. Gupta, A., Lamba, H., Kumaraguru, P., Joshi, A.: Faking sandy: characterizing and identifying fake images on twitter during hurricane sandy. In: Proceedings of the 22nd IC on WWW, pp. 729–736 (2013)

9. Hochreiter, S., Schmidhuber, J.: Long short-term memory. Neural Comput. 9(8), 1735–1780 (1997)

10. Jin, Z., Cao, J., Guo, H., Zhang, Y., Luo, J.: Multimodal fusion with recurrent neural networks for rumor detection on microblogs. In: Proceedings of the 25th ACM ICM, pp. 795–816 (2017)

11. Jin, Z., Cao, J., Zhang, Y., Zhou, J., Tian, Q.: Novel visual and statistical image features for microblogs news verification. IEEE Trans. Multimedia 19(3), 598–608 (2016)

12. Karpathy, A., Fei-Fei, L.: Deep visual-semantic alignments for generating image descriptions. In: Proceedings of the IEEE CVPR, pp. 3128–3137 (2015)

13. Keras: Repeatvector layer. https://keras.io/api/layers/reshaping_layers/repeat_vector/ Accessed in 2021

14. Keras: Timedistributed layer. https://keras.io/api/layers/recurrent_layers/time_distributed/ Accessed in 2021

15. Kim, Y.: Convolutional neural networks for sentence classification. In: Proceedings of EMNLP, pp. 1746–1751. ACL, Doha, Qatar (2014)

16. Kwon, S., Cha, M., Jung, K.: Rumor detection over varying time windows. PloS one 12(1), e0168344 (2017)

17. Lin, D., Lv, Y., Cao, D.: Rumor diffusion purpose analysis from social attribute to social content. In: 2015 IALP, pp. 107–110. IEEE (2015)

18. Ma, J., Gao, W., Wei, Z., Lu, Y., Wong, K.F.: Detect rumors using time series of social context information on microblogging websites. In: 24th ACM ICIKM, pp. 1751–1754 (2015)

19. Van der Maaten, L., Hinton, G.: Visualizing data using t-sne. J. Mach. Learn. Res. 9(11) (2008)

20. Newman, M.L., Pennebaker, J.W., Berry, D.S., Richards, J.M.: Lying words: predicting deception from linguistic styles. Pers. Soc. Psychol. Bull. 29(5), 665–675 (2003)

21. Pascanu, R., Mikolov, T., Bengio, Y.: On the difficulty of training recurrent neural networks. In: ICML, pp. 1310–1318 (2013)

22. Pennington, J., Socher, R., Manning, C.D.: Glove: Global vectors for word representation. In: Proceedings of the 2014 EMNLP, pp. 1532–1543 (2014)

23. Ruchansky, N., Seo, S., Liu, Y.: Csi: A hybrid deep model for fake news detection. In: Proceedings of ACM CIKM, pp. 797–806 (2017)

24. Shu, K., Mahudeswaran, D., Wang, S., Lee, D., Liu, H.: Fakenewsnet: A data repository with news content, social context, and spatiotemporal information for studying fake news on social media. Big Data 8(3), 171–188 (2020)

25. Shu, K., Sliva, A., Wang, S., Tang, J., Liu, H.: Fake news detection on social media: a data mining perspective. ACM SIGKDD Explor. 19(1), 22–36 (2017)

26. Simonyan, K., Zisserman, A.: Very deep convolutional networks for large-scale image recognition. In: Proceedings of ICLR (2015)

27. Wu, K., Yang, S., Zhu, K.Q.: False rumors detection on sina weibo by propagation structures. In: 2015 IEEE 31st ICDE, pp. 651–662. IEEE (2015)
28. Zhou, X., Wu, J., Zafarani, R.: SAFE: similarity-aware multi-modal fake news detection. In: Lauw, H.W., Wong, R.C.-W., Ntoulas, A., Lim, E.-P., Ng, S.-K., Pan, S.J. (eds.) PAKDD 2020. LNCS (LNAI), vol. 12085, pp. 354–367. Springer, Cham (2020). https://doi.org/10.1007/978-3-030-47436-2_27

Correction to: Automated Machine Learning for Satellite Data: Integrating Remote Sensing Pre-trained Models into AutoML Systems

Nelly Rosaura Palacios Salinas, Mitra Baratchi ⓘD,
Jan N. van Rijn ⓘD, and Andreas Vollrath

Correction to:
Chapter "Automated Machine Learning for Satellite Data: Integrating Remote Sensing Pre-trained Models into AutoML Systems" in: Y. Dong et al. (Eds.):
Machine Learning and Knowledge Discovery in Databases,
LNAI 12979, https://doi.org/10.1007/978-3-030-86517-7_28

The given name and surname of the author Nelly Rosaura Palacios Salinas have been wrongly attributed in some parts of the original publication. This has now been corrected.

The updated version of this chapter can be found at
https://doi.org/10.1007/978-3-030-86517-7_28

© Springer Nature Switzerland AG 2021
Y. Dong et al. (Eds.): ECML PKDD 2021, LNAI 12979, p. C1, 2021.
https://doi.org/10.1007/978-3-030-86517-7_32

Author Index

Alguacil, Antonio   102
Ali, Muhammad Intizar   20
Antaris, Stefanos   463
Arcelin, Bastien   135
Armand, Stéphane   182
Arriaza, Romina   463
Azri, Abderrazek   497

Bar, Kfir   271
Baratchi, Mitra   447
Bauerheim, Michael   102
Berndl, Emanuel   52
Blair, Philip   271
Blondé, Lionel   182
Bourrie, David   151
Breslin, John G.   20

Candido Ramos, Joao A.   182
Chawla, Nitesh V.   335
Chen, Li   151
Chen, Zhiguang   87
Chouragade, Ankur   218
Cieslak, David   335
Conan-Guez, Brieuc   135
Couceiro, Miguel   135

Darby, Paul   151
Darmont, Jérôme   497
Ding, Wenbiao   285
Dong, Kaiwen   335
Du, Yunfei   87

El-Azab, Anter   118
Eliav, Carmel   271

Fan, Cuncai   118
Färber, Michael   251
Favre, Cécile   497
Fu, Weiping   285

Gorniak, Adrian   481
Goyal, Vikram   381
Granitzer, Michael   52

Groh, Georg   481
Guyon, Isabelle   36

Hagenmayer, Daniel   251
Harbi, Nouria   497
Hasanaj, Fiona   271
Hassani, Marwan   234
Hingerl, Johannes   481
Huang, Guanjie   68

Jacob, Marc C.   102
Janiszewski, Piotr   351
Johnsten, Tom   151

Kabra, Anubha   365
Kalousis, Alexandros   182
Kanebako, Yusuke   3
Kapoor, Rajiv   365
Kashima, Hisashi   431
Kimball, Sytske K.   151
Krause, Franz   251
Kuniyoshi, Fusataka   319

Lango, Mateusz   351
Le, Thai   415
Lee, Dongwon   415
Li, Kejiao   168
Li, Zhenghong   302
Liang, Mingfei   201
Lin, Leyu   201
Liu, Yuecheng   168
Liu, Zhiyue   302
Liu, Zitao   285
Lu, Kai   335
Lu, Xiaotian   431
Lu, Yutong   87

Ma, Fenglong   68
Maciejewski, Henryk   399
Madan, Manchit   218
Maruhashi, Koji   431
Miwa, Makoto   319
Mohania, Mukesh   381

Moreau, Stéphane   102
Mosca, Edoardo   481

Napoli, Amedeo   135
Nasim, Md   118
Nguyen, Anna   251
Noûs, Camille   497

Okajima, Seiji   431
Ozawa, Jun   319

Palacios Salinas, Nelly Rosaura   447
Patel, Pankesh   20
Peng, Lu   151
Pennerath, Frédéric   135
Pinto, Wagner Gonçalves   102
Pitre, Boisy   151

Rafailidis, Dimitrios   463
Rappin, Eric   151

Sharma, Arushi   365
Shi, Wencheng   87
Stefanowski, Jerzy   351
Sudharsan, Bharath   20
Sun, Zhenlong   201
Szyc, Kamil   399

Takebayashi, Tomoyoshi   431
Takeuchi, Koh   431
Theobald, Claire   135
Tian, Hao   168
Tolmachev, Arseny   431
Tsukamoto, Kazuki   3

Tu, Wei-Wei   36
Tzeng, Nian-Feng   151

van den Berg, Sophie   234
van Rijn, Jan N.   447
Vempati, Sreekanth   218
Venktesh, V.   381
Vollrath, Andreas   447

Walkowiak, Tomasz   399
Wang, Fan   168
Wang, Jiahai   302
Wendlinger, Lorenz   52
Wich, Maximilian   481
Wu, Zhongqin   285

Xia, Xin   335
Xie, Ruobing   201
Xie, Tianyi   415
Xu, Fengyang   87
Xu, Guowei   285
Xu, Zhen   36
Xue, Yexiang   118

Yamamoto, Tatsuya   431
Yu, Haomin   201
Yuan, Xu   151

Zeng, Hongsheng   168
Zhang, Bo   201
Zhang, Maosen   118
Zhang, Xinghang   118
Zhang, Yihe   151
Zhou, Bo   168

Printed in the United States
by Baker & Taylor Publisher Services